Iron Age Archaeology and Trauma from Aymyrlyg, South Siberia

Eileen M. Murphy

BAR International Series 1152
2003

Published in 2016 by
BAR Publishing, Oxford

BAR International Series 1152

Iron Age Archaeology and Trauma from Aymyrlyg, South Siberia

ISBN 978 1 84171 522 3

BAR Publishing is the trading name of British Archaeological Reports (Oxford) Ltd. British Archaeological Reports was first incorporated in 1974 to publish the BAR Series, International and British. In 1992 Hadrian Books Ltd became part of the BAR group. This volume was originally published by Archaeopress in conjunction with British Archaeological Reports (Oxford) Ltd / Hadrian Books Ltd, the Series principal publisher, in 2003. This present volume is published by BAR Publishing, 2016.

Printed in England

BAR
PUBLISHING

BAR titles are available from:

BAR Publishing
122 Banbury Rd, Oxford, OX2 7BP, UK
EMAIL info@barpublishing.com
PHONE +44 (0)1865 310431
FAX +44 (0)1865 316916
www.barpublishing.com

Contents

This work is dedicated to my parents,

Anne and Patrick Murphy

Acknowledgements

I would like to begin by acknowledging the major contribution made by Prof. Yuri Chistov of the Research Division of the Peter the Great Museum of Anthropology and Ethnography, St. Petersburg, without whose help and support I would not have learned of, or gained access to, the Aymyrlyg skeletal collection. I am also extremely grateful to him for helping me to organise my various trips to St. Petersburg and for all the kindness he showed me during my stays in that great city. I am indebted to his colleague Prof. Ilyia Gokhman for granting me permission to study the Aymyrlyg skeletal population, and for the advice he provided throughout my research. Dr. Anna Radzjun, also of that organisation, must be thanked for providing me with background information on the assemblage and access to the results of her own research on the skeletal population from Aymyrlyg. I would like to thank Mr. Andre Gromov, Dr. Valery Khartanovich, Dr. Slava Moyseev, Dr. Sima Sankina and Dr. Albert Shevchenko for making my work in the Research Division both educational and enjoyable. Sincere thanks must also go to Vasily, Svetlana, Pavel, Regina and Mouska the Cat for their cheery smiles during the many long hours I spent working in SredniyProspekt!

I am very grateful to my supervisor Prof. Jim Mallory of the School of Archaeology and Palaeoecology, Queen's University Belfast, for the advice and encouragement he provided me with throughout the course of my doctoral research. I must also express my gratitude to the Department of Education for Northern Ireland who generously funded my PhD research and the numerous fieldtrips to St. Petersburg which it entailed. In addition, I am extremely grateful to the Emily Sarah Montgomery Travel Scholarship Scheme of Queen's University Belfast, for providing additional support during two of my periods of Russian fieldwork.

I am also grateful to Prof. Jonathan Pilcher, School of Archaeology and Palaeoecology, Queen's University Belfast, who was my internal examiner, and I would also like to express my sincere thanks to my external examiner, Dr. Keith Manchester, for his valuable comments on my doctoral work and all his friendship and support since the completion of my PhD. I would also like to acknowledge Prof. Viktor Masson of the Institute of the History of Material Culture, St. Petersburg, for enabling me to gain access to the textual and photographic archives of the excavations at Aymyrlyg. I received assistance from Ms. Mary Harrah, Ms. Katya Schreiber, and Ms. Zhenya Vladimirovna, each of whom undertook some translations of Russian texts. I am very grateful to Ms. Libby Mulqueeny, School of Archaeology and Palaeoecology, Queen's University Belfast, for her creativity in preparing the frontispiece for the book and to Mr. Philip Barratt, of the aforementioned school for typesetting the text.

I made a lot of very good friends during the time I spent in St. Petersburg, and I would like to pay particular thanks to Mr. Dmitry Chistov, of whom I have particularly fond memories, especially sitting in the flat in Prospekt Prosvecheniya chatting and sipping 'Baltica Tri' during the 'White Nights' of 1996. I am also grateful to Ms. Ulia Menshikova for her stimulating cultural discussions and nights at the theatre, and to Ms. Katya Schreiber for her enlivening company. Sincere thanks must also go to those I stayed with in St. Petersburg, particularly Ms. Bella Rubejova, who looked after me so well during my first trip to the city in March 1995. Thanks must also go to the Chistov families and the Schreiber family for all the help and enjoyable social occasions they provided. I would like to thank my English friends, who I met in Russia on a Russian language course in 1995, particularly Mr. Kevin Bickerstaff, for all his encouragement and many words of wisdom, and Ms. Karen Parkes, for providing such good company.

Mum, Dad and my brother Bryan were extremely supportive and encouraging and I would like to express my sincere thanks to them. My husband Dr. Colm Donnelly was an unfailingly considerate and understanding partner throughout the course of my doctorate research, a willing and most helpful assistant during my periods of fieldwork in St. Petersburg, and most recently a dedicated editor of this text capably assisted by our daughter Abigail. Thanks must go to the Donnelly family for all their interest and encouragement over the past years. My furry friends Rua, Liath, Fairfax, Tomyris and Heathcliff cannot be forgotten for the role they played in enlivening the seemingly endless days spent sitting in front of my computer during the write-up of my thesis.

Finally, I would like to remember Dr. A. M. Mandelshtam for the many years of excavations he undertook at Aymyrlyg, and I hope he would have approved of my attempts to put flesh back onto the bones of the people that he excavated from this highly important Iron Age cemetery.

List of tables

List of figures

List of Plates

1. Introduction

The work included in this monograph represents an aspect of the author's doctoral thesis – *An Osteological and Palaeopathological study of the Scythian and Hunno-Sarmatian period populations from the cemetery complex of Aymyrlyg, Tuva, South Siberia* – that was undertaken in the School of Archaeology and Palaeoecology, Queen's University Belfast, in the period between 1994 and 1998 (Murphy 1998). The objective of the thesis was to elucidate the nature of the health, diet and lifestyles of the two Iron Age populations buried at the cemetery complex of Aymyrlyg, Tuva, south Siberia, through an osteological and palaeopathological examination of their skeletal remains. A multidisciplinary approach was adopted which saw the integration of archaeological, documentary, and environmental evidence with the data derived from the skeletal analysis. During this work a rich array of traumatic lesions were identified among the remains, injuries which shed light on the everyday activities, occupations and warfare practices of the two population groups. The current text provides an in-depth account of the palaeopathological evidence for trauma, while placing it in its archaeological context.

The cemetery complex of Aymyrlyg is located in the Ulug-Khemski region of the Autonomous Republic of Tuva, south Siberia. It was excavated in the period between 1968 and 1984 by archaeologists of the Sayano-Tuvinskaya expedition team from the Institute for the History of Material Culture of St. Petersburg. The director of the expedition for the period between 1968 and 1978 was Dr. A. M. Mandelshtam but, following his death, Dr. E. U. Stambulnik continued with research excavations until the mid 1980s. When the excavation programme had concluded the skeletal remains were stored in the Research Division of the Peter the Great Museum of Anthropology and Ethnography, St. Petersburg, where a proportion of the remains were later subjected to craniometrical and osteometrical analyses (Bogdanova and Radzjun 1991). A detailed osteological study of the skeletal remains, however, was still required, as was a programme of palaeopathological analysis. With the kind permission of Professors Yuri Chistov and Ilyia Gokhman of the above institution the fieldwork for this thesis was undertaken in the period during 1995 to 1997 in order that these two objectives might be realised.

South Siberia and Mongolia are amongst the regions of the Old World with the most ancient traditions of pastoralism (Vainshtein 1980, 51). The analysis of the skeletal remains from Aymyrlyg provided a rare opportunity for the examination of a substantial corpus of skeletal remains (809 individuals) of semi-nomadic pastoralists from the vast Eurasian steppe-lands. The research represented one of the first palaeopathological studies of an archaeological population from south Siberia to have been undertaken and, as such, it has made a major contribution to our understanding of life and death in Iron Age Central Asia.

A number of academic papers have already been published on aspects of the completed thesis. These include a case study on a possible case of neurofibromatosis (Murphy *et al* 1998), and an overview of the evidence for disability based on the more severe developmental defects apparent among the remains (Murphy 2000a). Papers have also been written on the evidence for secondary burial practice (Murphy 2000b; Murphy and Mallory 2000; 2001), scalping (Murphy *et al* 2002), trepanation (Murphy 2003), and a case-study of an individual who appeared to have died as a consequence of a particularly frenzied attack (Murphy forthcoming a). An overview of the findings of the analysis of the Scythian period population has been published in Russian (Murphy 2001), and a Russian version of the results for the Hunno-Sarmatian period group is forthcoming (Murphy forthcoming b).

Throughout the text the corpus of skeletons from Aymyrlyg is divided into two distinct groups, those from the Scythian period (or Uyuk Culture) and those from the Hunno-Sarmatian period (or Shurmak Culture). This division has been made by Russian academics on the basis of differences in funerary monuments and material culture between the two groups, combined with information contained in Chinese and Classical texts on the inhabitants of the region. Russian archaeologists consider the Uyuk Culture in Iron Age Tuva to have belonged to the Scythian World (see Section 2.3). This culture has been approximately dated to between the 8th and the 2nd century BC and the majority of the Scythian period burials at Aymyrlyg are considered to date from the 3rd to 2nd centuries BC. Historical texts indicate that there was a spread of influence of the Hsiung-nu Empire into Tuva during the late 3rd and early 2nd century BC, and that this led to cultural and/or population changes at this time (Mandelshtam 1983, 331; 1992, 183, 185). Russian archaeologists have postulated that the Shurmak Culture developed during the 3rd to the 1st century BC as a result of a complicated and prolonged process that occurred in a climate of rapid political change when new population groups moved into Tuva. These Hunno-Sarmatian period people are considered to be related to the Hsiung-nu of the historical texts (Stambulnik 1983a, 34; Mandelshtam and Stambulnik 1992, 196) (see Section 2.4). It has further been suggested that the Hunno-Sarmatian period population at Aymyrlyg belongs to the earliest phase of this cultural change, when the Scythian period culture was ending and the new culture was gaining ground. As such, it is believed that the Hunno-Sarmatian period burials incorporate elements of both old and new cultures (Stambulnik 1983a, 39).

The monograph is divided into eight chapters, with two appendices at the end of the text. Appendix 1 contains data pertaining to the preservation of the remains and the minimum number of bone values that were employed during their examination, while Appendix 2 consists of a gazetteer of the skeletal remains included in the analysis which displayed evidence for trauma. Chapter 2 provides a background to life in Scythian and Hunno-Sarmatian period Tuva, information that has been drawn from archaeological evidence and the main historical sources. The archaeological evidence pertaining to funerary monuments and material culture is described in detail since the majority of this information is derived from archaeological reports written in Russian and generally unavailable to Western academics. Chapter 3 contains all archaeological information relating to the excavations at the cemetery complex of Aymyrlyg; again, this information is provided in a comprehensive manner since it is derived from both published and unpublished archaeological reports and archives written in Russian and not available in English. Chapter 4 details the basic methodology followed in the skeletal analysis that is of relevance for understanding the profile of the individuals that displayed evidence for trauma.

Chapter 5 provides an account of traumatic lesions apparent in the remains, including fractures, dislocations and soft-tissue injuries, while Chapter 6 concentrates on the evidence for weapon trauma evident in the two groups. The chapters begin with a general introduction which is then followed by a separate examination of the Scythian period population and the Hunno-Sarmatian period population. Both chapters conclude with an overall summary of the findings, in which information from the two population groups is compared and contrasted. In Chapter 7 the summary information from Chapters 5 and 6 is reviewed in light of the evidence derived from the archaeological, ethnographical and historical records. Relevant osteological and palaeopathological information from sites around the world is also called upon for comparative purposes. The text culminates in Chapter 8 where the overall findings of the analysis are imparted.

2. Life in Scythian and Hunno-Sarmatian period Tuva

2.1 Introduction

The Autonomous Republic of Tuva has an area of approximately 170, 500 square kilometres (Vainshtein 1980, 47), and is bounded to the north by the Sayan range and to the south by the Tannu-ola. Although both ranges have given the territory a fortress-like and secluded topography, the greater penetrability of the Tannu-ola from the south, relative to the Sayan in the north, has resulted in greater linguistic, cultural and political penetration from Mongolia (Ewing 1981, 175-176). In regions of central, western and southern Tuva 70,000 square kilometres of steppe-land is encompassed in the Tuvinian (Khemchik and Ulug-Khem) and Ubsu-Nur depressions, the mountains of south-west Tuva and the southern slopes of the western Sayan Mountains. The main cattle-grazing lands have been located in the central, western and southern regions of Tuva for centuries (Vainshtein 1980, 47). A large number of rivers flow through Tuva, and the Yenisey originates in the territory and runs for more than 200 km within the Republic. The upper reaches of the Yenisey are known as the Ulug-Khem or Great River (*ibid.*). The climate of Tuva has been described as 'sharply continental' with a snow cover of only 10 - 30 cm. This lack of heavy snow cover enables animals to pasture throughout the winter (Bokovenko 1995a, 261).

2.2 Previous archaeological research in Tuva

The archaeological monuments of Tuva were first mentioned in the late 19th and early 20th century accounts of early Russian scholars such as D. A. Klements, I. R. Aspelin, A. O. Gheikel, I. G. Grane and S. R. Mintslov. These scholars described the remains which were visible on the landscape without understanding the culture or time period to which they belonged (Mannai-Ool 1970, 4). Archaeological research was initiated at a much later stage in Tuva compared to the rest of south Siberia, and the objectives of the initial research were to develop the work which had been already undertaken in the Minusinsk Basin. The first excavations were undertaken in 1915-1916 by A. V. Adrianov in the northern area of Tuva in the vicinity of the Uyuk River. This research was followed by the work of S. A. Teplouhov who excavated over 160 Tuvinian monuments, forty-five of which were funerary monuments of the Uyuk Culture of the Scythian period (c. 8th - 2nd century BC), in 1926-1927 and 1929 (*ibid.*).

The systematic study of the archaeological monuments of Tuva began after its entry into the Soviet Union in 1947 and the next major series of excavations was undertaken in that year by S. V. Kiselev and the Sayano-Altaisky expedition team of the Russian Academy of Science. During this research a number of *olenniye* (deer) stones, tumuli and stone kurgans were recorded. In 1953 and 1954 archaeological investigations in Tuva were continued by the Institute of Ethnography of the Russian Academy of Science under the direction of A. D. Grach. Concomitantly, S. I. Vainshtein of the Tuvinsky Regional Museum also began a programme of field survey and excavation in Tuva (*ibid.*, 5). Excavations have regularly occurred in Tuva since 1955 and have included the work undertaken by S. I. Vainshtein and M. H. Mannai-Ool of the Tuvinsky Scientific Research Institute of Language, Literature and History, L. R. Kyzlasov (1955-1960, 1962) and the Tuvinsky archaeological expedition team of the Moscow State University, and L. P. Potapov (1957-1966) and the Tuvinsky archaeological and ethnographical expedition team of the Institute of Ethnography of the Russian Academy of Science (*ibid.*). In the period between 1959 and 1963 Y. I. Sunchugashev undertook extensive research in Tuva on Scythian period mining and metallurgy.

The Sayano-Tuvinsky expedition team from the Institute of the History of Material Culture, Russian Academy of Science, St. Petersburg, initiated a programme of excavations in 1965 in the region of the Sayano-Shushensky hydro-electric power station. In the period between 1965 and 1972 the leader of the expedition was A. D. Grach, who was succeeded in 1972 by S. N. Astahov (Mandelshtam 1992, 178). This program of excavations included the examination of the Aymyrlyg cemetery complex, which was systematically excavated in the period from 1968 to 1984 under the directorship of A. M. Mandelshtam and E. U. Stambulnik (see Chapter 2). In the mid-1980s N. A. Bokovenko excavated the Scythian period burial site of Badanka IV, while in 1993 he also excavated the Scythian period burial of Teplaya. Both sites were situated in the Usa Basin of the Sayans (Bokovenko 1994a, 48-49). More recently a series of excavations have been undertaken by V. A. Semenov in the vicinity of the Mongun-Taiga mountains at a number of sites dating from the Early Scythian period to the Turkic period (Semenov 1996, 25-29).

2.3 The Scythian World and the Scythian period in Tuva

The term Scythian World is applied to a group of archaeological cultures which date from approximately the 8th to the 2nd centuries BC (but see below and Sections 2.3.1 and 3.2), and are located in the zone of the steppes, forest-steppes, foothills and mountains of the Ukraine, Russia, Kazakhstan, Mongolia and the northern parts of China (Clenova 1994, 499). The Greeks referred to the nomads of Eurasia and Central Asia as the Scythians, while the Persians designated all nomadic tribes of the Eurasian steppes (including the Scythians) as the Saka (Abetekov and Yusopov 1994, 24). The cultures covered by the term Scythian World are wide ranging (Table 1).

The history of the Central Asian nomads is inseparable from that of the nomadic and semi-nomadic tribes of the Eurasian steppes. Their material culture was markedly similar, as too were the political and economic policies which they followed (Abetekov and Yusopov 1994, 23). The common material culture apparent among all of the tribes of the Scythian World is known as the Scythian Triad, which consists of weapons, horse harnesses and objects decorated in the Animal Style of artwork (Moskova 1994, 231). Although the artefacts of the

Cultures of the Scythian World
Scythian Cultures of the North Black Sea region, the Kusk Voronez region and the northern area of the Caucasus
Sauromatian Cultures from the Don river to the Urals
Tasmola and Kulazurga Cultures of Kazakhstan
Tagisken, Uygarak and Sakar-Caga Cultures in the Aral Sea region
Saka Cultures of Semirec'e, Kazakhstan, Kirgizia and the Pamirs
Pazyryk Culture of the High Altai
Bolserecenskaya Culture of the North Altai
Tagar Culture of the Minusinsk Basin
Uyuk Culture of Tuva
Ulangom-type Culture of western Mongolia
Slab-tomb Culture of eastern Mongolia

Table 1: Scythian period - A list of the main cultures included in the Scythian World (Clenova 1994, 499).

Scythian Triad display marked similarities, other components of the Scythian World (such as dwellings, burial customs, ceramics and adornments) differ considerably between the various cultures. Consequently, it is not possible to envisage a single Scythian Culture but rather a variety of cultures within a Scythian World (Clenova 1994, 500-501).

The majority of information on the Scythian period in Tuva has been obtained from cemetery excavations, particularly in the central area of the territory. In addition, some information has been gleaned from a study of Scythian period mining activities (Mandelshtam 1992, 179). The Scythian period culture in Tuva is called the Uyuk Culture after the region - the Uyuk river basin - where the first scientific excavations of monuments of this period occurred (Mannai-Ool 1970, 8). The Uyuk Culture is not confined solely to Tuva and it included the north-western part of Mongolia to the south and the eastern outlying regions of the Altai. The Uyuk Culture was bordered by two other Scythian period cultures, the Pazyryk Culture to the west, and the Tagar Culture to the north (Mandelshtam 1992, 179).

The dating of the Scythian period in Tuva is extremely complicated and a number of different chronological systems have been proposed, including those advocated by Kyzlasov (1958), Vainshtein (1958), Grach (1971), Mannai-Ool (1970) and Gryaznov (1983). Mandelshtam (1992, 179) used a corrected version of the Kyzlasov (1958) chronology for the dating of the Scythian period in Tuva. In general this period is considered to have lasted from the 8th to the 2nd century BC. In the dating system advocated by Mandelshtam (1992, 179) the Tuvinian Scythian period is divided into three phases - the 'Initial Stage' (which is also known as the Arzhan Culture - 8th - 7th century BC), the 'Early Stage' (7th - 6th century BC) and the 'Late Stage' (5th - 3rd century BC) (Mandelshtam 1992, 179; Bokovenko 1995a, 260). Russian archaeologists have identified a change in Iron Age Tuvinian culture during the 3rd - 2nd century BC and this is considered to be the time when the Hunno-Sarmatian period commences (Mandelshtam 1992, 179).

2.3.1 Initial Scythian period - Arzhan Culture (8th - 7th century BC)

The Arzhan Culture is named after the Arzhan Kurgan which was situated in the Uyuk depression of Tuva and was excavated by M. P. Gryaznov and M. K. Mannai-Ool in the period between 1971 and 1974 (Bokovenko 1995b, 259-260). Dendrochronology and radiocarbon dating have ascribed the Arzhan Kurgan to the end of the 8th century BC (Bokovenko 1994b, 32). As such, it is considerably older than the generally recognised Scythian remains throughout the rest of the steppe region. The dating of the Arzhan Kurgan has been used to support the argument for an eastern Central Asian origin of the Animal Style of artwork and the entire Scythian World (Askarov *et al* 1992, 470).

2.3.1a Funerary monuments and practices

The burials of the Arzhan Culture consisted of a stone cist, in which the body was buried lying on the ground surface, with a dome-shaped cairn used to cover the tomb. The cairn was surrounded by a circular stone fence (Bokovenko 1995b, 265). The Arzhan Kurgan is the most elaborate burial of the Initial Stage of the Scythian period. The kurgan was a huge circular structure with a diameter of approximately 120 m and a height of approximately 4 m. The kurgan had been built on the ground surface and had a reinforcing wall, and it has been estimated that it would have taken approximately 1500 individuals a period of seven or eight days to construct the monument. The careful manner in which the individual elements of the tomb had been fitted together indicated that the structure was probably built in a single session (Rolle 1989, 42). Beneath the stone cairn there was a complicated wooden structure which consisted of seventy burial chambers, each separated from one another by partitions of larch logs. The timber chambers were positioned radially around a central burial of an elderly male and female. The individuals in the central burial had been dressed in fur clothing and were adorned with lavish decorations. A number of attendants, six harnessed riding horses and seven elderly servants had been buried in the radiating chambers (Bokovenko 1995b, 266-267). All of the skeletons had been buried on their left sides with markedly flexed arms and legs. The only exception was a single individual who had been buried on his or her right side. A variety of orientations of the burials were evident, but a north-western orientation was most frequent (Mandelshtam 1992, 181). It has been estimated that approximately 160 horses had been buried in the kurgan, and that another 300 horses had been consumed at the funeral feast (Gryaznov 1980 as quoted in Abetekov and Yusupov 1994, 28).

2.3.1b Clothing and adornments

Fragments of clothing composed of sable fur and a variety of woollen cloths were recovered from the central burial. The woollen materials are thought to have originated from the Near East. One example displayed two colours and was decorated with single rhomboids and rows of rhombic shapes. Another was composed of four colours and was decorated with rows of stepped pyramids, while another consisted of material of a single colour decorated with doubled rhombic symbols (Mandelshtam 1992, 182).

Only a small number of items of jewellery were recovered from the Arzhan Kurgan. These consisted of a torque (which had been manufactured from a narrow strip of sheet gold), gold earrings (which had been ornamented with turquoise insets) and a number of rectangular gold plates. The gold plates had probably been used as belt decorations (Bokovenko 1995b, 271).

2.3.1c Weaponry

The weapons present in the Arzhan Kurgan consisted of two bronze daggers, several bronze pointed axes and a number of bronze and bone arrowheads. Both of the daggers had two blades, and a narrow wing-shaped hilt with a pommel. One of the pommels was composed of a depiction of a wild boar in bronze whose legs extended onto the hilt, and the other was roll-shaped. The pointed axe was large and a socket was present on the superior surface for the attachment of a plume. The bronze arrowheads were two-feathered, rhombic, leaf-shaped or bullet-shaped. The bone arrowheads generally had four edges and were rhombic in section, or had three edges and were bullet shaped (Mandelshtam 1992, 182).

2.3.1d Horse accoutrements

A large number of horse accoutrements were recovered from the Arzhan Kurgan. Bits with stirrup-like endings and ring endings were identified. All of the cheek-pieces had three perforations and were mostly composed of bronze. A number of curved cheek pieces, round and rectangular plates originally attached to harnesses, and pendants used to decorate bridles were also found. Bronze pendants were generally stirrup-shaped or ring-shaped, while pendants of horn, horse teeth and wild boar teeth were also present (*ibid.*). A number of other artefacts decorated in the Animal Style of artwork were associated with horse skeletons and had probably been used to decorate the horse trappings. These included a large plate which was decorated with a curled-up predator of the cat family - possibly a panther - and it is thought to have been attached to the trappings located on the breast of a horse. In addition, a number of elaborate bronze pommels and buckles covered in sheet gold were recovered (Schiltz 1994, 251).

2.3.1e Summary

Several of the objects retrieved from the Arzhan Kurgan (including a number of the arrowheads and the bronze cheek pieces) displayed parallels to objects recovered from the Black Sea region and the archaeologists considered these to be pre-Scythian in date. Other objects, including the bits with stirrup-like endings, were similar to those from pre-Scythian and Scythian period sites in the western part of the steppe territory. Radiocarbon dating has indicated that the Arzhan Kurgan probably dates to the late 8th - early 7th century BC (Mandelshtam 1992, 182).

2.3.2 Early Scythian period (7th - 6th century BC)

2.3.2a Funerary monuments and practices

The majority of funerary monuments from the Early Stage of the Scythian period Uyuk Culture in Tuva are those of the ordinary classes of society, characterised by stone and earth kurgans containing multiple burials and situated in small groupings. In some instances pairs of stone kurgans aligned north to south are conjoined. In the majority of cases the mounds appear to have been constructed after a certain number of individuals had been buried in the ground. In some instances, however, the burials of children are found situated at the edge of the mound, suggesting that they were buried after its construction. The central grave beneath the mound invariably contained the remains of an adult, while burials of children and adults might be situated around the circumference. Typical burials of this period were found in the Hemchik-Bom V cemetery, Tuva. The burials generally contained individuals although double burials containing the remains of an adult and a child have been discovered. The orientation is generally westerly (*ibid.*, 183).

Rows of stone kurgans covering single graves are also characteristic of this period. The graves may consist of an individual burial in a ground pit, such as those found at Zubovka, or collective burials in either log house tombs (e.g. Chinge II) or ground pits (e.g. Kuilug-Hem I). In some cases a stone cist may be present in the ground pit. In general, the individuals are buried on their side with arms and legs flexed and the head orientated to the west or the north-west. In central Tuva rectangular stone-works are commonly found overlying the burials, as was the case at Aymyrlyg (*ibid.*, 183).

2.3.2b Clothing and adornments

The remains of leather belts associated with sets of iron or bronze circular plates were recovered from several tombs, such as those at Ust-Hadynnyg. Some of the plates were decorated with animal motifs. Buckles, clasps and buttons similar to those recovered from the Arzhan Kurgan have also been found in Early Scythian period burials (*ibid.*, 184).

Though relatively few items of jewellery were present, those pieces which have been discovered include bronze torques, bronze earrings and bronze plates composed in the shape of a wild boar in a standing position. In addition, a number of stone and glass beads are known, including a Carnelian bead with a figure-of-eight design etched on its surface and a glass bead with eye-shaped designs. Both of these beads were considered by Mandelshtam to have been imported from the West (*ibid.*).

2.3.2c Weaponry

A small number of daggers and pointed axes and a considerable number of arrowheads were contained in the tombs. One of the daggers had a leaf-shaped blade, a narrow wing-shaped guard with curved endings and a mushroom-shaped pommel. Another dagger had a kidney-shaped guard and a short conical-shaped pommel. The pointed axes all had points which were circular in section. One of the pointed axes had a butt of bronze fashioned in the form of a pair of mountain goats. The majority of arrowheads were made from bronze, and displayed a variety of morphologies including two-feathered and leaf-shaped, and three-feathered and trilobate. The bone arrowheads were generally trilobate and tanged, while a number of arrows with both a wooden head and shaft were discovered in several of the tombs (*ibid.*, 184).

2.3.2d Horse accoutrements

The majority of Early Scythian period horse accoutrements from Tuva are represented by bits and the remains of harnesses. All bits were composed of bronze and had stirrup-shaped endings (*ibid.*, 183). During this period two-holed rod-shaped mouth bits began to be used, indicative that a new method of fastening the cheek-pieces to the bit had been invented. The cheek-pieces would have been inserted into the holes (and not secured by looping as had been the case in the Arzhan period) thereby enabling better control of the horse by the rider (Bokovenko 1995b, 274). The majority of cheek pieces were made from bronze, had two perforations and were Y-shaped. Other horse accoutrements included badges which would have been attached to the cheek-pieces. Hoards of horse accoutrements were recovered from several of the kurgans, most notably one of the mounds at Aldy-Bel. The horse fittings were amongst the objects recovered from the burials which the archaeologists found to be most suitable for dating purposes through a close study of their typologies and developmental sequence (Mandelshtam 1992, 183).

2.3.2e Tools

All of the knives recovered from the Early Scythian period in Tuva had a single blade. Some had clearly defined handles while other examples were more poorly defined. The latter were generally straight bodied with a perforation at the end of the handle. Some examples were discovered with ring-shaped pommels. Knives with clearly defined handles displayed a variety of pommels including one which was fashioned in the form of a wild boar in a standing position. The majority of awls were square in section and some appear to have had wooden or bone handles (*ibid.*, 184).

2.3.2f Household objects

Ceramics were rarely recovered from burials of the Early Scythian period and only one complete jar is known. When ceramics were encountered they tended not to be positioned within the graves but were located near them, as at Hemchik-Bom V, or discovered in other locations beneath the mound (*ibid.*).

2.3.2g Personal items

The majority of personal objects recovered from burials of the Early Scythian period consisted of mirrors and bone combs. Practically all of the mirrors displayed a rim around their edge and had a loop-shaped or button-like handle. All of the bone combs were elongated with finials which had been decorated with geometric or animalistic designs (Bokovenko 1995b, 275).

2.3.2h Summary

Early Scythian period grave goods in Tuva are considered to display all of the features of the classic 'Scythian Triad' and, on the basis of their typology, the grave goods recovered from this phase of the Scythian period are considered by Russian archaeologists to be later than those of the preceding Arzhan Culture (Mandelshtam 1992, 183).

2.3.3 Late Scythian period (5th - 3rd century BC)

The greater number of monuments of the Uyuk Culture in Tuva belong to the Late Scythian period. One of the most thoroughly excavated sites to contain burials characteristic of this period is the cemetery complex at Aymyrlyg, discussed in greater detail in Chapter 3 (*ibid.*, 185).

2.3.3a Funerary monuments and practices

Burial beneath a kurgan became infrequent relative to earlier phases of the Scythian period and an increase in the number of burials in log house tombs is characteristic of the Late Scythian period. The numbers of individuals buried within the log house tombs increased, with as many as fifteen individuals buried within a single tomb. In addition, individual or multiple burials in small and large stone cists became common. Large stone cists and log house tombs were generally positioned in chains across the landscape, while the smaller stone cists with individual burials either concentrated around larger tombs or were constructed in their own separate chain-like subgroups. The individual cist burials which were located near the chains of log house tombs and larger cists were invariably the burial places of children (*ibid.*, 185).

In the interior of the multiple tombs the skeletons were generally positioned in several rows. The orientation of the skeletons in the tombs varied and, in some cases, the orientation of individuals positioned in the same row was found to differ. The burials of children were also commonly recovered from within both log house tombs and large cist burials (*ibid.*).

2.3.3b Clothing and adornments

Fragments of preserved material indicate that the individuals would have worn an inner layer of clothing composed of woollen cloth, or more rarely of silk, and an outer layer of clothing made from leather and fur. The outer clothing appears to have been composed of sections of material of varying sizes sewn together leaving thick seams. In some cases the adjoining seams may have served a decorative purpose (*ibid.*, 189).

Conical-shaped objects of sheet gold located near the skulls of some individuals have been interpreted as the remains of head-dresses. Long leather boots tied at the bottom with a strap, short leather boots and felt footwear were all worn. The remains of a variety of belt types were evident. Broad belts composed of thick leather with areas for attaching straps were probably used to carry heavy objects such as weapons. Narrow belts made from thin leather were also recovered and these displayed areas for the attachment of lighter objects such as pendants. In some cases the leather appeared to have been covered by a layer of fabric. The remains of narrow belts made from thin strips of leather and fragments of wooden girdles were also retrieved. The straps for attaching objects to the belt appear to have been joined to the belt by iron rings. Belts were also associated with a variety of bronze buckles, some of which depicted zoomorphic motifs on their anterior surfaces. A number of iron and bone belts were also recovered. Large hooks were sometimes attached to the belts for suspending a quiver. A number of belt buckles and plates were probably used for decorating the belts. In those cases where the individual was wearing a number of narrow belts these were sometimes decorated with rows of small spherical-shaped bronze buckles which had been sewn on. Bronze and iron bow-tie shaped belt buckles with spiral motifs were associated with the broader belts. Bronze buckles and plates with depictions of feline predators were also used to decorate belts (*ibid.*, 189-190).

A number of horn and bone clasps were probably associated with clothing and footwear. The majority of the clasps were curved in shape and decorated with stylised animal head motifs, though clasps which had been carved in zoomorphic shapes were also found. Some clasps were manufactured from sheep metapodia and horse sesamoid bones (*ibid.*, 190).

A wide variety of personal adornments were recovered from burials of the Late Scythian period in Tuva. These include earrings, torques, pectorals, beads, rings and pins. Beads were used to make necklaces, bracelets and for decorating clothes. Pins appear to have been used to decorate the hair. The majority of earrings were manufactured from bronze, although examples in gold and silver and (to a lesser extent) iron have also been recovered. The simplest earrings were ring-shaped, while in more complicated forms a pendant was attached to the ring. Torques were made from bronze wire, twigs or pipes covered over with gold foil. Beads were manufactured from stone, gold, bronze, bone and glass. Pins were extremely common and were composed of bronze, iron, and bone, with iron pins occurring most frequently. The majority of pins displayed pommels, some of which were large and elaborate. In some cases the upper part of the pin was covered with sheet gold (*ibid.*, 191-192).

Clothing adornments included belt buckles and plates and pendants. The decorative belt buckles were made from thin sheet gold, on the surface of which geometric or zoomorphic designs had been carved. A number of belt buckles had been manufactured in the form of animal heads. Forms depicting a feline predator were most common, although rarer examples were fashioned in the form of the head of a mountain goat or the head of a ram. A few relatively large belt buckles were in the form of a bird with extended wings. A wide variety of bronze, bone and iron pendants were also probably used to decorate clothing. A number of pendants had been made from the teeth of musk-deer and marals, and from the canines of wild boars. In some cases the initial perforation in the tooth had broken and new perforations had been carved (*ibid.*, 192).

2.3.3c Weaponry

Weaponry did not change greatly from the Early to the Late Scythian period. Daggers, pointed axes and arrowheads remained common, though a wide variety of formats were apparent. The majority of the daggers were made of bronze, with only a small number of iron daggers having been recovered. Occasionally daggers with a bronze handle and an iron blade were identified. The majority of bronze daggers had a bow-tie shaped guard, accompanied by a hone-shaped, cigar-shaped, disc-like or zoomorphic pommel. The majority of blades were flat, although a number of triangular-shaped blades were present. Decoration was rare and, when encountered, was generally composed of geometric designs. A number of other daggers had narrow, triangular-shaped guards and hone-shaped pommels. The iron daggers were of two principal forms, either with bow-tie shaped guards and hone-like, disc-like or zoomorphic pommels, or with straight guards and disk-like or ring-like pommels. The daggers would have been worn in a leather scabbard, a section of which was also reserved for carrying a knife. The scabbards were rarely decorated, although at least one was encountered which had a zigzag design painted on it using a red pigment (*ibid.*, 186).

The majority of pointed axes displayed a socket for attachment to a shaft and were composed of bronze rather than iron. The pointed axes were of a variety of sizes and were generally straight or slightly curved, while the butts were round, hexagonal or oval in shape. The majority of iron pointed axes displayed rounded butts. All of the pointed axes would have been attached to a wooden shaft approximately 80 cm in length. The shaft was fastened to the socket using a wooden or a bronze wedge, and a bronze pommel was attached to the bottom of the wooden shaft. The pointed axes would have been carried in leather containers of similar shape to the axes (*ibid.*).

A large number and variety of bronze, bone and wooden arrowheads were discovered, with the various forms probably used for different purposes. Bone arrowheads with blunted or rounded points, for example, were probably used for hunting birds and fur-bearing animals (*ibid.*, 187), while the bronze arrowheads were probably used in combat (Bokovenko 1995b, 277). Bronze arrowheads took three principal forms - three-edged, three-feathered and three-edged and three-feathered - while bone arrowheads were divided into three classes - two-feathered, three-edged and six-edged. Wooden arrowheads were recovered in smaller quantities, although at least two completely wooden arrows were generally found in each quiver. As was the case with the bone arrowheads it has been suggested that these were used for hunting birds and fur-bearing animals. Occasionally arrows were found which were associated with bone whistles. Arrows were stored in long cylindrical quivers which were composed of leather or birch bark and leather (Mandelshtam 1992, 187). Bows that have survived have a length of up to 1 m, while arrows had a probable length of approximately 50-60 cm. The occurrence of arrowheads of a variety of sizes may indicate that both large and small bows were used (Bokovenko 1995b, 277).

2.3.3d Horse accoutrements

Horse fittings were only occasionally recovered from burials of the Late Scythian period. In the majority of cases the horse fittings were associated with the remains of horse skulls and metapodia located on the roofs covering log house tombs, and this has led to the suggestion that these are the remains of horse skins placed there following a burial service. Bronze bits with ring-shaped endings and rod-shaped cheek-pieces with two perforations are characteristic. Occasionally, bronze bits with ring-shaped and triangular-shaped endings and bronze cheek-pieces with distinctly curved endings have been recovered. While bronze predominates as the metal from which bits were manufactured, one burial contained an iron bit and iron cheekpieces. Other objects which may have been used in the bridle include bronze and bone buckles, rings, buttons, badges and blocks (*ibid.*, 187-188).

2.3.3e Tools

The most common tools found associated with Late Scythian period burials were knives and awls, which were often found together in both male and female burials. Knives were most frequently composed of bronze although several iron knives have also been identified. The blades had a variety of shapes, but handles with a loop-shaped perforation at their upper ends were commonly encountered. The iron knives were of similar morphology to the bronze knives, and both types would have been worn in a leather scabbard. The majority of knife scabbards also displayed a perforation for holding an awl (*ibid.*, 188).

The majority of awls were also manufactured in bronze. They had square or rectangular sections, and had a thin tang on which a sheep phalanx was attached as a handle. Iron awls generally had a similar morphology to bronze awls. A number of perforated sharpening stones of various sizes were also discovered and these were probably used for sharpening a wide range of different bladed objects (*ibid.*).

Bronze needles with elongated eyes were commonly recovered from female graves. The needles were stored in small cylindrical bone or leather cases, and sets contained needles of a variety of sizes. The needles would have been used with thin sinews and woollen threads. Occasionally bronze points were found associated with the needles and it is thought that they may have been used as borers in leather working. Spindle whorls of stone and pottery have been recovered from a small number of burials (*ibid.*, 188).

2.3.3f Household objects

A large number of ceramic and wooden vessels were present in the burials of the Late Scythian period in Tuva. The majority of the pottery vessels were handmade, although they may have been made on a moveable base. The clay contained large mineral inclusions, while the surface of the vessel was generally covered by a thin layer of the same clay thereby producing a burnished effect. The vessels were fired unevenly and this resulted in the lower parts of their bodies displaying irregular, dark stains. The pots had a variety of different morphologies, although handless varieties of pitchers and pots were most common. Although rarely decorated, some painted designs were apparent on a number of the vessels (*ibid.*, 192).

Round and oval wooden dishes and bowls of various size were the most common wooden vessels encountered, although bowls and cups with handles were relatively rare. When wooden vessels were recovered from burials they were generally found to have been stacked one inside the other (*ibid.*). A few small spherical-shaped bronze cauldrons with horizontal handles are also known and analogies with similar Scythian period finds in the Pazyryk kurgans of the Altai are indicative that the bronze vessels may have been used to inhale incense during ritual ceremonies (Bokovenko 1995b, 279).

2.3.3g Personal items

A number of bronze mirrors with button-shaped central handles and mirrors with loop-shaped handles were common. Rarer examples with rectangular and trapezoid-shaped handles were also used. Some handles were formed from zoomorphic depictions. Both males and females would probably have carried mirrors in round or oval-shaped leather pouches which would have been hung from their belts by straps. Some of the containers displayed signs of decoration in the form of appliqués and elaborate seams (Mandelshtam 1992, 190).

Bone combs were similar to those from the Early Scythian period and displayed carved zoomorphic figures, while wooden combs were sometimes composed from a single piece of wood, but generally they were composite in form. A single iron comb is considered to have been used for ceremonial purposes. The anterior surface of its handle was decorated in relief with vertical spiral shapes which were covered over with gold sheet. Combs appear to have been an exclusively female personal item (*ibid.*, 190-191).

Small cylindrical and conical-shaped cases made from horn are thought to have been used to contain either valuable objects or substances associated with female toiletries. The majority of the cases displayed skilful carvings depicting animals (*ibid.*, 191).

A wide variety of leather bags appear to have been predominantly used for keeping combs, threads and sinews. The majority of the leather bags were round or oval in shape, although rectangular bags with an outer pocket, sturdy barrel-shaped bags, and oval and trapezoid shaped folding bags were also used (*ibid.*).

2.3.3h Summary

A notable increase in the number of iron artefacts is apparent in the Late Scythian period and it has been suggested that this reflects an attempt to replace bronze with iron as the principal metal used in society. Iron duplicates of bronze artefacts were frequently encountered. A notable increase in the number and variety of ceramic vessels was also evident. As was the case for both the Initial and the Early phases of the Scythian period Culture, the 'Scythian Triad' was clearly apparent in the Late Scythian period in Tuva (Mandelshtam 1992, 186).

2.3.4 Society and social structure

Anthropologically, the Scythian period inhabitants of Tuva are thought to have been heterogeneous, with skulls displaying a combination of Caucasian and Mongoloid characteristics having been identified from cemeteries of this date (Vainshtein 1980, 39). The linguistic background of the tribes of Tuva's steppe-lands is undefined, but it is generally considered that the people would have spoken an Iranian language (*ibid.*). The recovery of well preserved remains of clothing in the Pazyryk kurgans of the Altai, which were contemporaneous with the Uyuk Culture, probably indicate that the population in Tuva at this time wore short leather or fur caftans, short boots, and felt or leather pointed caps. The trousers would have been made of woollen materials, with leather or fur having been utilised during the winter (Rudenko 1970, 83-98).

It is generally considered by Russian academics that the social organisation of the steppe tribes was based on a clan and tribal structure. The occurrence of burials in distinctive groups has been interpreted as representing the burial grounds of members of different clans (Abetekov and Yusupov 1994, 26). Khazanov (1975, 337) was of the opinion that Scythian social organisation had three main features. First, the root of the social structure were clans and tribal connections which were supported by fictitious ideals of genealogical kinship. Second, the clan and tribal structure of Scythian society was further subdivided at all levels. Third, this form of social structure enabled the development of a military organisation. The Classical author Lucian stated that the Pontic Scythians were divided into a royal clan who are thought to have been priests or members of a military aristocracy, and the 'eight-legged ones' who were ordinary herdspeople with a pair of oxen and a cart. Some Russian researchers have postulated that this division correlates with the partitioning of society into a three class or caste groups, a phenomenon which is thought to have been characteristic among all the ancient Indo-Iranians. The Classical sources also indicate that the Scythians had slaves although, in general, Russian academics have regarded this practice as having been limited in scale (Abetekov and Yusupov 1994, 28).

Khazanov (1975, 338) has postulated that Scythian society was composed of a number of different levels, the lowest tier of which appeared to have been occupied by slaves who did not, however, contribute greatly to society. He has suggested that the next strata of society contained a variety of classes of dependent people, who may have included impoverished nomads and royal servants, while the succeeding level comprised of the 'free' population of nomads. Khazanov (1975, 338) regarded the second highest level of Scythian society to have been composed of the religious elite and the nobility, with the highest level of society occupied by the royal clan. He stated that the historical sources infer that power in the royal family was patriarchal.

Mandelshtam (1992, 194) stated that the archaeological evidence derived from Tuvinian burials of the Uyuk Culture suggested that the social structure of these peoples was characterised by a primitive communal system. He interpreted the occurrence of groups of funerary monuments in chains across the landscape as evidence for a stable tribal organisation. In this scheme the main unit of the society was considered to have been a large patriarchal family which would have included several generations linked by mutual origin, and headed by the eldest male. Mandelshtam (1992, 194) was of the opinion that this form of social structure had not developed until the Early Scythian period, however, and the Arzhan Kurgan of the Initial Scythian period in Tuva represented evidence of a stratified society, with the two central burials probably those of aristocrats. It has been suggested that the 'central' system of the distribution of burials in the kurgans of the Early Scythian period reflect the initial orchestration of a primitive communal society. A subsequent development of the system and the strengthening of unity within the family grouping has been proposed on the basis of the increasing number of individuals buried in log house tombs, and in the existence of several rows of individuals laid out within these structures. Researchers have also suggested that monogamy and exogamy prevailed within Uyuk society (*ibid.*, 194).

Differences in the wealth and possibly social standing of the tribal family units are considered to be apparent in the burials of the Uyuk Culture, with relatively rich and poor graves ascribed on the basis of the accompanying grave goods. There were, however, no extreme differences in the wealth of these units and this has been interpreted by Russian academics as evidence that society was generally egalitarian (*ibid.*).

2.3.5 Lifestyle and economy

Pastoral nomadism probably emerged in the steppe-lands at the end of the 2nd millennium BC and the beginning of the 1st millennium BC, and differed from the generally stationary and complex economy of the preceding Bronze Age (Vainshtein 1978, 128; Khazanov 1984, 94). Nomadic pastoralism can only develop as an economic, cultural and social phenomenon on the basis of a number of prerequisites, including a suitable geographical environment, the existence of livestock which are suitably adapted to the region, the availability of riding animals and wheeled draft vehicles, a degree of property differentiation which involves private familial ownership of the livestock, and the opportunity for the division of labour between the nomads and agriculturists (Khazanov 1978, 119). It appears that all the prerequisites for a nomadic economy had been in place for a considerable length of time and that the final stimulus in the formation of pastoralist nomadism was the exceptionally dry climate which occurred at the start of the 1st millennium BC (Khazanov 1984, 95).

Artefacts recovered from Scythian period funerary monuments in Tuva indicate that the economy of the highland-steppe peoples was based upon a semi-nomadic form of pastoralism which was combined with land-cultivation, hunting and gathering. The Scythian period tribes of the mountain-steppe regions of Tuva would have made 'vertical' shifts seasonally, with the construction of permanent structures undertaken during the winter months (Vainshtein 1980, 51). This form of economy would have involved regular repeated seasonal migrations within the borders of a relatively defined territory (Mandelshtam 1992, 193). The existence of tribal burial grounds in pastoralist groups is connected with a semi-nomadic lifestyle, or with a fixed cycle of migrations which includes long stops at particular locations. Herdspeople who were constantly on the move for the majority of the year, and who did not stop for long periods of time at regular camping-places, did not have tribal burial grounds. The distribution of large tribal burial grounds in Tuva of Scythian period date indicates that cyclic migration with fixed routes and set winter camp sites existed amongst the steppe-mountain pastoralist tribes (Vainshtein 1980, 96). Presumably, herds would have been pastured in the mountains during the summer and in the more low-lying land during the winter (Bokovenko 1995a, 255). Ethnographic and historical evidence indicates that in nomadic societies the pastures, unlike the livestock, were communally owned and used by the tribe, since nomadism can only exist if the pastures are used extensively (Markov 1978, 309).

It is generally agreed by anthropologists that 'pure nomadism' cannot exist since the animal products provided by a herd are not sufficient for the nomads to survive upon alone and they are therefore required to practice a certain degree of agricultural activity or to have economic relationships with agriculturists (Hildinger 1997, 6). Semi-nomadic pastoralism is characterised by an economy which involves extensive pastoralism and secondary agriculture (Khazanov 1984, 19). The concentration by nomads on herding activities did not end the herder's need for agricultural produce. The nomad and the agriculturist always remained in contact, and in some cases the nomadic population subjugated the agriculturists (Johnson 1969, 3). Anthropological studies have indicated that once grain products are adopted by a nomadic population they will come to form a major component of the diet. In most nomadic societies breakfast consists of a thick soup or porridge made from mashed grain. It has been suggested that the Scythian nomads would also have eaten such porridges (Gavrilyuk 1987, 27). Archaeological evidence in the form of imprints of grain on ceramic vessels indicates that, by the 4th century BC at least, some of the Scythian tribes were familiar with millet, double-grained wheat, pellicle barley and naked barley (*ibid.*). Other cultivated plants available in the Scythian period included rye, seed oats, peas, lentils, sweet peas, beans, broad beans, rape, flax and hemp (Rolle 1989, 110). The place of agriculture in the Tuvinian Uyuk Culture is, however, unclear. The remains of millet and possible querns have been recovered from a number of burials and it is considered that the population may have practised occasional agriculture but that they more probably obtained agricultural products by exchange processes with neighbouring farming tribes (Mandelshtam 1992, 193).

The remains of animal bones recovered from the Uyuk burials have enabled estimates to be established for the composition of herds, and this has shown that sheep, goats, horses, and occasionally cattle and dogs were present. Camels would probably also have been bred (Mannai-Ool 1970, 91). Yak wool was recovered from the Pazyryk Kurgans in the High Altai which were dated to the second half of the 1st millennium BC (Rudenko 1970, 55). It is possible, therefore, that yak was also exploited in Tuva during the Scythian period. The Tuvinian Scythian period diet would probably have followed the pattern characteristic of nomadic pastoral peoples, principally depending upon dairy products. In addition, their nomadic pastoral economy evidently depended on having draught animals to facilitate their movements from one place to another (Vainshtein 1980, 59).

Faunal remains recovered from Scythian period tombs of the mountain-steppe tribes indicate that sheep and goats were the predominant meat animals maintained by these groups (Vainshtein 1980, 52), while the large number of pommels and handles manufactured from sheep bone indicates the maximal usage of products derived from these animals (Mannai-Ool 1970, 91). Ethnographical studies of late 19th and early 20th century Tuvinian tribes indicate that the mountain-steppe tribes predominantly maintained herds of sheep and goats, with sheep numbering approximately half of the total number of animals. Horses were next in frequency, followed by cattle, with camels and yaks forming less than 1% of the livestock (Vainshtein 1980, 54).

Sheep are equally suited to horses in travelling long distances (Vainshtein 1980, 59) since they can endure shortages of fodder better than other animals. In addition, both sheep and goats reproduce relatively quickly and the lambs and kids only require a few days special attention. Sheep, and goats to a lesser extent, would have provided high-quality meat, milk, skins and wool (*ibid.*, 61), while cattle would have provided meat, milk, skins, and (in the case of oxen) traction power for transportation. During the seasonal shifts oxen would pull carts filled with the belongings of the herdspeople (*ibid.*, 72). The main role of the horse was as a means of transport and as a pack animal. In addition, the horse provided milk, meat, leather and hair (Vainshtein 1980, 65). Although mare's milk is generally of inferior quality to the milk of the other main domesticates it is ideal for the preparation of an alcoholic drink called 'koumiss'. Mare's milk ferments quickly since it is rich in sugar, low in fat and contains the necessary organisms to trigger fermentation. Koumiss has a high calorie and vitamin content and a supposedly pleasant taste (Gavrilyuk 1987, 29). Although the Scythian semi-nomadic tribes maintained large flocks of sheep they attached most importance to their herds of horses, which were carefully reared and trained for riding (Talbot Rice 1966, 11-12). Pig was not a suitable animal for nomadic pastoralism since the pig stomach has a blind gut which acts as a cellulose fermenting system and does not function as efficiently as the composite stomach of ruminants. In addition, pigs require large amounts of water and would not have been capable of travelling the long distances required in the nomadic way of life (Bökönyi 1994, 26).

Hunting was probably of secondary economic importance in the Uyuk Culture. The recovery of pendants made from maral, musk-deer and wild boar teeth suggests that larger animals were hunted. The remains of furs and specific forms of arrowheads also indicate, however, that smaller fur-bearing animals - sable, ermine, squirrel, hare, fox and various rodents - were hunted. Fish vertebrae, which had been used as beads and buttons, have also been retrieved from Scythian period burials suggesting that some fishing was undertaken (Mandelshtam 1992, 193). Gathering wild natural foods probably played a role in the economy, and a leather bag containing cedar nuts recovered from one of the kurgans of the Kazylgan Burial may be indicative that the gathering of plant materials was practised (Mannai-Ool 1970, 93).

The main form of dwelling of the nomadic tribes of the Uyuk Culture would have been the transportable yurt which was made from felt, leather or skins. The severe winters, however, resulted in the tribes constructing polygonal wooden dwellings similar to those used by the present day tribes of this region until relatively recently (Bokovenko 1995b, 280). The discovery of ceramic vessels is not contradictory of a nomadic lifestyle. The ceramics could have been stored at stopping places and seasonal camps. Alternatively, the pots could have been carried by the nomads in their wagons (Okladnikov 1970, 204).

2.3.6 Weaponry and warfare

A large variety of weaponry was contained in the burials of the Uyuk Culture, indicative of the important role which warfare played in society. The weapons include pointed axes, daggers, and bows and arrows (see Sections 2.3.1c, 2.3.2c and 2.3.3c). The main objectives of warfare in the Uyuk Culture would probably have been to capture better pastures, valuable objects, animals and slaves. The latter would have been used as a work-force for husbandry of the tribal herds (Mannai-Ool 1970, 90).

The main body of the Scythian military organisation was cavalry. The Scythian horse-riders used saddles and this gave them an advantage over other tribes who did not. Although no examples have been found in any Uyuk burials, saddles were present in the contemporary Pazyryk burials of the Altai (Rudenko 1970, 129-139). It is not know if the Scythians used stirrups, but a vase from the Chertomylk Kurgan in the Ukraine depicts a saddled horse which had a thong hanging from the girth and this does seem to resemble a leather stirrup. Metal stirrups have not, however, been recovered from any burial of Scythian period date (Vernadsky 1943, 52). The Scythians are considered to have been very successful in warfare because of their tactic of assaulting the enemy with large groups of mounted archers. The bow-warriors were skilled horse-riders and they had even perfected the art of shooting backwards while riding at full gallop (Sulimirski 1970a, 394). It is interesting to note that late 19th and early 20th century Tuvinians were also reported as being capable of hitting an animal using a bow and arrow when on horseback at full gallop (Vainshtein 1980, 170). The steppe nomads would have used great numbers of horses when they were on the march, often as many as three or four per person. This system allowed them to ride a different horse each day, thereby resting tired horses but also making movement over great distances possible (Hildinger 1997, 17).

The bow was the weapon predominantly used by the Scythian period cavalry. The Scythian bow was generally short with a double curve and was extremely suitable for use on horseback. The range of Scythian period arrows is considered to have been up to approximately 122 m (*ibid*., 53). The written sources also relate that the Scythian arrowheads were poisoned. According to accounts written by Theophrastus, Aristotle and Pliny, the Scythians caught a type of small snake which they then killed and left to decompose. At the same time they also filled vessels with human blood and buried them in dung until the blood began to putrefy. The putrefying blood was then mixed with the remains of the decomposing snake to form a deadly poison that could be coated on arrowheads (Rolle 1989, 65). Pointed axes, daggers and knives would probably have been used in hand-to-hand combat (Mannai-Ool 1970, 90).

The existence of female warriors in the steppes during the Iron Age is heavily supported by evidence from the archaeological record, although this has not prevented some academics from remaining sceptical about their existence. The majority of Scythian graves in earlier modern studies were ascribed as 'male' or 'female' on the basis of preconceived academic notions concerning the types of grave goods expected for each gender (e.g. spindle whorls and combs for females and weapons for males), rather than the scientific determination of the skeleton's sex (Taylor 1994, 395). Contrary to these outdated notions, a significant number of burials of definite female warriors have now been identified in the funerary record. The majority of burials of armed females have been recovered amongst those of the ordinary classes of the population, rather than in tombs of the upper strata of society. In a number of cemeteries burials of female warriors represented as much as 37% of the burials. In the majority of cases the women were only buried with arrows, as was the case for the males from the ordinary level of society, although some female burials have been found to contain a greater range of weaponry including arrows, spears and, occasionally, a sword (Melyukova 1990, 106).

The burial of a Scythian female warrior was discovered in Kurgan Twenty at Cholodny Yar on the left bank of the Yyasmin River. The remains of a female skeleton and a young adult male, who was lying at the feet of the woman, were recovered from within the kurgan. Grave goods associated with the woman included two iron lance points, a wooden quiver with forty-seven three-feathered arrowheads and two bronze knives. The grave dates to the 4th century BC. Another female warrior was recovered from Kurgan Sixteen of Akkermen 1 in the Rayon Vasilevka. The remains of a young woman were recovered associated with a quiver (which contained twenty arrows), a number of lances and a heavy fighting belt which was covered in strips of iron. In addition, both of these female warriors displayed evidence of battle trauma (Rolle 1989, 88). A female skeleton with armour, arrowheads, a slate discus and an iron knife was recovered from mid-19th century excavations on the Terek river in the Caucasus (Taylor 1994, 395). Excavations at the Scythian 'royal' kurgan of Chertomlyk identified the remains of four probable female warriors from amongst fifty warrior graves. One of the women had an arrowhead embedded in her spine, and another had been buried with a large shield. Approximately forty burials of Scythian female warriors are known from the area to the west of the Don, while 20% of warrior burials recovered from the east of the Don, in the area Herodotus referred to as Sauromatae, have been of females (*ibid*.). In 1990 the remains of an adult male and an older teenage female of the Pazyryk Culture were recovered from a kurgan in the Ukok Plateau of the High Altai. Both individuals had been buried with identical weapons - battle-axes, knives and bows (Polosmak 1994, 95).

2.3.7 Technology

It is generally accepted that the steppe nomads lived in portable felt tents known as yurts. The skilled carving and building techniques evident in the remains of log house tombs suggest, however, that permanent buildings may also have been used, possibly at the site of their winter camps (see Section 2.3.5). The log house tombs were carefully constructed with timber floors, walls and roofs, and all display the carpentry skills which would have been required to construct a log cabin. In addition, the stone cist burials, particularly those with dome-shaped roofs, displayed evidence of skilled stone working (Mandelshtam 1992, 193).

The wide array of metal-work apparent in the Scythian period burials of Tuva suggest a skilled knowledge of metallurgy. The Sayan Mountains are known to be rich in copper ore (Vernadsky 1943, 41). In addition, a number of ancient copper mines have been discovered in Tuva in the Hovy-Aksy region, on the river On-Kazhaa and in Kizil-Torg near the tribute of the river Bouren-Hem. Evidence of mine shafts and quarries were discovered at these sites, as well as the remains of mining tools. Moulds, crucibles, pits and forges have been recovered in close proximity to the mines, indicating that copper smelting and the casting of bronze objects was undertaken nearby. Ceramics associated with the forges and the occurrence of log house tombs in the vicinity of the Hovy-Aksy mines have been taken as evidence that the mines were operated by people of the Uyuk Culture (Mannai-Ool 1970, 43-44).

The evidence for the iron mines of Tuva has been less extensively investigated. Iron mines have been discovered near the upper part of the river Buren-Hem, near the mountain of Sharbak, in the region of Alty-Khurak, and along the Chevarylyk, Aii-Ezim and Kurgat-Oii rivers. Pottery associated with the mines dated their use to both the Scythian and the Hunno-Sarmatian periods (*ibid.*, 44). The large numbers of gold objects recovered from the Scythian period burials suggests that gold mining was also undertaken in Tuva, but no gold mines have as yet been discovered (Mandelshtam 1992, 193).

2.3.8 Art

A large number of objects recovered from Scythian period burials are decorated in the Animal Style of art so characteristic of all Scythian period tribes (see Sections 2.3, 2.3.1, 2.3.2 and 2.3.4). Stone stelae, known as deer or *olenniye* stones, are also characteristic of the Scythian period. The *olenniye* stones have a height of up to approximately 4 m, and display a variety of inscribed decorations - ceremonial belts from which weapons are suspended, bows in cases, knives and various depictions of animals, predominantly deer. *Olenniye* stones have been found in Mongolia, Transbaikalia, the Altai, Kirgizia, Kazakhstan, the Urals, the Ukraine and the Caucasus. Approximately twenty *olenniye* stones have been recovered from Tuva where they are generally found near kurgans of the Uyuk Culture (Mannai-Ool 1970, 24). A fragment of an *olenniye* stone was recovered from the Arzhan Kurgan (see Section 2.3.1) (Bokovenko 1995b, 272).

A number of rock paintings which have been dated to the Scythian period on stylistic grounds are also known from Tuva. Deer and other ungulates are most commonly portrayed in the rock art. Mountain goats, animals with wings sprouting from their backs, feline predators, birds with extended wings, snakes, wolves and foxes were also commonly portrayed. In addition, battle and hunting scenes were depicted (*ibid.*, 32-33).

2.3.9 Religious and ritual beliefs

The evidence for the burial rites associated with the dead of the Uyuk Culture enables one to attempt to decipher the motives behind these rituals. The practice of inhumation and the burial of the dead with a number of grave goods suggests that the people believed that 'life' in the next world was similar to that on earth. The dead were buried with a number of household items, winter clothes and food and drink, perhaps for their journey into the next life. The conservation of decomposed bodies and bones prior to burial may also suggest the importance of being buried in one's ancestral tomb, while the careful positioning of the body on its side with flexed legs probably held some religious or ritual significance (Mandelshtam 1992, 196).

The occurrence of spirals, rosettes and circles with a central point on a variety of objects has been interpreted as indicative of a solar cult. Zigzag motifs, which were sometimes combined with spirals, have been interpreted as water, and may be indicative that water also had a religious or ritual significance. In some funerary monuments a central fire was present which may have been used to purify the tomb during its construction or before a burial, and the presence of fires within the tombs may be indicative of the ritual importance attached to fire (*ibid.*). A possible solar temple of the Uyuk Culture was discovered in Tuva at Ulug-horum where there is a large, circular stone structure, the interior of which was divided into segments by thirty-two rays, to form a wheel. The wheel and the circle are considered as classic solar symbols. The walls of the interior were decorated with images of horses, goats and rams (Antonini 1994, 293).

The occurrence of a wide variety of animalistic depictions on Scythian period objects has been interpreted as evidence of the importance the people of the Uyuk Culture placed on these animals and the natural world as a whole. Takats (1961, 722 as quoted in Raevskii 1982, 7) stated that the Animal Style of art had a magical and totemistic character which was principally related to hunting. Iatsenko (1971, 131-132 as quoted in Raevskii 1982, 7-8) was of the opinion that the magical and totemistic nature of the zoomorphic art was related to the warlike existence of the Scythians. She postulated that the portrayal of birds of prey and large cats was to provide the warrior and his or her equipment with the strength, agility and vigilance of these animals. Pendants made from the bones and teeth of wild animals may have been regarded as amulets (Mandelshtam 1992, 196) and it is known that up until the 20th century Tuvinian tribes used maral teeth and Cowrie shells as protective amulets (Mannai-Ool 1970, 97).

A number of Scythian artefacts have been recovered which are considered to depict mythological scenes, including the famous electrum vessel from the Ukrainian Kul'-Oba burial. Although it has been argued that such scenes simply depict everyday life the presence of these illustrations on apparently ritual vessels might suggest otherwise (Raevskii 1982, 14-15).

2.3.10 Historical sources

The principal historical sources which provide us with an insight into the lifestyles of the Scythian tribes are the *Histories* of Herodotus and the *Pseudo-Hippocratic Writings* of an unknown Greek doctor. Herodotus, writing in the 5th century BC, devoted Book IV of his *Histories* to discussing the Scythians, while the *Pseudo-Hippocratic Writings* includes some interesting and rather unflattering views on the health of the Scythian populations in the treatise on *Airs, Waters, Places*, probably written in the second half of the 5th century BC. The author of the text is referred to as Pseudo-Hippocrates. It is generally considered, however, that the descriptions of the author were biased by his alien sensitivities to the nomadic way of life (Rolle 1989, 54). A number of other Classical writers, such as Strabo, Lucian, Hesiod and Homer also refer to the Scythian tribes. In addition, accounts of Scythian activity in the Near East are recorded in the Assyrian Annals and form the basis of some of the Hebrew prophecies (Phillips 1972, 129).

Herodotus' work is considered to provide the most informative descriptions of the non-literate societies of his time. There has been considerable disagreement, however, over the reliability of his account but early 20th century archaeological discoveries have generally corroborated with Herodotus' story, especially when he described the burial rites associated with the death of a 'Royal Scythian' (see Section 2.3.10c) (Taylor 1994, 393-394).

2.3.10a Lifestyle and economy

Although he evidently did not admire the Scythians, Herodotus tells us that he did respect their success at self-preservation. He stated that their lifestyle dictated that no one could invade their territory without defeat, and if the Scythians wanted to avoid engaging with the attackers they could and would do so. He attributed this advantage to their nomadic lifestyle. They did not live in fortified towns but roamed around in wagons. They were all accustomed to fighting on horseback with bows and arrows, and they were dependent on their cattle for food rather than agriculture (Sélincourt and Burn 1972, 286). Herodotus tells us that pigs were never used for sacrificial purposes and that they were not bred within the Scythian territory (*ibid.*, 291).

Pseudo-Hippocrates also informs us that the Scythians were nomads who lived in wagons. The lighter wagons had four wheels but some had six, and all of the wagons were enclosed by felt walls, were built like houses, were divided into two or three sections, and provided protection against rain, snow and wind. The wagons were drawn by two or three yokes of hornless oxen. The Scythian females lived in the wagons, while the males rode on horseback followed by their herds. They would stay in one area and only moved on when there was no longer enough grass to feed their herds (Chadwick and Mann 1978, 163).

Herodotus indicated, however, that not all Scythians tribes were pastoralists. He stated that the Callipidae who lived to the west of the mouth of the Borysthenes (Dnieper) and their eastern neighbours the Alizones both had a similar lifestyle to that of the Scythians. They also grew grain for food, however, as well as onions, leeks, lentils and millet. According to Herodotus the agricultural Scythians, who grew grain for export, lived to the north of the Alizones (Sélincourt and Burn 1972, 277).

According to Pseudo-Hippocrates the Scythians ate boiled meat, and drank mare's milk, from which they also made cheese (Chadwick and Mann 1978, 163). Herodotus tells us that the Scythians blinded their slaves, who were then responsible for collecting the mare's milk (Sélincourt and Burn 1972, 271). Hesiod, who lived no later than the 8th century BC, also makes reference to the mare-milking Scythians (Sulimirski 1970b, 395).

2.3.10b Health and hygiene

Pseudo-Hippocrates states that the Scythians were the least prolific of all races, and attributes this to the harsh environment in which the they lived. The people were all of similar physique since they ate the same sorts of food, wore the same clothes, breathed moist thick air, drank water from the snow and ice and did no hard work. The Scythians were heavy and fleshy, and their joints were covered with flesh. Both males and females were described as being fat and hairless, with the two sexes resembling one another (Chadwick and Mann 1978, 164).

To remove moisture from the body the Scythians cauterised themselves on the shoulders, arms, wrists, chests, hips and loins, which was thought to make their bodies more sinewy and stronger. Pseudo-Hippocrates was of the opinion that they were flabby and stout because they were not wrapped in swaddling clothes as infants and were not accustomed to horse-riding when they were children. The writer also thought that they were too inactive as children, with the male children spending most of their time sitting in wagons, while the girls were described as being 'amazingly flabby and podgy'. The Scythian people also had ruddy complexions because of the cold environment, which caused their fair skins to become reddened and burned (*ibid.*, 164-165).

Pseudo-Hippocrates stated that the males were generally lacking in sexual desire because of their poor physical constitution and that they were so exhausted from horse-riding they were 'weak in the sexual act when they do have intercourse'. The females were infertile because of their 'fatness and flabbiness', while 'the opening of the womb was sealed by fat which does not permit insemination'. In addition, the Greek writer stated that the Scythian women 'menstruate infrequently and little' (*ibid.*, 165).

Rich male Scythians became impotent and effeminate, and were called 'Anarieis'. The Scythians believed this transformation to have occurred as a result of a divine visitation, and these men were revered within society. Pseudo-Hippocrates was of the opinion that the rich males became effeminate as a result of developing varicose veins and lameness from horse-riding. The Scythian cure for this was to cut the vein which ran behind each ear, resulting in haemorrhage and weakness. Pseudo-Hippocrates believed that the cure resulted in impotence and that the males then became gradually effeminate, although a differential diagnosis argued that the males spent so much time on horseback, wore trousers and did not 'handle their private parts', all of which led to no desire for sexual intercourse (*ibid.*, 165-167).

Herodotus also refers to the effeminity of the Scythians and states that the Scythians who robbed the temple of Aphrodite at Ascalon in Syria were punished by the goddess and inflicted with what is known as the 'female disease'. He related that the disease was still present in the Scythian population in his time, with those affected known as 'Enarees' (Sélincourt and Burn 1972, 84). It is interesting to note that amongst the Siberian tribes of the early 20th century it was common for male shamans to take on the characteristics of female shamans and, although more rare, for female shamans to take on the features of males. The phenomenon was also recorded by travellers in the 18th and 19th centuries (Czaplicka 1914, 248). In some cases the male shamans simply began to dress like women and undertake female tasks. In more extreme cases, however, the 'male' shaman began to 'feel himself a women', sought a male lover and occasionally got married (*ibid.*, 249).

Herodotus informs us that the Scythians did not take baths in water but rather washed their bodies in a vapour-bath. The vapour-bath consisted of a frame-work of three sticks joined together at the top, over which were stretched pieces of woollen cloth. A dish with red-hot stones was placed inside the structure. Once the individual was inside the tent-like structure hemp seeds were thrown onto the stones, and a vapour with narcotic qualities was produced. The women also made substances which would be the equivalent to modern-day face-masks. These were composed of ground cypress, cedar and frankincense which was mixed with a small amount of water to form a thick paste. The paste was applied to their faces and bodies, and left on for a day. Herodotus states that when the paste was removed the skin was 'clean, glossy and fragrant' (Sélincourt and Burn 1972, 295).

2.3.10c Funerary monuments and practices

Herodotus described the funerary rites associated with the death of a 'Royal Scythian' in great detail. First, the corpse had its abdomen cut open and the internal organs were removed. The cadaver was then stuffed with various aromatic plants, including galingale, parsley-seed and anise, and was sewn up. The whole body was then covered with wax and placed on a wagon for a tour through the territories of the various Scythian tribes. The groups which received the dead royal mutilated their bodies as a sign of grief. They would cut a piece from their ear, shave their heads, make circular incisions on their arms, slash their foreheads and noses, and stick arrows through their left hands. The king was then buried in the country of the Gerrhi. The corpse was placed on a mattress in a large rectangular pit. Spears were placed in the ground on either side of the mattress to support a roof of withies which were laid on wooden poles. Members of the king's household were buried with him including one of his concubines, his butler, his cook, his groom, his steward and his chamberlain, all of whom had been strangled. In addition, horses, gold cups and an array of other valuable items were also buried with the king. When the burial ceremony was complete a huge mound of earth was constructed above the tomb (Sélincourt and Burn 1972, 293-294).

A year after the death of the king fifty of his best servants and his finest horses were strangled and gutted and their bodies stuffed with chaff. A number of wheels were cut in half and attached in pairs to stakes which had been driven into the ground. Poles were driven horizontally through each horse which was then mounted on a set of the wheels. A pole was driven through the back of the neck of each servant who was then mounted onto a horse (*ibid.*, 294).

Herodotus additionally informs us of the burial rites associated with the death of an ordinary member of Scythian society. The closest relatives lay the body in a cart and brought it to visit their friends. Each of the families which were visited served their guests, including the corpse, a meal. The 'visits' were undertaken for a period of forty days after which the corpse was buried (*ibid.*). It is interesting to note that it has been recorded that the Siberian Gilyak and Ainu tribes also included the corpse in funerary feasts in the early 20th century (Czaplicka 1914, 152-153).

Herodotus also described the funerary rituals associated with the Issedones. When a man's father died, his kinsfolk brought sheep to his house which were to be offered as sacrifices. The sheep and the body of the dead man were cut into joints and sliced up. The sheep and the human flesh were then mixed together, served and eaten. The dead man's head was treated separately, with the hair having been stripped off and the endocranium cleaned out. The skull was then gilded and preserved as a sort of sacred image (Sélincourt and Burn 1972, 279). Herodotus recounted a similar story of the Massagetae. When a man reached old age his relatives threw a party in which the individual and a number of cattle were slaughtered. His flesh was then boiled and eaten. Anyone who died of disease, however, was not eaten but buried, and it was considered a misfortune not to have been sacrificed in this manner (*ibid.*, 128). It is interesting to note that Bendann (1969, 272) recorded the belief in Siberia in modern times that when a person died he or she continued their existence in the afterlife in the same physical condition. Consequently, parents often requested that their children kill them before their bodies began to physically decline.

2.3.10d Religious beliefs

Herodotus recounted that the Scythians worshipped Hestia, their main deity, Zeus, and Earth, whom they believe was the wife of Zeus. Deities of secondary importance included Apollo, Celestial Aphrodite, Heracles and Ares (Sélincourt and Burn 1972, 289). Animal sacrifice by strangulation was common. The animal was then skinned, the flesh was stripped from the bones and the meat was boiled in a cauldron under which a fire was kindled using the bones of the dead animal. Horses were the most common sacrificial animals, although cattle were also sacrificed (*ibid.*, 290).

Ares was the only deity for whom statues, altars and temples were constructed (*ibid.*, 289). Each district had a temple for worshipping Ares which consisted of a huge mound of brushwood, on top of which was a square platform. An ancient iron sword was placed at the summit of the monument to represent the deity. Each year horse and cattle were sacrificed to the sword. In addition, prisoners of war were also sacrificed to the deity. One man was chosen out of every one hundred, wine was poured over his head, and his throat was cut over a bowl. The bowl was then carried to the platform on top of the woodpile and the blood was poured over the sword. While this was occurring another ceremony was enacted on the ground beside the brushwood pile. This involved severing the right hands and arms of the slaughtered prisoners and tossing them into the air (*ibid.*, 290-291).

Herodotus stated that there were a large number of soothsayers in Scythia who divined using willow rods. In addition, the effeminate class of males known as the 'Enarees' (see Section 2.3.10b) divined using the inner bark of a lime-tree (*ibid.*, 292).

2.3.10e Weaponry and warfare

Herodotus recounted that it was Scythian custom for each man to drink the blood of the first man he killed. The heads of all the enemies who had been killed in battle were brought to the king. Only if a warrior presented the king with a head did he receive a share of the loot. The head was carefully processed - the skin was stripped from the head by making a circular cut around the ears and then shaking out the skull. The flesh was scraped off the skin using an ox rib. Once the flesh had been removed from the skin, it was worked in the warrior's fingers until it was supple and fit to be used as a sort of handkerchief. The handkerchiefs were then hung on the bridle of the warrior's horse as symbols of battle prowess (*ibid.*, 291).

Herodotus also related that many Scythian warriors sewed a number of scalps together to make cloaks. In addition, he stated that in some cases they removed the skin and nails of the right hands and arms of the dead enemies to cover their quivers. Cases were known where a complete body was flayed, and the skin was stretched on a wooden frame which the warriors carried around with them (*ibid.*).

According to Herodotus the skullcaps of the worst enemies of the Scythians were covered with rawhide, and in some cases the endocranial surface was gilded for use as a drinking vessel. The skulls of warring kinsfolk were supposed to be treated in the same manner (*ibid.*).

2.3.10f The role of females in society

Herodotus relates a story which involved the union of Scythian male warriors with Sauromatae female warriors (Amazons). The Scythian warriors had fought with a group of Amazons (whom they had presumed were males) in defence of their property. Once the Scythians discovered that the Amazons were female they wanted to join with them, and the two camps were united. The Scythians wanted to bring the Amazons back to their families, but the Amazons replied that they were too different to the Scythian females in that they were riders and warriors, whereas the Scythian females 'stayed at home in their wagons occupied with feminine tasks' and never hunted. Eventually the two groups compromised and settled together in a different part of the country. Here the Amazons were said to have maintained their traditional lifestyle - hunting on horseback, participating in war, and wearing the same clothes as males (*ibid.*, 307-308).

Pseudo-Hippocrates also wrote of the female warriors of the Sauromatae, who were described as a special race of Scythians. The women rode horses, shot arrows and threw javelins from horseback, and fought in campaigns for as long as they were virgins. Custom dictated that they did not lose their virginity until they had killed three enemies and offered the prescribed ritual sacrifices. Once a woman had married she did not ride again unless through military necessity. Pseudo-Hippocrates related that the Sauromatian women had no right breast since their mothers had applied a specially-heated iron to the breast when they were children which prevented it from developing. The objective of this practice was so that all the strength required to develop the breast would go instead into the development of the right arm and shoulder (Chadwick and Mann 1978, 162-163).

Both accounts of the Amazons suggest that in the 'Scythian' tribes' females were subservient to males and performed only domestic roles in society. Herodotus also recounts, however, that in the Massagetae tribe Queen Tomyris ruled following the death of her husband. He also indicates that she was responsible for co-ordinating the battle against the aggressor King Cyrus of the Medes (Gera 1997, 187-204). Another non-subservient female, the Sacian Queen Zarinaea, is described in the accounts of Diodorus Siculus. She was described as a warrior queen who fought on horseback and was ambitious in war (*ibid.*, 84-100). Herodotus also informs us that men and women held equal authority in the tribe of the Issedones (Sélincourt and Burn 1972, 279).

2.4 The Hsiung-nu and the Hunno-Sarmatian period in Tuva

During the late 3rd and early 2nd century BC several significant social changes occurred among the nomadic tribes of the steppes. In the west the Scythians were succeeded by the Sarmatians, while in the east the Hsiung-nu (Huns), who are often presumed to be the prehistoric ancestors of the later Huns, had emerged as a strong nomadic power. As a result this period in the history of the steppe nomads is referred to as the Hunno-Sarmatian period (Zadneprovskiy 1994, 457). Due to their rising prosperity from pastoralism, the development of an iron industry and military prowess, the twenty-four Hsiung-nu tribes dramatically increased in strength, resulting in the emergence of the powerful Hsiung-nu Empire (Ishjamts 1994, 153). According to the historical sources the Hsiung-nu initially engaged in a lengthy struggle against other nomadic tribes, particularly the Wu-sun and the Yüeh-chih, as well as with the Chinese (Litvinsky and Guang-da 1996, 29), and for several centuries the Hsiung-nu Empire acted as a rival power to the Han Empire of China. At its greatest extent the nomad empire is thought to have stretched from

Korea to the Altai, and from the border of China to Transbaikalia (Phillips 1965, 112). At the end of the 3rd and the beginning of the 2nd century BC the populations of Tuva, the Transbaikalia area, the Minusinsk Basin, and the Altai are all considered to have experienced the impact of Hsiung-nu military force (Davidova 1996, 159).

The Chinese sources recount that in 201 BC the *shan-yü* Motun conquered the country of Dinlin, which was located in the Minusinsk Basin on the Yenisey river. The Dinlin are considered to have been part of the Tagar Culture and are mentioned in the written sources as being among the acquired 'possessions' of the Huns (Mannai-Ool 1970, 107; Sulimirski 1970b, 112). It is generally considered that the Hunnish tribes would also have established rule to the south of the Tagar Culture which, by this time, would have also included the region of Tuva. The archaeological data has been used to support this theory but it also indicates that there was not an immediate transformation from the preceding Uyuk Culture of the Scythian period. It has been postulated, therefore, that the new culture developed as a result of a complicated and prolonged process that occurred in a climate of rapid political change which involved the movement into Tuva of new groups of people (Mandelshtam and Stambulnik 1992, 196). The population of Tuva is thought to have established greater ethnic links to the Minusinsk Basin during this time, and in the 1st century BC part of the population of Tuva and the Gianguny tribe, which had been forced out of north-western Mongolia by the Huns, moved to the Minusinsk Basin and formed the basis of the Tashtyk Culture (Mannai-Ool 1970, 107).

The Hunno-Sarmatian period culture in Tuva is referred to as the Shurmak Culture. The older Scythian period Uyuk Culture in Tuva does not appear to have totally disappeared from the archaeological record, however, and a large proportion of its characteristics - its funerary monuments, burial rituals and grave goods - seem to have became assimilated into the new Shurmak Culture (*ibid*., 108). Three main categories of archaeological monuments are generally encountered in the Hunno-Sarmatian period in Tuva. The first group displays characteristics of the preceding Uyuk Culture, and are thought to represent the monuments of the native Tuvinian population. The second group contains monuments which are considered to have been constructed by the new Hsiung-nu population, and are mainly concentrated in the southern regions of Tuva. Both of these two monument types are considered characteristic of the Early Phase of the Hunno-Sarmatian period in Tuva. The third category of monuments are those which display the fully developed characteristics of the Late Hunno-Sarmatian period. These monuments were eventually to become typical for much of Tuva in the later stages of this period, although a number of local differences were apparent (Mandelshtam 1992, 196-197).

2.4.1 Early Hunno-Sarmatian period (3rd - 1st century BC)

The Early Hunno-Sarmatian period is considered by Russian academics to have been an age of transition when the native Uyuk Culture gradually became assimilated into the new Hsiung-nu Culture. The burial grounds ascribed to this phase of the Hunno-Sarmatian period have been dated on the basis of typological similarities of their grave goods with those recovered from neighbouring territories. It has not proved possible, however, to determine the exact period when the Shurmak Culture originated (*ibid*., 197). The two main Tuvinian cemeteries to have revealed burials of the Early Stage of the Hunno-Sarmatian period are Bay-Dag II and Aymyrlyg XXXI. Since the funerary monuments and grave goods from Aymyrlyg XXXI are described in detail in Chapter 3 (see Section 3.3), the current discussion concentrates predominantly on the monuments and material culture of the Bay-Dag II cemetery.

2.4.1a Funerary monuments and practices

The number of Uyuk style burials - log house tombs and stone cists - which definitely date to the Hunno-Sarmatian period is relatively small. While the architecture of these tombs is distinctly of a Scythian period style, they have been found to include arrowheads and pottery vessels of Hunno-Sarmatian typology and this has led to the suggestion that they represent burials from the transitionary phase. A number of secondary graves placed in burial structures of the Uyuk Culture have also been identified. The first variety of these consist of stone cists in which the skeleton is positioned lying on its side and has flexed legs. Burials of this kind have been recovered from the cemeteries of Bay-Dag, Hovy-Aksy, Ulyg-Oymak, Aymyrlyg and Karadag II. The morphology of the burial structure is identical to that for the Uyuk Culture, but the accompanying grave goods are characteristic of the Hunno-Sarmatian period culture. The second form of secondary grave consist of rectangular or oval pits, in which the skeleton is positioned lying on its side with flexed legs. Burials of this kind have been discovered in the cemeteries of Bay-Dag II, Turan III, Uspenskoye, Hemchik-Bom I, Chinge II, and Aymyrlyg. In some cases the pits were lined with boulders, or stone slabs were positioned along the walls. In general, very few grave goods are associated with these burials, with clay vessels, iron knives and iron buckles having only occasionally been recovered. A single case of a skeleton which was lying in typical Uyuk Culture fashion - on its side with flexed arms and legs - in a wooden coffin was recovered from the cemetery of Aymyrlyg, but its presence in a wooden coffin would suggest it belonged to the Hunno-Sarmatian period (*ibid*.).

The Bay-Dag II cemetery is located on the left bank of the Yenisei river to the east of the town of Shagonar. The early phase of the cemetery is considered to date to the Hunno-Sarmatian period. There were two main forms of funerary monument - large trapeziform-shaped mounds, and large and small circular mounds. The trapeziform-shaped mounds attained a length of up to 27 m. The diameter of the circular mounds ranged from 10-15 m for the larger forms, and 4-6 m for the smaller varieties. A rectangular pit, which was generally orientated from south to north or, more rarely, from south-east to north-west, was located beneath both the trapeziform and round structures. The size of the pit was dependent on the size of the overground monuments. The pits of the larger monuments generally contained a trapeziform-shaped coffin chamber, inside which a coffin was positioned on stone slabs. The space between the walls of the pit and the walls of the coffin chamber were lined with stones (*ibid.*, 197-198).

The bodies were positioned in an extended supine position with the head orientated to the north or the north-west. The remains of animals were generally positioned near the legs of the skeletons and, in one case, a complete sheep skeleton was recovered from beneath the coffin. Grave goods were located in the coffin near the skeleton, in the coffin chamber or in the space between the coffin chamber and the coffin. All of the coffins recovered from the large monuments had been richly decorated with iron plates and thin strips of gold attached with iron nails. In addition, some of the coffins had been covered with silk. The coffins recovered from the smaller monuments were also decorated but to a much lesser extent (*ibid.*, 198). For further discussion see Section 3.3.1.

2.4.1b Clothing and adornments

The remains of clothing from the Bay-Dag II cemetery consisted mostly of belt buckles. These displayed a variety of forms - ring-shaped, rectangular and teardrop-shaped - and generally had a long, moveable lug although, in some cases, the lug was stationary. One of the teardrop-shaped buckles was found to be covered with gold foil. A number of iron nail-shaped pendants and a number of beads composed of a variety of materials were also recovered from the burials (*ibid.*). For further discussion see Section 3.3.3e.

2.4.1c Weaponry

Weapons recovered from the Bay-Dag II site included the remains of a single bladed iron sword with a ring-shaped pommel, fragments of iron daggers and the remains of long iron knives. In addition, large numbers of brackets from bows and a variety of iron arrowheads were recovered. The arrowheads were generally elongated, and three-feathered with a diamond-shaped head. In some cases the arrowheads were associated with bone whistles. The remains of the bone bow brackets suggest that the bows were of large and complex construction (*ibid.*, 198). For further discussion see Section 3.3.3a.

2.4.1d Horse accoutrements

The horse accoutrements recovered from Bay-Dag II were all made from iron. The bits appeared to have been ring-shaped with additional diamond-shaped loops for the attachment of the bridle. The cheek-pieces were rod-shaped (*ibid.*). Information on the horse accoutrements from the Aymyrlyg XXXI cemetery was not provided by the excavators in the excavation summary accounts.

2.4.1e Tools

Details of the tools recovered from the Group XXXI burials at Aymyrlyg are provided in Section 3.3.3b.

2.4.1f Household objects

All of the clay and bronze vessels recovered from the Bay-Dag II burial ground were typical of the Hunnish culture. The clay pots displayed a variety of sizes, and all of the pots had well-defined rims. The neck of the pots were generally decorated with a wavy ornament, while the bodies of the pots displayed vertical burnishing. A number of narrow, vase-shaped clay pots were also recovered. These had a high neck and a well-defined rim, and displayed vertical burnishing and horizontal curved stripes on the upper part of the pot bodies. The bronze vessels were of similar morphologies and displayed similar decorations to the clay pots. Several of the bronze vessels displayed handles which were attached to the rim (*ibid.*). For further discussion see Section 3.3.3c.

2.4.1g Personal items

Tweezers were recovered from several of the burials at Bay-Dag II. In some cases two pairs of tweezers of different size were recovered from a single grave. Some of the tweezers were plain while others were decorated. A number of small bone spoons, which are also considered to have been toilet items, were recovered from some of the burials (*ibid.*). For further discussion see Section 3.3.3f.

2.4.1h Summary

The Early Phase of the Hunno-Sarmatian period in Tuva is characterised by the occurrence of a wide variety of funerary monuments, and the existence of a number of different burial rites. In addition, some of the burials displayed a combination of features typical of both the Uyuk Culture and the Shurmak Culture. These traits have been considered by Russian archaeologists as evidence for the existence of an unstable society in the early stages of development, with an apparently considerable amount of interaction between the native Uyuk Culture of the Scythian period and the newcoming tribes of the Hunno-Sarmatian period Shurmak Culture who had just entered Tuva at this time (Stambulnik 1983a, 39).

2.4.2 Late Hunno-Sarmatian period (1st century BC - 5th century AD)

The majority of evidence for the Late Phase of the Hunno-Sarmatian period Shurmak Culture has been obtained from cemetery excavations, although a small amount of research has also been undertaken on nomadic camp-sites and mines of this period. The remains of burials from the Late Phase of the Hunno-Sarmatian period have been discovered in all areas of Tuva (Mandelshtam and Stambulnik 1992, 200).

2.4.2a Funerary monuments and practices

The majority of the cemeteries consist of a small number of mounds or only a single mound. In addition, most of these burial mounds are small. The kurgans may be dispersed or they may be positioned together in compact groups near Scythian period burials mounds. In a restricted number of cases the mounds appear to have been deliberately positioned in chains, although in the majority of cases no such deliberate system is evident and the chain alignment is most characteristic of the larger groups of kurgans. Burials may be placed beneath individual mounds or large numbers of burials may occur beneath a single large kurgan. At the Kokel cemetery, for example, the number of burials beneath a single mound ranged from five to 141 burials. Some of the graves had cut into others indicating that the cemetery had been used for a long period of time. In addition to the burial monuments of the Shurmak Culture, one of the most specific features is the existence of ritual structures associated with funeral rites. The ritual structure and the funerary monument often form a single complex (*ibid.*).

Seven forms of burial construction have been identified on the basis of their surface characteristics:

1. Round stone kurgans under which are one or, more rarely, up to four burials. In some cases the mound consists of a circular border in which the internal central area is paved with stone. The organisation of the burials associated with such structures is diverse. In some cases the body may have been placed in a pit covered with a wooden board or branches. In others it may have been left lying on a board within the pit covered over with more boards. Elsewhere the body may have been buried in a pit with stone or slab-lined walls or one which had stones at the corners on top of which an overhead cover was supported. Cases were recorded where bodies were buried in wooden coffins, some of which were also surrounded by stone slabs. One or more burials could be placed in a large rectangular coffin, while bodies could be interred in miniature log house tombs and stone cists. Finally, a number of individuals were buried on the ancient ground surface or in pits with an alcove, the entrance of which was filled in with stones. Burials beneath a single mound can display uniform or diverse features and a variety of orientations one cutting into another (*ibid.*, 200-201).

The remains of clay pots were found beneath several of the kurgans. These had been dug into the ground and were surrounded by a square 'box' of slabs which were positioned near or in a burial. Pottery sherds or spreads of ash have been discovered on the ground surface at the margins of a number of barrows. It has been suggested that the pots and spreads of ash are associated with ritual activities which occurred around the time of burial (*ibid.*, 201).

2. Rectangular or square stone kurgans covering from one to four burials. The burials can be found in simple ground pits, pits where the skeleton is lying on boards or within a wooden coffin and the pit lined with stones, and in stone cists.

3. Low oval-shaped mounds covering an individual burial in a ground pit. This class of funerary monument is rare.

4. Circular stone borders, inside the interior of which is a grave. The majority of burials are in ground pits, but burials in wooden coffins and stone cists have also been recovered.

5. Low irregularly-shaped stone mounds. Burials of all varieties have been recovered from beneath such structures.

6. Enigmatic constructions. The appearance of the ground constructions which would have overlain individual graves beneath the large kurgans with multiple interments is uncertain, although it is considered that they would have been circular or rectangular in shape. Practically all of the variations in burial form discovered under kurgans with single burials have also been discovered under the large multi-burial barrows. Burials in coffins are, however, most commonly found beneath these structures. The remains of ritual structures associated with the burial rites are commonly found adjacent to these large multi-grave barrows, and include the remains of funerary pyres (some of which are associated with a ceramic vessel), fragments of individual metal objects or pottery sherds, and rows of stones which are often associated with pottery sherds and other objects (*ibid.*).

7. Burials which do not have any surface signs, and which are usually only identified by the existence of a small surface depression. The majority of these burials are found in coffins or lying in pits lined with branches and covered over with boards (*ibid.*).

The remains of structures associated with funerary rituals have been classed into three main categories:

1. Circular barrows, under the centre of which is a clay vessel which has been embedded in the ground. The pot is frequently surrounded by a small square box made from stone slabs. In some cases charcoal spreads, pottery sherds and animal bones are recovered from the ground surface at the periphery of the mound.

2. Circular barrows, covering a circular, oval or rectangular demarcated area. A clay pot is embedded in the ground at the centre of the structure. On rare occasions a spread of charcoal is also present in the central area. In some cases the demarcated area also contains a small square box composed of stone slabs in addition to a pot. Ash and charcoal are generally present at the bottom of this box.

3. Slab arrangements in a rectangular form, which are sometimes surrounded by a rectangular stone fence. A clay pot embedded into the ground is usually found to be buried beneath one of the edges or corners of the rectangular structure.

The majority of interments in the burials of all variations are single, although a number of double and collective interments have been discovered. Burials of two individuals are generally found to contain the remains of a male and a female individual. In these double burials the male and the female can either be found within a single burial chamber or in separate burial chambers where the individuals are placed in separate coffins. Fewer double graves containing individuals of the same sex have been recovered, and the majority of these have been found to consist of the burials of two males who were placed in separate burial structures, usually coffins. The burial of two females are rare, as are a small number of burials which contain a female and a child, or a female, male and two children. A variety of other collective burials have been recovered but, in general, these are also rare (*ibid.*, 201-202).

The majority of skeletons are placed in an extended supine position, with one or both arms flexed and the hands positioned near the pelvis. Less frequently the skeleton is positioned on its side, or lying in a supine position with the knees raised. The orientation of the burials varies, but an orientation to the north-west, north or west is most common (*ibid.*, 202).

2.4.2b Clothing and adornments

In general, only small fragments of leather, woollen or silk materials were recovered from the burials. A male burial of Late Hunno-Sarmatian period date from the cemetery of Kara-Dash, however, was found dressed in an almost completely preserved set of clothes. The individual was wearing a leather gown, a leather hat with side flaps and a pair of low leather boots which had been tied with leather belts at the ankles (*ibid.*, 203).

Fragmentary remains of leather, wool and wicker belts were recovered from several burials. In the majority of cases several buckles were associated with the belts. The buckles were predominantly made from iron and displayed a variety of shapes - rectangular, trapeziform, oval and lyre-shaped. A variety of rings were also recovered which probably served as buckles. Occasionally bronze and bone buckles were recovered. The majority of bronze buckles were small and decorated, while most of the bone buckles were trapeziform in shape. A number of metal and bone pendants were also associated with the remains of belts and had probably been used for decorative purposes. Small hooks associated with the belts had probably been used for attaching various objects to the belts (*ibid.*).

Only a small amount of jewellery was recovered from burials of the Late Phase of the Hunno-Sarmatian period Shurmak Culture. Small round plates with a central perforation made from iron, bronze, bone or shell were frequently used to decorate head-dresses or clothing. A variety of shapes of buckles/plates made from sheet gold were probably also used for decorating clothing. In some cases the plates displayed geometric designs in relief. Pendants in the form of birds with outstretched wings, and pendants in the form of miniature 'Scythian' cauldrons were recovered from several of the burials (*ibid.*).

Earrings, torques and beads were retrieved from several burials. The three main classes of earrings were simple rings sometimes associated with a leaf-shaped gold pendant, spiral-shaped fine strips of gold, and thin oval and rectangular shaped sheets of gold decorated with geometrical motifs. The torques were generally small and composed of a twisted strip of sheet gold with fastenings at either end. Beads were only occasionally recovered. The beads displayed a variety of shapes and had been composed of Carnelian, limestone and blue glass. In some cases bronze rings, Cowrie shells and maral teeth were associated with beads (*ibid.*).

2.4.2c Weaponry

Items of weaponry have been recovered from nearly all burials of the Late Phase of the Hunno-Sarmatian period Shurmak Culture, particularly in the large multi-burial kurgans. The complete weaponry of a warrior probably consisted of a sword, a dagger, a spear and a quiver containing arrows, although swords were only occasionally present in the burials. In addition, miniature versions and models of weapons have also been found. The miniature swords consist of iron knife-like implements with ring-shaped ends and decorated wooden hilts. These objects were generally placed in wooden scabbards which displayed a variety of shapes and ornamentation, and some of the scabbards are similar to those typical for Sarmatian burials. Models of swords were completely manufactured from wood (*ibid.*, 202).

Daggers were rarely recovered from the burials although spearheads have occasionally been discovered. The spears were all socketted with either a leaf-shaped, triangular-shaped or knife-shaped end. Smaller versions of spearheads probably represent the remains of darts (*ibid.*).

Bows were of the complex Hunno type, with a length of 1.3 - 1.5 m. They were fitted with a number of horn brackets, either two pairs of end brackets or three middle brackets. The remains of wooden models of bows of a variety of size and displaying a variety of skill in manufacture were retrieved from a large proportion of the burials. In many cases the model bows had been painted in red and black. Arrows were generally made from birch or willow wood. They had a length of up to 0.7 m. The end of the arrow which was not attached to the arrowhead was thickened, and had a slit cut into it for placing the arrow on the bow string. In some cases two painted belts of red were positioned above the slit. A large proportion of the arrows were associated with barrel-shaped bone whistles (*ibid.*).

A large variety of arrowheads were present in the burials, the majority of which had been manufactured from iron, although bronze and bone examples were occasionally discovered. Some of the quivers contained arrowheads of wood. All of the bronze arrowheads had shafts and were either diamond-shaped or had a tiered head. Iron arrowheads also had shafts and displayed a number of principal forms - three-feathered; three-bladed arrowheads with a diamond-shaped end; three-bladed arrowheads with a leaf-shaped end; three-bladed arrowheads with a hexagonal-shaped end; three-bladed split arrowheads; flat arrowheads with an oval or square section; small flat spade-shaped arrowheads and flat diamond-shaped arrowheads (*ibid.*, 202-203).

In general, only the bases of the quivers were preserved. These consisted of wooden discs which displayed depressions for the attachment of the remainder of the quiver. Some of the quivers were made from birch bark. The quivers would have been hung from the belt with the aid of small iron hooks (*ibid.*, 203).

The only remains of armour discovered in the burials were perforated metal plates which are thought to have been part of coats of mail (*ibid.*).

2.4.2d Horse accoutrements

Only a small number of horse accoutrements have been found in the burials. Iron horse-bits and miniature copies of the bits have been discovered in several burials. All of the bits were ring-like, and examples composed of two sections prevailed. Some of the annulated bits were provided with freemoving additional rings. The majority of cheek-pieces had been made from iron, and were propeller-like in shape, with the blades having been made from either oval, leaf-shaped or trapeziform pieces of iron. The remains of thin iron plates may have been used to decorate the foreheads of the horses (*ibid.*).

2.4.2e Tools

The majority of tools recovered from the burials were knives and, more rarely, awls. In some cases several knives were recovered from a single burial. The knives might be found lying together as a group, or else situated at various points surrounding the skeleton. There were three main classes of knives. Some examples did not have a well defined handle, but displayed a straight back and a ring-shaped ending. Others had a well defined handle, a straight back and a ring-shaped ending, while a final group were tanged with a wooden or bone handle, and had a clearly curved back. The knives appear to have been carried in wooden scabbards, which in some cases displayed curved decorations, painted with red pigment (*ibid.*).

All of the awls were rod-like and round in section, and the majority had hafted wooden handles. In some cases whetstones and razors were associated with awls. Wooden beaters and miniature wooden beaters were also identified, and stone and ceramic spindle whorls accompanied a number of female burials (*ibid.*). In some cases fire-making implements were recovered from the burials. These consisted of a small wooden board with a number of perforations and a spindle-shaped rod. They appear to have been stored in leather bags (*ibid.*, 203).

2.4.2f Household objects

The occurrence of a large quantity of clay and wooden vessels in the large funerary mounds is characteristic of the Shurmak Culture. All of the ceramic vessels were hand-made and displayed evidence of uneven firing. The surfaces were generally a dark grey colour, with a smooth or burnished appearance. Four main types of clay vessels were apparent - vase-shaped, pot-shaped, flasks and mugs. Mugs were only occasionally included in the burials, and the pot-shaped vessels were invariably retrieved only from the larger burial mounds. The vase-shaped vessels were generally uniform in shape, but with a variety of ornamentation. The pot-shaped vessels appeared to be smaller versions of the bronze vessels characteristic of the Scythian period. Other, more rare, forms of ceramic vessels included bowls, miniature flasks and infant feeding cups (*ibid.*, 204).

A diverse range of wooden vessels were recovered from the burials. These consisted of oval and square-shaped dishes with four legs (which were probably used as dish-tables), bowls, plates, barrels, mugs, beakers, pots, ladles, scoops, beaters and spoons. Some of the wooden vessels had been decorated with geometric designs which had been painted using black and red pigments (*ibid.*).

The remains of bronze vessels which were similar to ceramic pots of both Scythian and Hunno-Sarmatian period type have been recovered from several of the large burial mounds. Small iron vessels characteristic of the Shurmak Culture were commonly recovered from the burials, and these were in the form of small bowls with two vertical handles supported on ring bases or three legs, and goblet-shaped vessels with ring bases. The function of the metal vessels is not known, although it is considered that they may have had a ritual purpose (*ibid.*).

2.4.2g Personal items

The remains of bone combs were often associated with female burials, and sets which contained miniature awls, a variety of knives and scrapers were occasionally discovered. These have been interpreted as either toiletry items or miniature versions of larger tools. The remains of Chinese mirrors and combs are also found in these sets. The sets of objects were generally contained in rectangular wooden boxes (*ibid.*, 203).

2.4.2h Summary

A wide variety of funerary monuments and funerary rituals are characteristic of the Late Phase of the Hunno-Sarmatian period. In addition a wide array of grave goods are considered characteristic of this period, with only the occasional object or structure reminiscent of the material culture of the earlier Uyuk Culture. In many cases the burials from the Late Phase of the Hunno-Sarmatian period are, however, located near those of the Scythian period as if they had been deliberately positioned next to these sites (*ibid.*, 200).

2.4.3 Society and social structure

The Chinese sources imply that the Hsiung-nu were hairy and had prominent noses (Phillips 1965, 113), and previous craniometrical research on skulls ascribed to Hsiung-nu tribes have generally been found to display a mixture of Mongoloid and European characteristics (Maenchen-Helfen 1973, 367-374). The royal tribes and the *shan-yü* of Hsiung-nu society are regarded by some to have borne Iranian names, and Hsiung-nu words recorded by the Chinese chroniclers are thought to have been derived from an Iranian language similar to that of the Saka (Harmatta 1994, 488).

Although the Hsiung-nu Empire is considered to have been composed of a large number of different peoples, they are thought to have divided themselves into twenty-four major tribes, each of which may have consisted of kinship groups, clans and patriarchal families. The oldest and most prominent tribes are regarded as the Hu-yen, the Hsü-pu and the Lan. Once Mo-tun became the *shan-yü* in 209 BC his noble family, the Lüan-ti, became the most distinguished group of all (Ishjamts 1994, 158). The tribal nobility are considered to have formed an aristocratic elite, while the ordinary members of society were probably all relatively poor. A large number of slaves are thought to have worked in agriculture, the manufacture of crafts and in pastoralism. It is generally argued by Russian academics that the Hsiung-nu society was in the process of transforming from a tribal to a class system, and that the Hsiung-nu Empire was representative of a form of class-based state organisation (*ibid.*).

Mandelshtam and Stambulnik (1992, 205) postulated that the social structure of the Hunno-Sarmatian period Shurmak Culture in Tuva was characterised by the destruction of the system of communal relationships. They believed that the temporary submission of Tuva to the Hsiung-nu State and the introduction of new ethnic groups into the region at this time enabled this process to occur. If this theory is accepted then it is probable that all levels of society would have been affected by these events. It has also been argued by academics that the traditional social structure of the Uyuk Culture underwent a transition during the development of the Shurmak Culture, and that this transition is reflected in the relative infrequency with which collective burials occurred during this time (*ibid.*). In addition, the differences between rich and poor burials were considered to be more apparent in the burials of the Shurmak Culture than had been the case for the Uyuk Culture, with the richest burials of the former culture containing large numbers of objects made from precious metals. The occurrence of large kurgans which covered a number of individual interments, however, has been interpreted as evidence that tribal structure was still strong (*ibid.*).

Military organisation is thought to have played a major social role in the Shurmak Culture. Evidence in support of this theory is provided by the widespread occurrence of weapons in a large number of burials, and the existence of skeletons with evidence of battle trauma. The majority of weapons were associated with male skeletons, while their diversity is thought to be indicative that a number of different classes of warriors existed, and that these warriors held different roles in the structure of society (*ibid.*).

2.4.4 Lifestyle and economy

The Hsiung-nu were excellent stock-breeders but they were also familiar with agriculture and possessed several permanent fortified settlements where they produced handicrafts (Harmatta 1994, 488). The Shurmak Culture in Tuva had a similar semi-nomadic pastoralist economy to that of the preceding Uyuk Culture but with a greater degree of mobility (Vainshtein 1980, 52). As was the case for the Scythian period, the existence of tribal burial grounds in Tuva of Hunno-Sarmatian period date indicates that cyclic migration continued in this era, with the tribes moving through their land along fixed routes and overwintering at set camp sites (*ibid.*, 96).

The remains of food debris from the burials indicates that sheep composed the greatest proportion of the herds, although goats, horses and cattle may also have been kept (Mandelshtam and Stambulnik 1992, 204). The remains of sheep from these archaeological contexts usually consisted of the bones of the leg (or, in some cases, the remains of several legs), vertebrae, ribs, sacra and skulls. The recovery of a bronze plaque from the Minusinsk area of Hunno-Sarmatian period date with a depiction of two standing camels suggests that the Hsiung-nu also used camels (Maenchen-Helfen 1973, 173). Occasionally horse skulls were recovered. The occurrence of a large proportion of weaponry in the burials, which was typical for horse-riders, may be indicative of the important role which the horse played in society (Mandelshtam and Stambulnik 1992, 204).

The majority of the Hsiung-nu, including the *shan-yü*, lived in yurts all year round. The left side of the entrance to the yurt was for males, while the right side was used by females (Ishjamts 1994, 164). The predominance of cattle of a small breed, the types of household objects, and the lack of burial in log house tombs has been interpreted by academics as evidence that the mobility of the Shurmak Culture was greater than that of the preceding Uyuk Culture. It has further been suggested that there was a transition from a semi-nomadic to a nomadic form of pastoralism at this time (*ibid.*). There is no definite evidence, however, that this transition occurred. Indeed, as stated above, the positioning of kurgans of the Shurmak Culture near those of the preceding Uyuk Culture may be indicative of the continued use of traditional winter camp-sites. Millet and the remains of other cereals have been commonly recovered from burials of this period, and this might indicate that a certain level of agriculture was being practised. It is probable that this agricultural activity took place in the vicinity of the winter camp-sites and that a proportion of the population may have stayed to tend and guard the crops during the year. It has been suggested by academics that the reason for the development of agriculture may have been related to the subjugation of the Tuvinians by the Hsiung-nu, and it may have been Hsiung-nu policy to have agriculture practised in their new territory so that each region would be economically independent (*ibid.*).

2.4.5 Weaponry and warfare

The Hsiung-nu were renowned for their military capability. When the *shan-yü* Mo-tun reorganised the Hsiung-nu army in 209 BC he replaced the heavily armed horse-riders with light cavalry armed with long composite bows. This resulted in the creation of a military force with much greater manoeuvrability. He subjected the army to strict discipline and introduced major improvements in strategy and tactics. Elements of the new army's technology, such as the composite bow and the stirruped saddle, then became adopted throughout the steppe lands (Ishjamts 1994, 163).

The Hsiung-nu females were skilled horse-riders and also used bows and arrows, enabling them to accompany the males in defending the children and older people of the tribe during an enemy attack. Women also participated in teaching the children how to handle the bow and arrow (*ibid.*, 164). In general, weaponry accounts for a large proportion of the grave goods accompanying Hunno-Sarmatian period burials (see Sections 2.4.1c and 2.4.2c).

2.4.6 Technology

In the Hunno-Sarmatian period the nomadic tribes of Tuva, like the Huns of Central Asia, had developed forms of professional as well as domestic craftworking. Advances in the extraction of iron ore resulted in the majority of tools and weapons being manufactured from iron, although bronze was still used to a lesser extent. It has been suggested that the Hunno-Sarmatian period tribes in Tuva would not have been capable of practising metal-founding and that their high quality iron and bronze artefacts were acquired as a result of trade with the agriculturists of East Asia and, in particular, China. Analysis of the chemical composition of these objects has, however, indicated that, with the exception of mirrors, the objects were unlikely to have been imported (Vainshtein 1980, 199), and the uniformity apparent in the design of some items of weaponry may indicate that they were centrally produced in Tuva (Mandelshtam and Stambulnik 1992, 204).

Leather, wool and wood-working were probably developed as domestic crafts. Some of the craftwork of the tribes of the Hunno-Sarmatian period was of such high quality and so suitable to a nomadic lifestyle that it was still being practised by the nomadic people of Tuva until recent times. The embroidered felt rugs recovered from the Hsiung-nu burial grounds in Noin Ula, Mongolia, are, for example, directly comparable to those produced by modern Tuvinians. In addition, many of the wooden vessels recovered from Hunno-Sarmatian period burials are similar to those which were used by the Tuvinians in the late 19th and early 20th centuries (Vainshtein 1980, 200).

A set of leather-working tools recovered with a female burial in the Kokel burial ground dates to the Late Phase of the Hunno-Sarmatian period. This find suggested that leather-working was regarded as a female task in Hunno-Sarmatian times, as was the case among the Tuvinian population of the late 19th and early 20th centuries. Remarkably, the tools recovered from the burial were nearly identical to those used in recent times by Tuvinians and included a scraper and a curved iron blade with a wooden handle (*ibid.*).

2.4.7 Art

Information on the art of the Hsiung-nu has been chiefly obtained from the Noin Ula tombs of Mongolia and the Ordos bronzes. The furniture included in the Noin Ula tombs included fabrics embroidered with Chinese silk and decorative motifs, as well as ornamentation which probably derived from Hellenistic north-west India. The Animal Style of art apparent in the Noin Ula tombs, however, has been regarded as signalling the decline of the style, since the animals depicted have lost their dynamism and are dull compared to those of the Scythian Animal Style (Antonini 1994, 294-295). The Animal Style apparent in the Ordos bronzes, however, is more flamboyant and appears to show links with Shang and Zhou metalworking in China (*ibid.*, 295).

2.4.8 Religious and ritual beliefs

The Hsiung-nu are thought to have believed in animism and totemism and regarded the sun, moon and other heavenly bodies as deities, making sacrifices to the heavens, the earth, spirits and ancestors. During the rule of Mo-tun, however, Shamanism became the state religion. The chief shaman was selected from among the sorcerers and served the *shan-yü* and his relatives (Ishjamts 1994, 165). The Hsiung-nu traditionally launched a military expedition during a full moon, the outcome of which relied upon the prophecies of the shamans, sorcerers and soothsayers, who used ram scapulae for predicting the future. The bone was placed in a fire and the future was read from the lines which developed in the fire-cracked bone (*ibid.*).

2.4.9 Historical sources

The majority of our literary information on the Hsiung-nu is derived from the Chinese Chronicles. The *Shih chi*, or the *Historical Records*, which were started at the end of the 2nd century BC and completed between 74 and 49 BC contain particularly detailed accounts of the Hsiung-nu and their relations with the Han Empire of China. Other chronicles which contain information on the Hsiung-nu are the *Ch'ien Han Shu*, or the *History of the Former Hans*, which was compiled during the 1st century AD, and the *Heou Han Shu*, or the *History of the Later Hans*, which was completed in the later part of the mid 5th century AD (Vernadsky 1943, 77).

2.4.9a History

The *Shih chi* records that in the late 3rd century BC the Hsiung-nu tribal confederation was headed by a chief named T'ou-man who held the title of *shan-yü* which means 'the greatest' or 'the best'. Legend states that he had two sons from different wives. To ensure that his youngest and favourite son inherited the throne he gave his eldest son Mo-tun as a hostage to the Yüeh-chih, hoping that they would kill him. Mo-tun managed to steal a horse, however, and returned home. His father was so impressed with his bravery that he put 10,000 families under his control. Mo-tun set about training his cavalry in the arts of war and ordered all his troops only to shoot their arrows after he had shot his own whistling arrow. Failure to obey his command resulted in instant death. Once he was satisfied that they were adequately trained, Mo-tun and his warriors shot and killed his father during a hunting trip. Mo-tun then proclaimed himself the *shan-yü* in 209 BC, after which the Hsiung-nu state rapidly became a powerful nomadic empire (Yü 1990, 120).

In 198 BC a treaty of 'peace and alliance' was agreed between the Hsiung-nu and China. The Han emperor officially acknowledged that the Hsiung-nu had power on a similar level to that of his own empire, and that its lands surrounded the northern borders of China. The ruler of the Hsiung-nu acknowledged that the Chinese emperor had sovereignty over all of the territory behind the Great Wall. The treaty also entailed giving the Han emperor's daughter in marriage to the *shan-yü*. In addition, the Han emperor was to send lavish gifts to the *shan-yü* each year, including silks, fabrics, crafts, wine, rice, gold and money. This was regarded by the Hsiung-nu as a form of tribute (*ibid.*, 122).

Regular trade occurred between the Chinese and the Hsiung-nu from the time of Mo-tun, with the Hsiung-nu exchanging cattle, wool and furs for Chinese goods. After 129 BC the Han emperor Wu-ti changed the policy of his predecessors and launched a number of military campaigns against the Hsiung-nu. The Hsiung-nu, however, defeated the Chinese. In the period between 70 and 60 BC the Hsiung-nu were weakened as a result of internecine wars between the various leadership factions who wanted to become the *shan-yü*, and in AD 48 the Hsiung-nu split into a northern and a southern faction as a consequence of increased internal disagreements. The southern tribes gradually fell under Han rule, while the northern tribes attempted to defend their independence. In AD 93, however, they were attacked by an alliance of the Chinese, the southern Hsiung-nu, the red-haired, blue-eyed giants of the north who were known as the Ting-Ling, and the Hsien-pi from the east. Although a large proportion of the northern Hsiung-nu, including the *shan-yü*, were killed, a proportion of the Hsiung-nu, particularly the western branch, did not surrender. The Hu-yen tribe assumed leadership and marched westwards. They have become known to history as the Western Huns and their descendants later reached Afghanistan, India and the Roman Empire (Ishjamts 1994, 155).

2.4.9b Society and social structure

The ruler of the Hsiung-nu emperor was the *shan-yü*, who also referred to himself as '*ch'eng-li kut'u*' which can be translated as 'son of heaven'. The considerable power of the *shan-yü* was hereditary. It is also possible to identify several classes of officials in the administrative organisation of the Hsiung-nu. These consisted of nobles who were divided into eastern and western groups or senior and junior groups. The first class consisted of the Chu-ch'i princes, with the Eastern Chu-ch'i prince reputed to be the *shan-yü* heir apparent, although his right to succeed to the throne was disregarded on many occasions. The second class consisted of the Lu-li princes, and these high-ranking nobles were always members of the *shan-yü's* clan (Watson 1961, 163).

As well as the aristocracy who were relatives of the *shan-yü* there was also the service nobility that performed administrative tasks and were referred to as the *Ku-tu-hou*. In addition, there was the clan nobility which consisted of princes who were affiliated with the clan chiefs or elected elders (*ibid*.).

The *Shih chi* records that anyone who drew their sword a foot from the scabbard during times of peace was condemned to death, while anyone who stole from another was punished by the confiscation of their property. Minor offences were punished by flogging and major ones by death. No one could be kept imprisoned while awaiting sentence for a period longer than ten days, and the number of prisoners awaiting trial at any one time never exceeded 'a handful' (*ibid*., 164).

In the 2nd century BC a renegade Chinese eunuch named Chung-hsing Shuo taught the *shan-yü* how to write official letters to the Chinese court on a wooden tablet, 31 cm long and decorated with broad stamps and long seals (Watson 1961, 171). The same sources also indicate, however, that when the Hsiung-nu made notes or wanted to send a message they made cuts on a piece of wood. They also mention a 'Hu script' (Ishjamts 1994, 165).

2.4.9c Lifestyle and economy

In the *Shih-chi* the Hsiung-nu are described as a people who wandered from place to place pasturing their animals. It is stated that they predominantly raised horse, cattle and sheep but that they also maintained rarer animals including camels, asses, mules and wild horses. They moved around in search of water and pasture, had no walled cities or permanent buildings, and did not practice agriculture (Watson 1961, 155).

Everyone in Hsiung-nu society ate the meat of domestic animals and wore clothes made from hide, or wraps made from felt or fur. The young men were given the highest quality food, while the old got whatever was left over. The Hsiung-nu were said to respect the young and strong, and despise the old and weak (*ibid*., 155-156). Later in the *Shih chi*, however, the Han eunuch Chung-hsing Shuo, who was an ally of the Hsiung-nu, explained to a Han envoy that the Hsiung-nu were a warrior people. As such, the old and weak were not capable of fighting, and the best food was therefore reserved for the younger members of society who were responsible for defending the nations. As a consequence of this system both the old and young lived in relative harmony and security (*ibid*., 171).

2.4.9d Funerary monuments and practices

The *Shih chi* records that the Hsiung-nu used an inner and outer coffin for burial, with accessories of gold, silver, clothing and fur. It is also stated that they did not construct burial mounds, plant trees on the graves or wear mourning clothes. On the death of a ruler the favoured ministers and concubines were obliged to accompany their master in death. The source states that these individuals often numbered in their hundreds or even thousands (*ibid*., 164).

It is also recorded in the *Shih chi* that after a man's father died he married one of his stepmothers. In addition, when a brother died the surviving brothers would take his widows for their own wives (*ibid*., 156). Later in the *Shih chi*, however, the Han eunuch Chung-hsing Shuo, an ally of the Hsiung-nu, explained to a Han envoy that the reason for this practice was that the Hsiung-nu 'hated to see the clan die out' (*ibid*., 172). It is interesting to note that Czaplicka (1914, 117) reported that in the early 20th century it was not unusual for an Altaian widow to marry her brother-in-law.

2.4.9e Religious beliefs

The *shan-yü* considered himself to have been 'born of heaven and earth, brought forth by the sun and moon'. The Chinese sources record that 'at daybreak the *shan-yü* sets out from camp to worship the rising sun, at nightfall to worship the moon'. In the fifth lunar month the Hsiung-nu nobility met at the *shan-yü's* headquarters to make sacrifices to their ancestors, the heavens, the earth and the spirits. Each year they congregated three times at the shrine of the moon where, on the day of the 'dog' during the first, fifth and ninth months, they made offerings to the heavenly spirit (Ishjamts 1994, 165).

In one Chinese source the story is related of how a treaty of alliance was made between the Han and Hsiung-nu rulers. Following the conclusion of the pact the Chinese dignitaries, the *shan yü* and the clan elders 'went up to the Hun's Mountains by the east side of the River No-Shui, and impaled a white horse. The *shan-yü* took a costly sword and moistened its tip with wine; they drank the dedicated wine from the skull of a Yüeh-chih lord who had been killed by the *Shan-yü* Laoshan'. This account has been interpreted as evidence that a sword cult existed among the Hsiung-nu (Rudenko 1970, 291).

2.4.9f Weaponry and warfare

A reference in the *Shih-chi* indicates that the Hsiung-nu actively encouraged their children to learn how to hunt from an early age. It is stated that the young boys started off by learning how to ride sheep and shoot birds and rats with a bow and arrow. When slightly older they were taught how to shoot foxes and hares which they used for food. As a result of this practice all of the young men knew how to use a bow and arrow and were able to form an armed cavalry in periods of war (Watson 1961, 155). It is also recorded that during times of peace the Hsiung-nu were content to herd their flocks and make their living by hunting but that during times of war they would take up their arms and go on plundering and marauding expeditions. Bows and arrows were used for long-distance attack, while swords and spears were used at close range (*ibid*.).

The *Shih chi* states that the Hsiung-nu attacked during a full moon, and withdrew their troops when the moon waned. Following a battle, the Hsiung-nu who had decapitated their enemies or taken prisoners were presented with a cup of wine and allowed to keep any of the spoils which they had captured. Prisoners were used as slaves. Anyone who succeeded in recovering the dead body of a comrade who had died in battle was allowed to keep the dead individual's property (*ibid*., 165). During an attack on the Talas fortress by the Chinese troops in the 1st century BC, Hsiung-nu women fought alongside their male counterparts and a group of Roman mercenaries in defence of the citadel. It is also stated that the women were the last to leave their positions (Ishjamts 1994, 164).

3. The cemetery complex of Aymyrlyg

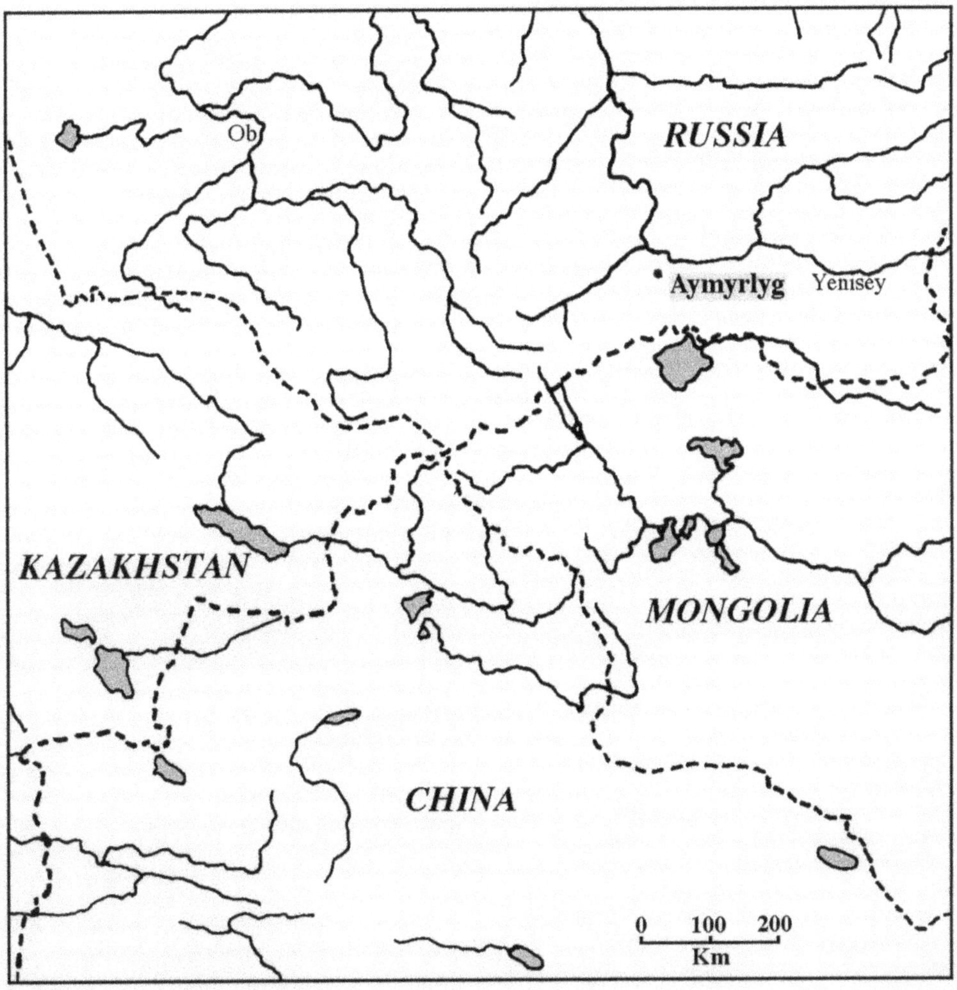

Figure 1: Location map of the cemetery of Aymyrlyg.

3.1 Introduction

The cemetery complex of Aymyrlyg is located in the Ulug-Khemski region of the Autonomous Republic of Tuva, south Siberia (Fig. 1). It is situated in close proximity to the Sayansky ravine of the Yenisey river, on the left bank of the Chaa-Hol river, where it extends for a distance of over ten kilometres. The Aymyrlyg burial ground is considered to be the largest archaeological monument located in the flooding zone of the Sayano-Shushenskoye hydro-electric power station (Mandelshtam 1983, 25). By 1974 Aymyrlyg was the most completely and systematically investigated monument of Scythian period date in Tuva (Mandelshtam 1975, 220).

The first scientific excavations to have been undertaken at Aymyrlyg occurred in 1964 under the direction of Dr. M. H. Mannai-Ool (*ibid.*). The cemetery complex was then excavated in the period between 1968 and 1984 by archaeologists of the Sayano-Tuvinskaya expedition team from the Institute for the History of Material Culture, St. Petersburg. The director of the expedition for the period between 1968 and 1978 was Dr. A. M. Mandelshtam but, following his death, Dr. E. U. Stambulnik continued the research until the mid 1980s. Concomitantly, burials of the Turkic and later periods were excavated by Dr. B. B. Ovchinnikov of the Urals University.

Aymyrlyg was used as a burial ground for a considerable period of time with the earliest burials dating to the Bronze Age and the latest burials dating to at least the 18th century AD (Mandelshtam 1983, 26). The majority of the burials from the excavations undertaken by Mandelshtam were from the Early and Late Phases of the Scythian period (c. 8th - 2nd century BC), while most of the burials from the later years of the programme under the direction of Stambulnik were from

Burial group	Location	Years of excavation	Predominant period
I	northern	1968, 1969	Scythian
II	northern	1968, 1969, 1972	Scythian
III	northern	1968, 1969, 1972	Scythian
IV	northern	1968, 1969, 1972, 1975	Scythian
V	northern	1968, 1969	Late Medieval
VI	northern	1969, 1970, 1971	Scythian
VII	southern	1970, 1971, 1972, 1974, 1975	Scythian
VIII	southern	1970, 1971, 1972, 1976	Scythian
IX	central area	1974, 1975, 1976	Scythian
X	southern	1971	Scythian
XII	southern	1972, 1975, 1976	Scythian
XIII	southern	1972	Bronze Age
XX	**	1975	Scythian
XXI	**	1975	Scythian
XXIII	**	1975	Scythian
XXV	**	1977	Scythian
XXXI	southern	1977, 1979, 1980, 1981	Hunno-Sarmatian
B*	south-western	1968	Turkic
B	northern	1975, 1976	Scythian
C	northern	1974	Scythian
D	northern	1974	Scythian
E	northern	1974	Scythian
G	northern	1971, 1975	Scythian

Table 2: Aymyrlyg - Summary of the excavations as detailed in the archive reports. The excavation reports for 1973, 1978 and post-1981 were not available for consultation.

**There are two distinct groups referred to as 'B' in the archive reports.*

*** Information on the location of Groups XX, XXI, XXIII and XXV is not provided on the site plan or in those archive reports which were available for consultation in the Institute of the History of Material Culture, St. Petersburg.*

the Hunno-Sarmatian period (c. 3rd century BC - AD 2nd century). Details of the burial groups which were excavated each year at Aymyrlyg are provided in Table 2.

3.2 Scythian period

Although some of the Scythian period burials recovered from Aymyrlyg are considered to date to the earlier phases of the Uyuk Culture in the 7th - 6th century BC (Mandelshtam 1992, 183), the majority of burials are thought to date to the Late Phase of this culture during the 3rd - 2nd century BC (*ibid.*, 185). On the basis of the findings at Aymyrlyg, Mandelshtam (1983, 33) was of the opinion that the Scythian period in Tuva should be extended, at least partially, into the 2nd century BC.

3.2.1 Funerary monuments

3.2.1a Location of burials

Scythian period burials were found in all areas of the site, with the exception of the southern quadrant, where a compact Turkic period cemetery was located. There were two main characteristics relating to the location of the Scythian period burials. There appeared to have been a preference for the construction of the funerary monuments at the edge of the river terrace, while the monuments were distributed in chains which ran along the riverside. These chains generally formed the core of each group of burials, with later burials being concentrated around each central chain (*ibid.*, 26).

3.2.1b Morphology of funerary monuments

The funerary monuments of the Scythian period at Aymyrlyg displayed markedly uniform surface structures but had diverse interior features. The exterior of each tomb was generally composed of a low, rectangular-shaped wall which comprised stone slabs sunk edgewise into the ground. The size of the wall was dependent on the size of the underlying grave, which in turn was reliant on the number of individuals who were buried within the tomb (*ibid.*).

The most characteristic interior structure used in the Scythian period funerary monuments was the rectangular log house tomb (Plate 1). The walls of the log house tomb were composed of timber, and a ceiling of thinly chopped logs was generally present. The wooden structures were placed within deep pits dug into the ground. A typical log house tomb consists of three internal levels, but tombs with two to four levels have also been encountered. Invariably the numbers of individuals buried within an Aymyrlyg log house tomb was considerable, with as many as fifteen skeletons being recovered from individual examples (*ibid.*).

The stone cist was the second form of Scythian period funerary monument which was commonly encountered at Aymyrlyg. The cists were generally rectangular in shape and were composed of large stone slabs. The roof of the cist comprised numerous layers of flat slabs, and was apparent on the ground surface in those cases where the cist had not been buried deep into the earth. The majority of cists were small, with room only for the burial of a single individual (Plate 2). There were also, however, larger cist burials which contained the remains of two or three individuals (Plate 3) (*ibid.*).

Both the log house tombs and the larger cists of each burial group were positioned in chains. The smaller cists, however, either comprised separate chains and subgroups or were concentrated around the log house tombs and the larger cist burials. The majority of the smaller cists contained the remains of subadult individuals (Plate 4) (*ibid.*).

Several other forms of Scythian period funerary monuments were identified in the archaeological record in addition to the site's characteristic log house tombs and stone cist burials.

Plate 1: Scythian period - Log House Tomb VI. 6 (Photographic Archive of the Institute for the History of Material Culture, St. Petersburg).

Plate 2: Scythian period - Burial of a single adult individual in Cist 1 from Group V (Photographic Archive of the Institute for the History of Material Culture, St. Petersburg).

Plate 3: Scythian period - Burial of two adults in a large stone cist of unknown context (Photographic Archive of the Institute for the History of Material Culture, St. Petersburg).

Plate 4: Scythian period - Burial of a young child in a stone cist of unknown context (Photographic Archive of the Institute for the History of Material Culture, St. Petersburg).

These monuments were of notably different morphology, and tended to be located separately from the main tombs in each group chain. Some had stepped pits with rectangular surface cairns while others had pits with walls which were lined with stone slabs. Burial cists which had been roofed with dome-like constructions of stone were also noted. In addition, several individuals had been buried in simple ground pits. Variations were also evident in the southern part of the cemetery complex and cases existed where the tomb's stone cairn was surrounded by a stone slab border of circular plan. Examples of log house tombs composed of squared beams of wood as opposed to logs were also recorded (*ibid.*).

The majority of the funerary monuments were dated on the basis of the typology of the artefacts contained within each tomb, and the greater majority were considered to date from the 5th to the 2nd century BC. The variety of artefacts contained within the tombs suggested that the cemetery complex spanned the entire Scythian period and developed without interruption into the subsequent Hunno-Sarmatian period (*ibid.*, 33).

3.2.2 Treatment of the bodies

The burial rite was generally uniform, with the majority of individuals having been buried lying on their left sides and with their legs flexed, though in some cases skeletons were encountered where the individual had been buried on his or her right side. The arms were generally extended anterior to the body, but in several instances they were flexed. Several skeletons which were lying in a prone position were discovered, all of which had flexed legs (*ibid.*). In conjunction with evidence for the burial of complete corpses, the remains of disarticulated skeletons in approximate correct anatomical position were recovered (Plate 1). These remains indicate that the bodies of these individuals were not buried immediately in the tomb (Murphy 2000b; Murphy and Mallory 2000; 2001). The burials in the log house tombs were generally positioned lying side by side in one or more rows. In some cases, however, the burials were overlying one another (Mandelshtam 1970, 184).

The majority of the Scythian period individuals at Aymyrlyg were orientated to the west, with some deviations to the north and the south. Other orientation were, however, encountered (Mandelshtam 1983, 33). Several of the log house tombs contained rows of individuals and, in some cases, the individuals present in a single row displayed a variety of orientations. The location and orientation of subadult individuals also showed a notable deviation from the characteristic western orientation, while the orientation of individuals buried within stone cists appeared to differ to those of the individuals buried in the log house tombs of the same burial group. This diversity of orientations apparent amongst the Scythian period individuals might suggest that the choice of orientation was dictated by a complex system of social behaviour (*ibid.*).

3.2.3 Grave goods

Despite considerable robbing of the tombs, a large proportion of the original grave goods still accompanied the burials. A comparison of the grave goods present in tombs that had been robbed against those from intact tombs indicated that a considerable proportion of the original grave goods were still preserved in the robbed tombs (Mandelshtam 1969, 190). Two components of the Scythian Triad (see Section 2.3) were represented among the grave goods; the characteristic weapons and artefacts decorated with the Animal Style of artwork. The third component - horse accoutrements - were, however, only occasionally recovered (Mandelshtam 1983, 33).

3.2.3a Weaponry

The weaponry recovered from the tombs consisted of bronze and iron daggers, bronze and iron pointed axes (Plate 5), and bronze, bone and wooden arrowheads. A large proportion of the weaponry had been robbed, as was evidenced by the occurrence of empty sheaths associated with oxide stains, and the existence of organic arrowshafts with signs that the arrowheads had been removed (*ibid.*, 28).

The bronze daggers recovered from the tombs had similar bow-tie shaped guards, but displayed a variety of pommels of different morphology. The various dagger pommels were oval-shaped, disc-shaped, or heart-shaped. In addition, several daggers had bronze pommels in the form of a pair of animal heads which were turned to face each other. The iron daggers had a similar morphology to the bronze examples, although several had different forms of guards including those which were straight, those which were angular and those which were designed to form the shape of a pair of animal heads (*ibid.*). A bone scabbard associated with a dagger was recovered during the 1972 excavations (Mandelshtam 1973, 228).

Bronze pointed axes had a narrow pointed working edge and a butt with a central hole for the attachment of the organic shaft. The main difference in morphology among these artefacts occurred in the shape of the butt. This could be rounded, oval, hexagonal or flat in shape. A single bronze pointed axe displayed a butt which had been designed in the form of a pair of mountain goats, and had an opening positioned opposite the shaft-hole for the location of a plume. The iron pointed axes were generally extensively corroded and it was difficult to ascertain variations in their morphological characteristics (Mandelshtam 1983, 30). The first socketed axe (bronze) to have been recovered from a Tuvinian burial was retrieved during the 1975 season of excavations (Mandelshtam 1976, 259).

The majority of bronze arrowheads were designed to attach to an organic shaft. A variety of arrowheads of different morphologies were apparent, with the majority being trilobate. A large number of bone arrowheads were also recovered, the majority of which were designed to attach to a shaft and were trilobate in morphology. Several sets of quivers were found to contain a number of wooden arrowheads, and it has been suggested that these could have been utilised for hunting small animals and birds. No iron arrowheads were recovered (Mandelshtam 1983, 30).

3.2.3b Horse accoutrements

Although rarely discovered within the tombs, horse bits represented the majority of horse accoutrements discovered dur-

Plate 5: Scythian period - Bronze pointed battle axe (chekan), bronze awl and unidentifiable metal object from Log House Tomb II. 6 (Photographic Archive of the Institute for the History of Material Culture, St. Petersburg).

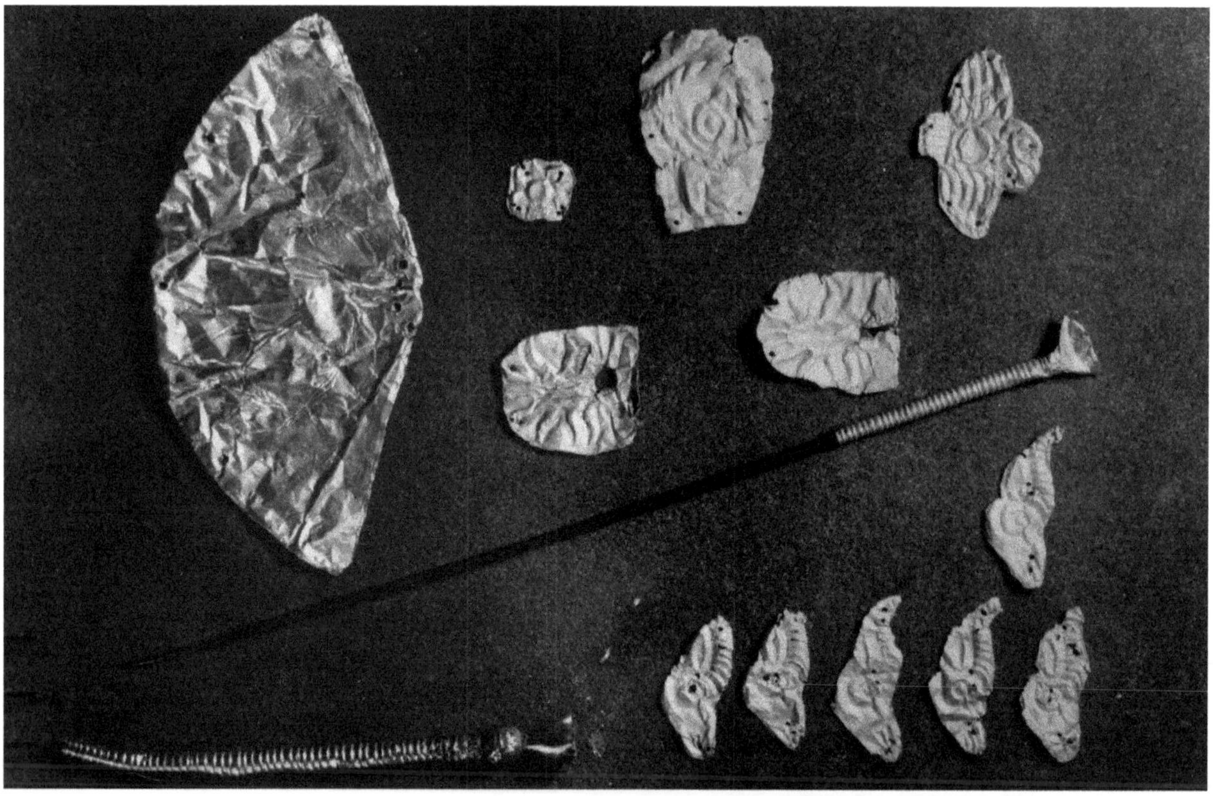

Plate 6: Scythian period - Gold foil decorated in relief and pins covered with gold foil recovered from Log House Tomb II. 3 (Photographic Archive of the Institute for the History of Material Culture, St. Petersburg).

ing the excavations. All of the bits were composed of bronze, had a ring-like morphology, and were invariably associated with horn cheek-pieces. In some cases bone clasps and blocks were recovered in close proximity to the bits and cheek-pieces and it is considered that they may represent the remains of bridles. The horse accoutrements, in conjunction with horse skulls and metapodia, were generally recovered on the roofs of the log house tombs. These findings suggest that horse skins, with the bit, cheek-pieces and bridle attached to the skull, were left lying on top of the tomb, perhaps as part of the funeral ceremony (*ibid.*).

3.2.3c Tools

Knives, awls (Plate 5) and needles represented the majority of utilitarian tools recovered from the tombs. All the bronze and iron knives and awls had a single, straight or slightly curved blade with a poorly formed handle. A few of the bronze knives displayed ring-shaped handles, with a central tear-shaped perforation. The majority of iron knives did not display perforations in their handles, and ring-shaped handles were only occasionally encountered. The bronze awls were generally square or rectangular in section. Several awls displayed nail-like heads, and some had been provided with a small handle made from an animal phalanx. A proportion of the iron awls had bone handles, while some of them were composed solely of an iron rod (*ibid.*, 31).

All of the needles were composed of bronze, and they were sometimes recovered in sets. A knife and an awl in a single leather scabbard were commonly recovered beside male and female skeletons. In many cases a hole had been perforated in the scabbard to hold the awl. A proportion of knives were recovered in scabbards with daggers. It is possible that the knives associated with daggers may have had a different function than the knives associated with awls. Several wooden discs and perforated boards which were used to start fires were also recovered from the cemetery complex (*ibid.*).

3.2.3d Household objects

The majority of clay vessels recovered from the site had been hand-made, although the morphology of several pots suggested that they had been constructed on a revolving base. The vessels were generally unevenly (and in some cases incompletely) fired and this had resulted in the appearance of dark stains on the exterior of the vessels. Decoration on the pots was minimal with grooves or, more rarely, incised geometric designs being apparent on the upper part of the body of the pot. In some cases spiral shaped rolls of clay had been attached to the middle level of the pot, while horse-shoe shaped rolls had been attached to the neck of the pot beneath the rim. Several pots were also recovered which had been painted with curvilinear figures and spirals (*ibid.*, 32). The majority of pottery vessels were handleless jars (Plates 1 and 3), though jars and cups with handles, and pots with low, broad necks were recovered. A number of bowls and cask-like flasks were also found accompanying the burials (*ibid.*).

Wooden vessels were generally poorly preserved. The majority of wooden containers appeared to have been round and oval-shaped low dishes and bowls which, in some cases, had a flat, horizontal handle. Small bowls and cups were also recovered. An incised geometric design was apparent on several of the wooden containers (*ibid.*). In some instances the wooden vessels appeared to have been grouped together into specific sets of objects within the tomb (Mandelshtam 1971, 217). A vessel which was composed of birch bark was recovered during the 1971 season of excavation (Mandelshtam 1972, 281).

3.2.3e Foodstuffs

In general, foodstuffs were not preserved. The exception to this was the occurrence of sheep scapulae or sacra lying on oval dishes, which suggests that the dead were buried with joints of meat. In several cases wooden spoons were recovered positioned in wooden vessels, and this has been interpreted as indicative that some form of soup or porridge-like food was eaten by the Scythian period population (Mandelshtam 1983, 32).

3.2.3f Clothing and adornments

The remains of clothing were generally preserved only as small scraps of fabric. In the majority of cases the fabric had survived as a consequence of having been buried in close proximity to bronze artefacts (*ibid.*). The remnants of clothing suggested that the individuals had worn an outer layer of clothes made from leather and fur, and an inner layer of clothes made from a woollen fabric. When clothing seams were encountered their morphology suggested that they had been markedly solid, and had in some cases played a decorative role (*ibid.*).

Belts which were principally composed of leather were commonly encountered. Some of the belts were relatively thick and wide, and were probably used for suspending a variety of objects, including weapons. Bronze buckles and bronze and iron hooks were commonly associated with the remains of belts. Straps were also used for suspending objects such as knives and small bags from the belts. These would have been fastened to the belt by bronze, iron or bone buttons and were in some cases decorated with a set of bronze or iron rings (*ibid.*). Some of the belts were decorated with semi-spherical or bow-tie shaped bronze badges (*ibid.*, 32).

Earrings were only occasionally recovered from the burials. The simplest earrings were composed of a ring with overlapping ends, while more complicated varieties consisted of a chain on the end of which hung a pendant. All of the earrings which were recovered were composed of gold or bronze. It has been suggested that any iron earrings buried with the individuals would have been completely oxidised and consequently not preserved (*ibid.*, 31).

Bracelets composed of bronze rods and covered with sheet gold were occasionally recovered, while beads and pendants were frequently encountered. Glass, spherical eye-shaped beads, and spherical or cylindrical-shaped stone beads composed of red sandstone or carnelian were commonly retrieved from the burials. The majority of pendants were composed of bronze and bone and consisted of a bar with a bell. In addition, bronze pot-shaped pendants were recovered. Perforated teeth from marals and musk deers, wild boar canines and a small number of Cowrie shells were also encountered in the burial chambers (*ibid.*, 32). It is probable that the Cowrie shells were imported from the region adjacent to the Indian Ocean.

Miniature belt buckles composed of thin gold sheets may have been used to decorate the clothing and headgear (Plate 6). Some of the miniature buckles displayed relief patterns of animals including a feline predator, a mountain goat and a ram (*ibid.*).

3.2.3g Personal objects

A large number of iron, bronze and bone pins, bone and wooden combs, bronze mirrors and small leather bags were associated with the Scythian period burials. All of the mirrors were circular and all were contained in leather cases. Some of the mirrors had a prominent tear-shaped handle, while others had a small side handle (*ibid.*). A bronze mirror was recovered from the 1969 excavations on which an engraving of a mountain goat was apparent on the back of the mirror (Mandelshtam 1970, 184). Some of the heads of the pins were covered with gold foil (Plate 6) (*ibid.*). A small leather bag with an appliqué design was recovered during the 1971 excavations (Mandelshtam 1972, 281).

3.2.3h Ritual objects

Several objects were recovered which were interpreted as having a ritual or ceremonial function. Bronze pots with two handles were discovered containing burnt stones. These were interpreted as having been used for herb burning, and may have involved the burning of narcotic plants for social or ritual purposes. Birch bark portraits of human beings are also considered to have had a ritual significance (Mandelshtam 1983, 32).

3.2.3i The Animal Style of art

A variety of objects of both ceremonial and utilitarian function were found to display artwork in the Animal Style. These objects included combs, mirrors, pommels and clasps. The animals represented comprised ungulates and feline predators, portrayed in both stationary or dynamic poses. More complicated scenes of landscapes, or displays of animals in attacking or fighting poses were also recovered. The Animal Style of decoration was most commonly found on objects manufactured from bone or horn (*ibid.*, 30). A large stone slab which displayed a number of carvings of mountain goats on one of its sides was discovered during the 1974 excavations (Mandelshtam 1975, 219).

3.3 Hunno-Sarmatian period

The burials in Group XXXI at Aymyrlyg were classified as dating to the Hunno-Sarmatian period (c. 3rd century BC - AD 2nd century) on the basis of the various styles of tombs and the grave goods which were buried with the individuals (Stambulnik 1983a, 34). The Group XXXI cemetery is referred to as Podgorka in some reports (Bogdanova and Radzjun 1991, 55). It was observed, however, that the Hunno-Sarmatian period burials in Aymyrlyg differed considerably to those from the Scythian period at the site and also other Hunno-Sarmatian period burials in Tuva, such as those from Kokel. The principal differences were apparent in the diversity of funerary rites associated with the bodies and in the extensive variety of features evident in the subterranean sections of the tombs (Stambulnik 1983a, 34). It was concluded that the Hunno-Sarmatian period cemetery at Aymyrlyg existed at the time when the Scythian period culture was concluding and the Hunno-Sarmatian culture was commencing (*ibid.*, 39). This chronological border is considered to occur between the 3rd and the 2nd centuries BC, a time when the area was part of the rapidly developing Hsiung-nu Empire (see Section 2.4) (Mandelshtam and Stambulnik 1992, 196).

3.3.1 Funerary monuments

3.3.1a Location of burials

Group XXXI is a compact Hunno-Sarmatian period cemetery located at the base of the southern slope of the Mon mountain near the Chaa-Hol river. It is situated in the southern region of the Aymyrlyg cemetery complex (Stambulnik 1978, 13).

3.3.1b Morphology of funerary monuments

The surface signs of the tombs had undergone extensive destruction as a consequence of torrential rain flowing down the mountain slope. In those cases where the surface signs were still evident they consisted of low, near-rectangular shaped cairns, surrounded by partially preserved circular stone borders (Stambulnik 1983a, 34). In general, the diameter of the circular enclosure did not exceed the length of the burial pit (Mandelshtam and Stambulnik 1992, 198). The graves were situated in rows which were aligned from south-west to north-east (Stambulnik 1983a, 34).

In total, approximately 200 burials were excavated, though many of these had been disturbed by rodents and foxes. The majority of burials consisted of a single individual (Plate 7), although several double and triple burials were encountered (Plates 8 and 9). All of the burial pits were rectangular in shape with a depth of less than one metre. Several burials were encountered, however, which had a depth of between one and two metres. The burial pits were generally orientated from south-east to north-west, although some deviations in orientation were apparent (*ibid.*).

Burial in a composite wooden coffin or, less frequently, in a log or a wooden block is characteristic for the Hunno-Sarmatian period. The tombs were classified into three classes:

A - Ground pit
B - Stone cist
C - Ground pit with an alcove

The tombs of each class were then further subdivided:

Class A

1. Body lying at the bottom of a ground pit (12 examples).
2. Body lying at the bottom of a ground pit. Stone slabs line the two narrow walls of the pit which was roofed with logs propped on the stone slab side walls (3 examples).
3. A twig or birch bark mat was placed at the bottom of the pit. The body was covered with twigs or birch bark (2 exam-

ples).
4. A wooden block or a log was positioned at the bottom of the pit (3 examples).
5. A wooden coffin was positioned at the bottom of the pit (19 examples) (Plates 7 and 8).
6. A large outer wooden coffin was positioned at the bottom of the pit. Stones were placed on top of the coffin (2 examples).

Class B
7. The burial consisted of a stone cist (75 examples). In some cases the walls of the cist were incomplete and consisted of overlapping stones (6 examples).
8. A wooden block was positioned in a pit which was partially lined with stone slabs (23 examples).
9. A wooden coffin was positioned inside a stone cist (26 examples) (Plate 9).
10. A large outer coffin was positioned inside a stone cist (10 examples).
11. An individual burial with an inner wooden block and a large outer wooden coffin was positioned inside a stone cist (1 example).
12. An individual burial with an inner wooden coffin and a large outer wooden coffin was positioned inside a stone cist (2 examples).

Class C
13. A pit with an alcove which was sealed with stone slabs (1 example).
14. A pit with an alcove which contained a stone cist and was sealed with stone slabs (1 example).

A total of fourteen variations in the plan of the inner grave were apparent (*ibid*., 34-37). In addition, each of the basic formats of burial outlined above showed a great variation in detail, with coffins, for example, being composed of long boards, short slats, or thin rounded wooden bars (*ibid*., 37). The diversity of features apparent in the burial structures has been interpreted as being indicative of an unstable cultural group. This social instability has been attributed to the onset of the Hunno-Sarmatian period which saw a new group of people inhabiting Tuva from the late 3rd/early 2nd century BC onwards. The diversity of graves indicates that the Scythian period inhabitants of Tuva interacted with the newcomers and it is suggested that this resulted in a blending of the two cultures (*ibid*., 39).

3.3.2 Treatment of the bodies

The extended, supine position is most characteristic for burials of the Hunno-Sarmatian period. The burials from the Early Phase of the Hunno-Sarmatian period at Aymyrlyg revealed a wide number of burial positions. Some individuals had been buried lying on their side with flexed legs, while others had been buried in a supine position with their knees bent upwards, and some were lying in the expected extended supine position characteristic for this period. The diversity in body position appeared to correlate with a particular type of burial plan (see Section 3.3.1b) (*ibid*.).

The majority of the individuals were orientated to the northwest, but deviation in the orientation of some bodies was apparent (Mandelshtam and Stambulnik 1992, 199).

3.3.3 Grave goods

Only a small proportion of the Hunno-Sarmatian period burials had been robbed and a large variety of grave goods were recovered from the tombs (Stambulnik 1978, 20).

3.3.3a Weaponry

The majority of weapons recovered from the burials were arrowheads, bone bow brackets, spearheads and darts. The majority of arrowheads were shafted and had been manufactured from iron, although some bone arrowheads were also recovered. The typology of the iron arrowheads took a number of formats. There was a corpus of flat, lozenge-shaped examples which had been attached directly to their shafts. A second group comprised those which were flat with a triangular-shaped anterior aspect and a rectangular-shaped posterior section. These arrowheads had also been attached directly to their shafts. A third identifiable group of arrowheads consisted of those which were flat or trapeziform-shaped with blunt tips. A final group had trilobate-shaped tips. In addition, a number of bone whistles, either attached to arrowshafts or lying separately from the arrows, were also discovered in the tombs. The spearheads were generally bung-shaped, flat or trilobate in format (Mandelshtam and Stambulnik 1992, 200).

Bows usually had seven horn brackets - two pairs of end brackets with a hollow for the bow string, and three middle brackets. In one of the burials the partly preserved remains of a bow were recovered, with a full imprint of the bow preserved on the wall of the coffin. The bow was composite in form and had a length of approximately 1.5 metres (*ibid*.).

3.3.3b Tools

The majority of tools recovered from the burials consisted of iron knives and awls. Sets which contained a knife and an awl were typically found. In some cases several knives were found associated with a skeleton. These could be positioned together in a single group or located at various points around or on the body. The typology of the knives took a number of formats. There was a corpus of examples with a ring-shaped pommel, a straight back and a poorly defined handle. A second group comprised those which had a ring-shaped pommel, a straight or concave back and a well defined handle. A third identifiable category consisted of those which had a wooden or bone handle and were tanged with a straight or convex back. A final group comprised tanged and double-bladed knives which would probably have had wooden or bone handles (Mandelshtam and Stambulnik 1992, 199).

The majority of awls retrieved were rod-shaped with a flat or circular cross-section. They were generally associated with a wooden or, more rarely, a bone handle. Several awls were recovered which were rod-shaped, circular in cross-section and had a ring-shaped pommel (*ibid*.). A wooden triangular-shaped knife with a straight back and a number of knife-shaped objects with ring-shaped pommels and well defined handles were interpreted as files (*ibid*.). Disc-shaped stones and ceramic spindle whorls were also recovered from the burials. In

Plate 7: Hunno-Sarmatian period - Single adult individual buried in a wooden coffin of unknown context (Photographic Archive of the Institute for the History of Material Culture, St. Petersburg).

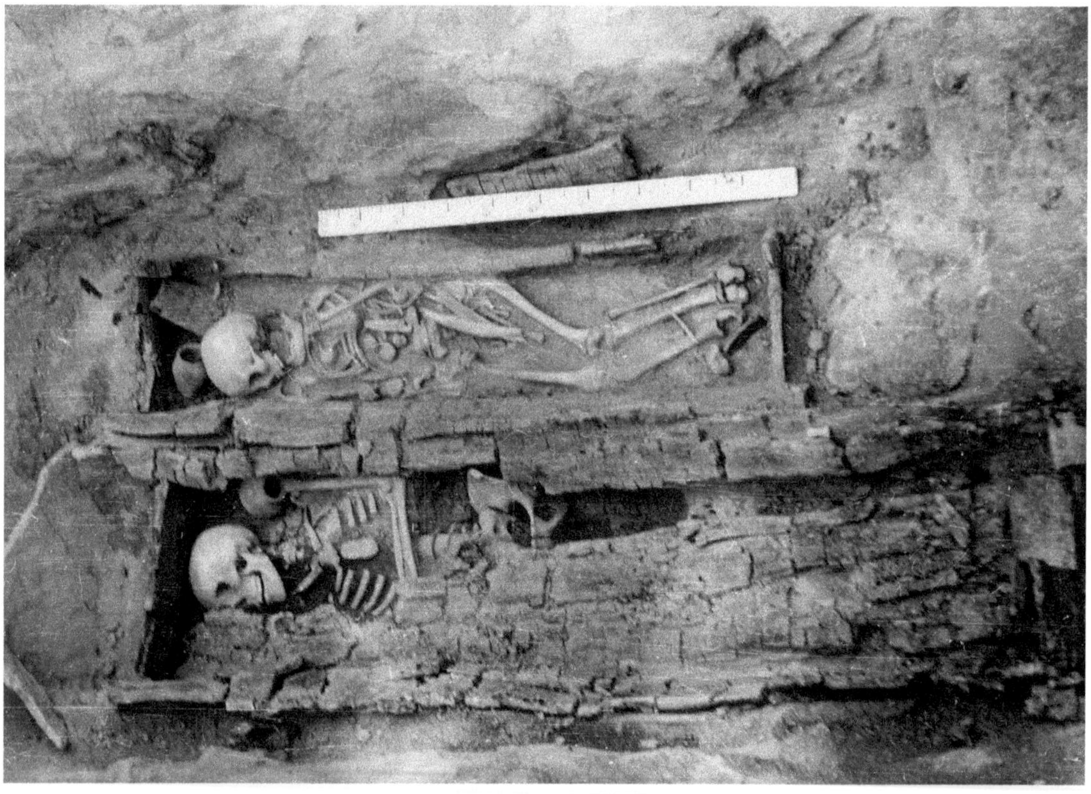

Plate 8: Hunno-Sarmatian period - Adult female and 5.5 - 7.5 year old child buried in a pair of adjoining coffins in Context XXXI. 85 (Photographic Archive of the Institute for the History of Material Culture, St. Petersburg).

Plate 9: Hunno-Sarmatian period - Pair of adult individuals possibly buried in wooden coffins and contained in a large stone cist (Context XXXI. 30) (Photographic Archive of the Institute for the History of Material Culture, St. Petersburg).

one case a wooden spindle was preserved in association with a ceramic spindle whorl (*ibid.*).

3.3.3c Household objects

Pottery recovered from the tombs showed parallels with objects discovered from Hunno-Sarmatian period cemeteries in Transbaikalia and Mongolia, while a small proportion of the pottery vessels were similar to those from the Late Hunno-Sarmatian period burial ground at Kokel which is also located in Tuva. These predominantly occurred in tombs of the fifth or ninth types (see Section 3.3.1b). A proportion of the Hunno-Sarmatian period Aymyrlyg pottery vessels also showed parallels with those of Scythian period date from the same site (Stambulnik 1983a, 38). The majority of pots were vase-shaped vessels of varying size, although a number of pots and goblets of a variety of shapes were recovered (Plates 8 and 9). In some cases the vessels had vertical handles which were fixed to the rim of the vessel (Mandelshtam and Stambulnik 1992, 199).

A wide variety of wooden vessels were recovered which included oval and round dishes, cups, scoops, pots with handles, pots without handles, and beaker-shaped vessels. In addition, wooden ladles and birch bark boxes with wooden bases were recovered (*ibid.*). A number of miniature spoons with long handles (which were probably used for measuring) were also retrieved (Stambulnik 1983a, 39).

3.3.3d Foodstuffs

The available archival information suggests that Skeleton XXXI. 45 was the only individual buried with plant material, and it is stated in the report that plant seeds were recovered from the burial (Stambulnik 1981, 2). A large number of burials contained the remains of sheep bones which may indicate that the individuals were buried with joints of mutton (Stambulnik 1980, 1981 and 1983). In Burial XXXI. 97 a sheep scapula was discovered lying on an oval wooden dish (Stambulnik 1983b, 1). In Burial XXXI. 100, Skeleton 1 was associated with a sheep skull and an iron knife, which were lying together on an oval wooden dish (*ibid.*, 3). Six vertebrae and two ribs from a horse were associated with Skeleton XXXI. 119 (*ibid.*, 12). This finding may indicate that, in addition to mutton, horse meat had been consumed by the Aymyrlyg population.

3.3.3e Clothing and adornments

The remains of clothing can be reconstructed on the basis of fragments of leather, silk and woollen fabrics which were recovered from the burials. In one of the burials fabric was draped across the skull, including the facial area, and extended to the thoracic vertebrae. It is probable that the fabric represented the remains of a shroud (Mandelshtam and Stambulnik 1992, 199). Occasionally the leg bones of the individuals were covered with leather which probably represented the remains of leather footwear (Stambulnik 1980, 10).

Traces of fur discovered beneath the pelvis of Skeleton XXXI. 120 probably came from an outer garment, partially composed of fur (Stambulnik 1983b, 13).

Fragments of leather belts were generally preserved lying adjacent to metal buckles. Each individual usually had a number of belt buckles of either uniform or multivariate shapes and forms. In some cases the positioning of the belt buckles were indicative that the individual had been wearing two belts. In cases where the individual had been wearing two belts there was an external belt with large buckles and an internal belt with smaller buckles. The belt buckles were classified into two main groups - those with a mobile catch and those which had an immobile hook. The buckles with a mobile catch were generally round, oval, or rectangular in shape. In some cases this form of buckle had a figure-of-eight morphology or was lyre-shaped. The buckles with an immobile hook were generally circular in shape with a round or a rectangular hole. A number of iron rings were also recovered which would probably have functioned as either buckles or pendants (Mandelshtam and Stambulnik 1992, 199).

A number of perforated square and circular-shaped plates with small hooks also appear to have been associated with belts. A group of these plates were manufactured from horn, gold and bronze and may be indicative of the high status of the wearer. These plates were generally large in size and could occur singly or in pairs. Some of the larger plates displayed floral artwork or depictions of animals (*ibid*., 200).

Jewellery was only occasionally recovered from the burials. Gold items were rare, with the majority taking the form of badges or plates which would have been originally sewn onto the clothing. Some very elaborate gold objects were, however, retrieved and it would seem that semi-precious stones were set into these items. A number of pendants made from genuine and fake Cowrie shells were found. Several pot-shaped pendants were also retrieved from the burials. A variety of cylindrical and keg-shaped beads made from colourful semi-precious stones were discovered. A small carving of a ram which had been covered with gold foil displayed a socket in its base which suggests that it was hafted onto a shaft. This object may also have been used as an adornment, or it may have had ritual or social significance (*ibid*.).

3.3.3f Personal objects

Fragments of mirrors were identified accompanying several burials. Of probable Far Eastern origin the majority of the mirrors were found to date to the 1st century BC or later (Stambulnik 1983a, 38). Round or oval-shaped birch bark boxes containing a variety of toiletry items - tweezers, small knives, scraper-like objects, fragments of mirrors and combs - were frequently recovered (Mandelshtam and Stambulnik 1992, 199). In most cases the mirror fragments were wrapped in soft silk covers. Some of the boxes also contained miniature bone or, more rarely, wooden spoons which were probably used for toiletry purposes. Several of the boxes were decorated with stylistic designs. A number of examples depicted animals, while entire scenes were portrayed on a small number of the boxes. At least two of the birch bark boxes displayed symbols which had been scratched in a row onto the surface of the box. These symbols have been interpreted as an unknown written language (*ibid*., 199).

3.3.3g Art

A number of highly decorated bronze objects were present in the tombs including a number of rectangular waist-belt plates which had been gilded and depicted a variety of relief pictures of running and winged horses. In addition, a pendant was discovered which displayed a standing deer with its head turned backwards. A near-rectangular badge with a stylised floral decoration was also recovered. All of these objects were previously undiscovered in Tuva, but analogous objects from other Central Asian locations indicates that they were of probable Hunno-Sarmatian period date (Stambulnik 1983a, 39).

3.4 Previous osteological analyses on the skeletal remains from Aymyrlyg

A detailed craniometrical and osteometrical analysis was undertaken on a number of Hunno-Sarmatian period individuals from the Group XXXI cemetery complex at Aymyrlyg (Bogdanova and Radzjun 1991, 55-100). The results of the craniological and osteological analysis of the Hunno-Sarmatian period skeletons from Aymyrlyg were considered to indicate that the individuals were of mixed Caucasian-Mongoloid extraction. Both craniometrical and osteometrical differences were observed between the individuals who had been buried in stone boxes and those who had been interred in ground pits. The individuals who had been buried in ground pits were considered to display more pronounced Mongoloid facial characteristics and had a greater estimated living stature relative to those who had been buried in stone cists. It has been suggested that the results are indicative that a heterogeneous population group was burying its dead at Aymyrlyg during the Hunno-Sarmatian period (*ibid*.).

4. Osteological methodology

4.1 Post-excavation treatment of the skeletal remains

The skeletal remains from the Scythian and Hunno-Sarmatian periods from Aymyrlyg were excavated by Dr. A. M. Mandelshtam and Dr. E. U. Stambulnik in the period between 1968 and 1984, as part of the Sayano-Tuvinskaya expedition of the Institute of the History of Material Culture, St. Petersburg (see Section 3.1). The post-excavation treatment of the skeletal remains in the period prior to 1995 is not known by the current author, but it is understood that an unknown proportion of the Scythian and Hunno-Sarmatian period skeletons were not curated by the anthropologists due to their lack of archaeological context (Prof. Ilyia Gokhman 1996, pers. comm.).

When the current author undertook the analysis of the skeletal remains in the period between 1995 and 1997 all surviving skeletons from Aymyrlyg were curated in the Research Division of the Peter the Great Museum of Anthropology and Ethnography, St. Petersburg. The majority of skeletons had been washed, but it was necessary to remove surface soil from the remains of several individuals using a soft brush. Any skeletal elements which were fragmentary had been reconstructed using wax by the anthropologists.

The remains of the Scythian period adult individuals had been accorded an anthropological number with the prefix '6862-', which was followed by a number consecutively given to the curated individuals by the anthropologists. The remains of Scythian and the majority of Hunno-Sarmatian period subadult individuals were labelled with the prefix '6863-', similarly followed by a consecutive number allocated to the individual during curation. The remains of all Hunno-Sarmatian period adults and several Hunno-Sarmatian period subadult individuals were labelled as 'XXXI-', which was succeeded by the burial number given to each skeleton by the archaeologists during the excavation. Consequently, establishing the archaeological context for the Hunno-Sarmatian period individuals was relatively easy to achieve. To obtain information on the archaeological contexts for the individuals labelled '6862-' and '6863-', however, it was necessary to identify any details which had been written in indelible ink by the archaeologists on the skeletal remains of the individuals. In addition, a list of anthropological numbers and archaeological contexts provided by Dr. Anna Radzjun of the Research Division of the Peter the Great Museum of Anthropology and Ethnography gave further details on the individuals. It was not possible, however, to successfully match all of the anthropological numbers to the archaeological record, and this resulted in the exclusion of several individuals from the current skeletal analysis.

The skeletal remains were stored in rectangular wooden boxes, with the cranial and post-cranial remains of the majority of Scythian and Hunno-Sarmatian period individuals curated separately. Prior to the onset of the analysis of the skeletons it was necessary, therefore, to create a catalogue of all human remains from Aymyrlyg contained in the storeroom at the Research Division of the Peter the Great Museum of Anthropology and Ethnography. This system enabled the remains of each individual to be examined as a single entity.

4.2 Recording of the skeletal remains

The sex, age-at-death, estimated living stature, and post-cranial metrics were recorded for all adult individuals from Aymyrlyg. In addition, a sample of adult individuals from both the Scythian and Hunno-Sarmatian periods were examined for cranial and post-cranial non-metric traits. The only osteological feature of the subadult individuals to be methodically determined was the age-at-death. If an individual displayed an unusual or a prominent non-metric trait this was also recorded. All individuals were subjected to a rigorous palaeopathological examination of both their skeletal and dental remains. Cultural modifications (such as discoloration due to the association of the remains with metal artefacts in the tomb, defleshing cut marks and cranial deformations) were also recorded for all individuals.

All osteological and dental characteristics were documented on a series of proforma recording sheets devised by the author and adapted from the standard recording sheets utilised by osteoarchaeologists in the Department of Archaeology and Prehistory, University of Sheffield and the Department of Archaeological Sciences, University of Bradford. All palaeopathological conditions and cultural modifications apparent in the skeletal remains were recorded in a series of laboratory notebooks. In addition, a detailed photographic record of palaeopathological lesions, unusual non-metric traits, signs of cultural modification, and evidence of soft-tissue preservation was created using a portable photographic studio conveyed to the Research Division of the Peter the Great Museum of Anthropology and Ethnography by the author.

4.3 Sex determination

The determination of the sex of an individual is affected by the age-at-death, race and state of preservation of the skeleton. Prior to puberty it is not possible to accurately determine the sex of a subadult skeleton. If sex-related features were evident in the remains of older subadult individuals the sex was noted in the gazetteer entry for that individual. Only the adult skeletons from Aymyrlyg, however, were methodically examined for the purposes of sex determination.

In general, when a complete skeleton is present the sex can be established with 95-100% accuracy. If, however, only sections of the skeleton are present, then the accuracy of sex determination diminishes. The pelvis alone provides 90-95% accuracy, the skull provides 80-90% accuracy, while the long bones enable the sex to be established with 80-90% accuracy (Krogman and Iscan 1986, 189; Ubelaker 1989a, 53-55). The human remains from Aymyrlyg ranged from being in a very poor state of preservation to an excellent state of preservation. In general, as many sex determination techniques as possible were used in the study. The combination of morphological characteristics and metrical data was usually sufficient to enable a definite identification of sex. In cases where the remains of the individuals were not sufficiently preserved to enable a definitive determination of sex the skeletons were categorised as 'possible female', 'possible male', 'probable female' or 'probable male'. Individuals were classified as being of unidentifiable sex when no sex related characteristics were apparent.

It is important to remember that ethnic characteristics can obscure the morphological traits generally used for sex determination. Consequently, it is necessary to have a group of skeletons from the same ethnic group for which a pattern of relative male and female characteristics can be established. Since the individuals from both the Scythian and Hunno-Sarmatian periods at Aymyrlyg were all classed as being of a Caucasian and Mongoloid admixture (Bogdanova and Radzjun 1991, 60; also see Section 3.4), for the purposes of sex determination the group may be considered as fairly homogenous.

Of all of the bones of the skeleton the pelvis is considered to provide the most accurate information for the determination of sex (Ferembach *et al* 1980, 517; Krogman and Iscan 1986, 200). The morphology of the pelvis is different in males and females since the female pelvis is specially adapted for pregnancy and childbirth. Table 3 contains details of the features of the pelvic bones which were used for determining the sex of the individuals from Aymyrlyg.

As stated above, sex determinations made solely on the basis of the bones of the skull are not as accurate as those ascertained from a study of the pelvic bones. In general, the male skull is more robust with larger muscle markings than those visible on a female skull. The morphological traits of the skull examined for sex determination in the remains from Aymyrlyg are listed in Table 4.

When determining the sex of the individuals from Aymyrlyg metrical analysis of the long bones was used only as a supplementary method to the pelvic and skull sex determination techniques. The metrical information obtained from the long bones, and used for the purposes of sex determination is provided in Table 5.

Characteristic	Difference between males and females
Sciatic notch	In females the notch is wide and forms an angle of c. 60°. In males the notch is narrower with an angle of c. 30°.
Subpubic angle	A sub-pubic angle greater than or equal to 90° is indicative of the female sex.
Subpubic concavity	Female concavity is large and obvious, while in males (if present) it is very slight.
Ischio-pubic ramus	Medial area of ischio-pubic ramus is flat in males. Ridge occurs at the site in females.
Ventral arc	Ventral arc is a ridge on the anterior surface of the pubis which is only present in females.
Preauricular sulcus	Sulcus is more commonly present in females.
Obturator foramen	The obturator foramen is small and triangular shaped in females and larger and oval shaped in males.
Pelvic brim	Iliac crests are less angulated in females than in males.
Acetabulum	Acetabulum is larger in males.
Ilium auricular surface	In females the auricular surface is raised above the surface of the ilium and there is a wide post-auricular space.
Sacral curvature	Sacrum is more curved in males than females.

Table 3: Features of the pelvic bones used for sex determination (after Ferembach et al 1980, 518; Krogman and Iscan 1986, 208-216; Buikstra and Ubelaker 1994, 16-19).

Characteristic	Difference between males and females
Supraorbital ridges	More prominent in males.
Mastoid processes	Larger in males.
Posterior zygomatic arch	Posterior end of zygomatic process extends as a well-defined ridge for a distance past the external auditory meatus in males.
External occipital protuberance	Larger and more prominent in males.
Anterior mandible	Chin is more square in males while rounded with a medial point in females.
Orbital rims	Upper margin of the orbit is blunt in males and sharper in females.
Frontal/parietal bossing	The female skull retains the childhood characteristics of frontal and parietal bossing into adulthood

Table 4: Features of the skull used for sex determination (after Ferembach et al 1980, 523; Krogman and Iscan 1986, 192-193; Buikstra and Ubelaker 1994, 19-20).

Measurement	Female (mm)	Male (mm)	Author
Femur (vertical head diameter)	< 43	> 48	Stewart (1979, 100-101)
Femur (bicondylar breadth)	< 74	> 76	Pearson (1917-1919, 56)
Humerus (vertical head diameter)	< 43	> 47	Stewart (1979, 100)
Radius (maximum head diameter)	< 21	> 23	Berrizbeitia (1989, 1210)
Scapula (width of glenoid cavity)	< 26	> 29	Krogman and Iscan (1986, 227)
Clavicle (maximum length)	< 138	> 150	Krogman and Iscan (1986, 228)

Table 5: Adult long bone measurements used for the determination of sex. The measurements (mm) in the second column are the values below which the bone can be confidently assigned to the female sex. The measurements in the third column are the values above which the bone can be assigned to the male sex. Values which are intermediate cannot be assigned to either sex.

4.4 Determination of the age-at-death

The determination of the age at which an individual died involves the observation of morphological features in the skeletal remains and the comparison of the information with age-related changes recorded for modern-day populations of known age. This is followed by the estimation of any possible variabilities which may exist between the unknown and the recent population on which the comparative data is based (Ubelaker 1989a, 63). There are two main periods in the life of a human; development, which occurs approximately between birth and 25 years of age, and degeneration, which occurs roughly from the age of 25 years onwards. It is easier to determine the age-at-death of children and adolescents than it is to obtain an age estimation for fully developed adults, since the growth and maturation of a subadult skeleton follows a reasonably predictable sequence (Chamberlain 1994, 13). It is more difficult to make an accurate assessment of the age-at-death of an adult since once the skeleton has completed growth the age-related changes associated with degeneration do not progress at a predictable rate (Saunders 1992, 8).

4.4.1 Subadults

The subadult individuals were grouped into the following age categories:

Infant	< 2 years
Child	2 - 6 years
Juvenile	6 - 12 years
Adolescent	12 - 17 years

Age-at-death determination of subadults was principally undertaken on the basis of the tooth eruption and developmental sequence devised by Ubelaker (1989a, 64), and the tooth calcification method of Moorrees *et al* (1963) modified by Smith (1991). In addition, the epiphyseal fusion data of Ferembach *et al* (1980, 530-532) and Ubelaker (1989a, 75) were also used for the purposes of age determination. The epiphyseal method, however, is generally considered less reliable than the dental methods since it is known that there is a marked difference in the timing of epiphyseal fusion between males and females (Ubelaker 1989b, 56). It is not possible to determine the sex of subadult individuals, and age-at-death estimations obtained using only the epiphyseal fusion ageing method must be considered as approximate values. Consequently, age-at-death estimates based solely on the epiphyseal fusion method were used only when dental remains were not preserved. The diaphyseal length values of Ubelaker (1989a, 70-71), which were obtained from the Native American Arikara population group, were also employed as a means of determining age-at-death. As was the case for the epiphyseal fusion technique, however, the diaphyseal length method is considered to be less reliable than the dental age determination methods. The main problems with the method arise due to the fact that the bone growth rates will differ widely both among populations of the same race and between different racial groups.

Each individual was placed into a relevant age category. In some cases the established age-at-death lay in the boundary between two of the age groups. When this occurred the individual was placed in the category where the most significant overlap in years was noted. As many as possible of the various ageing techniques discussed in this section were applied to the remains of each subadult individual. In cases where it was possible to apply multiple ageing methods to an individual the estimated age-at-death obtained from the more reliable ageing techniques was given priority.

4.4.2 Adults

It is more difficult to determine accurately the age-at-death of an adult skeleton than subadult remains. The majority of adult ageing techniques are based on degenerative changes which occur in the skeleton as the individual increases in age. The age-related changes associated with degeneration do not progress at a predictable rate. Consequently, it is not possible to obtain precise age-at-death values for adult individuals, and it is necessary to ascribe an age-at-death band to each adult individual. The following age-at-death categories were applied to the adult individuals from Aymyrlyg:

Young adult	17-25 years
Adult	25-35 years
Adult	35-45 years
Mature adult	45+ years

As many as possible of the various ageing techniques discussed in this section were applied to the remains of each adult individual. When it was possible to apply a multifactorial approach to an individual the estimated age-at-death obtained from the more reliable ageing techniques (such as the Suchey-Brooks pubic symphysis ageing method) was given priority.

The methods employed for age-at-death determination of the adults included the examination of the state of fusion of the sternal ends of the clavicles (McKern 1970, 46) and the centra of the first and second sacral vertebrae (*ibid.*, 48), and the state of eruption of the third molars (*ibid.*, 43). The tooth attrition standards for age determination devised by Brothwell (1981, 72) were also employed for the purposes of age-at-death determination. An examination of the morphological changes of the pubic symphysis is regarded as one of the most reliable methods for determining the age-at-death of adult skeletons (Buikstra and Ubelaker 1994, 21), and the Suchey-Brooks ageing method was followed (Brooks and Suchey 1990, 237). To ensure accuracy the morphological characteristics of the pubic symphyses of the adult individuals from Aymyrlyg were compared to Diane France Suchey-Brooks male and female age determination epoxy resin/plaster casts (Klepinger *et al* 1992, 763). The auricular surface ageing technique of Lovejoy *et al* (1985), the cranial suture methodology of Meindl and Lovejoy (1985) and the conditions of the sternal ends of the ribs were also examined for the purposes of establishing age-at-death (Iscan *et al* 1984; Iscan *et al* 1985).

4.5 Estimation of living stature

The estimation of living stature was determined for adult individuals using the mathematical methods devised by Trotter and Gleser (1952, 1958) and Trotter (1970). The greatest lengths of the long bones were measured, and the values were applied to their regression formulae. The most accurate estimations of stature are obtained from measurements obtained from the long bones of the lower limb, with the combined femur-tibia formula generally providing the most accurate estimation. Consequently, where possible, the combined femur-tibia formula was used to determine the estimated stature of the individuals from Aymyrlyg.

5. Trauma

5.1 Introduction

Trauma may be defined as any injury or wound that affects the skeleton and/or soft tissues of the body (Roberts 1991, 226). In general, trauma affects the skeleton in four main ways - partial or complete break of bone, subluxation or dislocation of bone, a disruption of the nerve and/or blood supply and the artificially induced development of an abnormal shape or contour in bone (Ortner and Putschar 1981, 55). The evidence of trauma among a population group can provide important information on the lifestyle of the individuals, such as their material culture, economy, living environment, occupation and the levels of interpersonal violence they may have been exposed to (Roberts and Manchester 1995, 65). The skeletal lesions of trauma are indicative of gross injury since everyday cuts, scratches, bumps and bruises will not be visible in the palaeopathological record (*ibid.*, 66).

A fracture may be defined as a partial or complete break in the continuity of bone and/or cartilage, and each fracture is associated with damage to the adjacent soft tissues (Resnick *et al* 1995, 2570). Fractures which occur a short time before or at the time of death will display no evidence of healing and are difficult to identify in skeletal remains since their morphology can be identical to post-mortem damage (Roberts and Manchester 1995, 67). Consequently, the majority of fractures described throughout the present chapter either displayed a degree of healing or were completely healed. Classification of the state of healing associated with the fractures is based on the descriptions of Resnick *et al* (1995, 2575). The majority of fractures encountered in the Scythian and Hunno-Sarmatian period populations were either transverse or oblique simple fractures in which the skin would have remained intact (*ibid.*, 2570), although several examples of crushing or compression fractures were also apparent. The only fractures evident in subadult individuals were depressed fractures of the crania. The most common fractures to arise in immature bone are greenstick, or incomplete, fractures (Adams 1978, 8) which may be difficult to identify in archaeological skeletal remains. Consequently, the prevalence rates of fractures present in each type of bone, with the exception of the cranial vaults, were calculated on the basis of the number of observable complete adult bones only (see Appendix 1). The frequencies of depressed cranial fractures were based on the number of near-complete and complete subadult and adult crania available for examination.

Radiographic facilities were not available during the analysis of the skeletal remains, and the descriptions of traumatic lesions among the populations are based solely upon a morphological examination of the bones. Since such a large number of traumatic lesions of various manifestations were apparent the information on these injuries is presented in two separate chapters (Chapters 5 and 6). Only traumatic lesions with no clear associations with weapons or surgical implements are included in the current chapter. Injuries inflicted by weapons, or sustained during surgical procedures, such as amputations or trepanations are discussed in Chapter 6.

5.2 Scythian period individuals

5.2.1 Bone trauma

5.2.1a Fractures of the appendicular skeleton

A total number of 1710 complete adult long bones were examined for evidence of fractures, of which 0.9% (15/1710) displayed lesions. Details of the individuals with fractures among the bones of the appendicular skeleton are provided in Table 6, and it was found that a greater proportion of males (69%; 9/13) than (31%; 4/13) females displayed the injuries. A summary of the frequencies of fractured bones in the appendicular skeleton is provided in Table 7.

The ulna (2.2%) was the long bone which most frequently displayed fractures, followed by the radius (1.6%) and clavicle (1.3%). When each individual with evidence of one or more fractures is examined separately one can attempt to determine the possible aetiology of the various fractures.

5.2.1a (i) Fractures of the clavicles

Skeleton I. 3. Sk. 4, Skeleton VIII. 54. Sk. 5 and Skeleton XX. 10. Sk. 2 all displayed fractured clavicles. In all three cases the right clavicle was fractured, and two of the lesions were located at the acromial ends, while one was situated in the midshaft region. Two of the affected individuals were male, and one was female. Fractured clavicles usually arise as a result of the person falling onto their shoulder or outstretched hands (Adams 1978, 114), and Skeleton I. 3. Sk. 4, a 25-35 year old male, also displayed fractures of the right ulna and radius (see Section 5.2.1a (v)), which may also have occurred as a result of a fall. Skeleton VIII. 54. Sk. 5, a 35-45 year old female, also displayed a fractured right zygomatic (see Section 5.2.1b (ii)), and it is possible that she had suffered a heavy fall onto her right shoulder and the side of her head.

5.2.1a (ii) Fractures of the humeri

Skeleton XX. 8. Sk. 2, a 45+ year old male, displayed a compression fracture of the right humeral head (see Plate 18), which was associated with a dislocated right glenohumeral joint (see Section 5.2.1d). An anterior dislocation of the glenohumeral joint is commonly associated with a compression fracture of the postero-lateral aspect of the humeral head, a feature which is generally referred to as the Hills-Sachs Lesion. The humeral head trauma occurs as a result of its impaction against the anterior margin of the glenoid fossa (Resnick and Goergen 1995, 2695), and it is probable that this action was responsible for the fractured humeral head evident in the Aymyrlyg case. A Colles' fracture of the left radius (see Section 5.2.1a (iv)), a fracture of the right fibula (see Section 5.2.1a (vii)) and an ossified haematoma of the left fibula (see Section 5.2.1c) were also evident in the remains of the indi-

Context	Age	Sex	Bone	Location	Type	Shortening	Alignment	Apposition	Healing
B. 8. Sk. 1	35-45	M	L ulna	distal	simple transverse	-	good	100%	healed
I. 3. Sk. 4	25-35	M	R radius	distal	simple transverse	+	c. 40° post.	100%	healed
			R ulna	distal	simple transverse	-	c. 30° lat.	100%	heaed
			R clavicle	acromial	simple oblique	+	good	100%	healed
IV. 4. Sk. 3	35-45	M	L tibia	distal	simple oblique	+	good	90-100%	healed
			L fibula	proximal	simple oblique	+	c. 20° ant.	50%	healed
VI. 16. Sk. 1	35-45	F	L ulna	distal	simple transverse	+	good	100%	healed
VII. 5	35-45	M	L fibula	distal	simple oblique	+	good	50%	healed
VIII. 54. Sk. 5	35-45	F	R clavicle	acromial	simple oblique	+	good	100%	healed
IX. 19. Sk. 1	25-35	M	L tibia	distal	simple oblique	+	c. 5° post.	100%	healed
XX. 7. Sk. 4	17-25	M	L radius	proximal	simple oblique	+	c. 10° post-med.	50%	healed
			L ulna	distal	simple oblique	+	c. 45° lat.	75%	healed
XX. 8. Sk. 2	45+	M	L radius	distal	simple transverse	+	c. 30° post.	90-100%	healed
			R humerus	proximal	compressi-on	+	-	100%	healed
			R fibula	distal	simple oblique	+	c. 10° post.	90-100%	healed
XX. 10. Sk. 2	25-35	M	R clavicle	midshaft	simple oblique	+	good	90-100%	healed
XXI. 4. Sk. 3	35-45	F	L radius	distal	simple transverse	+	c. 5° post.	90-100%	healed
			L ulna	styloid	simple transverse	+	c. 50° post.	50%	healed
XXIII. 4	25-35	M	L MC3	midshaft	simple oblique	+	good	100%	healed
			L MC4	midshaft	simple oblique	+	good	100%	healed
			R navicular	anterior surface	compressi-on	-	good	-	healed
XXIII. 8	45+	F	R MC4	midshaft	simple transverse	+	good	100%	healed
			R MC5	midshaft	simple transverse	+	c. 45° ant.	100%	healed
			L ulna	distal	simple transverse	+	c. 10° med.	100%	healed

Table 6: Scythian period individuals - Fractures of the appendicular skeleton. The degree of displacement referred to in the 'Alignment' column relates to the distal part of the fractured bone (R = right, L = left, MC = metacarpal, + = present, - = not applicable).

Bone	No. fractures	No. bones	% fractures
clavicle	3	227	1.3
humerus	1	283	0.4
ulna	6	267	2.2
radius	4	255	1.6
tibia	2	326	0.6
fibula	3	246	1.2
metacarpal	4	216	1.8
navicular	1	48	2

Table 7: Scythian period individuals - Prevalence of fractures in the bones of the appendicular skeleton.

vidual, and it is probable that all of the injuries were obtained as a result of a serious fall.

5.2.1a (iii) Night-stick fractures of the ulnae

Three individuals (Skeleton B. 8. Sk. 1, Skeleton VI. 16. Sk. 1 and Skeleton XXIII. 8) displayed night-stick fractures of the ulna, in which only the diaphysis of the ulna was fractured. In all three cases the fractures had occurred at the distal ends of the diaphyses of the left ulnae. Night-stick fractures commonly arise during an assault when the victim raises his or her forearm to protect the head from being struck (*ibid.*, 2732). Two of the individuals with night-stick fractures were female, and Skeleton XXIII. 8, a 45+ year old female, also displayed fractured 4th and 5th (boxer's fracture) metacarpals (see Section 5.2.1a (ix)) and a fractured rib (see Section 5.2.1b (vi)). Fractures of the metacarpals generally arise as a result of a fall onto a hand or from a blow upon the knuckles as obtained in boxing (Adams 1978, 182), and it is probable that she had procured the fractures in an incident of violence at some stage during her life. Skeleton B. 8. Sk. 1, a 35-45 year old male, also displayed an ossified haematoma and evidence of weapon trauma (see Section 6.2.1), which again may suggest that his fractured ulna had been attained during an incident of violence.

5.2.1a (iv) Colles' fractures of the radii

Two individuals displayed Colles' fractures which involve the fracture of the distal end of the radius with dorsal displacement. Colles' fractures are generally attained as a result of fall on outstretched hands and are more common in older individuals, particularly females (Resnick and Goergen 1995, 2734). Skeleton XX. 8. Sk. 2, a 45+ year old male, also displayed a fractured right fibula (see Section 5.2.1a (vii)), a dislocated right glenohumeral joint with an associated fractured humeral head (see Sections 5.2.1a (ii) and 5.2.1d) and an ossified haematoma on the left fibula (see Section 5.2.1c). It is probable that all of the injuries had been attained as a result of a fall. In Skeleton XXI. 4. Sk. 3, a 35-45 year old female, the Colles' fracture evident on the left radius was also associated with a fractured left ulnar styloid process, and it is probable that the injuries were all sustained as a result of a fall onto the left hand.

5.2.1a (v) Fractures of the diaphyses of both the radii and ulnae

Two individuals displayed fractures of the diaphyses of both the radius and the ulna. Injuries which involve both bones of the forearm generally result from direct blows to the arm which may have been attained during a fall or a fight (*ibid.*, 2732). In Skeleton I. 3. Sk. 4, a 25-35 year old male, the fractures were situated at the distal ends of both bones, while in Skeleton XX. 7. Sk. 4, a 17-25 year old male, the fracture on the radius was positioned more proximally than that evident on the ulna (Plate 10). Skeleton I. 3. Sk. 4 also displayed a fractured right clavicle (see Section 5.2.1a (i)), and it is probable that he had fractured the bones of his right forearm as a result of a fall onto the right side of his body.

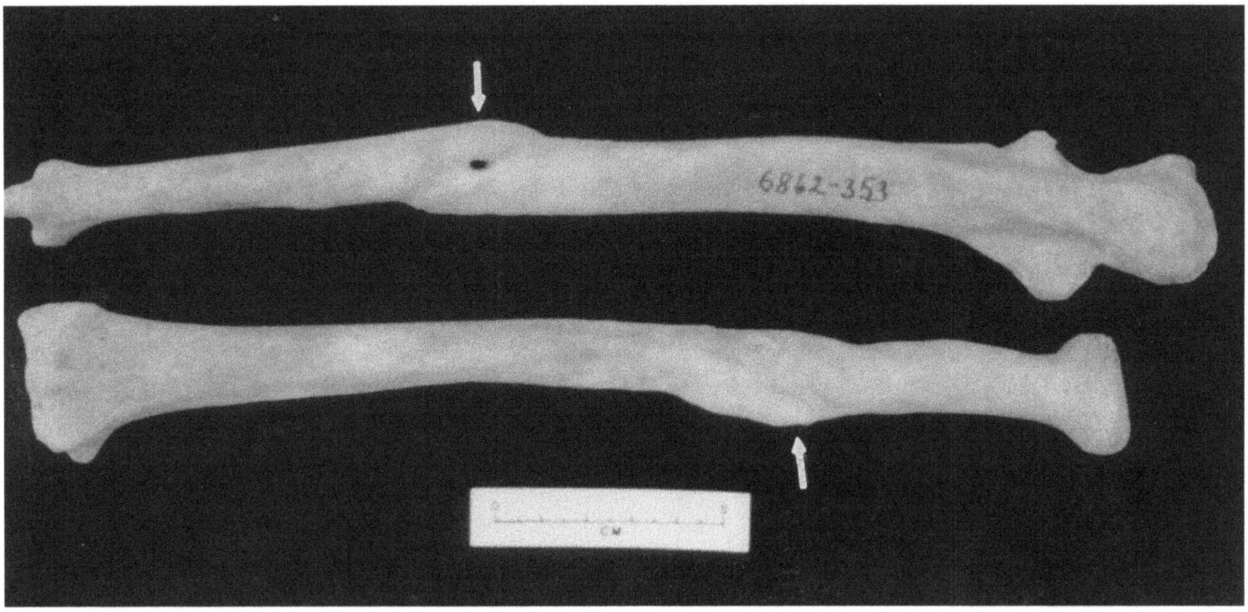

Plate 10: Scythian period - Fractured right ulna and radius in Skeleton XX. 7. Sk. 4.

5.2.1a (vi) Fractures of the tibiae not associated with fractured fibulae

Skeleton IX. 19. Sk. 1, a 25-35 year old male, displayed a fracture of the left distal tibia only. This type of injury is a relatively uncommon finding (Apley and Solomon 1988, 315), and generally arise as a result of either direct or indirect forces. Since the fracture was oblique in the Aymyrlyg case it had probably been sustained as a result of the action of an indirect force, such as a twisting injury or a fall (Resnick and Goergen 1995, 2777).

5.2.1a (vii) Fractures of the fibulae not associated with fractured tibiae

Two individuals displayed isolated fractures of the diaphysis of the fibula. The shaft of the fibula is rarely fractured without an associated fracture of the shaft of the tibia, or an injury to the ankle, and such fractures generally arise as a result of direct trauma to the bone (Adams 1978, 257). Skeleton VII. 5, a 35-45 year old male, displayed a fracture of the diaphysis of the left fibula alone. In Skeleton XX. 8. Sk. 2, a 45+ year old male, a fracture was present on the right fibula, and the individual also displayed a dislocated right glenohumeral joint (see Section 5.2.1d), a fractured right humeral head (see Section 5.2.1a (ii)), a Colles' fracture of the left radius (see Section 5.2.1a (iv)) and an ossified haematoma of the left fibula (see Section 5.2.1c). It is probable that these injuries were all procured as a result of a fall, during which the fibula had been subject to direct trauma.

5.2.1a (viii) Fractures of both the tibiae and fibulae

Skeleton IV. 4. Sk. 3, a 35-45 year old male, displayed a fractured left tibia and fibula. An oblique fracture was present at the proximal end of the fibula and the distal end of the tibia. It is therefore probable that the injuries were due to the action of a rotational force (*ibid.*, 248), and that they were obtained as a result of a fall.

5.2.1a (ix) Fractures of the bones of the hands and feet

Skeleton XXIII. 8, a 45+ year old female, displayed fractured 4th and 5th (boxer's fracture) right metacarpals. Fractures of the metacarpals generally arise as a result of a fall on the hand or a blow upon the knuckles as obtained in boxing (Adams 1978, 182). It is highly probable that the individual was right-handed and, since the fractured metacarpals were from the right hand, it is probable that she was dealing a blow when she sustained the injuries. In addition, the presence of a night-stick fracture of the left ulna (see Section 5.2.1a (iii)) and a fractured rib (see Section 5.2.1b (vi)) may also suggest that she had attained the injuries as a consequence of physical violence.

Skeleton XXIII. 4, a 25-35 year old male, displayed fractured left 3rd and 4th metacarpals, and a crushing fracture of the right navicular. The occurrence of a fracture in the navicular is relatively rare (Resnick and Goergen 1995, 2795), and this case is particularly unusual since none of the other bones of the right ankle or foot present displayed evidence of trauma. All of the injuries had probably been obtained as a result of a fall, although the individual also displayed possible evidence of weapon trauma (see Section 6.2.1).

5.2.1a (x) Evidence of treatment and healing

All of the fractures of the appendicular skeleton present in this corpus of individuals from Aymyrlyg were healed, and this indicated that they had been sustained by the individuals long before their death (see Table 6), with all lesions displaying 50% or greater apposition of the fracture parts. In addition, the alignment of the fractured bones was generally good with only slight angular deviation. It is therefore possible that some form of splints were used to treat the fractures, although the majority of the affected bones displayed a degree of shortening which may suggest that the population was unfamiliar with the manipulative techniques required to disimpact the parts of the fracture. It is interesting to note that two cases of fractured tibiae were evident; both individuals would probably have been incapacitated for at least a number of weeks and dependent for their welfare on other members of society during this time. This finding would suggest that these individuals were accorded some degree of life support during their periods of incapacitation.

5.2.1b Fractures of the axial skeleton

5.2.1b (i) Depressed fractures of the cranial vaults

Depressed fractures generally occur when bone fails at, or in close proximity to, the site of an impact, and a linear fracture arises when the bone rebounds after the trauma. The bone at the site of impact flattens to conform with the shape of the surface with which it has impacted (Gordon *et al* 1988 as quoted in Roberts 1996b, 132). It was possible to examine 473 cranial vaults from adult and subadult individuals, 1.7% (8/473) of which displayed one or more depressed fractures (Table 8).

All fractures were healed and had generally rounded appearances, which may suggest that they were made using blunt implements. The smallest dimension of the injuries was 1.5 mm, and the greatest was 43 mm. In all, 24% of the population with depressed fractures were female, 38% were males and 38% were subadults. The majority of the injuries were present on the parietals (56%; 5/9), while 33% were situated on the frontals and 11% were located on the occipitals.

The frequencies of depressed fractures in the observable frontal bones, parietals and occipitals is presented in Table 9.

Bone	No. fractures	No. bones	% fractures
frontal	3	471	0.6
parietal	5	929	0.5
occipital	1	453	0.2

Table 9: Scythian period individuals - Prevalences of depressed fractures in the bones of the adult and subadult cranial vaults.

The head is the main target of attack during episodes of interpersonal and intergroup violence (Roberts and Manchester 1995, 79), and the left fronto-parietal region of the cranium is the position where an injury would commonly be

Context	Age	Sex	Type of fracture	Location	Shape	Size (mm)	Healing
G. 5. Sk. 1	45+	F	depressed	L frontal	circular	6x6	healed
II. 4. Sk. 3 (Plate 11)	11.5-12.5	-	depressed	Mid frontal	oblique line	43x12	healed
VII. 5. Sk. 6	25-35	?M	2 x depressed	L parietal	circular	18.5x18.5	healed
					oval	37x12	healed
VIII. 12. Sk. 5	35-45	?M	depressed	L parietal	oval	31.5x17	healed
X. 2. Sk. 1 (Plate 12)	17-25	?F	depressed	L occipital	circular	20x20	healed
X. 3	14-18	-	depressed	R parietal	oval	15x5	healed
X. 13. Sk. 6	15-18	-	depressed	R parietal	transverse line	17x1.5	healed
XXIII. 13 (i)	25-35	?M	depressed	L frontal	circular	17x17	healed

Table 8: Scythian period individuals - Details of individuals with depressed fractures of the cranial vault (L = left, R = right).

afflicted in hand-to-hand combat when facing a right-handed opponent (Wakely 1997, 26). Consequently, the majority of the depressed fractures apparent in Scythian period crania had probably been deliberately inflicted during conflict. Only Skeleton XXIII. 13 (i), a 25-35 year old male, displayed evidence of multiple fractures, and a fractured left zygomatic and right nasal bone were also present in the remains (see Section 5.2.1b (ii)). It is therefore probable that the depressed fracture on the left frontal bone was obtained during an incident of interpersonal violence, and the presence of possible healed weapon trauma in the remains is a further indication that he had been involved in interpersonal violence (see Section 6.2.1).

5.2.1b (ii) Fractures of the facial skeleton and mandible

As stated above (Section 5.2.1b (i)) the majority of cases of skull trauma occur as a result of intentional blows struck by an opponent during conflict. Consequently, the presence of fractures in the facial bones and mandible may also be considered indicative of interpersonal or intergroup violence. Five members of the Scythian period population displayed fractures of the facial bones, excluding the nasal bones (Table 10).

The majority of the population with injuries to the facial skeleton were male (80%; 4/5), and only 20% (1/5) of females displayed such trauma. In Skeleton VIII. 54. Sk. 5, an adult female, a fractured right clavicle (see Section 5.2.1a (i)) was also apparent, and it is possible that both injuries were sustained as a result of a fall rather than interpersonal or intergroup violence. Skeleton G. 7. Sk. 5, a 25-35 year old male, also displayed possible weapon trauma (see Section 6.2.1), and Skeleton XXIII. 13 (i), a 25-35 year old probable male, displayed a depressed fracture on the frontal bone (see Section 5.2.1b (i)), a fractured right nasal bone (see Section 5.2.1b (ii)), and possible evidence for weapon trauma (see Section 6.2.1). It is probable, therefore, that both of these individuals had attained their injuries as a result of interpersonal or intergroup violence.

Thirteen Scythian period adults displayed fractures to their nasal bones. Some of the individuals had a single fractured nasal bone, while both nasal bones had been fractured in other skeletons (Table 11). The prevalence of these lesions in the population group was 3.4% (23/675).

Fractured nasal bones are common among archaeological populations and probably arise as a consequence of interpersonal violence (Roberts and Manchester 1995, 80). It is interesting that 92% of the Scythian period individuals with these injuries were male (12/13), and only one female displayed this trauma. This finding may suggest that males were more frequently involved in interpersonal and intergroup violence than females.

Context	Age	Sex	Type	Location	Healing
G. 7. Sk. 5	25-35	M	Simple oblique	L mandible	healed
II. 3 (ii) (Plate 13)	adult	?M	Depressed	L orbit	advanced
VIII. 54. Sk. 5	35-45	F	Simple transverse	R zygomatic	healed
XXIII. 10. Sk. 1 (Plate 14)	25-35	M	Simple transverse	L zygomatic	healed
XXIII. 13 (i)	25-35	?M	Simple transverse	L zygomatic	healed

Table 10: Scythian period individuals - Details of fractures present in the facial bones and mandibles (L = left, R = right).

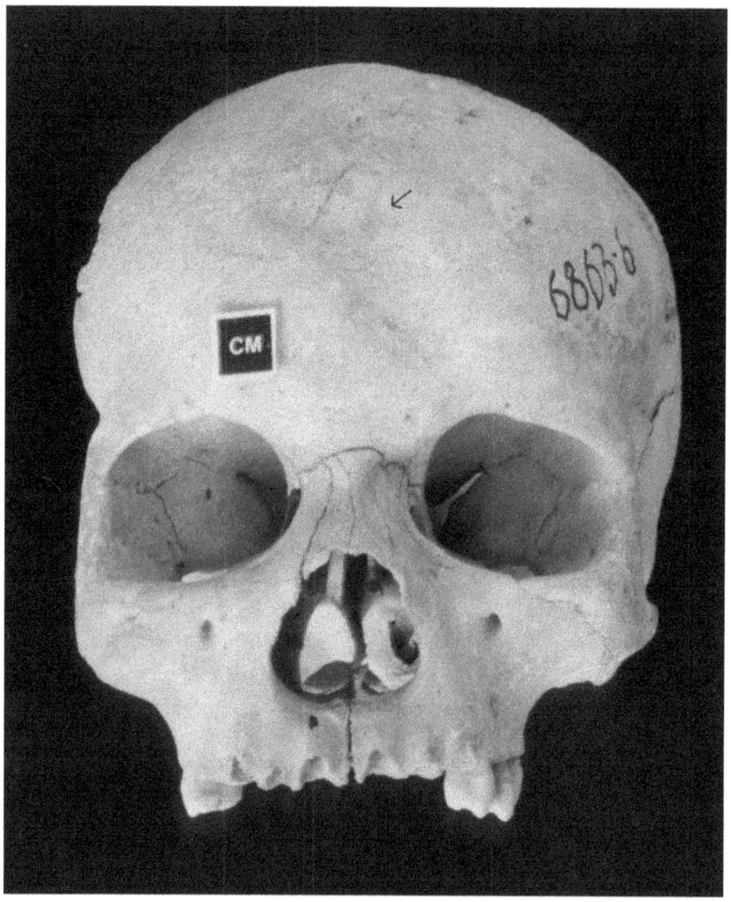

Plate 11: Scythian period - Depressed fracture on the frontal bone of Skeleton II. 4. Sk. 3.

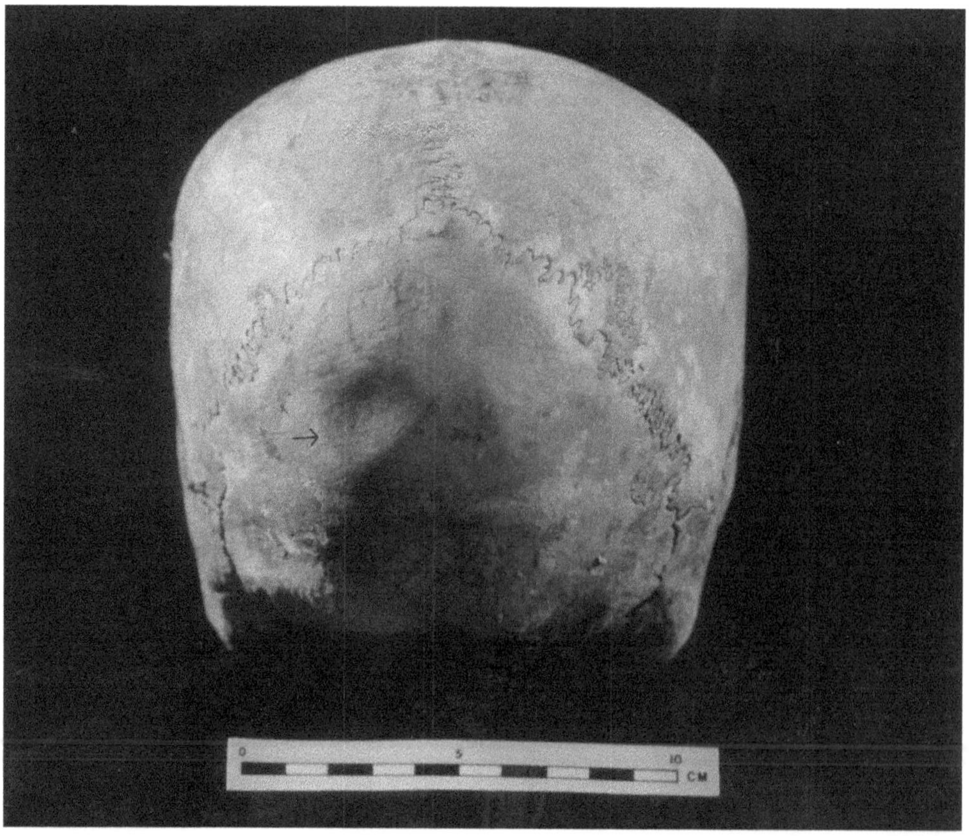

Plate 12: Scythian period - Depressed fracture on the occipital of Skeleton X. 2. Sk. 1.

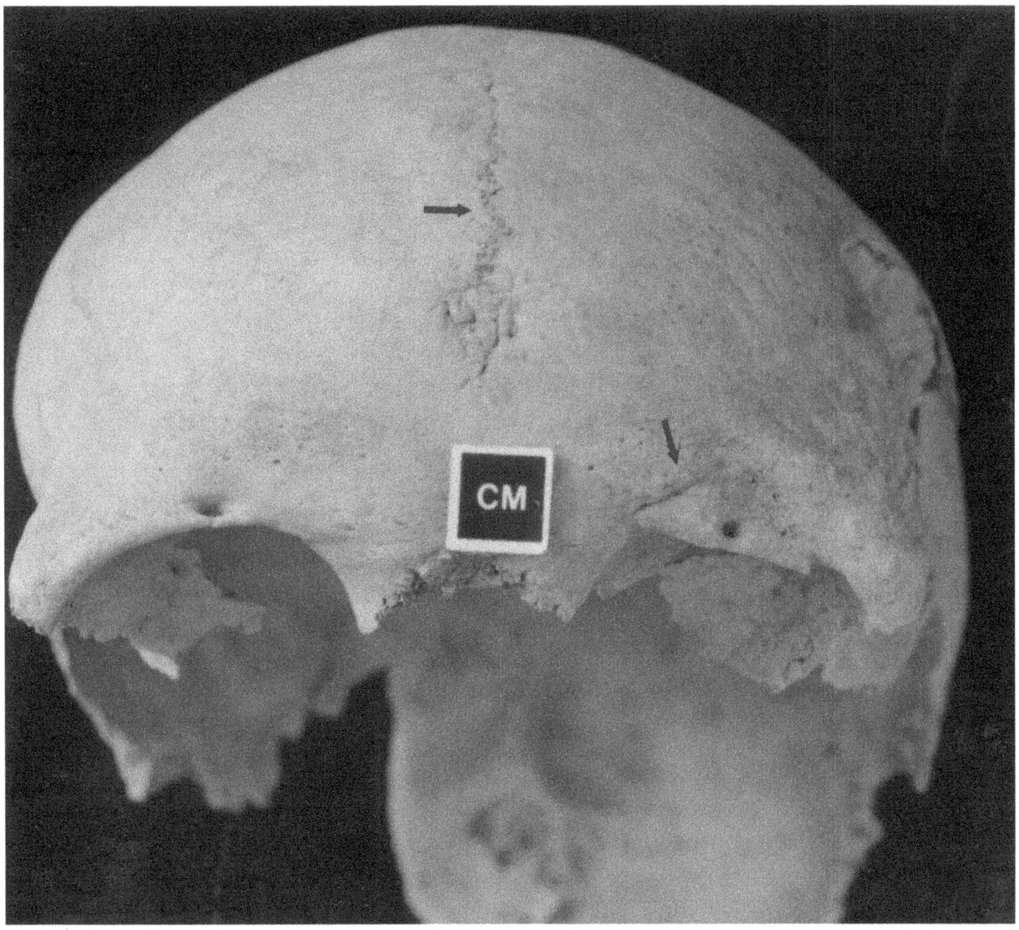

Plate 13: Scythian period - Fracture of the superior orbital margin in Skeleton II. 3 (ii).

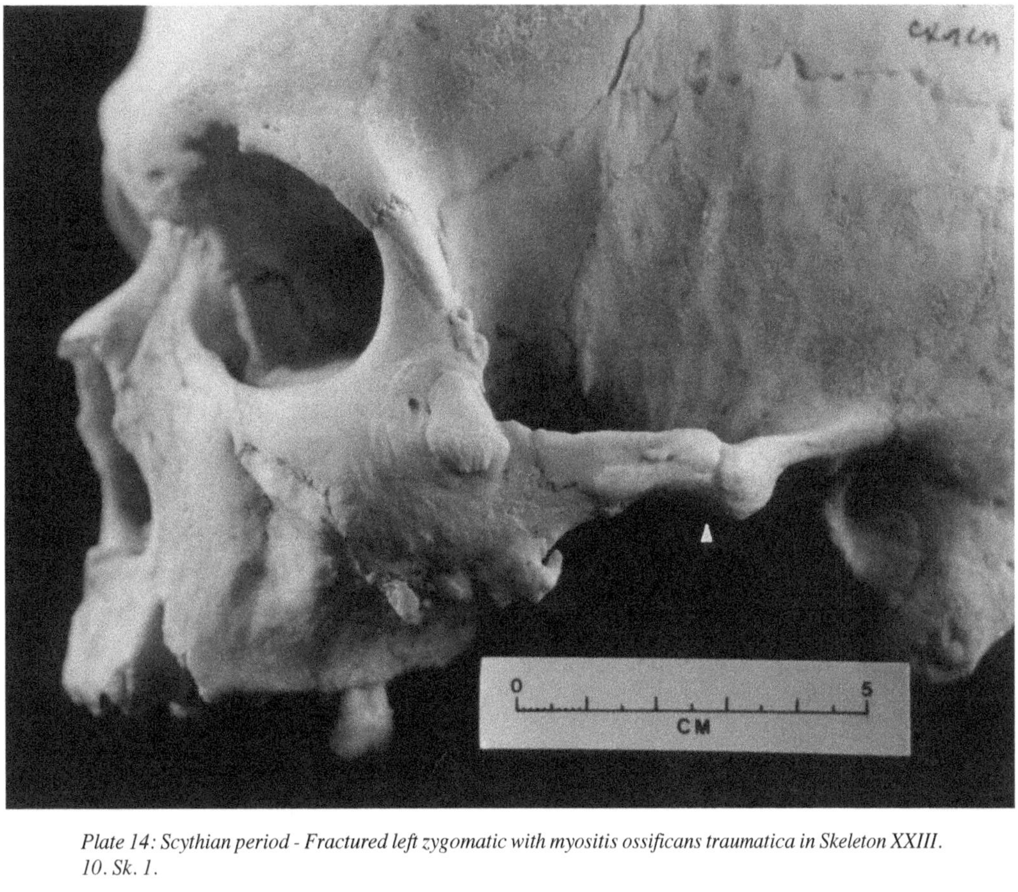

Plate 14: Scythian period - Fractured left zygomatic with myositis ossificans traumatica in Skeleton XXIII. 10. Sk. 1.

Context	Age	Sex	R		L	
			Type	Healing	Type	Healing
III. 16 (ii)	35-45	?M	Simple oblique	healed	Simple transverse	healed
III. 18	25-35	F	Simple oblique	healed	Simple oblique	healed
VI. 5. Sk. 6	25-35	?M	Simple oblique	healed	Simple transverse	advanced
VI. 13. Sk. 8	25-35	M	Simple oblique	early	Simple transverse	early
VII. 1 (iii)	35-45	?M	Simple transverse	advanced	Simple transverse	advanced
VII. 3	17-25	?M	Simple oblique	advanced	Simple transverse	advanced
VII. 5. Sk. 1	25-35	?M	Simple transverse	advanced	Simple transverse	advanced
VIII. 10 (ii) (Plate 15)	35-45	?M	Simple transverse	early	Simple oblique	mid
VIII. 12. Sk. 7 (i)	35-45	?M	Simple transverse	healed	Simple oblique	healed
VIII. 35. Sk. 1	35-45	?M	Simple transverse	advanced	Simple transverse	advanced
XX. 9	25-35	M	Simple transverse	mid	No fracture	-
XXIII. 13 (i)	25-35	?M	Simple oblique	mid	No fracture	-
XXIII. 19. Sk. 3	35-45	M	No fracture	-	Simple transverse	mid

Table 11: Scythian period individuals - Details of fractured nasal bones.

A summary of the frequencies of fractures of the facial skeleton and mandible is provided in Table 12.

Bone	No. fractures	No. bones	% fractures
mandible	1	202	0.5
orbit	1	702	0.1
zygomatic	3	642	0.5
nasal bone	23	675	3.4

Table 12: Scythian period individuals - Frequencies of fractured facial bones in the population.

The nasal bones were evidently most susceptible to fracture (3.4%), while injuries of the zygomatics, orbits and mandibles were relatively uncommon among the Scythian period population.

5.2.1b (iii) Compression fractures of the vertebrae

Skeleton XXI. 4. Sk. 3, a 35-45 year old female, displayed a compression fracture of the right side of the body of the 1st lumbar vertebra. The thoracolumbar region is the most susceptible area of the spine to traumatic compression fractures, and it is also a common site of fractures caused by minor trauma in an individual with osteoporosis (Pathria 1995, 2876). A Colles' fracture of the left radius and a fracture of the styloid process of the left ulna (see Section 5.2. 1a (iv)) were also present in the individual. In addition, the bones of the skeleton were generally osteopaenic, and it is probable that the female was suffering from osteoporosis. Consequently, the compression fracture of the 1st lumbar vertebra may have been secondary to osteoporosis.

5.2.1b (iv) Spondylolysis

Spondylolysis is a vertebral defect in which a lack of bony continuity occurs at the pars interarticularis and the defective area is bridged by fibrous tissue (Adams and Hamblen 1990, 190). The defect most commonly affects the lumbar spine, although it has been reported in cervical vertebrae, and clinical studies have indicated that the fifth lumbar vertebra is most frequently affected (67% of cases), followed by the fourth (15% - 30% of cases), and third lumbar vertebrae (1% - 2% of cases) (Resnick *et al* 1995, 2597). The lesion occurs in adults of modern populations with a frequency of 5% to 6% (Schmorl and Junghanns 1971, 88), although certain ethnic groups, such as the Japanese and the Inuit, have a higher incidence of the defect. Up to 50% of the Inuit may display spondylolysis, a finding that has been explained as a consequence of their difficult and strenuous physical existence (Merbs and Wilson 1960, 160). The defect occurs with higher frequency among athletes, especially those involved in gymnastics, diving, weight-lifting, pole-vaulting and American football, and the majority of cases in modern populations occur in males (Resnick *et al* 1995, 2597).

The aetiology of spondylolysis is uncertain, with genetic or traumatic causes both considered as possible aetiologies; it is generally accepted, however, that it is an acquired traumatic lesion which would have occurred at some stage between infancy and young adulthood. It has been suggested that the lesion may arise as a result of a fatigue fracture occurring as a consequence of repeated trauma, possibly due to bending and lifting in an upright position, although acute fracture at the pars interarticularis, after a single incident of trauma, probably accounts for a proportion of the defects (Roberts and Manchester 1995, 78). It is likely that, to a certain extent, genetic factors are also involved in the occurrence of the lesion, and families are known in which over 25% of individu-

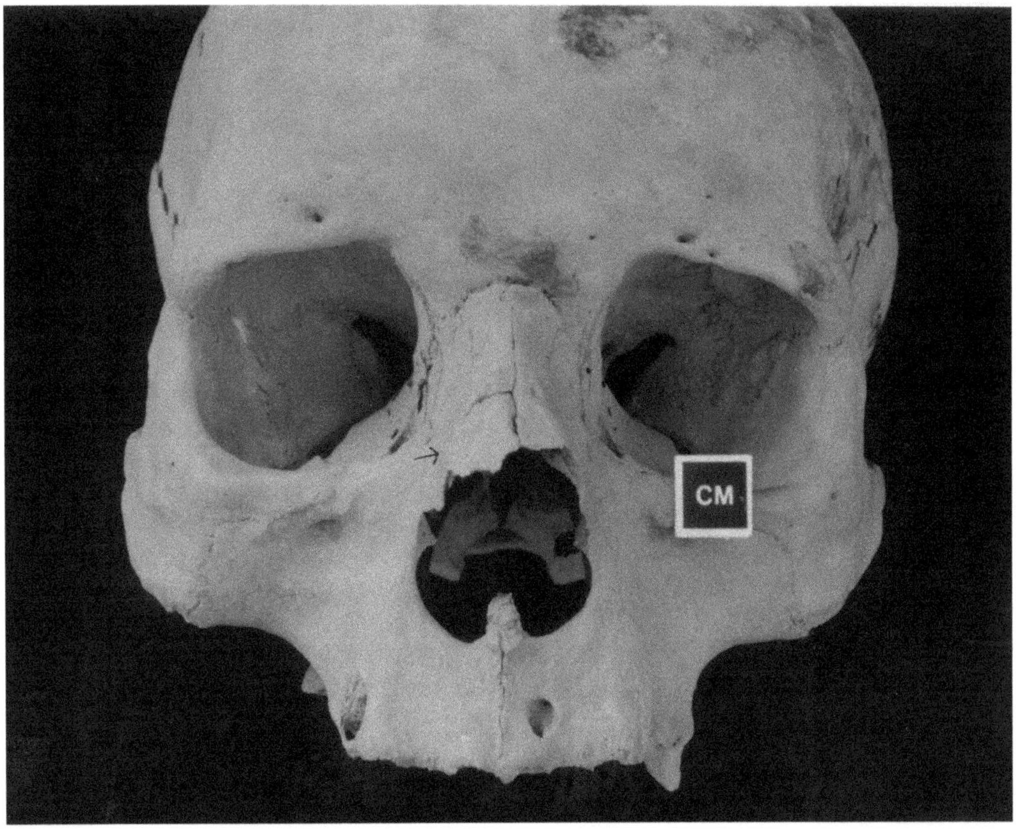

Plate 15: Scythian period - Fractured right nasal bone in Skeleton VIII. 10 (ii).

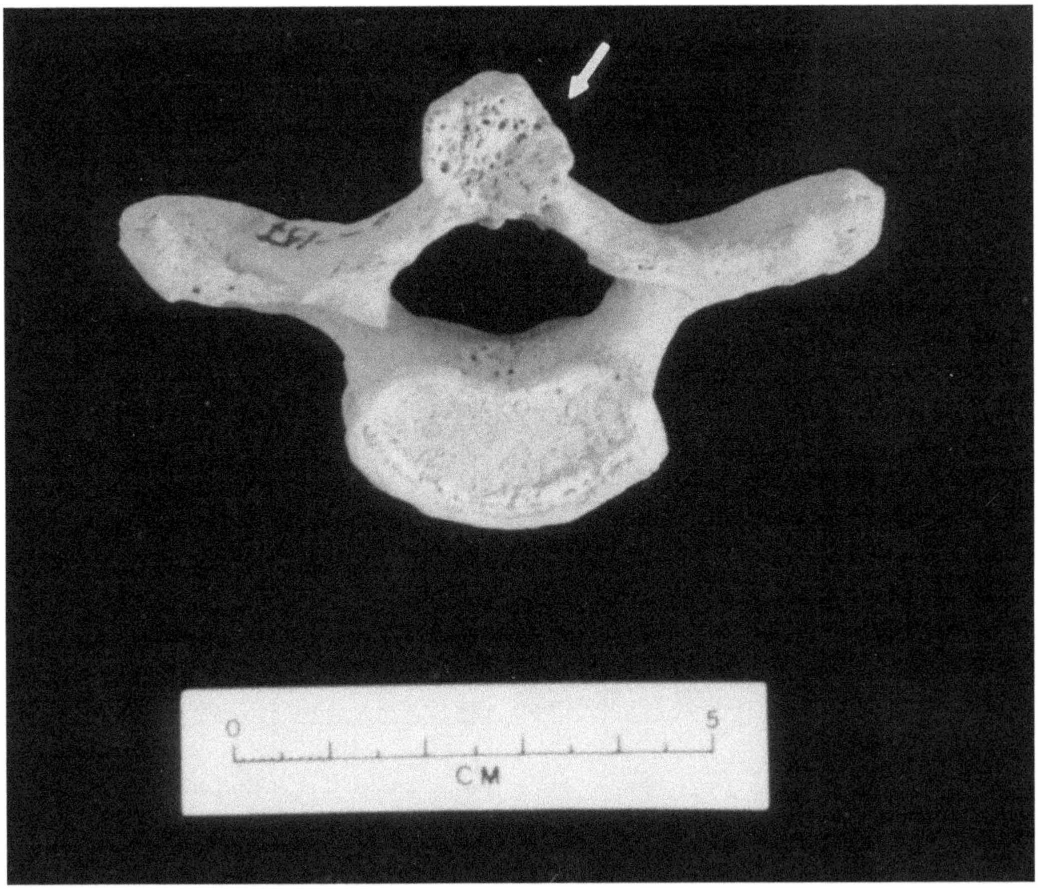

Plate 16: Scythian period - Clay shoveller's fracture of the first thoracic vertebra in Skeleton VI. 15. Sk. 6.

als display the lesion. In addition, other congenital malformations of the spine, such as cleft neural arches and transitional vertebrae, are common in individuals with spondylolysis. Although the defect itself may not be inherited it has been suggested that there is a genetic influence acting on the strength of the pars interarticularis that predisposes some individuals to stress fracture at this point (Resnick *et al* 1995, 2599). The defect may be asymptomatic, but lumbar back pain, tenderness, gait abnormality and neurologic problems can occur (*ibid.*, 2597).

Twelve Scythian period individuals displayed spondylolysis (Table 13). An average minimum number of ninety-two lumbar spines was calculated on the basis of the numbers of each lumbar vertebra present for the adult individuals. The results of this calculation indicated that 13% (12/92) of adults displayed the lesion in one of their lumbar vertebrae.

Context	Sex	Age	Vertebra	R	L	Paraxial mesoderm defects	Other trauma
D. 2 (i)	M	17-25	LV5	+	+	o	+
I. 15	F	17-25	LV5	+	+	o	+
III. 22 (ii)	F	17-25	LV5	+	+	+	+
VI. 12. Sk. 3	F	25-35	LV5	+	+	o	o
VI. 15. Sk. 4	F	35-45	LV5	+	+	o	+
VI. 16. Sk. 1	F	35-45	LV5	+	o	+	+
VI. 16. Sk. 4	M	17-25	LV5	+	+	o	+
VIII. 18. Sk. 4	F	17-25	LV5	+	+	o	+
XII. 4 (i)	M	25-35	LV5	+	+	+	+
XX. 9. Sk. 2	F	35-45	LV5	+	+	o	+
XX. 10. Sk. 4	M	25-35	LV3	+	+	+	+
XXI. 4. Sk. 3	F	35-45	LV4	+	+	o	+

Table 13: Scythian period individuals - Details of spondylolysis in the population. (+ = present, o = absent, LV = lumbar vertebra)

The fifth lumbar vertebra was affected in a notable majority of cases (83%; 10/12), with the third and fourth lumbar vertebrae only rarely affected. The majority of individuals displaying the lesion were female (67%; 8/12), and the prevalence based on age-at-death was fairly uniform with 42% (5/12) aged 17-25 years, 25% (3/12) aged 25-35 years, and 33% (4/100) having an age-at-death of 35-45 years. The lesion was incomplete and unilateral in only one case (Skeleton VI. 16. Sk. 1), in which the left side of the vertebra was normal and loss of bony continuity at the pars interarticularis was restricted to the right side of the vertebra.

Four cases also displayed developmental defects of the paraxial mesoderm. The right side of the posterior arch of the atlas was aplastic in Skeleton III. 22 (I), partial occipitalization was apparent in Skeleton VI. 16. Sk. 1, partial sacralization was present in Skeleton XII. 4 (i), and complete sacralization was evident in Skeleton XX. 10. Sk. 4. In addition, both Skeletons 1 and 4 from Log House Tomb VI. 16 displayed spondylolysis of LV5, and these findings may support the theory that an underlying genetic weakness of the pars interarticularis contributes to the development of the lesion. With the exception of Skeleton VI. 12. Sk. 3, all cases displayed lesions indicative of physical stress to the skeleton including Schmorl's nodes. It is therefore probable that spondylolysis is a further indicator of the arduous physical lifestyle of the Scythian period individuals from Aymyrlyg.

5.2.1b (v) Clay-shoveller's fractures

A single Scythian period adult male (Skeleton VI. 15. Sk. 6) displayed a clay-shoveller's fracture of the first thoracic vertebra (Plate 16). A clay-shoveller's fracture is an oblique fracture of the spinous process, which can occur in CV6, CV7 and TV1-TV3. The fracture arises as a consequence of avulsion by the supraspinous ligament (Pathria 1995, 2661). Isolated fractures of the spinous processes generally occur as a result of hyperflexion. A direct blow to the posterior spine can, however, also result in a fracture of the spinous process (*ibid.*). A clay-shoveller's fracture can be painful, but is generally harmless to the individual (Schmorl and Junghanns 1971, 261).

The name clay-shoveller's fracture is derived from the common occurrence of this type of fracture among Australian clay miners (Pathria 1995, 2861). It is possible, therefore, that the occurrence of a clay-shoveller's fracture in the vertebrae of an individual is indicative that he or she had been subject to repeated strenuous physical exertion which involved lifting heavy weights. It is interesting to note that Skeleton VI. 15. Sk. 6 also displayed prominent enthesopathies and Schmorl's nodes, further indicators of stress and strain to the skeleton. In addition, lesions with a similar morphology to squatting facets were present on the lateral condyles of the distal femora. The occurrence of these lesions may indicate that the individual was involved with the repeated lifting of heavy weights, an activity which put strain on his vertebral column and knees.

5.2.1b (vi) Rib fractures

The prevalence of fractures among the ribs of the Scythian period adults was 0.5% (3/571), and one or more fractured ribs were evident in two individuals (Table 14).

Context	Age	Sex	Type of fracture	Location	Healing
VI. 16. Sk. 2	35-45	M	Simple transverse	L 7th rib	advanced
			Simple transverse	L 8th rib	advanced
XXIII. 8	45+	F	Simple transverse	R 8th rib	mid

Table 14: Scythian period individuals - Fractures of the ribs (L = left, R = right).

The majority of fractures of the ribs occur as a result of a direct injury, such as a blow to the chest or a fall against a hard object (Adams 1978, 103). Skeleton VI. 16. Sk. 2, a 35-45 year old male, displayed fractures on the left 7th and 8th ribs, and no other signs of trauma were evident in his skeletal remains. Skeleton XXIII. 8, a 45+ year old female, displayed a fracture of the right 8th rib, a night-stick fracture of the left distal ulna (see Section 5.2.1a (iii)) and fractured 4th and 5th metacarpals (see Section 5.2.1a (ix)), and it is probable that she had procured all of the fractures in a violent incident during her life.

5.2.1c Ossified haematomas

Muscle contusion can arise as a consequence of direct trauma made by a blunt implement, which may result in the rupture of capillaries and the occurrence of bleeding between the fibres of the damaged connective tissue. This type of injury is generally followed by oedema and the development of an inflammatory mass (Resnick *et al* 1995, 2673), and the resultant soft tissue swelling may become replaced by a smooth bone swelling known as an ossified haematoma (Baker and Brothwell 1980, 83). Five Scythian period adults displayed evidence of these injuries (Table 15), and the prevalence of the lesions in the femora was 0.9% (3/333), 0.3% (1/328) in the complete tibiae, and 0.4% (1/246) in the fibulae.

Context	Age	Sex	Bone
B. 8. Sk. 1	35-45	M	L femur
D. 6. Sk. 1	45+	F	R tibia
G. 3. Sk. 2	25-35	F	R femur
III. 11	25-35	F	R femur
XX. 8. Sk. 2	45+	M	L fibula

Table 15: Scythian period individuals - Details of ossified haematomas in the population (L = left, R = right).

Skeleton B. 8. Sk. 1, a 35-45 year old male, also displayed a night-stick fracture of the left ulna (see Section 5.2.1a (iii)), and it is possible that the fracture and the ossified haematoma were sustained in a single incident of interpersonal violence. Skeleton XX. 8. Sk. 2, a 45+ year old male, also displayed a dislocated right glenohumeral joint (see Section 5.2.1d), a fractured right humeral head (see Section 5.2.1a (ii)), a Colles' fracture of the left radius (see Section 5.2.1a (iv)) and a fracture of the right fibula (see Section 5.2.1a (vii)), and it is probable that all of these traumatic lesions had occurred as a consequence of a fall.

5.2.1d Dislocations

A single example of a dislocated joint was present in the Scythian period population. In Skeleton XX. 8. Sk. 2, a 45+ year old male, the right glenohumeral joint was dislocated with the humerus completely displaced anteriorly (Plates 17-19), and a new joint surface, which displayed osteophytes, porosity and eburnation, was present on the antero-inferior surface of the coracoid process. The right humeral head had been subject to a compression fracture, and it displayed a flattened appearance with a porotic surface and marginal osteophytes. An area of porosity and eburnation was present on the supero-posterior surface of the anatomical neck, and large osteophytes, which were probably due to strain of the supraspinatus, infraspinatus and teres minor muscles, were positioned immediately inferior to the area of eburnation. An enthesopathy of the costoclavicular ligament was present in the right clavicle, and enthesopathies of the deltoid and pectoralis major were apparent in the right humerus. Osteophytosis was visible on the proximal and distal ends of the right ulna and the distal end of the right radius, and these lesions may have been secondary to the dislocation.

The glenohumeral joint is considered to be relatively unstable since its integrity is predominantly dependent on the capsular and ligamentous structures with which it is associated. In addition, the humeral head is approximately three times larger than the glenoid fossa with which it articulates (Resnick and Goergen 1995, 2695). Anterior dislocations are the most frequent dislocations of the glenohumeral joint, being seen in 95% of modern cases (*ibid.*), and in the majority of such cases the humeral head is displaced through a tear in the capsule and becomes positioned in the infraclavicular fossa. Consequently, the term subcoracoid dislocation is commonly used to describe the lesion (Adams 1978, 120), and Skeleton XX. 8. Sk. 2 appears to have suffered from this form of injury.

Anterior dislocations are commonly associated with compression fractures of the postero-lateral aspect of the humeral head, a feature which is generally referred to as the Hills-Sachs Lesion. The fracture occurs as a result of its impaction against the anterior margin of the glenoid fossa (Resnick and Goergen 1995, 2695), and it is probable that Skeleton XX. 8. Sk. 2 had been suffering from a typical Hills-Sachs Lesion (see Section 5.2.1a (ii)). In anterior dislocation the individual is commonly subject to severe pain, and the lateral outline of the shoulder may have a flattened appearance (Apley and Solomon 1988, 252).

The most common cause of anterior dislocation of the glenohumeral joint is a fall onto an outstretched hand or onto the shoulder itself (Adams 1978, 120). A Colles' fracture was present in the left radius (see Section 5.2.1a (iv)), a simple oblique fracture was visible at the distal end of the right fibula (see Section 5.2.1a (vii)) and an ossified haematoma was apparent on the left fibula (see Section 5.2.1c). It is possible that the individual had been subject to a severe fall in which he had dislocated his right shoulder, fractured the right humeral head, fractured the left radius, fractured the right fibula and injured the muscle of the left fibula. Since the individual was an older adult it is possible that he would have been suffering from underlying osteoporosis which could have contributed to the occurrence of multiple fractures. All of the lesions were in an advanced state of healing, and the presence of eburnation on the false joint surface of the dislocated shoulder may indicate that they had been sustained a considerable amount of time before his death.

5.2.2 Soft-tissue trauma

5.2.2a Myositis ossificans traumatica

Skeleton VIII. 39 (ii), an adult individual of unknown sex,

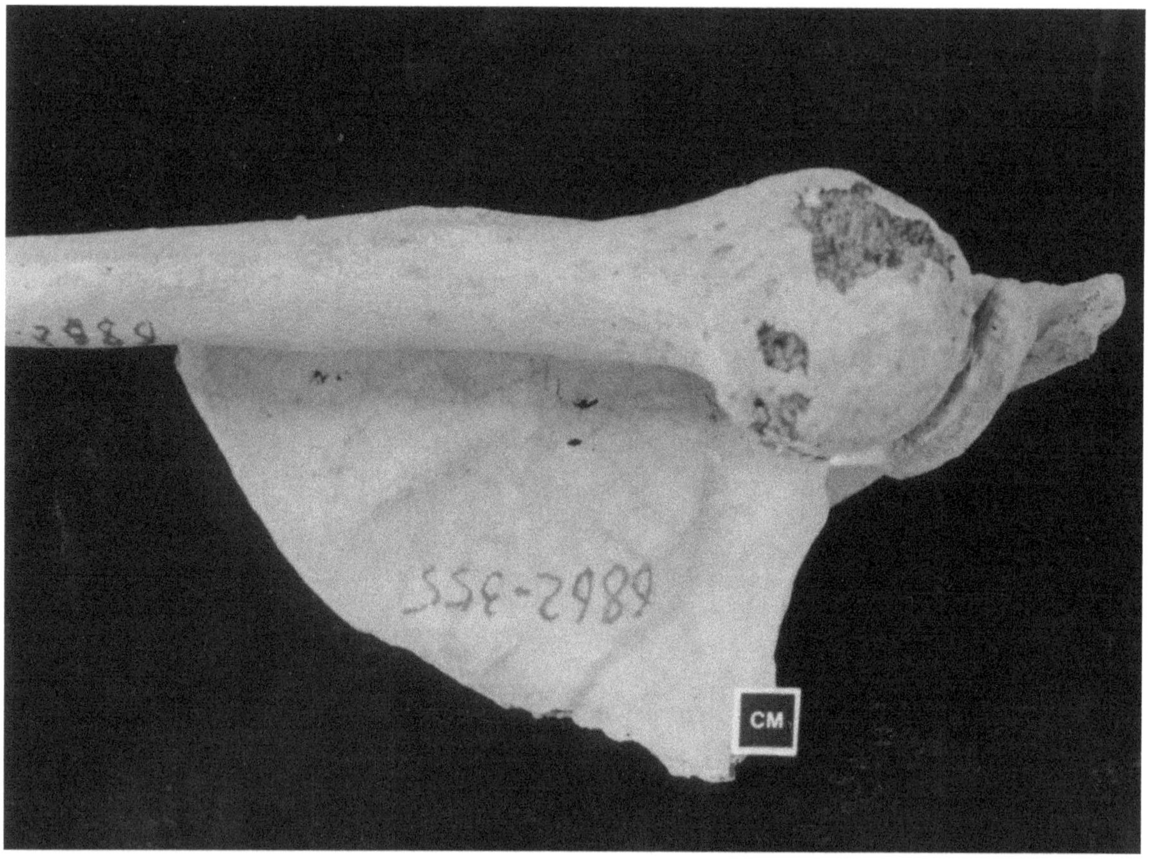

Plate 17: Scythian period - Dislocation of the right glenohumeral joint in Skeleton XX. 8. Sk. 2.

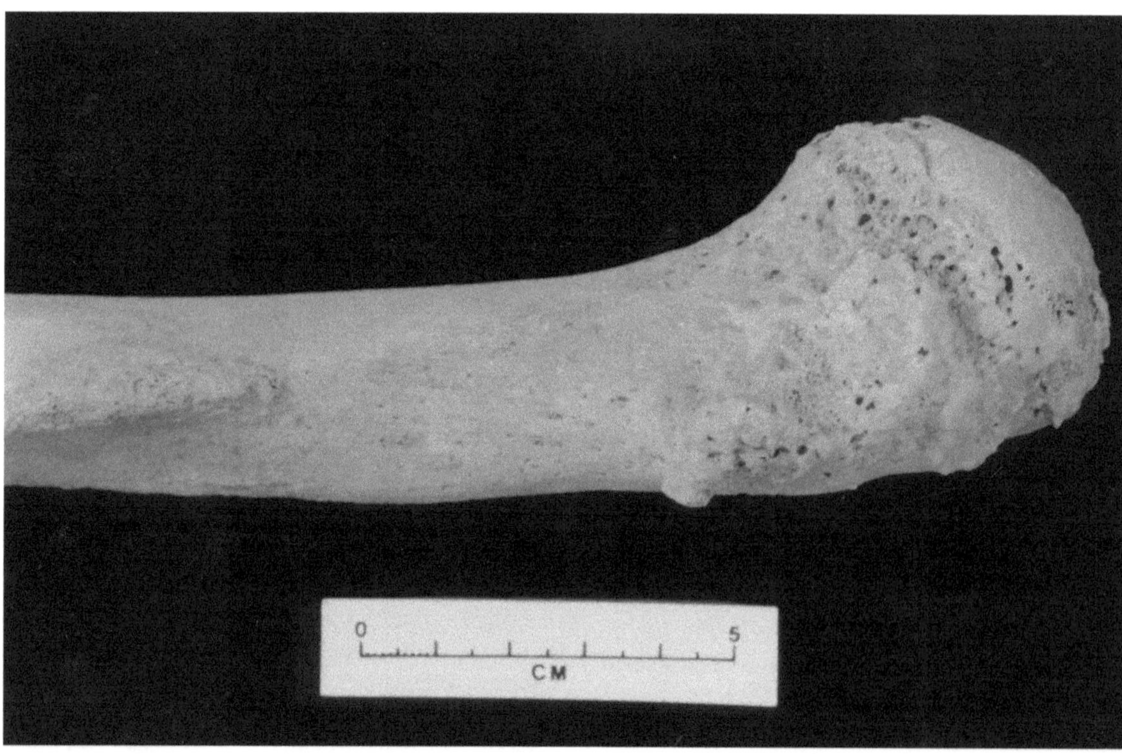

Plate 18: Scythian period - Compression fracture of the right humeral head associated with the dislocated right glenohumeral joint in Skeleton XX. 8. Sk. 2.

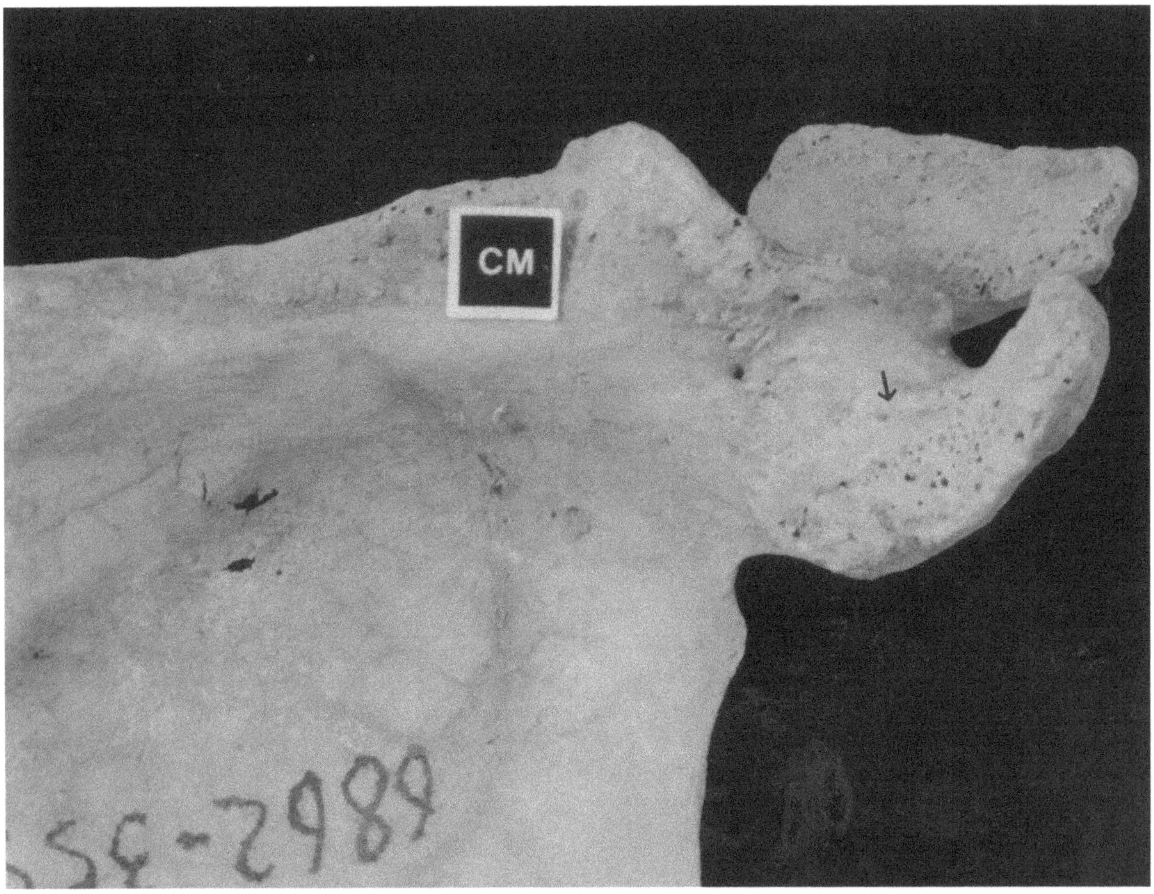

Plate 19: Scythian period - False joint surface associated with the dislocated right glenohumeral joint in Skeleton XX. 8. Sk. 2.

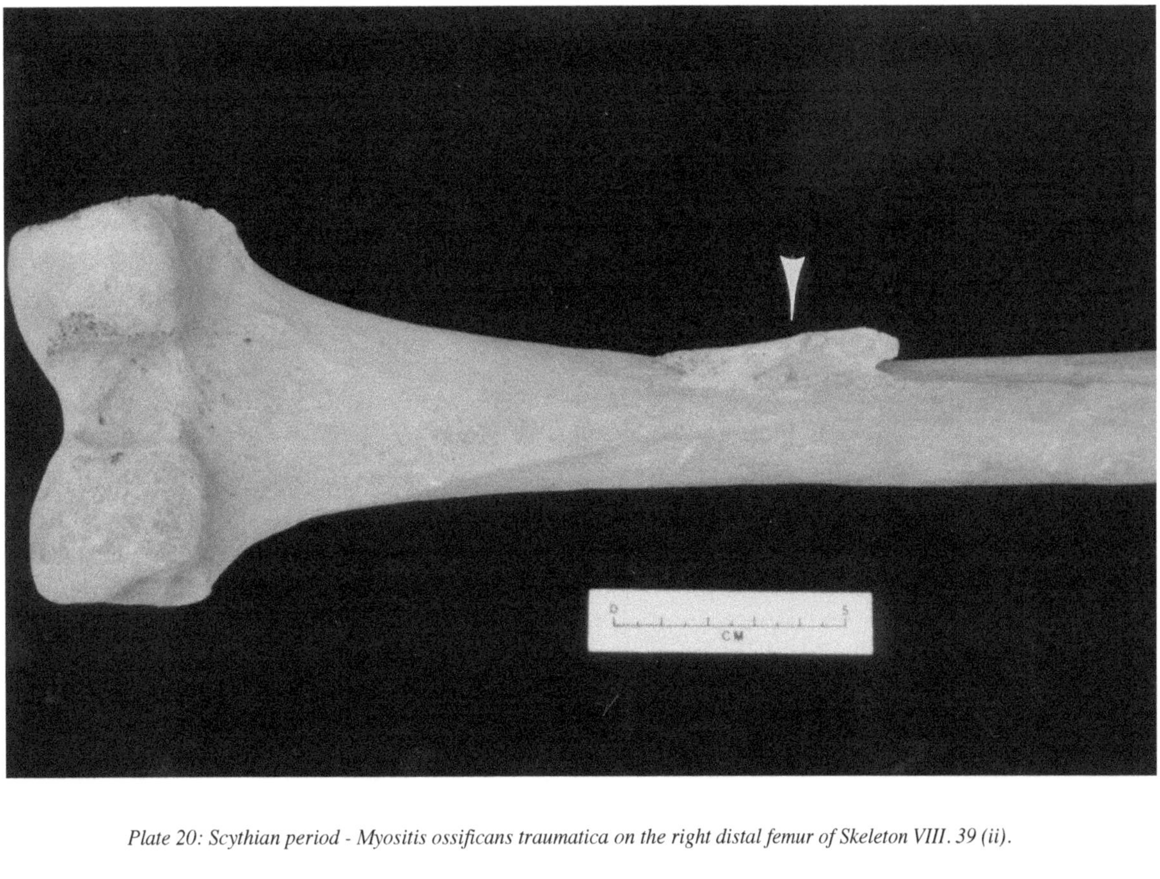

Plate 20: Scythian period - Myositis ossificans traumatica on the right distal femur of Skeleton VIII. 39 (ii).

displayed a spur-like bony process on the medial supracondylar line of the distal right femur (Plate 20) which measured 49 mm supero-inferiorly, 3.3 mm medio-laterally and 10 mm antero-posteriorly. The exostosis may have been indicative of myositis ossificans of part of the insertion of the adductor magnus. Myositis ossificans traumatica generally occurs as a sequela to trauma, and the sites most commonly affected are those most susceptible to trauma including the elbow, the thigh and the buttocks. Post-traumatic ossification generally arises after an incident of severe trauma, but it is also known to occur as a result of repetitive incidents of minor trauma which can be occupationally induced (Resnick and Niwayama 1995, 4577). The functions of the adductor magnus are to stabilise the hip, adduction and extension of the hip, and medial rotation of the thigh (Romanes 1986, 149), and it is probable that the lesion had occurred as a consequence of muscle strain during one of the above motions.

5.2.2b Intracranial haemorrhaging

Skeleton D. 7. Sk. 3, a 2-3 year old child, displayed a localised area of increased vascular impressions at the midpoint of the right parietal (Plate 21). The affected area was associated with the middle meningeal vessels, and measured approximately 60 mm by 60 mm. In the majority of cases of trauma to the head the individual displays evidence of increased intracranial pressure, and in some cases an intracranial haematoma may be present (Roberts 1996b, 130). It is possible that the increased vascularisation was the result of intracranial haemorrhaging, possibly due to direct trauma, although no evidence for fractures was visible on the bones of the skull.

Context	Age	Sex	Bone	Muscle involved
B. 7. Sk. 1	25-35	F	R tibia	tibial collateral ligament
B. 7. disturbed	adult	M	L tibia	tibial collateral ligament
E. 9. Sk. 2	25-35	F	R humerus	pectoralis major
G. 6. Sk. 4 (i)	45+	F	L calcaneus	bifurcate ligament
III. 3	17-25	M	L fibula	extensor digitorum longus
VI. 3. Sk. 3	35-45	M	L femur	short head of biceps
VIII. 4. Sk. 12	17-25	M	R tibia	popliteus
XXI. 4. Sk. 4 (Plate 22)	17-25	M	L femur	adductor magnus
XXIII. 4	25-35	M	L tibia	tibialis anterior
XXIII. 11. Sk. 3	25-35	M	L tibia	iliotibial tract
XXIII. 19. Sk. 1	17-25	F	L femur	capsule attachments of hip

Table 16: Scythian period individuals - Details of the enthesophytes in the population (L = left, R = right).

5.2.2c Enthesophytes

Enthesophytes can occur at the sites of musculotendinous and ligamentous attachments, and they arise as a consequence of chronic repetitive irritation, bleeding and inflammation, periostitis or microavulsions (Pavlov 1995, 3246). Enthesophytes were apparent in the remains of eleven Scythian period individuals (Table 16).

The majority of enthesophytes occurred in males (64%; 7/11), and 36% of females displayed the lesions. Tibiae were most commonly affected (46%; 5/11), followed by femora (27%; 3/11), humeri (9%; 1/11), fibulae (9%; 1/11) and calcanei (9%; 1/11). The prevalence of enthesophytes in the population group is provided in Table 17.

Bone	No. enthesophytes	No. bones	% enthesophytes
humerus	1	283	0.4
femur	3	333	0.9
tibia	5	326	1.5
fibula	1	246	0.4
calcaneus	1	161	0.6

Table 17: Scythian period individuals - Prevalence of enthesophytes in the population.

The greatest frequency of enthesophytes was apparent in the tibiae (1.5%; 5/326) and femora (0.9%; 3/333), and it is therefore probable that the lifestyle of the individuals had resulted in more physical stresses to the bones of the lower body relative to those of the upper skeleton.

5.3 Hunno-Sarmatian period individuals

5.3.1 Bone trauma

5.3.1a Fractures of the appendicular skeleton

A total number of 1593 complete adult long bones were examined for fractures, 0.8% (12/1593) of which displayed the lesions. Details of the individuals with fractures of the appendicular skeleton are provided in Table 18. When the fractures of the long bones and the clavicles are considered it was found that a greater proportion of males (64%; 9/14) displayed the lesions relative to females (36%; 5/14). A summary of the frequencies of fractured bones from the appendicular skeletons of the adults is provided in Table 19.

The long bone with the greatest prevalence of fractures was the ulna (2.3%), followed by the radius (1.5%), the clavicle (0.8%) and the fibula (0.8%). When each individual with evidence of one or more fractures is examined separately it is possible to determine the possible aetiology of the various injuries.

5.3.1a (i) Fractures of the clavicles

Skeleton XXXI. 211A, a 35-45 year old male (Plate 23), and Skeleton XXXI. 214, a 35-45 year old female, both displayed fractures of the acromial ends of their clavicles. These injuries generally arise as a result of a fall onto an outstretched hand (Adams 1978, 114). Skeleton XXXI. 211A also displayed

Context	Age	Sex	Bone	Location	Type	Shortening	Alignment	Apposition	Healing
XXXI. 3	25-35	M	L ulna	distal	simple oblique	+	good	100%	healed
XXXI. 11. Sk. 1	35-45	F	L radius	distal	simple oblique	+	c. 40° postero-lat.	100%	healed
			L ulna	styloid process	simple transverse	+	good	100%	healed
XXXI. 15	35-45	F	L ulna	distal	simple oblique	+	c. 40° postero-lat.	80%	healed
XXXI. 30. Sk. 2	25-35	M	R MC5	proximal	simple oblique	-	good	100%	early
XXXI. 48	35-45	F	L radius	distal	simple transverse	+	c. 5° post.	100%	healed
XXXI. 80	35-45	M	L ulna	midshaft	simple oblique	-	good	100%	advanced
XXXI. 108	25-35	M	L ulna	distal	simple oblique	+	c. 20° lat.	100%	advanced
XXXI. 110	35-45	F	L fibula	distal	simple oblique	+	c. 10° ant.	50%	advanced
XXXI. 127 (i)	25-35	M	L fibula	distal	simple oblique	-	good	100%	healed
			L talus	lateral malleolus	compression	-	-	-	healed
			L calcaneus	posterior artic. facet	compression	-	-	-	healed
XXXI. 138	25-35	M	L ulna	midshaft	simple transverse	+	c. 30° ant.	50%	healed
XXXI. 139 (i)	25-35	M	L MC4	midshaft	simple oblique	+	-	100%	healed
			L MT1	proximal artic. surface	crushing	-	-	-	healed
XXXI. 211A	35-45	M	L clavicle	acromial	simple transverse	+	c. 45° post. (acrom)	90-100%	healed
XXXI. 212	35-45	M	R radius	distal	simple oblique	+	c. 45° postero-lat.	30%	advanced
XXXI. 214	35-45	F	R prox. hand PH4	midshaft	simple transverse	-	-	0%	unhealed
			R clavicle	acromial	simple oblique	+	good	100%	healed
			R radius	distal	simple transverse	+	c. 10° post.	90%	healed

Table 18: Hunno-Sarmatian period individuals - Fractures of the appendicular skeleton. The degree of displacement referred to in the 'Alignment' column relates to the distal part of the fractured bone (R = right, L = left, MC = metacarpal, MT = metatarsal, PH = phalanx, + = present, - = not applicable).

Bone	No. fractures	No. bones	% fractures
clavicle	2	237	0.8
ulna	6	257	2.3
radius	4	263	1.5
fibula	2	253	0.8
metacarpal	2	185	1.1
hand phalanx	1	220	0.4
metatarsal	1	209	0.5
talus	1	212	0.5
calcaneus	1	215	0.5

Table 19: Hunno-Sarmatian period individuals - Prevalences of fractured bones of the appendicular skeleton.

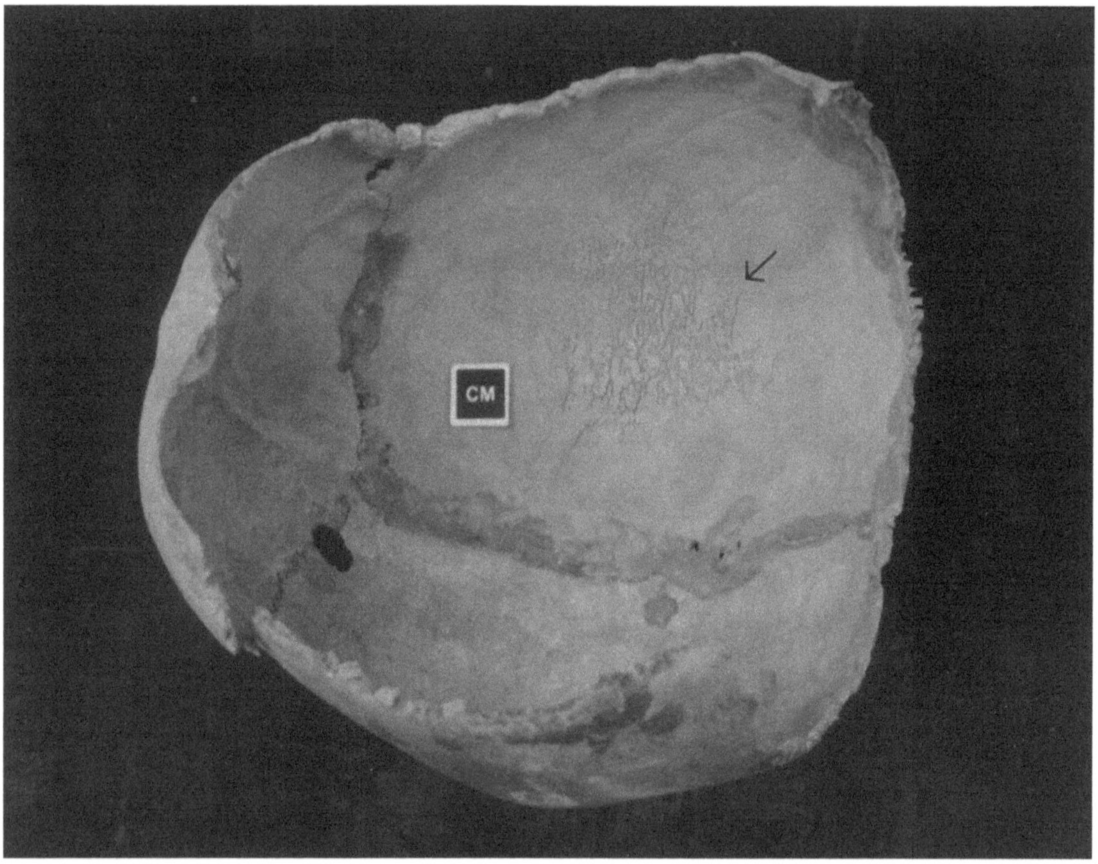

Plate 21: Scythian period - Localised area of increased vascular impressions at the midpoint of the right parietal in Skeleton D. 7. Sk. 3 which may have been due to intracranial haemorrhaging as a result of trauma.

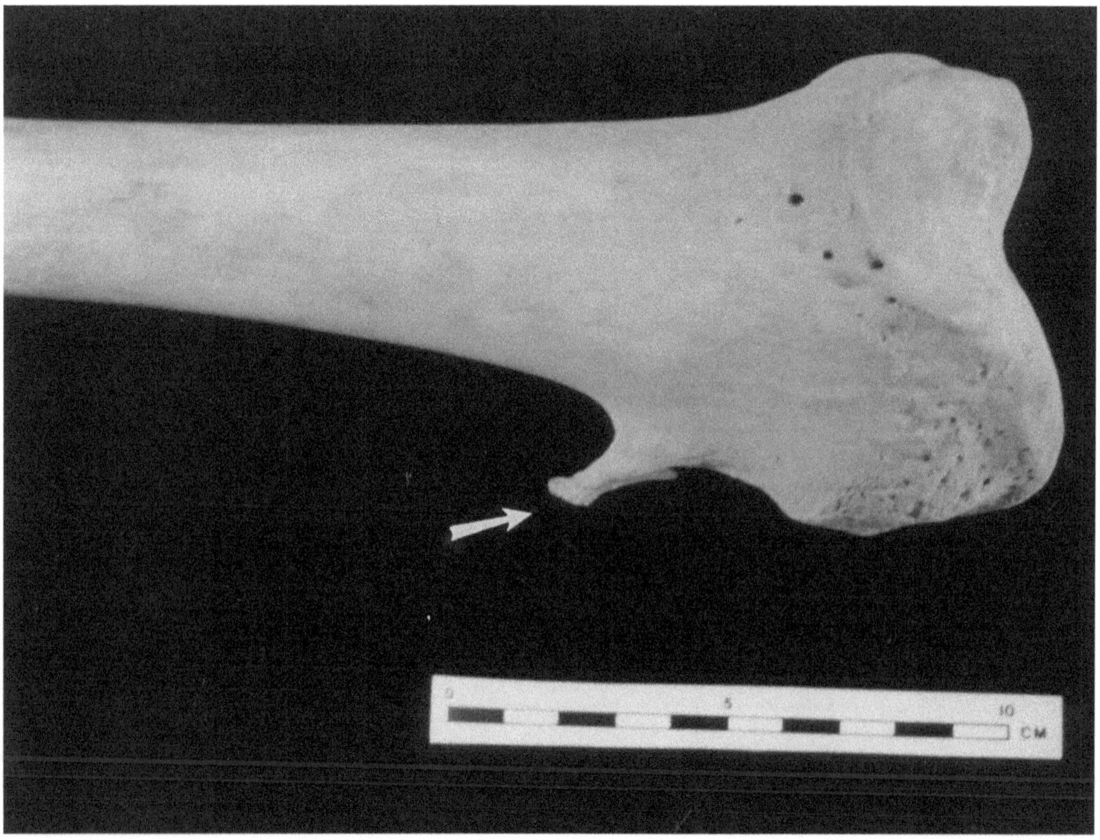

Plate 22: Scythian period - Enthesophyte on the distal left femur of Skeleton XXI. 4. Sk. 4.

fractured nasal bones (see Section 5.3.1b (i)), and it is therefore possible that the injuries may have been attained in an incident of interpersonal violence. A Colles' fracture of the right radius (see Section 5.3.1a (iii)) and an ununited fracture of the right fourth proximal phalanx (see Section 5.3.1a (v)) were also present in Skeleton XXXI. 214, and it is possible that all three injuries were obtained as a consequence of a fall.

5.3.1a (ii) Night-stick fractures of the ulnae

Six individuals (Skeleton XXXI. 3, Skeleton XXXI. 11. Sk. 1, Skeleton XXXI. 15, Skeleton XXXI. 80, Skeleton XXXI. 108 and Skeleton XXXI. 138) displayed night-stick fractures of the ulnae. In all cases the left ulna was affected, and the fractures were present at either the distal end or the midshaft of the diaphysis. The majority of affected individuals were male (67%; 4/6), and only two females displayed this injury. Night-stick fractures generally arise during assault when the victim raises his or her forearm to protect the head from being struck (Resnick and Goergen 1995, 2732). The left nasal bone was also fractured (see Section 5.3.1b (i)) and unhealed weapon trauma (see Section 6.3.1) were present in the remains of Skeleton XXXI. 3, while in Skeleton XXXI. 108 both of the nasal bones displayed fractures (see Section 5.3.1b (i)). Since fractured nasal bones are commonly attained during incidents of interpersonal violence, their occurrence in the remains of both these individuals may support the theory that the night-stick fractures of the ulnae were attained during incidents of conflict.

5.3.1a (iii) Colles' fractures of the radii

Four individuals (Skeleton XXXI. 11. Sk. 1, Skeleton XXXI. 48, Skeleton XXXI. 212 and Skeleton XXXI. 214) displayed Colles' fractures of the radii, which are commonly attained as a result of a fall onto an outstretched hand (*ibid.*, 2734). In Skeleton XXXI. 11. Sk. 1, a 35-45 year old female, the injury evident in the left radius was associated with a fracture of the styloid process of the left ulna. Skeleton XXXI. 212, a 35-45 year old male, also displayed a fractured right 9th rib (see Section 5.3. 1b (v)), and in Skeleton XXXI. 214 the Colles' fracture of the right radius was associated with a fractured right clavicle (see Section 5.3.1a (i)) and a fractured right fourth proximal phalanx (see Section 5.3.1a (v)). It is probable that all of the Colles' fractures were sustained due to falls onto outstretched hands.

5.3.1a (iv) Fractures of the fibulae not associated with fractured tibiae

Two individuals (Skeleton XXXI. 110 and Skeleton XXXI. 127 (i)) displayed isolated fractures of the diaphysis of the fibulae, which generally arise as a result of direct trauma to the bone (Adams 1978, 257). In Skeleton XXXI. 127 (i) the left talus and left calcaneus also displayed crushing fractures (see Section 5.3.1a (v)), an ossified haematoma was present on the right tibia (see Section 5.3.1c) and myositis ossificans traumatica was apparent on the right humerus (see Section 5.3.2a). It is probable that the individual had obtained all of the injuries as a result of a serious fall.

5.3.1a (v) Fractures of the bones of the hands and the feet

Skeleton XXXI. 30. Sk. 2 and Skeleton XXXI. 139 (i), both of whom were 25-35 year old males, displayed fractured metacarpals, which commonly arise as a result of falling onto the hands or from blows upon the knuckles obtained during boxing (*ibid.*, 182). In Skeleton XXXI. 30. Sk. 2 the right 5th metacarpal (boxer's fracture) was fractured, and it is probable that the injury had been procured during an incident of interpersonal violence. In Skeleton XXXI. 139 (i) the left 4th metacarpal was fractured while a crushing fracture was also present at the proximal end of the left 1st metatarsal. It is probable that the individual had attained both injuries as a result of a fall.

Skeleton XXXI. 214, a 35-45 year old female, displayed a fractured right fourth proximal hand phalanx. The fracture parts were ununited and the metacarpal, proximal phalanx and intermediate phalanx were ankylosed. Marked angulation was evident and the finger would have been claw-shaped. Angulation, rotation and displacement are commonly found in fractures of the metacarpals and phalanges (Resnick and Goergen 1995, 2751), and it is probable that marked angulation and rotation would have been associated with the fractured proximal phalanx in this individual. In modern treatment of fractured fingers long-term immobilisation of the injured finger is avoided since it can result in permanent stiffness (Adams 1978, 185). It is possible that the individual did not use her fourth finger due to the pain caused by the fracture, and this immobilisation may have contributed to the lack of unification of the fracture parts, and the eventual ankylosis of the metacarpal, parts of the proximal phalanx and the intermediate phalanx. The individual also displayed a fractured right clavicle (see Section 5.3.1a (i)) and a Colles' fracture of the right radius (see Section 5.3.1a (iii)), and it is probable that all of the injuries were sustained as a result of a fall.

Skeleton XXXI. 127 (i), a 25-35 year old male, displayed crushing fractures of the lateral malleolus of the left talus and the posterior articular facet of the left calcaneus. The individual also displayed a fracture of the left fibula (see Section 5.3.1a (iv)), an ossified haematoma of the right tibia (see Section 5.3.1c), and myositis ossificans traumatica on the right humerus (see Section 5.3.2a). It is probable that the individual had procured the injuries as a result of a fall, and he may have fallen heavily onto the lateral aspect of the left foot which would account for the presence of the crushing fractures.

5.3.1a (vi) Evidence of treatment and healing

The fractures apparent in the bones of the appendicular skeleton displayed various states of healing, ranging from early to completely healed (see Table 18). The alignment of the fractured bones was generally good with only slight angular deviation, although most of the fractured bones displayed a degree of shortening which may suggest that the population was unfamiliar with the manipulative techniques required to disimpact the parts of the fracture. The majority of the fractured bones displayed 50% or greater apposition of the fracture parts, although one case, however, displayed a poorly aligned Colles' fracture of the radius in which there was only 30% apposition of the fracture parts, and another individual displayed a fractured proximal hand phalanx which was completely ununited. It is probable that splints were used to treat fractures, some of which were more successful than others.

5.3.1b Fractures of the axial skeleton

5.3.1b (i) Fractures of the facial skeleton

In interpersonal and intergroup violence the head is the main focus of attack (Roberts and Manchester 1995, 79). Skeleton XXXI. 40, a 25-35 year old female, displayed a fracture of the right orbital margin as well as fractured nasal bones (see below). It is probable that these facial injuries had been procured in a single incident of interpersonal violence. Fractures of the orbit occurred in the Hunno-Sarmatian period population with a frequency of 0.4% (1/219).

Eight Hunno-Sarmatian period adults displayed fractures to their nasal bones. Some of the individuals had a single fractured nasal bone, while both nasal bones had been fractured in other skeletons (Table 20). The prevalence of these lesions in the population group was 6% (13/215).

Fractured nasal bones are common in archaeological populations and probably arise as a result of interpersonal violence (*ibid.*, 80). The majority of individuals displaying fractures of one or both nasal bones were male (75%; 6/8), and it is interesting that a greater proportion of the fractures were present on the left (50% 8/16) rather than the right nasal bones (31%; 5/16). This finding may be indicative that

Content	Age	Sex	R		L	
			Type	Healing	Type	Healing
XXXI. 3	25-35	M	No fracture	-	Simple oblique	healed
XXXI. 40 (Plate 24)	25-35	F	Simple transverse	early	Simple transverse	early
XXXI. 52 (Plate 25)	25-35	M	No fracture	-	Simple oblique	mid
XXXI. 108	25-35	M	Simple oblique	healed	Simple oblique	healed
XXXI. 137	35-45	M	Simple transverse	early	Simple transverse	early
XXXI. 149	17-25	F	Simple oblique	mid	Simple transverse	mid
XXXI. 189 (ii)	17-25	M	No fracture	-	Simple transverse	healed
XXXI. 211A	35-45	M	Simple transverse	healed	Simple oblique	healed

Table 20: Hunno-Sarmatian period individuals - Details of fractured nasal bones.

the individuals responsible for striking the blows to the face were predominantly right-handed, since a right handed opponent is most likely to strike the left side of the body during face-to-face combat.

Skeleton XXXI. 3 and Skeleton XXXI. 108, both 25-35 year old males, also displayed night-stick fractures of their left ulnae (see Section 5.3.1a (ii)), and Skeleton XXXI. 40, a 25-35 year old female, also displayed a fracture to the right orbital margin (see above). In addition, Skeleton XXXI. 3, Skeleton XXXI. 157 and Skeleton XXXI. 189 (ii) revealed evidence of weapon trauma (see Section 6.3.1). These lesions are all further indications that the individuals had been involved in interpersonal violence. In Skeleton XXXI. 211A, a 35-45 year old male, fractures were also present on the left clavicle and the left 5th, 6th and 7th ribs, and it is possible that the injuries in this case may have been due to a fall.

5.3.1b (ii) Compression fractures of the vertebrae

A total number of 2638 adult cervical, thoracic and lumbar vertebrae were examined for traumatic lesions, 0.7% (19/2638) of which displayed compression fractures. Of the total number of lumbar vertebrae examined, 1.5% (9/602) displayed the lesions, and the injuries were evident in 0.7% (10/1356) of thoracic vertebrae. Details of compression fractures among the population are provided in Table 21.

The majority of individuals displaying compression fractures to one or more vertebrae were male (86%; 6/7). The distribution of these injuries among the Hunno-Sarmatian period adults is presented in Figure 2. The fractures were concentrated at the midthoracic, thoracolumbar and lumbar regions of the spine. The majority of fractures of the vertebral bodies of the thoracic or lumbar vertebrae are caused by vertical forces

Context	Age	Sex	Location	Healing
XXXI. 13	35-45	M	TV12	healed
XXXI. 24a	45+	M	TV7	healed
			LV1	healed
XXXI. 28. Sk. 2	45+	M	TV8	healed
			TV12	healed
			LV1	healed
XXXI. 31b	17-25	M	TV11	healed
XXXI. 182a	17-25	M	TV5	healed
(Plate 26)			TV6	healed
			TV7	healed
XXXI. 209	45+	M	TV4	healed
			TV11	healed
			LV1	healed
			LV2	healed
			LV3	healed
XXXI. 214	35-45	F	LV2	healed
			LV3	healed
			LV4	healed
			LV5	healed

Table 21: Hunno-Sarmatian period individuals - Details of vertebral compression fractures present in the population (TV = thoracic vertebra, LV = lumbar vertebra).

acting through the long axis of the spinal column. The forces can act from above, as in situations where heavy weights fall on top of an individual, or from below as a result of a heavy fall onto the feet or the buttocks (Adams 1978, 96). Compression fractures of the midthoracic area are commonly found in individuals with osteoporosis, and are generally solitary or

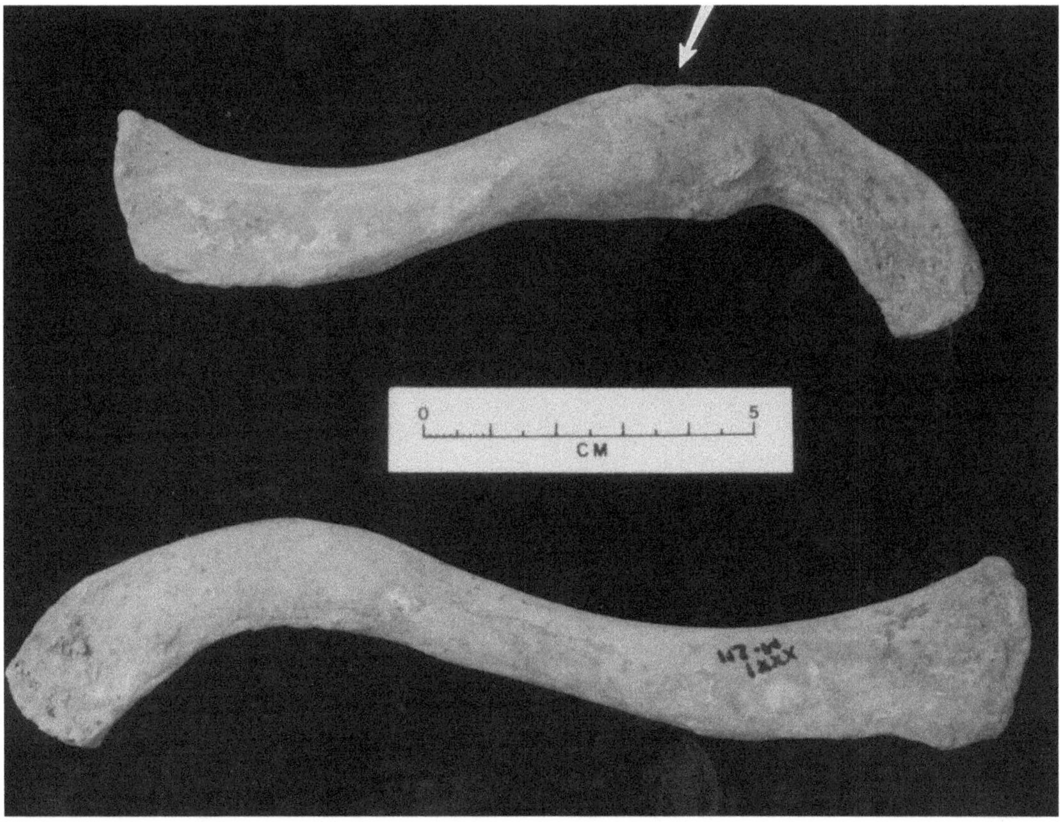

Plate 23: Hunno-Sarmatian period - Fractured left clavicle in Skeleton XXXI. 211A.

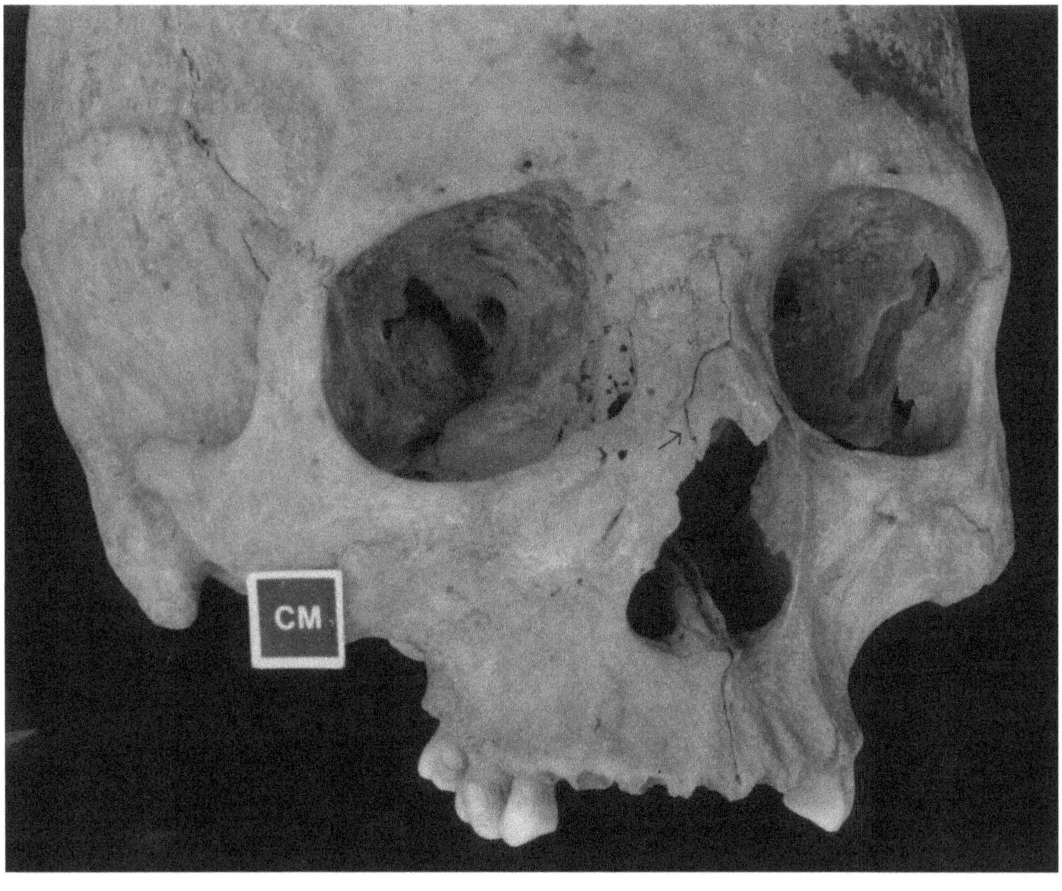

Plate 24: Hunno-Sarmatian period - Fractured nasal bones in Skeleton XXXI. 40.

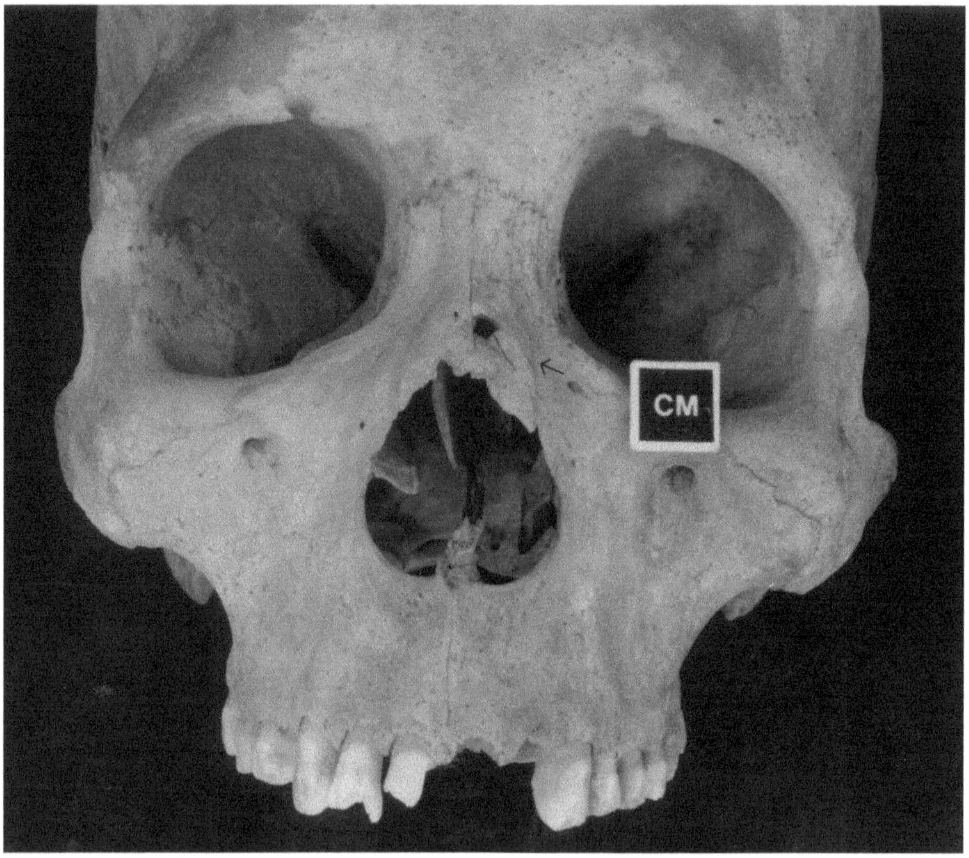

Plate 25: Hunno-Sarmatian period - Fractured left nasal bone in Skeleton XXXI. 52.

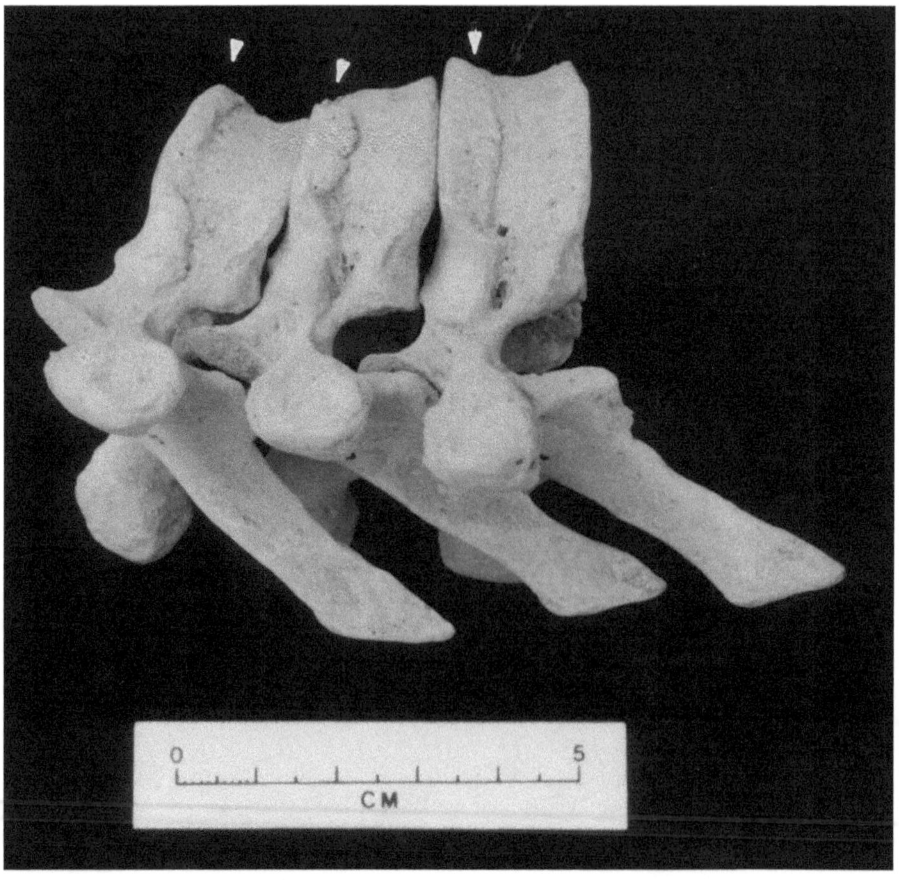

Plate 26: Hunno-Sarmatian period - Compression fractures of the fifth, sixth and seventh thoracic vertebrae in Skeleton XXXI. 182a.

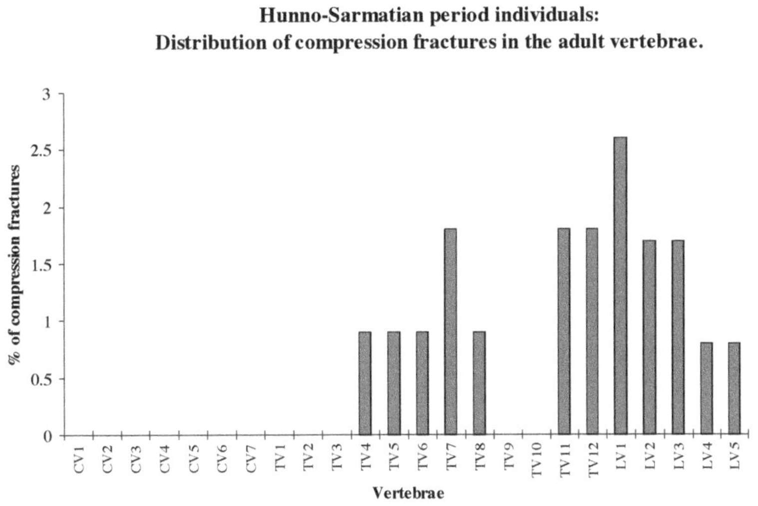

Figure 2: Hunno-Sarmatian period individuals - The distribution of compression fractures in the adult vertebrae.

disseminated throughout the spine. Solitary compression fractures associated with osteoporosis are rare above the 7th thoracic vertebra (Pathria 1995, 2869-71). It is possible, therefore, that the compression fractures apparent in Skeleton XXXI. 13, a 35-45 year old male, and Skeleton XXXI. 24a and Skeleton XXXI. 28. Sk. 2, both of whom were 45+ year old males, were related to osteoporosis, while the compression fractures apparent in the midthoracic spine of Skeleton XXXI. 182a, a 17-25 year old male, were probably due to trauma.

The thoracolumbar area is the most susceptible region of the spine to traumatic compression fractures (*ibid.*, 2876), and it is probable that the injuries apparent in Skeleton XXXI. 31b, a 17-25 year old male, Skeleton XXXI. 209, a 45+ year old male, and Skeleton XXXI. 214, a 35-45 year old female, were due to trauma. Skeleton XXXI. 31b was, however, considered to have had an erosive arthropathy, possibly Reiter's syndrome, and this condition may have been the underlying factor which caused the compression fracture.

Fractures were also present on the right 12th and left 5th to 11th ribs (see Section 5.3.1b (v)) and the sternal body (see Section 5.3.1b (vi)) of Skeleton XXXI. 209, and in Skeleton XXXI. 214 fractures were apparent on the right fourth proximal hand phalanx (see Section 5.3.1a (v)), the right radius (see Section 5.3.1a (iii)) and the right clavicle (see Section 5.3.1a (i)). The occurrence of multiple fractures in both of these cases supports the theory that the compression fractures of the vertebrae were caused by heavy falls.

5.3.1b (iii) Spondylolysis

Six Hunno-Sarmatian period individuals displayed spondylolysis (Table 22). An average minimum number of 120 lumbar spines was calculated on the basis of the numbers of each individual vertebra present for the adult individuals. The results of the calculation indicated that 5% (6/120) of the population were suffering from the condition.

In the majority of cases the fourth lumbar vertebra was affected (67%; 4/6), with the third and fifth lumbar vertebrae only affected in single cases. The majority of individuals displaying spondylolysis were female (67%; 4/6), and the prevalence based on the age-at-death of the individuals was fairly widespread with one adolescent (17%; 1/6), two young adults (33%; 2/6) and three 35-45 year old individuals (50%; 3/6) displaying the defect. In all cases the spondylolysis was bilateral and complete.

Context	Sex	Age	Vertebra	R	L	Paraxial mesoderm defects	Other trauma
XXXI. 2	M	17-25	LV4	+	+	o	+
XXXI. 100. Sk. 1	F	35-45	LV4	+	+	o	+
XXXI. 148	F	35-45	LV4	+	+	+	+
XXXI. 149	F	17-25	LV3	+	+	o	o
XXXI. 175	?F	15.5-18.5	LV5	+	+	o	+
XXXI. 212	M	35-45	LV4	+	+	o	+

Table 22: Hunno-Sarmatian period individuals - Details of spondylolysis in the population (+ = present, o = absent, LV = lumbar vertebra).

Only one of the individuals with spondylolysis displayed a developmental defect of the paraxial mesoderm; in Skeleton XXXI. 148 the neural arch of the twelfth thoracic vertebra was bifurcated. All of the other individuals with spondylolysis, with the exception of Skeleton XXXI. 149, displayed lesions indicative of stress and strain to the skeleton including Schmorl's nodes. This finding may suggest that the occurrence of spondylolysis is a further indication of the strenuous physical lifestyle of the Hunno-Sarmatian period population from Aymyrlyg.

5.3.1b (iv) Clay-shoveller's fractures

Skeleton XXXI. 100. Sk. 1, a 35-45 year old female, displayed a clay-shoveller's fracture of the 1st thoracic vertebra, and it is possible that the presence of this injury is indicative that the woman had been subject to repeated strenuous physical exertion which had involved lifting heavy weights. Spondylolysis (see Section 5.3.1b (iii)), Schmorl's nodes and prominent enthesopathies were also apparent, and are further evidence that the lifestyle of the woman had been responsible for the exertion of repeated physical stresses on her skeleton.

5.3.1b (v) Rib fractures

The prevalence of rib fractures among the Hunno-Sarmatian period adults was 2.3% (11/481), and details of the injuries present among the population are provided in Table 23.

Context	Age	Sex	Type of fracture	Location	Healing
XXXI. 20	35-45	F	simple transverse	L 2nd rib	early
XXXI. 203. Sk. 2	35-45	F	simple transverse	R 3rd rib	healed
XXXI. 209	45+	M	simple transverse	R 12th rib	healed
			simple transverse	L 5-11th ribs	healed
XXXI. 212	35-45	M	Simple oblique	R 9th rib	healed

Table 23: Hunno-Sarmatian period individuals - Fractures of the ribs (L = left, R = right).

Equal proportions of males (50%) and females (50%) displayed one or more fractured ribs. The majority of rib fractures occur as a result of direct injury, such as a blow to the chest, or a fall against a hard object (Adams 1978, 103). Skeleton XXXI. 20, Skeleton XXXI. 203. Sk. 2 and Skeleton XXXI. 212 all displayed a single fractured rib, and a Colles' fracture was present on the right radius of Skeleton XXXI. 212 (see Section 5.3.1a (iii)). The rib injuries in Skeleton XXXI. 20 and Skeleton XXXI. 203. Sk. 2 were probably due to direct blows on the chest, while the fractured rib in Skeleton XXXI. 212 had probably occurred as a result of a fall. Eight fractured ribs, compression fractures of several vertebrae (see Section 5.3.1b (ii)) and a fractured sternal body (see Section 5.3.1b (vii)) were visible in Skeleton XXXI. 209, a 45+ year old male, and it is highly probable that all of these injuries had been sustained as a result of a heavy fall.

5.3.1b (vi) Fractures of the sterna

Two Hunno-Sarmatian period adults displayed fractures of the sternal bodies providing a prevalence of 6% (2/32) for the injuries. Details of the fractures are provided in Table 24.

Context	Age	Sex	Type of fracture	Location	Healing
XXXI. 203. Sk. 1	35-45	F	simple transverse	sternal body	early
XXXI. 209	45+	M	simple transverse	sternal body	healed

Table 24: Hunno-Sarmatian period individuals - Fractures of the sternum.

Fractures of the sternum generally arise as a result of direct force, or vertebral compression of the thoracic cage with concomitant fracture of the thoracic spine (*ibid.*, 104). In Skeleton XXXI. 203. Sk. 1 only the sternal body was fractured which may indicate that the injury had probably been due to direct trauma (Plate 27), while in Skeleton XXXI. 209 eight ribs (see Section 5.3.1b (v)) and several thoracic and lumbar vertebrae were also fractured (see Section 5.3.1b (ii)), and it is probable that the injuries had occurred as a result of vertebral compression of the thoracic cage, possibly during a severe fall. In Skeleton XXXI. 203. Sk. 1 the fractured sternum was in the early stages of healing when the individual died, while in Skeleton XXXI. 209 all of the fractures had healed by the time of the individual's death. It is possible that the fractured sternal body in Skeleton XXXI. 203. Sk. 1 had caused obstruction of the airway, and eventually resulted in death due to pulmonary complications. Plaques of woven bone present on the rib heads may be indicative that the individual had been subject to emphysema or pneumonia before she died.

5.3.1c Ossified haematomas

Ossified haematomas generally occur as a result of direct trauma (Resnick *et al* 1995, 2673), and six Hunno-Sarmatian period adults displayed these lesions (Table 25).

Context	Age	Sex	Bone
XXIII. 3. entrance (i)	25-35	F	L fibula
XXXI. 2	17-25	M	L tibia
XXXI. 97	25-35	F	R tibia
XXXI. 127 (i)	25-35	M	R tibia
XXXI. 127 (iii)	25-35	M	R tibia
XXXI. 157 (Plate 28)	35-45	M	R femur

Table 25: Hunno-Sarmatian period individuals - Details of ossified haematomas in the population (L = left, R = right).

The prevalence of the lesions was 1.5% (4/273) for the tibiae, 0.4% (1/275) for the femora, and 0.4% (1/253) for the fibulae. The concentration of the haematomas on the bones of the legs, and in particular the lower leg bones, may indicate that these parts of the body were most susceptible to direct trauma. The majority of the lesions were visible in males (67%; 4/6).

Plate 27: Hunno-Sarmatian period - Fractured sternal body in Skeleton XXXI. 203. Sk. 1.

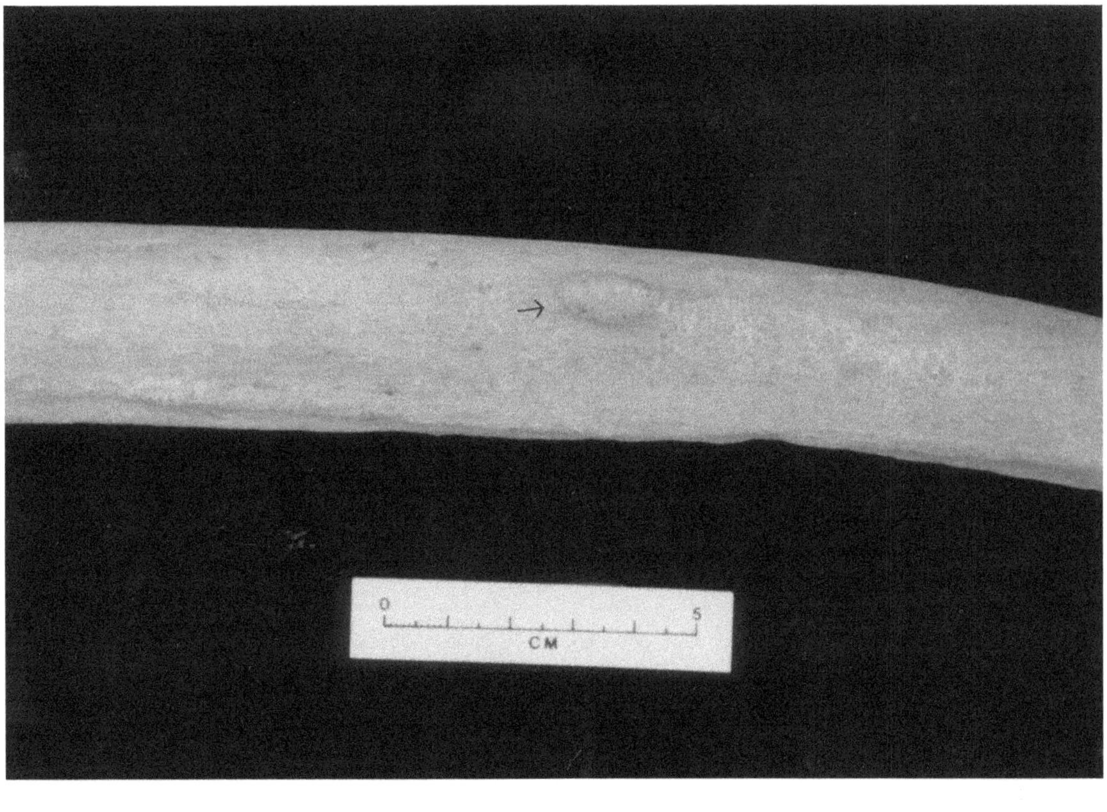

Plate 28: Hunno-Sarmatian period - Ossified haematoma on the midshaft of the right femur in Skeleton XXXI. 157.

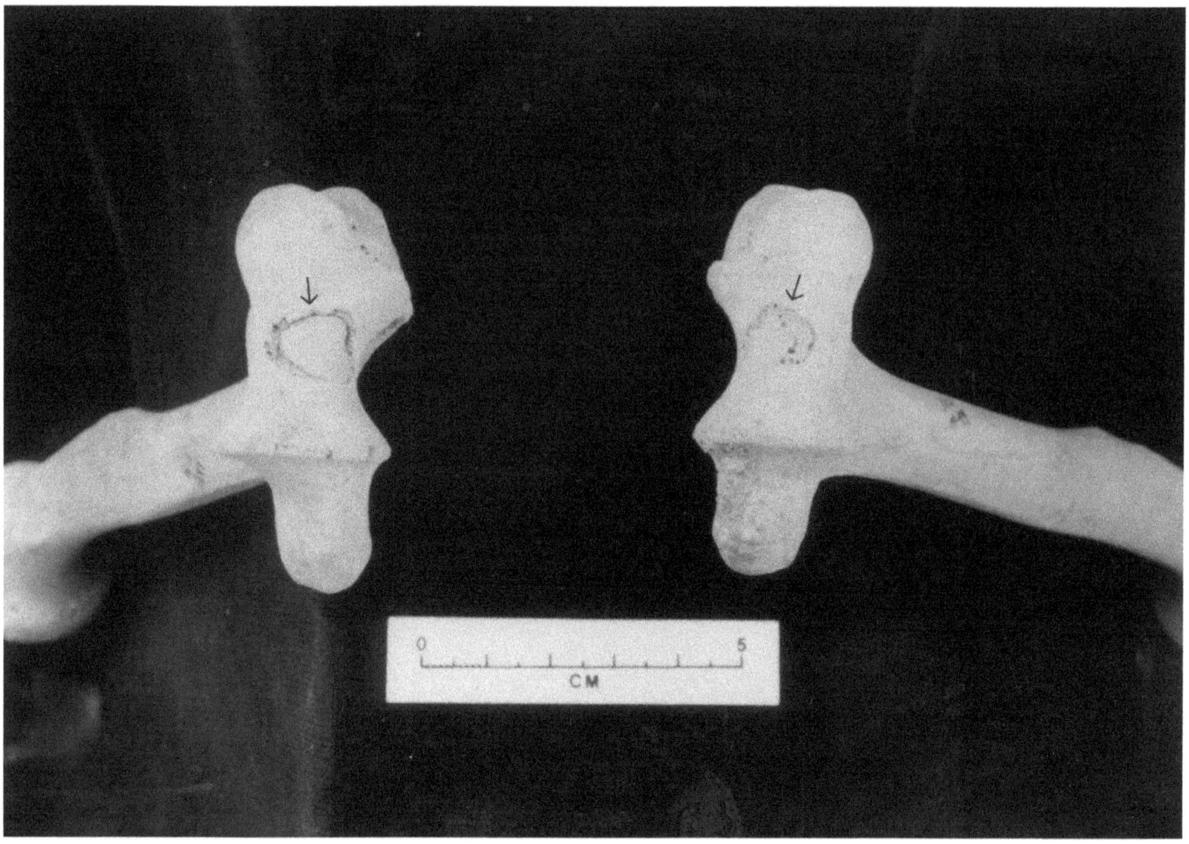

Plate 29: Hunno-Sarmatian period - Possible osteochondritis dissecans in the distal humeri of Skeleton XXXI. 99.

5.3.1d Dislocations

Skeleton XXXI. 15, a 35-45 year old female, displayed possible evidence for dislocation of the right knee, and the proximal end of the right tibia appeared to have been rotated by approximately forty-five degrees so that the lateral condyle was positioned more posteriorly, and the medial condyle was situated more anteriorly, than normal. Secondary degenerative joint disease had ensued. Dislocation of the knee is a rare injury which is due to the joint being exposed to high energy trauma, such as that attained in a fall from a considerable height, or by a severe blow to the knee (Resnick and Goergen 1995, 2773). It is possible that the individual had been subject to a rotatory (postero-lateral) dislocation, which is an uncommon injury associated with damage of the collateral ligaments. Rotatory dislocation of the knee can be irreducible due to intra-articular invagination of the medial capsule and the medial collateral ligament (*ibid*., 2774), and major complications with such injuries include damage of the popliteal artery and peroneal nerves (Adams 1978, 233).

The Aymyrlyg case also displayed a night-stick fracture of the left ulna (see Section 5.3.1a (ii)). The two injuries were not necessarily attained in the same incident, however, and she may have fallen from a height and damaged her right knee only, and fractured her left ulna in an earlier or later incident. Alternatively, it is indeed possible that both injuries were obtained in a single incident of interpersonal violence during which the left arm and right knee were dealt severe blows by an aggressor.

5.3.1e Osteochondritis dissecans

Osteochondritis dissecans is a localised disorder which affects convex joint surfaces. An area of subchondral bone becomes avascular and, in conjunction with its associated articular cartilage, may slowly detach from the surrounding joint surface to form a loose body within the joint (Adams and Hamblen 1990, 125). The joints most commonly affected by the disorder are the knees and elbows, although in some cases the femoral heads and tali may display the lesions (*ibid*., 126). Males are more likely to be affected by the disorder than females, and the lesion is more likely to be unilateral than bilateral (Resnick *et al* 1995, 2616). In modern clinical cases the disorder is generally presented in adolescents or young adult individuals. During the early stages of necrosis the individual may feel an aching sensation in the affected joint, and once the fragment completely separates from the articular surface the symptoms include recurrent sudden locking of the joint followed by sharp pain and effusion of a clear fluid into the joint (Adams and Hamblen 1990, 126).

Osteochondritis dissecans is generally considered to be the end result of an osteochondral fracture initially caused by shearing, rotatory or tangentially aligned impaction forces (Resnick *et al* 1995, 2613). Wells (1974, 367) recorded that that the lesion occurred in the knee or foot in 95% of all cases among Romano-British and Anglo-Saxon populations. He observed that when using the feet for very heavy work - weight bearing, thrusting and turning movements - the principal line of force is directed through the big toe, the first cuneiform, the navicular, and the head and trochlear surface of the talus to the distal articular surface of the tibia, the line along which the majority of osteochondritic dissecans lesions are observed. This observation supports the theory that trauma or stress is a major cause of the lesion (*ibid.*), but modern studies have also indicated, however, that there is likely to be a genetic predisposition for the condition, since it can occur throughout families, and an autosomal dominant mode of inheritance has been suggested. In addition, irregularities of ossification are considered as a possible aetiology (Resnick *et al* 1995, 2611). Another theory suggests that the lesion occurs as a result of ischaemia, in which the blood supply to the affected area is cut off, possibly as result of thrombosis of an end-artery (Clanton and DeLee 1982, 51).

Three Hunno-Sarmatian period individuals displayed lesions which were possibly indicative of osteochondritis dissecans (Table 26).

Context	Sex	Age	Location	Bilateral / unilateral	Other strains and trauma
XXXI. 99	F	25-35	Distal humerus	bilateral	o
XXXI. 111	M	25-35	Distal humerus	unilateral	Schmorl's nodes
					Enthesopathies
XXXI. 139 (i)	M	25-35	Distal femur	unilateral	Fractures LMC4, LMT1
					Schmorl's nodes
					Enthesopathies

Table 26: Hunno-Sarmatian period individuals - Details of osteochondritis dissecans in the population (L = left, o = not present, MC = metacarpal, MT = metatarsal) (vertebral osteophytosis and single enthesopathies were not included in the 'other strains and trauma' category).

It was possible to examine 272 adult distal humeri for evidence of osteochondritis dissecans among the Hunno-Sarmatian period population, and it was found to have occurred with a frequency of 1.1% (3/272). In total, 276 adult distal femora were available for examination, only 0.4% (1/276) of which displayed the lesions.

Circular lesions were present on the trochleae of both distal humeri (Plate 29) of Skeleton XXXI. 99, which had diameters of 8 mm and consisted of depressed areas of spongy bone with central discs of thin cortical bone. The circular fragments of bone may not have been completely necrotic since they were only partially detached from the trochleae. Clinical studies have shown that once the demarcating line between the avascular bone and the normal articular surface has developed it takes a further two to three months for the fragment to separate and form a loose body (Adams and Hamblen 1990, 234). In Skeleton XXXI. 111 a circular area, which had a diameter of approximately 5 mm and a porotic appearance, was present on the anterior aspect of the capitulum of the right humerus. Studies of modern populations have indicated that the symptoms of osteochondritis dissecans of the elbow initially involve an aching pain which is aggravated by use, followed by locking, clicking or intermittent severe pain, and eventual degenerative joint disease. In addition, at all stages of the condition a limitation of movement, particularly a loss of extension, is characteristic of the condition (Woodward and Bianco 1975, 38; Lindholm *et al* 1980, 248). In Skeleton XXXI. 139 (i) an oval shaped depressed lesion, which measured approximately 24 mm by 14 mm, was present on the lateral condyle of the left femur.

It is interesting to note that both Skeleton XXXI. 111 and Skeleton XXXI. 139 (i) displayed other traumatic lesions, including fractures, Schmorl's nodes and enthesopathies. These lesions may indicate that the individuals had practised active lifestyles which had resulted in considerable physical stress to their bodies and supports the theory that osteochondritis dissecans is related to trauma. Skeleton XXXI. 99 also displayed slight vertebral osteophytosis and an enthesopathy of the left Achilles' tendon, and it is possible that she too had an active lifestyle which had caused physical stress to her bones.

5.3.2 Soft-tissue trauma

5.3.2a Myositis ossificans traumatica

Skeleton XXXI. 127 (i), a 25-35 year old male, displayed a bony spur on the antero-lateral surface of the midshaft of the right humerus, which may have been indicative of myositis ossificans traumatica of part of the insertion for pectoralis major. Since this muscle is responsible for protracting the scapula and depressing the shoulder, as well as flexion, adduction and medial rotation of the shoulder (Romanes 1986, 115) it is probable that the muscle injury was sustained during one or more of these motions. The individual also displayed a fractured left fibula (see Section 5.3.1a (iv)), left talus and left calcaneus (see Section 5.3.1a (v)), and an ossified haematoma on the right tibia (see Section 5.3.1c), and it is probable that all of the injuries had been attained as a result of a heavy fall.

5.3.2b Enthesophytes

Skeleton XXXI. 64, a 25-35 year old female, displayed an enthesophyte on the left tibia, immediately inferior to the lateral condyle. The prevalence of enthesophytes in the observable adult tibiae was 0.4% (1/273).

5.4 Summary

Fractures were observed in 0.9% of all adult long bones from the Scythian period population group, with 69% of cases affecting males, and 31% females. The ulnae most frequently displayed the injuries (2.2%), followed by the radii (1.6%) and the clavicles (1.3%). The fractured clavicles had probably been caused by falls onto outstretched hands or onto the shoulders. Night-stick fractures of the ulnae may have occurred as a consequence of aggressive activities, and it was interesting to note that two of the three Scythian period cases were female, and that one of these women also displayed fractured metacarpals, also indicative of interpersonal violence. Colles' fractures of the radii may have been attained as a result of falls on outstretched hands, and in cases where the diaphysis of both the radius and ulna were fractured the injury had probably occurred as a result of direct blows to the forearms. A single Scythian period case displayed a fracture of the diaphysis of the tibia not associated with a fractured fibula, and it is probable that the trauma had occurred as a consequence of a twisting injury or a fall. Cases where only the fibula (and not the tibia) was fractured were probably the result of direct trauma, and the presence of multiple injuries in one of the affected individuals suggests that the direct trauma had occurred as a result of a fall. Fractures which involved both the diaphysis of the tibia and the fibula probably occurred as a result of a fall, and a variety of fractures of the bones of the hands and feet were present among the Scythian period populations, some of which may have been attained during falls, while others were characteristic of aggressive activities.

A variety of fractures were also apparent in the axial skeletons of the Scythian period population, and depressed fractures were present in 1.7% of the adult and subadult cranial vaults. In total, 38% of depressed cranial fractures had occurred in males, 38% in subadults and 24% in females. Most of the depressed fractures were located on the frontal bones (0.8%) or the parietals (0.7%), and it is probable that the majority of these injuries had occurred as a result of interpersonal violence. Several members of the Scythian period population also displayed fractures of the facial skeleton (excluding the nasal bones), the majority of which were male (80%) with only 20% of their female counterparts affected. Again, it is probable that a large proportion of these injuries were attained as a result of interpersonal violence. A frequency of 3.4% of fractured adult nasal bones was observed in the Scythian period population, 92% of which were present in males.

A single Scythian period older adult female displayed a compression fracture of a vertebra, a Colles' fracture of the left radius, a fracture of the styloid process of the left ulna and generalised osteopaenia, and it is probable that she had been suffering from underlying osteoporosis. A total frequency of 13% of Scythian period adults displayed spondylolysis in the lumbar spine, and the majority of cases were evident in females (67%), with only 33% of males affected. A single clay-shoveller's fracture of the first thoracic vertebra was apparent in the Scythian period population, and fractures were observed in 0.5% of adult ribs. A female with fractured ribs also displayed a night-stick fracture of the ulna and fractured metacarpals, and it is probable that she had obtained all injuries in a severe incident of interpersonal violence. Ossified haematomas were evident in 0.9% of femora, 0.3% of tibiae and 0.4% of fibulae. A single Scythian period individual displayed a dislocated right glenohumeral joint, with a fractured humeral head, a Colles' fracture, a fractured fibula and an ossified haematoma, and it is probable that the injuries had occurred as a result of a fall.

Several examples of soft-tissue trauma were apparent among the remains of the Scythian period population, and myositis ossificans traumatica was evident in the femur of a single individual. A localised area of increased vasculature was apparent at the midpoint of the endocranial surface of the right parietal of a child, which may have occurred as a result of intracranial haemorrhaging due to trauma. Prominent enthesophytes were visible in the remains of eleven individuals, the majority of which were male (64%). The greatest frequency of enthesophytes occurred on the tibiae (1.5%), followed by the femora (0.9%), which may indicate that the bones of the lower body had been subjected to greater physical stress than those of the upper extremities.

Fractures were observed in 0.8% of all adult long bones from the Hunno-Sarmatian period population group, and 64% of the affected individuals were male while 36% were female. The ulnae were the long bones which most frequently displayed trauma (2.3%), followed by the radii (1.5%), the clavicles (0.8%) and the fibulae (0.8%). Two Hunno-Sarmatian period individuals displayed fractured clavicles, one of which was probably sustained as a result of a fall and the other during an incident of interpersonal violence. Night-stick fractures of the ulnae were probably due to interpersonal violence, and 67% of individuals with these injuries were male. Colles' fractures of the radii may have occurred as a consequence of falling on outstretched hands. Two Hunno-Sarmatian period individuals displayed fractures of the fibulae not associated with fractured tibiae, and the presence of an ossified haematoma and myositis ossificans traumatica in one of the cases may indicate that all injuries had been obtained during a fall. Fractures present in the bones of the hands and feet had probably resulted from falls and interpersonal violence.

A variety of fractures were also apparent in the axial skeleton of the Hunno-Sarmatian period population, although no depressed cranial fractures were evident. A single Hunno-Sarmatian period female displayed a fractured orbit as well as fractured nasal bones, which were probably all sustained in an incident of interpersonal violence. The prevalence of fractured nasal bones among the Hunno-Sarmatian period adults was 6%, with 75% of cases male, and only 25% female.

Compression fractures were apparent in 0.7% of Hunno-Sarmatian period vertebrae, and the majority of cases occurred in males (86%). All of the compression fractures had occurred at the midthoracic, thoracolumbar and lumbar regions of the spine and, although the lesions in several cases may have been due to underlying palaeopathological conditions such as osteoporosis, it is probable that in the majority of cases they had been obtained as a result of heavy falls. A total of 5% of Hunno-Sarmatian period adults displayed spondylolysis in the lumbar spine, and the majority of affected individuals were female (67%), with only 33% of their male counterparts affected. A single Hunno-Sarmatian period female displayed a clay-shoveller's fracture of the first thoracic vertebra. Fractures were observed in 2.3% of adult ribs, with males and females being equally affected. The majority of individuals with rib fractures displayed other injuries characteristic of falls and it is probable that the rib fractures were also attained as a result of falling. Two Hunno-Sarmatian period individuals displayed fractured sternal bodies, one of which may have occurred as a result of a direct blow, while the other may have been obtained as a result of a heavy fall. Ossified haematomas were evident in 1.5% of tibiae, 0.4% of femora and 0.4% of fibulae. A single Hunno-Sarmatian period individual displayed possible evidence for a dislocated right knee, which may have occurred either as a result of a considerable blow to the knee or been due to a heavy fall. Osteochondritis dissecans was present in 1.1% of distal humeri and 0.4% of distal femora. The presence of other lesions indicative of physical stress in the remains of a proportion of these affected individuals may indicate that the lesions had occurred as a result of joint strain.

Soft-tissue trauma occurred infrequently in the remains of the Hunno-Sarmatian period population, and myositis ossificans traumatica was evident in the humerus of a single individual, while enthesophytes were apparent in 0.4% of adult tibiae.

Fractures were observed in 0.9% of all adult long bones from the Scythian period and 0.8% of all adult long bones from the Hunno-Sarmatian period, and the frequencies of appendicular fractures in both population groups were generally low. A greater proportion of males than females displayed fractures of the appendicular skeleton in both populations, and details of these fractures are provided in Figure 3. The ulna was most often affected in both groups, and the relatively high frequency of fractures of the Scythian period naviculars is probably misleading since only a small number of naviculars had been preserved in this group. The majority of appendicular fractures in both population groups had probably occurred as a result of accidents, particularly those which had involved falling, although the occurrence of night-stick fractures and boxer fractures also suggests that a proportion of the injuries may have been due to interpersonal violence.

Depressed fractures were apparent in 1.7% of Scythian period adult and subadult cranial vaults, whereas they did not occur in any of the Hunno-Sarmatian period vaults. Details of the skull fractures present in both population groups are provided in Figure 4, and it is evident that the Scythian period population displayed a wider variety of these injuries than the Hunno-Sarmatian period population. Skull fractures generally occurred in very small frequencies, with the exception of fractures of the nasal bones. The frequency of frac-

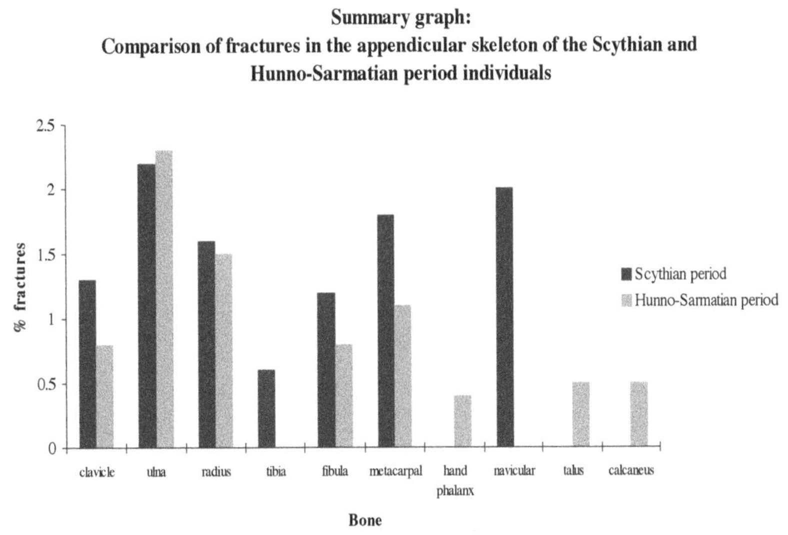

Figure 3: Summary graph - Comparison of the frequencies of fractures of the appendicular skeleton in the Scythian and Hunno-Sarmatian period individuals.

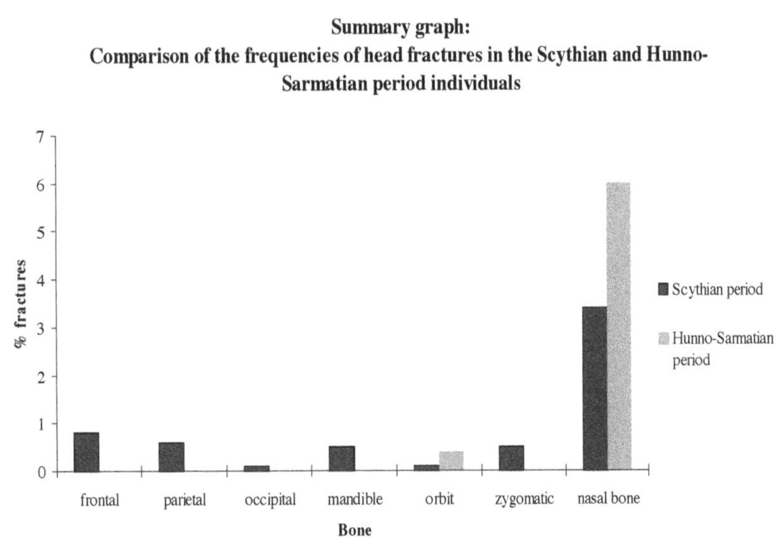

Figure 4: Summary graph - Comparison of the frequencies of fractures of the skull in the Scythian and Hunno-Sarmatian period individuals.

tured nasal bones in the Hunno-Sarmatian period population was greater than that for the Scythian period population group, and a larger proportion of males than females displayed the injuries in both populations. The difference, however, was more substantial between Scythian period males (92%) and females (8%) relative to their Hunno-Sarmatian counterparts. The occurrence of trauma to the skull, and in particular the nasal bones, is generally indicative of interpersonal violence, and it is probable that members of both population groups, particularly males, were exposed to interpersonal violence. The occurrence of depressed cranial fractures in Scythian period subadults, and traumatic skull injuries in females of both periods, however, indicates that this was not an exclusively male preoccupation.

Only a single Scythian period individual displayed a compression fracture of the vertebra, and this was probably related to osteoporosis; the lesions were, however, apparent in 0.7% of the Hunno-Sarmatian period adult population. Although some of these fractures were probably related to underlying palaeopathological conditions it is probable that the majority of fractures were due to falls. Spondylolysis occurred in the Scythian period population with a frequency of 13%, and in the Hunno-Sarmatian period population with a frequency of 5%. The majority of affected individuals in both population groups were female. Single Scythian and Hunno-Sarmatian period individuals displayed clay-shoveller's fractures, and rib fractures were visible in 0.5% of the Scythian period population and 2.3% of Hunno-Sarmatian period group. Ossified haematomas occurred infrequently in both population groups, and were concentrated on the lower leg bones, with the femora (0.9%) most frequently affected in the Scythian period population and the tibiae (1.5%) in the Hunno-Sarmatian period population. A Scythian period individual displayed a dislocated glenohumeral joint, while a Hunno-Sarmatian period individual displayed a possible dislocated knee.

Evidence of soft-tissue trauma occurred infrequently in both population groups, and only one Scythian period individual displayed evidence for myositis ossificans traumatica in a femur, while a single Hunno-Sarmatian period individual displayed the lesion in a humerus. A Scythian period child may have developed intracranial haemorrhaging secondary to an incident of direct trauma. A number of Scythian period individuals displayed large enthesophytes, mainly affecting the tibiae (1.5%) and the femora (0.9%). Prominent enthesophytes were also apparent in 0.4% of Hunno-Sarmatian period tibiae. It is possible, therefore, that the bones of the lower body would have been subjected to more physical stresses than those of the upper extremities in both population groups.

Although occurring with generally small frequencies, a variety of examples of skeletal and soft-tissue trauma were apparent among both populations. The morphological characteristics and positions of a large proportion of the traumatic lesions indicated that the individuals were susceptible to falling injuries, and this finding may reflect the semi-nomadic lifestyles of both populations which would have involved a considerable amount of horse riding and travelling in wagons. It is interesting to note that clinical studies have indicated the majority of modern horse riding injuries are caused by falls (Williams 1988, 140). The presence of numerous cases of spondylolysis and the two cases of clay-shoveller's fractures may suggest that both populations followed lifestyles which involved considerable physical exertion, and caused extensive stress to their skeletal systems. The occurrence of numerous fractures to the skull (particularly the nasal bones), night-stick fractures and boxer fractures is indicative of the occurrence of interpersonal violence in both population groups. This finding is also reflected in the numerous cases of unhealed weapon trauma apparent in both populations as will be seen in the following chapter.

6. Weapon trauma and surgery

6.1 Introduction

Trauma also includes those injuries inflicted by a weapon, or sustained during a surgical procedure such as an amputation or trepanation. The evidence for weapon trauma in a population group can provide important information on past societies. It is possible to determine the types of weapons which were used by a population, and the damage which these objects would have caused to the body. Information can be gleaned on the actions of the warriors during combat, and on the techniques of warfare practised by a society. In addition, it is possible to gain an insight into the levels of interpersonal and intergroup violence that a population would have been subjected to, thus providing an indication of political stability throughout the ages in different geographical locations.

Lesions sustained as a result of surgical procedures are a valuable source of information on the level of medical expertise present in a society. In addition, the attitudes of the members of a population to certain diseases and disabilities can be reconstructed. In many cases the lesions indicative of weapon or surgical trauma present in archaeological human remains are unhealed, and it is probable that the trauma was directly responsible for the death of these individuals. Consequently, an examination of skeletal lesions sustained as a result of weapon or surgical trauma potentially enables the reconstruction of the cause and the manner of death of an individual (Roberts and Manchester 1995, 83).

The prevalence rates for the weapon and surgical trauma evident in the individuals from Aymyrlyg are based on the numbers of observable adult and subadult bones, since immature and mature bones are affected in the same manner by these forms of trauma. It is difficult, however, to determine the overall prevalence of trauma in the Aymyrlyg population groups due to the differential preservation of the individuals, but in the present chapter the evidence for weapon trauma and surgery apparent in the remains is discussed.

6.2 Scythian period individuals

6.2.1 Weapon trauma

6.2.1a Pointed battle axe trauma

Twelve Scythian period individuals displayed evidence of weapon trauma which was probably inflicted using pointed

Context	Age	Sex	No. injuries	Location	Size (mm)	Healing	Identification
B. 8. Sk. 1	35-45	M	2	L parietal	13x18	o	definite
(Plate 32)				L parietal	15x17.5	o	definite
G. 7. Sk. 5	25-35	M	1	L parietal	4x4	+	probable
II. 4. Sk. 12	25-35	?M	3	L frontal	44x24.5	o	definite
				R frontal	64x25	o	definite
				R temporal	67x50	o	definite
II. 8 (iii) (Plate 33)	25-35	?F	1	L parietal	24x24	+	possible
VI. 6 (Plate 34)	4-5	-	1	L parietal	13.5x13	o	definite
VI. 9. upper (ii)	7-10	-	1	R parietal/ frontal	incomplete	o	possible
VII. 5. Sk. 4	25-35	M	1	L parietal	11.5x16.5	o	definite
IX. 3 (i)	17-25	M	4	Mid frontal	45x31	o	definite
(Plate 35)				L parietal	8x8	o	definite
				R parietal/ temporal	92.5x20	o	definite
				Lambda	34x22.5	o	definite
XXI. 4. Sk. 4	17-25	M	1	R frontal	5x5	o	definite
XXIII. 4	25-35	M	1	R frontal	25x20	+	probable
XXIII. 13 (i)	25-35	?M	1	L zygomatic	10x10	+	possible
XXV. 16. Sk. 4	17-25	M	3	R parietal	25x12	o	definite
				Mid frontal	26x18	o	definite
				Mid frontal	12.5x11.5	o	definite

Table 27: Scythian period individuals - Summary of pointed axe trauma in the adults and subadults (+ = healing present, o = healing not present, R = right, L = left).

axes or *chekany* (Murphy 2003) (Plates 30-35 and see Plate 5), resulting in a prevalence of 2.5% (12/473) for these injuries. A summary of the injuries caused by pointed axes is presented in Table 27.

Of the twelve individuals displaying evidence of trauma inflicted by a pointed axe, 75% (9/12) were male, 17% (2/12) were subadults, and 8% (1/12) were female. Twenty injuries were identified, 40% (8/20) of which were situated on the parietals, 35% (7/20) on the frontal, 5% (1/20) on the temporal, 5% (1/20) on the zygomatic, 5% (1/20) on the occipital, and two injuries had affected more than one bone (10%; 2/20). The preponderance of injuries on the parietals and the frontal bones suggests that the individuals were likely to have been directly facing their opponent during an episode of violence.

The individual with the injury apparent at the lambda (Skeleton IX. 3 (i)) also displayed trauma to the frontal bone, and the right and left parietals. It is possible that the injury on the occipital represents the last blow struck by the attacker, possibly when the victim was defeated, already fatally injured and lying in a prone position on the ground. The majority of injuries present on the parietals alone were situated on the left parietal (88%; 7/8). This finding is probably indicative of face-to-face combat in which the individuals who struck the blows were right handed. A summary of the prevalence of the injuries in each observable cranial bone in the population group is provided in Table 28.

Bone	No. injuries	No. observable	% injuries
frontal	6	471	1.3
parietal	9	929	1
occipital	1	453	0.2
temporal	2	848	0.2
zygomatic	1	860	0.1

Table 28: Scythian period individuals - Prevalence of pointed axe trauma in the adult and subadult cranial bones. A single injury which involved two bones is included in the count for both bones. If a bone displayed more than one injury it is only counted once for the purposes of determining the prevalence.

Eighty percent (16/20) of the pointed axe injuries were unhealed, attesting to the effectiveness of this type of axe as a lethal weapon. All four of the individuals - Skeleton G. 7. Sk. 5, Skeleton II. 8 (iii) (Plate 33), Skeleton XXIII. 4 and Skeleton XXIII. 13 (i) - who displayed signs of healing associated with an axe injury only displayed a single lesion. The two subadults (Skeleton VI. 6 (Plate 34) and Skeleton VI. 9. upper (ii)) with pointed axe injuries each displayed a single injury, and it is probable that the axe was able to perforate the immature cranial bones with ease, causing irreparable trauma to the brain. The four individuals (Skeleton B. 8. Sk. 1, Skeleton II. 4. Sk. 12, Skeleton IX. 3 (i) (Plate 35) and Skeleton XXV. 16. Sk. 4) with evidence of multiple axe injuries displayed trauma on several cranial bones suggesting that they and/or their assailant were not stationary during the combat.

Four of the individuals with evidence of trauma inflicted by pointed axes also displayed conventional fractures. In Skeleton B. 8. Sk. 1, a 35-45 year old male, a night-stick fracture was present on the left ulna (see Section 5.2.1a (iii)) and an ossified haematoma was apparent on the left femur (see Section 5.2.1c). In Skeleton G. 7. Sk. 5, a 25-35 year old male, the left side of the mandible was fractured (see Section 5.2.1b (ii)). In Skeleton XXIII. 4, a 25-35 year old male, fractures were apparent in the left 3rd and 4th metacarpals, and the right navicular (see Section 5.2.1a (ix)). In Skeleton XXIII. 13 (i), a 25-35 year old probable male, fractures were visible on the left side of the frontal bone, the left zygomatic, and the right nasal bone (see Section 5.2.1b (ii)). The lesions with no clear association with weaponry apparent in Skeleton B. 8. Sk. 1, Skeleton G. 7. Sk. 5 and Skeleton XXIII. 13 (i) are all probably indicative of interpersonal violence. It is probable, therefore, that these three males were subject to repeated aggressive activities, and they may have been warriors.

6.2.1b Blade trauma

Three Scythian period individuals displayed evidence of trauma inflicted by a bladed weapon. A summary of the blade injuries which were present in the remains of the individuals is provided in Table 29.

It is evident that the three individuals with possible evidence of blade trauma were all adult males. The injuries appear to have been made by the slashing motion of a large bladed weapon, probably a sword. None of the examples of trauma displayed signs of healing, thereby attesting to the effectiveness of the bladed weapons.

The sword cut apparent in the sacrum of Skeleton VI. 15 (ii) had completely separated the distal part of S5 from the remainder of the sacrum. Since there was no sign of weapon trauma on the innominates it is probable that the assailant had attacked the individual from behind. In Skeleton XXIII the supero-inferior transverse sword cuts on the right parietal and the right zygomatic were synonymous, and had been caused by a single blow (Plate 36). There were no signs of healing associated with the injuries and it is probable that they had caused fatal damage to the brain. The prevalence of sword trauma in the observable adult and subadult remains is provided in Table 30.

6.2.1c Decapitation

Various explanations have been proposed to explain the presence of decapitated individuals in archaeological contexts. The beheading may have been carried out for punitive reasons, as part of a rite of human sacrifice, or for the purposes of obtaining the head of a vanquished enemy as a trophy of warfare (Bush and Stirland 1987). Skeleton D/G, a 5.5-6.5 year old child, was the only Scythian period individual to display evidence for decapitation. Two chop marks were visible on the inferior surface of CV3. The posterior aspect of the right inferior articular facet had been cleanly chopped

Plate 30: Scythian period - Log House Tomb X. 1 containing the remains of at least two adult individuals with pointed axe trauma (Photographic Archive of the Institute for the History of Material Culture, St. Petersburg).

Plate 31: Scythian period - Log House Tomb XX. 7 which contains a pointed axe in situ (Photographic Archive of the Institute for the History of Material Culture, St. Petersburg).

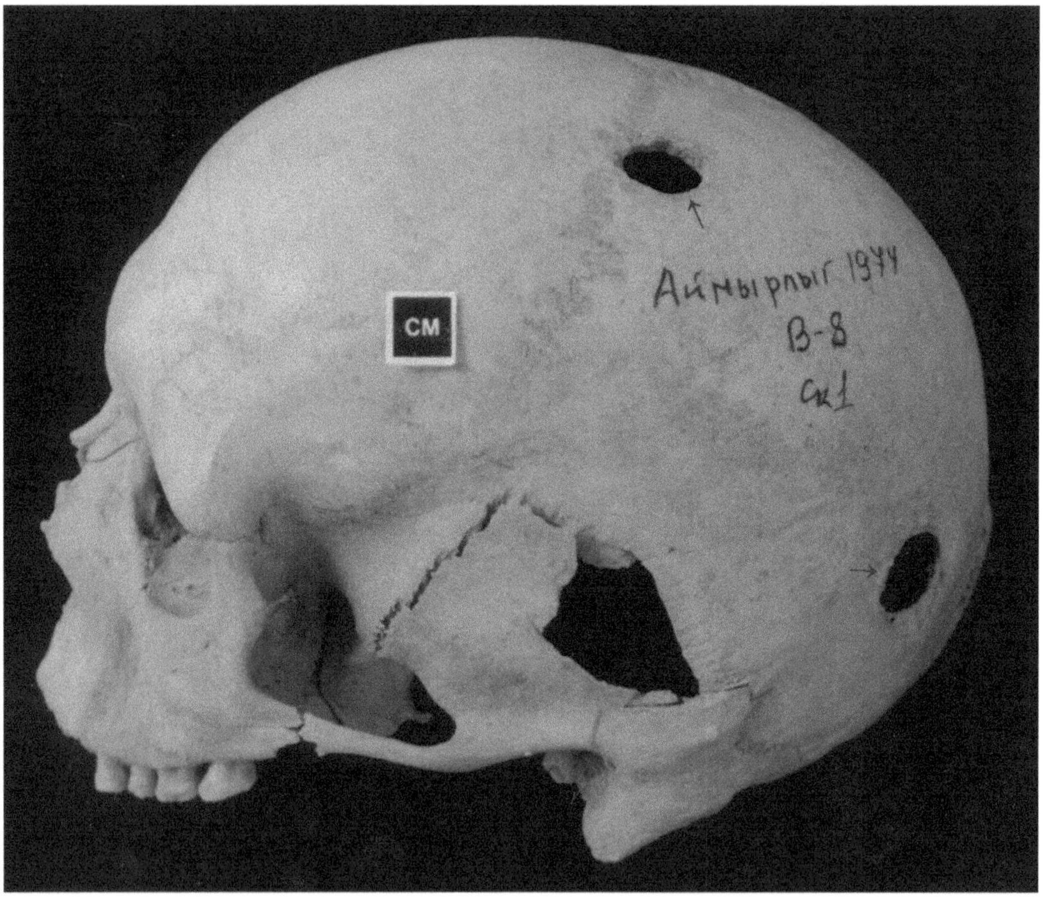

Plate 32. Scythian period - Pointed axe injuries in the left parietal of Skeleton B. 8. Sk. 1.

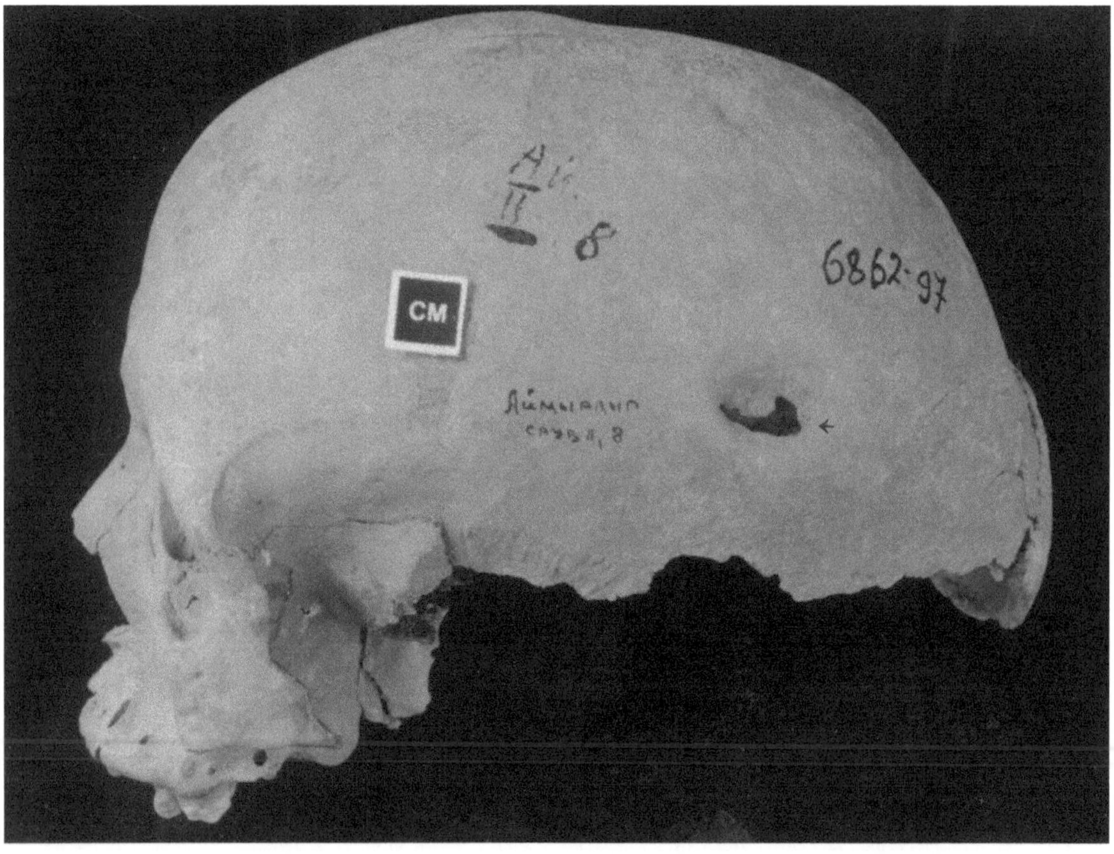

Plate 33: Scythian period - Possible healed pointed axe injury in the left parietal of Skeleton II. 8 (iii).

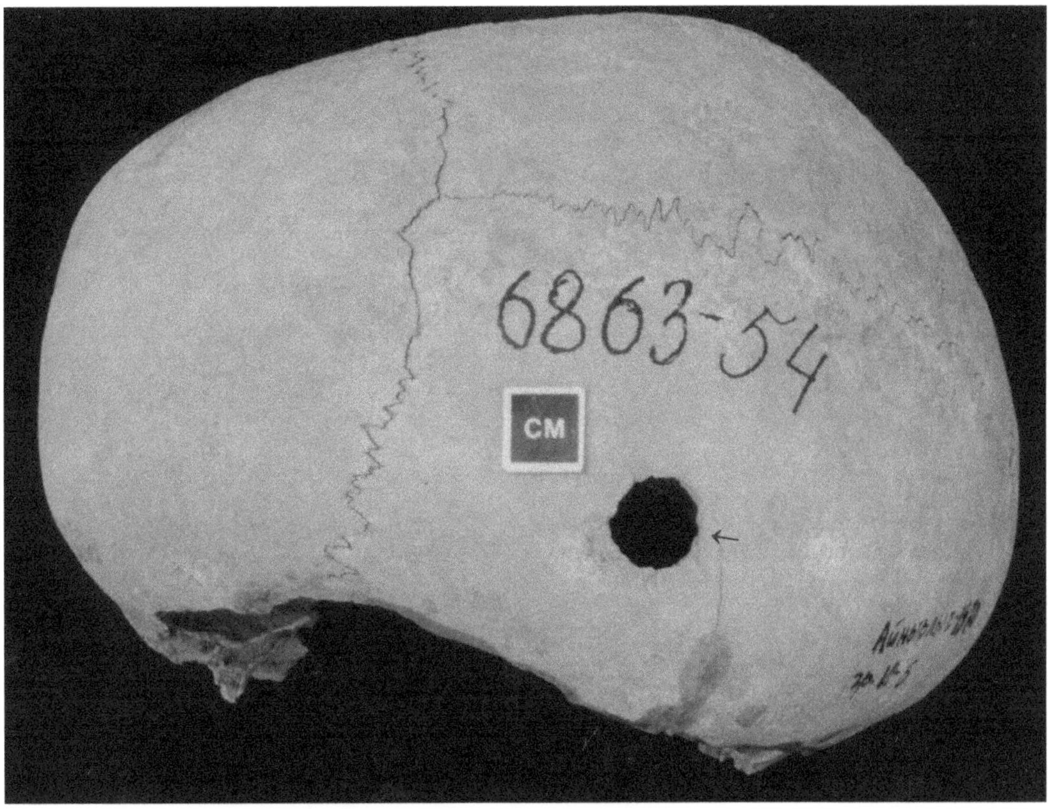

Plate 34: Scythian period - Pointed axe injury in the left parietal of Skeleton VI. 6, a 4-5 year old child.

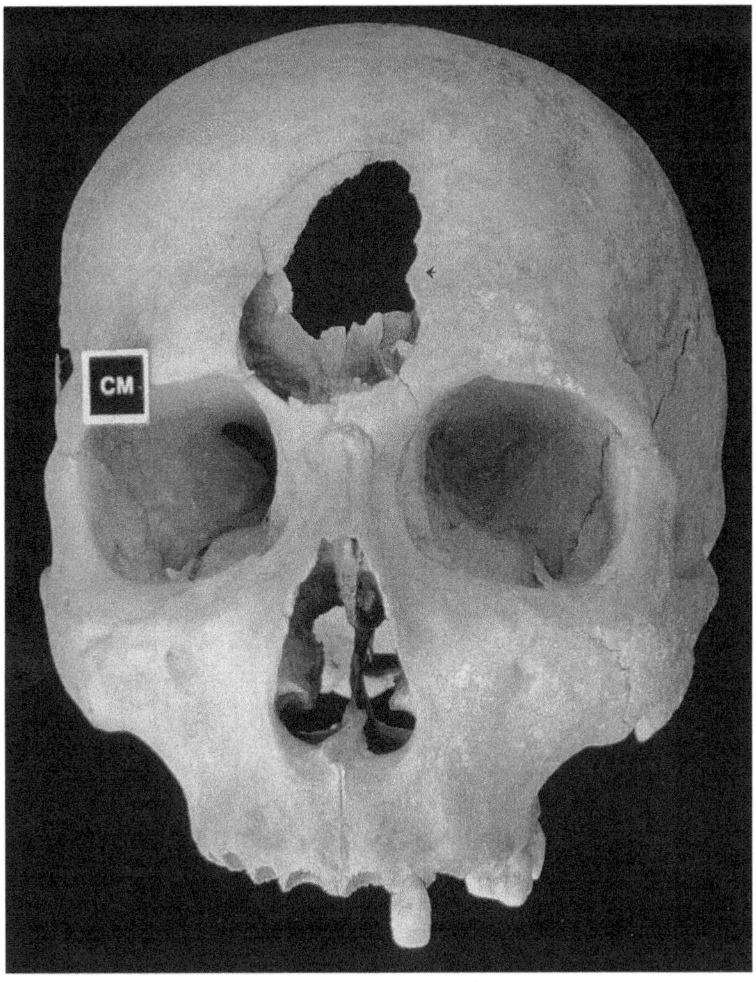

Plate 35: Scythian period - Pointed axe injury at the glabella of Skeleton IX. 3 (i).

Context	Age	Sex	No. injuries	Location	Size (mm)	Direction	Healing	Identification
II. 13. south	25-35	M	2	R humeral head	23 AP	oblique	o	possible
				R femur	20 SM-IL	oblique	o	possible
VI. 15 (ii)	35-45	M	1	sacrum	c 20 ML	transverse	o	probable
XXIII	25-35	?M	2	R parietal	59.5 SI	transverse	o	definite
(Plate 36)				R zygomatic	c 40 SI	transverse	o	definite

Table 29: Scythian period individuals - Details of sword injuries apparent in the population (+ = healing present, o = healing not present, AP = antero-posterior, SM = supero-medial, IL = infero-lateral, ML = medio-lateral, SI = supero-inferior, R = right, L = left).

Bone	No. trauma	No. observable	% trauma
humerus	1	860	0.2
femur	1	459	0.2
sacrum	1	130	0.8
parietal	1	929	0.1
zygomatic	1	860	0.1

Table 30: Scythian period individuals - Prevalence of sword trauma in the adults and subadults.

through. In addition, the left side of the inferior surface of the vertebral body displayed an oblique chop mark which extended to the medial margin of the left transverse foramen. Two chop marks were also apparent on the superior surface of CV4. A chop mark was visible on the medial margin of the left superior articular facet, and a further chop mark was present on the lateral margin of the right superior articular facet. It is probable that the cut marks on CV3 and CV4 were attained in a single incident and that the head and neck had been severed from the remainder of the body as a result of these blows. It is probable that the decapitation had been undertaken from behind the individual. The skull of the individual was not present and information concerning the archaeological context was not available, but it is possible that he or she had been buried without their head. Decapitation was observed in the Scythian period cervical vertebrae with a frequency of 1% (1/98).

Cut marks were present on the anterior surface of CVI and CV2 which may be indicative that the individual had his or her throat cut (see Section 6.2.1d) and that after this had occurred he or she was decapitated. It is impossible, however, to discern with any degree of certainty whether this procedure was undertaken for ritual or aggressive purposes. The possible evidence that the individual had his or her throat cut, and the evidence that the individual was a child would, however, tend to suggest a ritual purpose for the throat cutting and beheading.

6.2.1d Evidence for deliberate trauma to the throat

Skeleton D/G, a 5.5 - 6.5 year old child, displayed several small, transverse cutmarks on the anterior surfaces of CV1 and CV2 (Plate 37). The superior aspect of each cutmark had a bevelled edge suggesting that they had been made in an upward manner. The lesions may be indicative that the individual had his or her throat cut. Chop marks indicative of decapitation were also apparent on CV3 and CV4 (see Section 6.2.1c). After having had his or her throat cut it is possible that the individual was decapitated. It is impossible to know with any degree of certainty, however, whether this procedure was undertaken for ritual or aggressive purposes, though the evidence that this was a child might tend to suggest a ritual reason for the trauma. The evidence for possible throat cutting was observed in the Scythian period cervical spines with a frequency of 1% (1/98).

6.2.1e Scalping

Scalping can be defined as the forcible removal of all or part of the scalp from the head, and involves the incision of the skin overlying the skull with a sharp object in a circular pattern (Axtell and Sturtevant 1980, 467) to enable the easy removal of the soft tissues from the cranial vault (Hamperl 1967, 630). Scalping was commonly undertaken to retrieve human trophies as indicators of success and bravery in warfare in different cultures of a variety of periods throughout the world. It has also been suggested that scalping may have been undertaken for therapeutic reasons, or for magico-religious reasons (Owsley 1994, 337-338).

A Scythian period probable male, with an age-at-death of 25-35 years, displayed evidence for scalping (Murphy *et al* 2002). Unfortunately, the context for this case was uncertain and only the skull was available for analysis. Numerous small sharp cutmarks were apparent on the surface of the cranial vault, which ranged in length from 4 mm to 62 mm, and appeared to follow the circumference of the skull and may suggest that the objective of the scalper had been to carefully remove a circular area of skin from the superior and posterior aspects of the cranium. The cutmarks on the occipital and the posterior regions of the parietals ran in a horizontal direction, parallel to one another. The superior margins of the cutmarks were generally bevelled, suggesting that the person

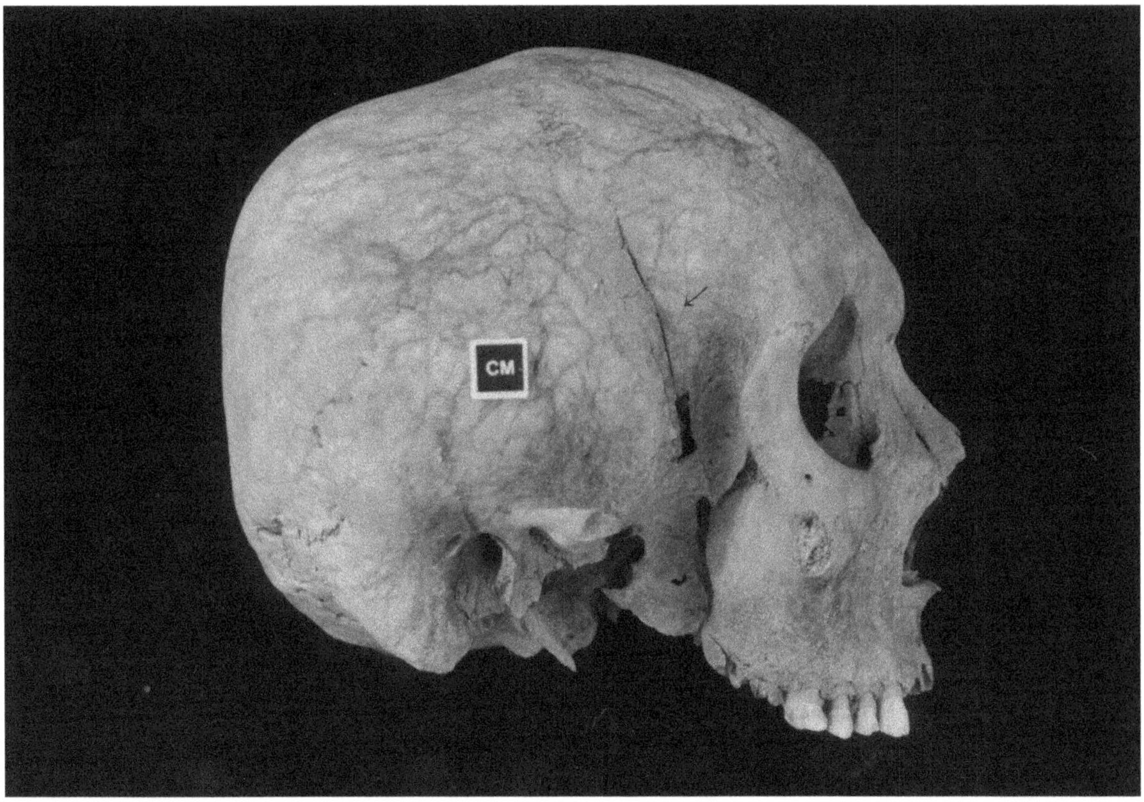

Plate 36: Scythian period - Sword trauma in the right parietal, temporal and zygomatic of Skeleton XXIII.

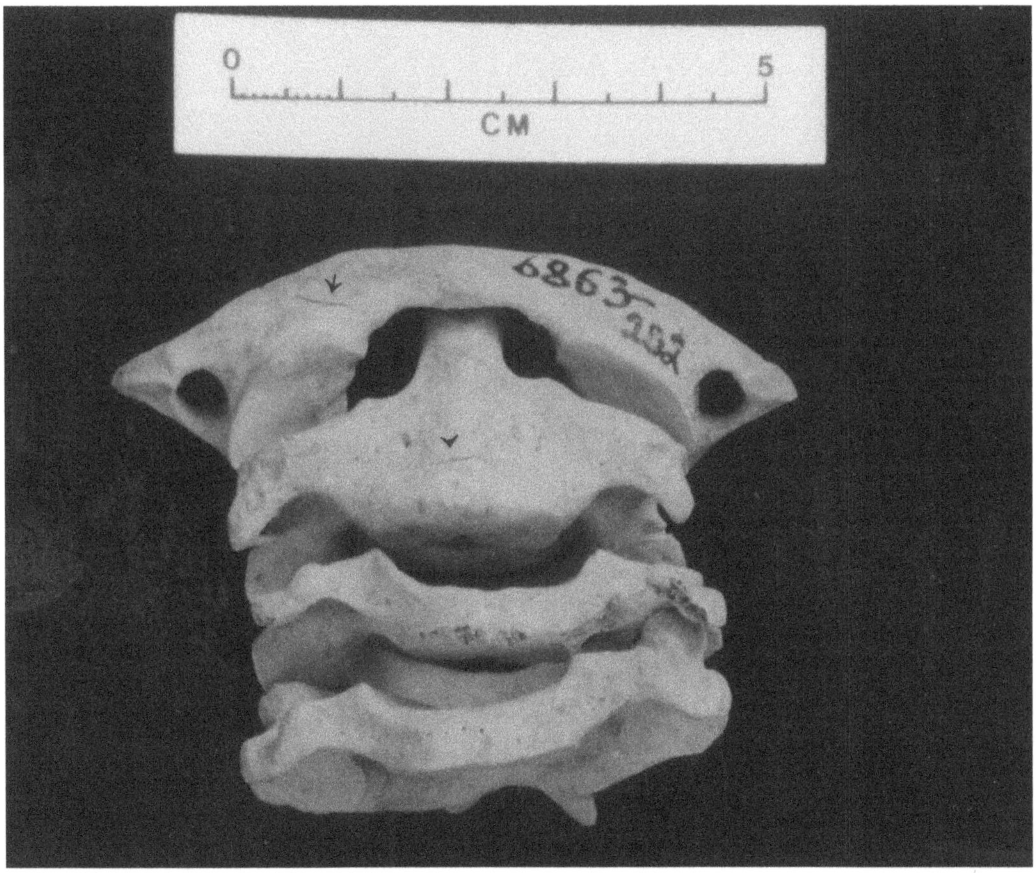

Plate 37: Scythian period - Cutmarks on the anterior surfaces of the first and second vertebrae of Skeleton D/G, a 5.5-6.5 year old child, which may be indicative that the throat was cut.

who had made the cut marks had been positioned superior to the skull, and that the skin had been peeled off in an anterior direction. No signs of healing were associated with the scalping cutmarks, and it is therefore probable that the procedure may have occurred just prior to death or in the peri-mortem period. Fractured nasal bones were also apparent.

6.2.2 Evidence for surgical procedures

6.2.2a Trepanation

Numerous examples of trepanations are known from the palaeopathological record, with the earliest cases dating to Mesolithic times (Rokhlin 1965, 290). In addition, the practice of trepanation has been widely observed by modern anthropologists throughout the populations of the world (Ackernecht 1967, 642). Trepanation of the cranium involves the removal of a section of the calvarium without damaging the underlying blood vessels, meninges and brain (Lisowski 1967, 651). Five forms of trepanation have been identified in archaeological human remains throughout the world. The trepanation can be undertaken by scraping (possibly using a piece of flint or a shell), gouging, boring and sawing (using a drill-like implement), sawing alone, and (those in which a small perforation is created) by using a drill (trephination) (Culebras 1993, 11). Lisowski (1967, 659) has summarised the motives behind trepanation operations into three categories. The operations may have been undertaken for a therapeutic reason to treat head injuries, including fractures and scalp wounds. The trepanation could also have been carried out for magico-therapeutic motives whereby the cause of symptoms, such as headaches, neuralgia or epilepsy, may have been regarded as a consequence of possession by evil spirits. The third of Lisowski's categories of motivation is magico-ritual, in which the trepanation was undertaken purely for ritual or magical purposes.

Two possible examples of trepanation were apparent in the Scythian period individuals, making a prevalence of 0.4% (2/473) for the population group (Murphy 2003). In Skeleton VIII. 21. Sk. 3, a 35-45 year old probable female, a well healed perforation which may have been due to a trepanation was present on the left side of the frontal (Plate 38). The lesion was near circular, measuring 23.5 mm by 24.4 mm, and had very smooth margins. The morphology of the lesion indicates that it was probably either undertaken using the scraping or gouging techniques. Although, the regular circular shape of the lesion would suggest that the gouging method was practised, the presence of a depressed area surrounding the perforation makes it more probable that the operation had been undertaken using the scraping technique. Studies on the healing of trepanations have indicated that little osseous regeneration occurs (*ibid.*, 666). It is, therefore, probable that the individual survived for a fairly long period of time following the operation since the margins of the perforation were smooth.

The majority of trepanation operations identified in archaeological skeletal remains have been undertaken on the left side of the cranium. It has been suggested that this is the case since the majority of traumatic lesions sustained as a result of interpersonal violence occur on this side of the head. This is thought to occur since the adversary would most probably have been right handed and positioned facing the victim (*ibid.*, 659) (see Section 6.2.1a). The only palaeopathological conditions present in the skull of Skeleton VIII. 21. Sk. 3 were related to the dentition. It is possible that the three dental abscesses apparent in the remains may have caused the individual to have suffered from headaches, prompting her to undergo a trepanation operation. This is, however, mere speculation and it is impossible to ascertain the motives which lay behind the operation on this woman.

In Skeleton II. 4. Sk. 6, a 12-14 year old adolescent, an oval perforation, measuring 18 mm medio-laterally by 8 mm antero-posteriorly, was present on the coronal suture immediately to the right of the bregma (Plate 39). The perforation had a smooth, bevelled margin which had a thickness of 8 mm. The bevelled margin had a glossy appearance which may indicate that it was made using a metal implement. Alternatively, the polished margins of the perforation may be suggestive of wear (Owsley *et al* 1994, 368). It was initially thought that the perforation represented a trepanation operation undertaken using the scraping technique while the individual was alive, but that they had not survived the procedure. The regularity of the lesion and the presence of shallow scrape marks over the entire surface of the cranial vault may, however, indicate that the perforation was made post-mortem.

Ethnographic studies in Africa have indicated that the motive behind posthumous trepanation operations was to obtain roundels of human skull. The roundels were generally circular in shape and were frequently perforated and polished so that they could be worn on necklaces. It was thought that the roundels had magico-religious properties and they were regarded as amulets or charms (Lisowski 1967, 659). It is possible, therefore, that Skeleton II. 4. Sk. 6 had been trepanned post-mortem for ritual purposes. The scrape marks were distributed randomly over the cranial vault and were fairly coarse in morphology. In scalping the incisions are generally made in a deliberate manner following the curvature of the cranium (Hamperl 1967, 630) (see Section 6.3.1e). It is possible, therefore, that the individual was scalped posthumously in a non-regular manner mainly to enable the retrieval of a roundel of bone. It is also possible that the scalp may also have been removed for specific purposes. An alternative explanation for the post-mortem perforation of the crania may be found in the work of Gillman (1876) who reported on a series of artificially perforated crania from Moundbuilder sites in Michigan. The perforations on these crania were all considered to have been made post-mortem and they were invariably positioned on the superior aspect of the cranium. The openings were generally situated at or near the bregma, were circular, and measured approximately 10-25 mm in diameter (*ibid.*, 243-244). Gillman (1876, 238-239) recounted that ethnographic studies of the Dyaks of Borneo had revealed that each village had a central house in which they kept the heads of their dead prisoners suspended by a string which passed through a perforation in the superior aspect of the cranium. It is possible that whoever had perforated the cranium of the individual from Aymyrlyg had intended to suspend the crania, rather than obtain a roundel of bone. It is probable that the posthumous perforation was made in the crania of the individual from Aymyrlyg for either ritual or warfare pur-

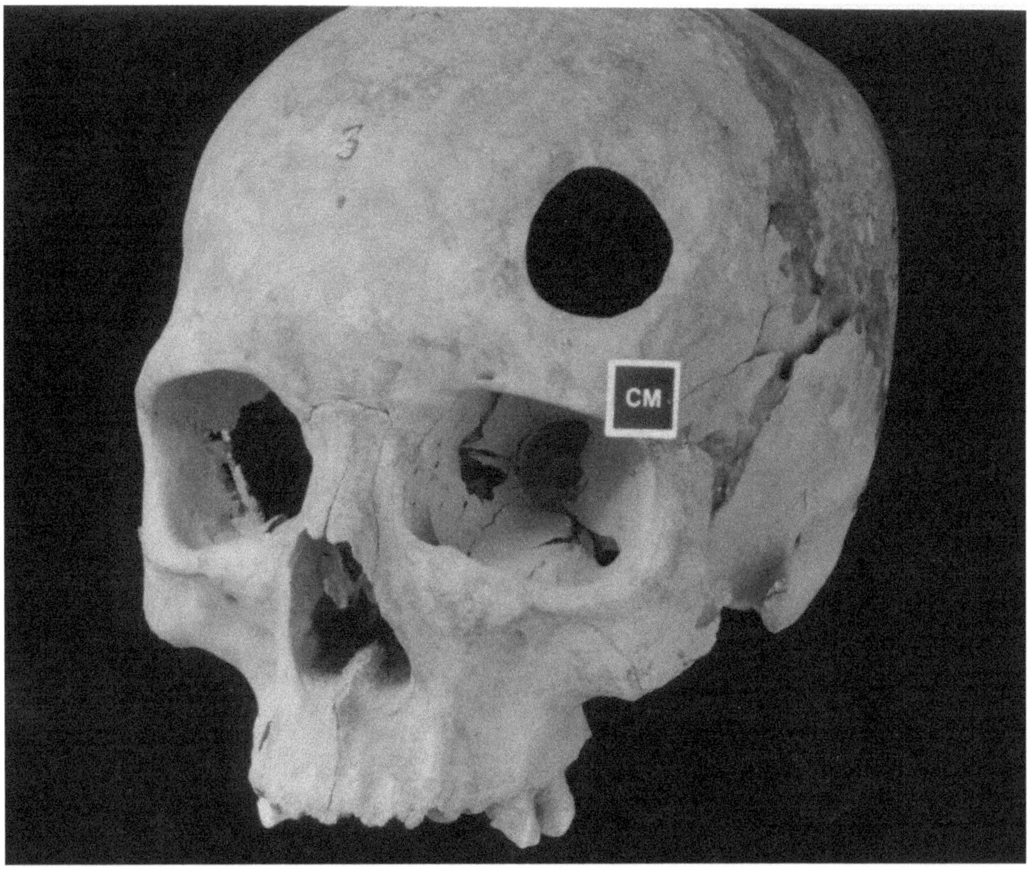

Plate 38: Scythian period - A healed trepanation on the frontal of Skeleton VIII. 21. Sk. 3.

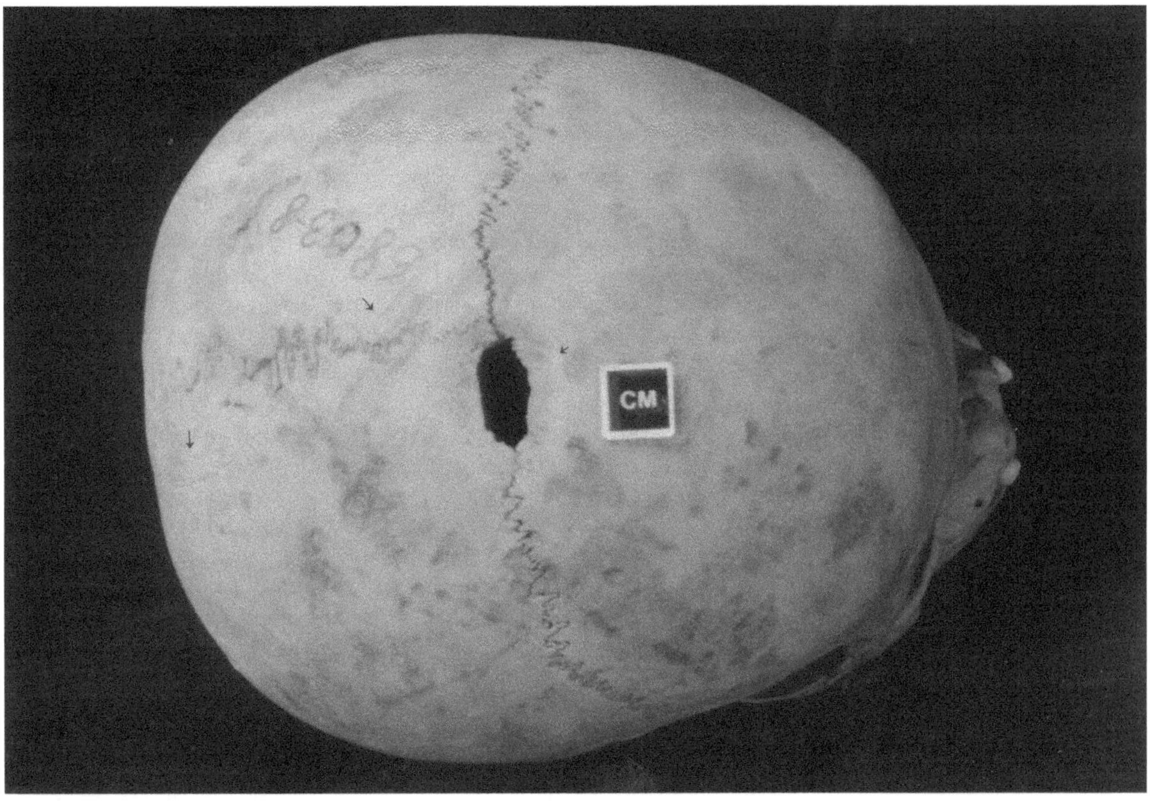

Plate 39: Scythian period - A post-mortem trepanation at the bregma of Skeleton II. 4. Sk. 6.

poses or as a combination of the two.

6.2.2b Amputation

Amputation of limbs is known to have occurred in the past for a variety of reasons. The amputation could have been voluntary and permitted by the person, perhaps as a surgical procedure or as a sign of grief, but amputation may also have been involuntary, with the individual's limb removed in the course of punitive action against a criminal or captive (Merbs 1989, 181). In addition, accidental amputation could occur during battle or in the course of everyday life (Roberts and Manchester 1995, 89). Evidence of amputation in the palaeopathological record is rare, probably due to the difficulty of identifying this form of trauma.

A possible amputation was apparent in the remains of Skeleton VI. 6. Sk. 4, a 25-35 year old male (Plates 40 and 41). The left forearm and hand were not present, and there appears to have been an amputation at the distal end of the humerus. The trochlea and the capitulum of the left humerus were not present with only the superior part of the medial epicondyle surviving. The inferior aspect of the distal end had a smooth and curved morphology. A bony nodule was present in the area of the olecranon fossa. The anterior surface of the distal end had a very porotic appearance. The metaphysis displayed pitted woven bone which was in the advanced stages of remodelling. The presence of new bone formation in the late stages of remodelling may indicate that the amputation was at an advanced stage of healing. The left humerus had a greatest length of 306 mm as opposed to 328 mm for the normal right humerus. The entheses on the left humerus were pronounced and similar to those present on the right humerus. This may suggest that the amputation had occurred quite shortly before the individual died, and that he had been actively using his arm until the amputation had occurred. Consequently, it is possible to postulate that he had not been suffering from any chronic disease of the left arm which may have warranted amputation.

It is not possible to determine the reason for the amputation with any degree of certainty. Since the entire forearm and hand had been removed, however, it is unlikely that the amputation was incurred in the process of punitive action. In addition, there was no evidence of a palaeopathological condition apparent in the remains of the individual which may have necessitated the amputation of a limb. This would suggest that the amputation was probably not undertaken for deliberate medical reasons. It is most likely, therefore, that the amputation arose as a consequence of a domestic accident or as a result of warfare. Given the evidence for interpersonal trauma and weapon injuries (see Sections 5.2 and 6.2.1) apparent in the Scythian period population the latter aetiology seems most plausible. Possible differential diagnoses include degenerative changes, infectious processes or an unhealed fracture of the distal humerus. Since the skeleton was in a good state of preservation (60-70% of bones present), however, it seems too great a coincidence that the left forearm and hand were not present while the left distal humerus displayed lesions indicative of an amputation.

6.3 Hunno-Sarmatian period individuals

6.3.1 Weapon trauma

6.3.1a Blade trauma

Twelve Hunno-Sarmatian period individuals displayed evidence of trauma inflicted by a bladed weapon. A summary of the blade injuries which were present on the remains of the individuals is presented in Table 31.

The majority of the individuals with evidence of blade trauma were adult males (75%; 9/12), with notably fewer female (17%; 2/12) and subadult individuals (8%; 1/12) displaying injuries. The majority of injuries had been made in a slashing motion using a very sharp, large bladed weapon, which was probably a sword. The only evidence for a blade injury healing was apparent in the cranium of Skeleton XXXI. 108, where two sword cuts on the right parietal were almost totally healed. Ninety-two percent (11/12) of the individuals with blade trauma displayed unhealed injuries. The majority of the corpus displayed multiple sword cuts (58%; 7/12), suggesting that the injuries were sustained during combat in which both individuals did not remain in a stationary position.

In Skeleton XXXI. 54 the injuries to the left parietal and frontal, the left clavicle and the left scapula had all been sustained in a single blow (Plates 42 and 43). The attacker had probably been standing in front of, and slightly to the right of, the victim, and the weapon must have been wielded with a great deal of force to have damaged four bones. In addition, the cranium had been almost cleaved in two. Numerous sword cuts were also present in Skeleton XXXI. 115 (Plates 46-48), Skeleton XXXI. 176 (Plates 49 and 50) and Skeleton XXXI. 189 (ii). In all three of these individuals the distribution of the injuries indicates that the victim and aggressor were non-stationary. The injuries present in Skeleton XXXI. 115 indicate that the fighters were probably facing each other for the majority of the encounter. Although the assailants were facing the victims at some stage during the course of the combat, the distribution of lesions apparent in Skeleton XXXI. 176 and Skeleton XXXI. 189 (ii) (Plates 51-53 and see Plates 60 and 61) suggests that they also struck blows when they were positioned posterior to the victims. This is evident by the occurrence of sword cuts on the neural arches of several lumbar vertebrae in Skeleton XXXI. 176, and on the neural arches of TV4, LV2 and S4 in Skeleton XXXI. 189 (ii). It is also possible that these two individuals had been overwhelmed by multiple assailants. Skeleton XXXI. 87. Sk. 1 and Skeleton XXXI. 157 also displayed injuries to the neural arches of several vertebrae. Neither of the two individuals displayed other examples of weapon trauma on their remains. It is probable, therefore, that in both cases the injuries were either sustained when they were fleeing from the enemy, or that the injuries were caused by the aggressors attacking the two individuals from behind.

It was surprising to find sword injuries in the remains of Skeleton XXXI. 59 (i), a 1-1.5 year old infant (Plates 44 and 45). The position on the injuries on the superior aspect of the left parietal and occipital suggest that they were made by an adult or an older subadult, who was standing above the infant. It is interesting that two females also displayed evidence of sword

Context	Age	Sex	No. injuries	Location	Size (mm)	Direction	Healing	Identification
XXXI. 42	35-45	M	1	R femur	7.5 ML	transverse	o	probable
XXXI. 54	17-25	M	3	L parietal/ frontal	102 AP	transverse	o	definite
(Plates 42 and 43)				L clavicle	17 ML	oblique	o	definite
				L scapula	42.5 ML	oblique	o	definite
XXXI. 59 (i)	1-1.5	-	2	L parietal	90 SI	oblique	o	definite
(Plates 44 and 45)				L parietal/ occipital	72 AP	oblique	o	definite
XXXI. 87. Sk. 1	17-25	??M	1	neural arch TV11 + V12	17 SI	oblique	o	definite
XXXI. 87. Sk. 2	35-45	M	1	R femur	8 SI	oblique	o	definite
XXXI. 108	25-35	M	2	L parietal	94 AP	oblique	+	possible
					24.5 AP	oblique	+	possible
XXXI. 115	35-45	F	6	L temporal/ zygomatic	68 AP	transverse	o	definite
(Plates 46-48)				L zygomatic	22 SI	transverse	o	definite
				R scapula	35 SI	transverse	o	definite
				L humerus	6 SI	oblique	o	definite
				L humerus	23 SI	oblique	o	definite
				L radius	14 SI x 8 ML	transverse	o	definite
XXXI. 157	35-45	M	2	neural arch TV4	7 SI	oblique	o	definite
				neural arch TV9 + TV10	15.8 SI	oblique	o	definite
XXXI. 172	25-35	M	1	R scapula	26.5 SI	oblique	o	definite
XXXI. 176	25-35	M	6	mid frontal	14 AP	oblique	o	definite
(Plates 49 and 50)				R parietal	14 AP	oblique	o	definite
				R humerus	PM damage	oblique	o	definite
				R radius	8 AP	oblique	o	definite
				neural arch LV1 + LV2	50 ML	oblique	o	definite
				neural arch LV4	45 ML	transverse	o	definite
XXXI. 185	35-45	F	1	R femur	23 AP x 12 SI	transverse	o	possible
XXXI. 189 (ii)	17-25	M	13	R frontal/ parietal	91 AP	transverse	o	definite
(Plates 51-53 and see Plates 60 and 61)				L mandible	10 AP	oblique	o	definite
				R mandible	15 AP	oblique	o	definite
				R scapula	5 AP	oblique	o	definite
				L humerus	20 AP	oblique	o	definite
				L humerus	20 ML	transverse	o	definite
				R 12th rib	10 SI	oblique	o	definite
				L 10th rib	35 SI	oblique	o	definite
				neural arch TV4	7.6 AP	oblique	o	definite
				Anterior body TV12	35 SI	oblique	o	definite
				neural arch LV2	35 ML	transverse	o	definite
				Sacrum (S4)	20 ML	transverse	o	definite
				L femur	2 SI x 21 AP	transverse	o	definite

Table 31: Hunno-Sarmatian period individuals - Summary of blade trauma in the population (+ = healing present, o = healing not present, AP = antero-posterior, ML = medio-lateral, SI = supero-inferior, R = right, L = left, TV = thoracic vertebra, LV = lumbar vertebra, S = sacral vertebra).

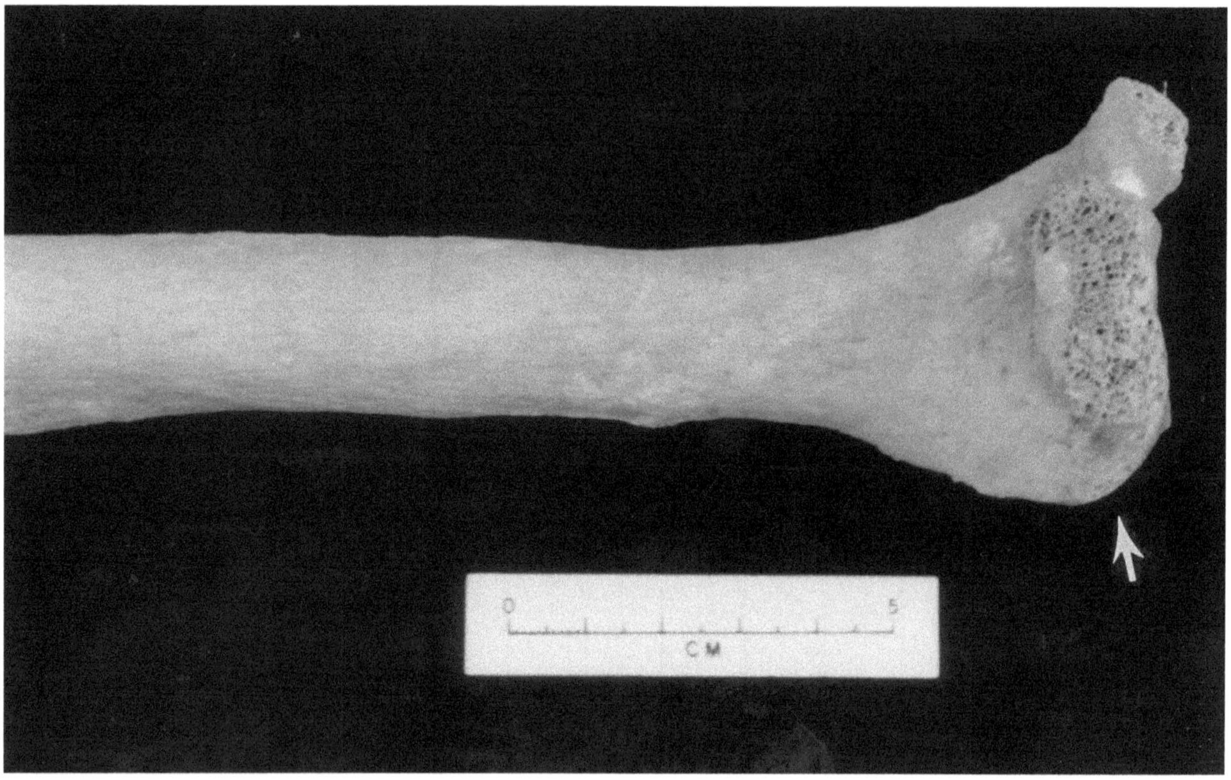

Plate 40: Scythian period - Anterior view of the possible amputation in the left distal humerus of Skeleton VI. 6. Sk. 4.

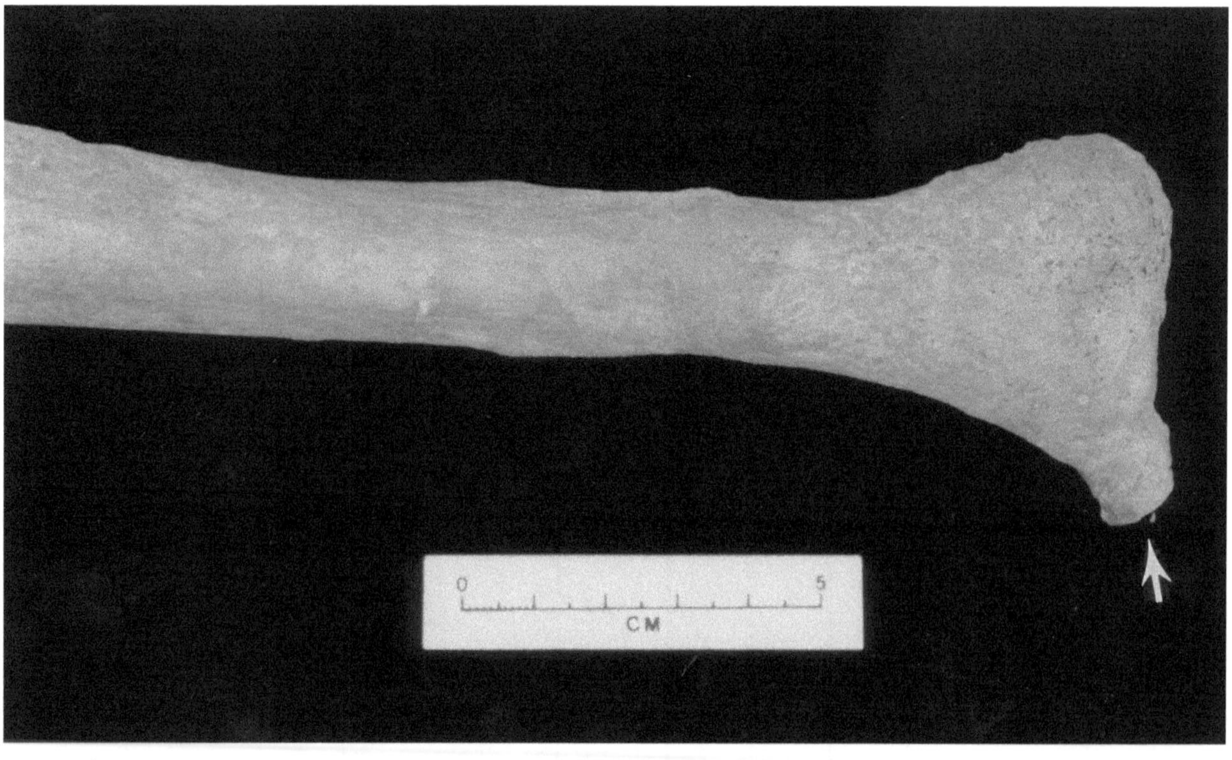

Plate 41: Scythian period - Posterior view of the possible amputation in the left distal humerus of Skeleton VI. 6. Sk. 4.

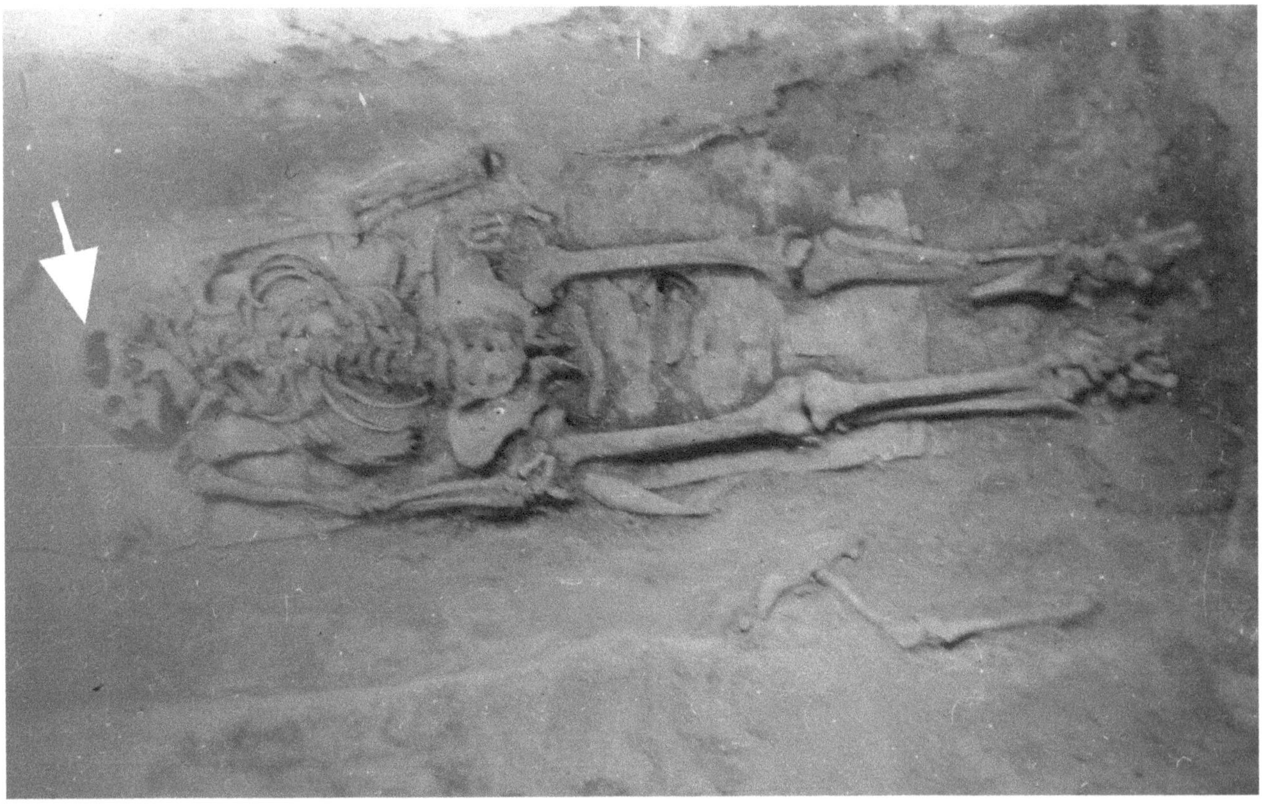

Plate 42: Hunno-Sarmatian period - Burial of Skeleton XXXI. 54 with the cranial injuries evident in the individual apparent in situ (Photographic Archive of the Institute for the History of Material Culture, St. Petersburg).

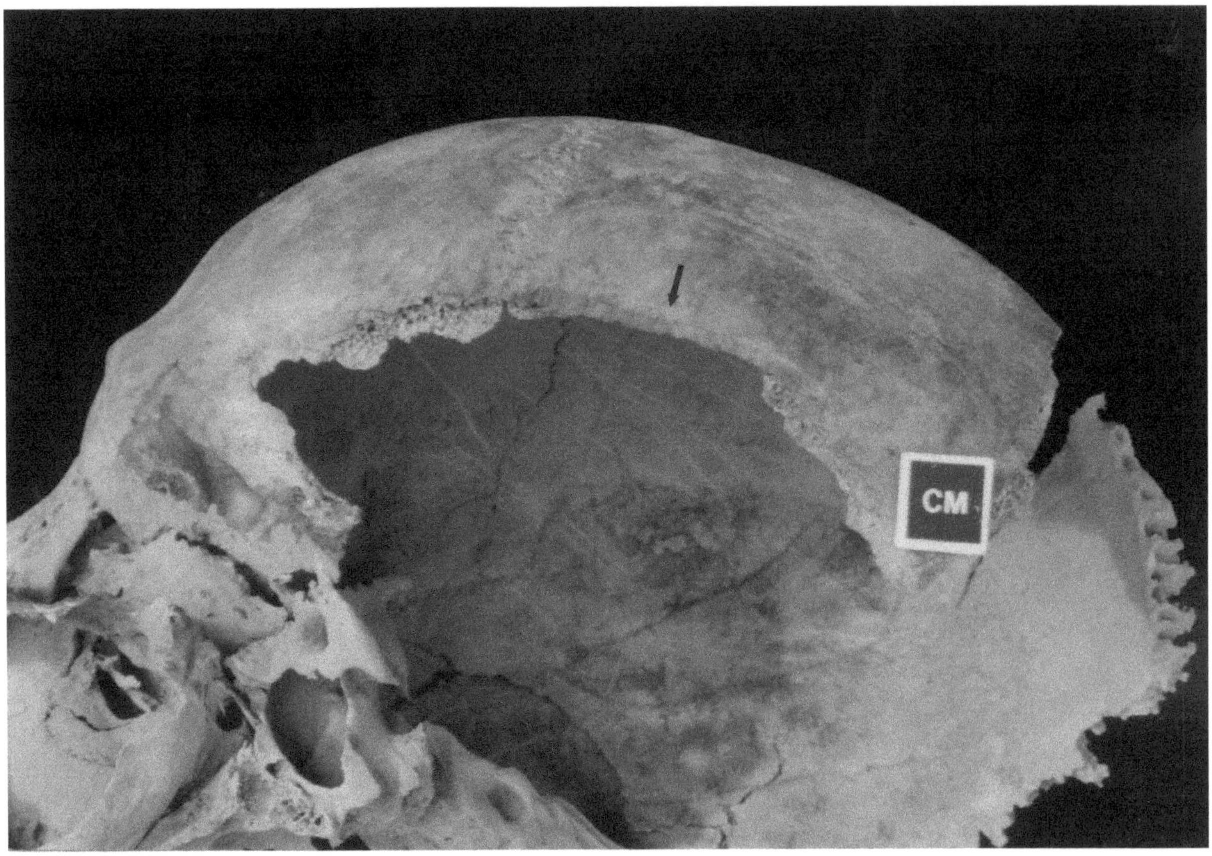

Plate 43: Hunno-Sarmatian period - Sharp sword cut in the Left frontal and parietal of Skeleton XXXI. 54.

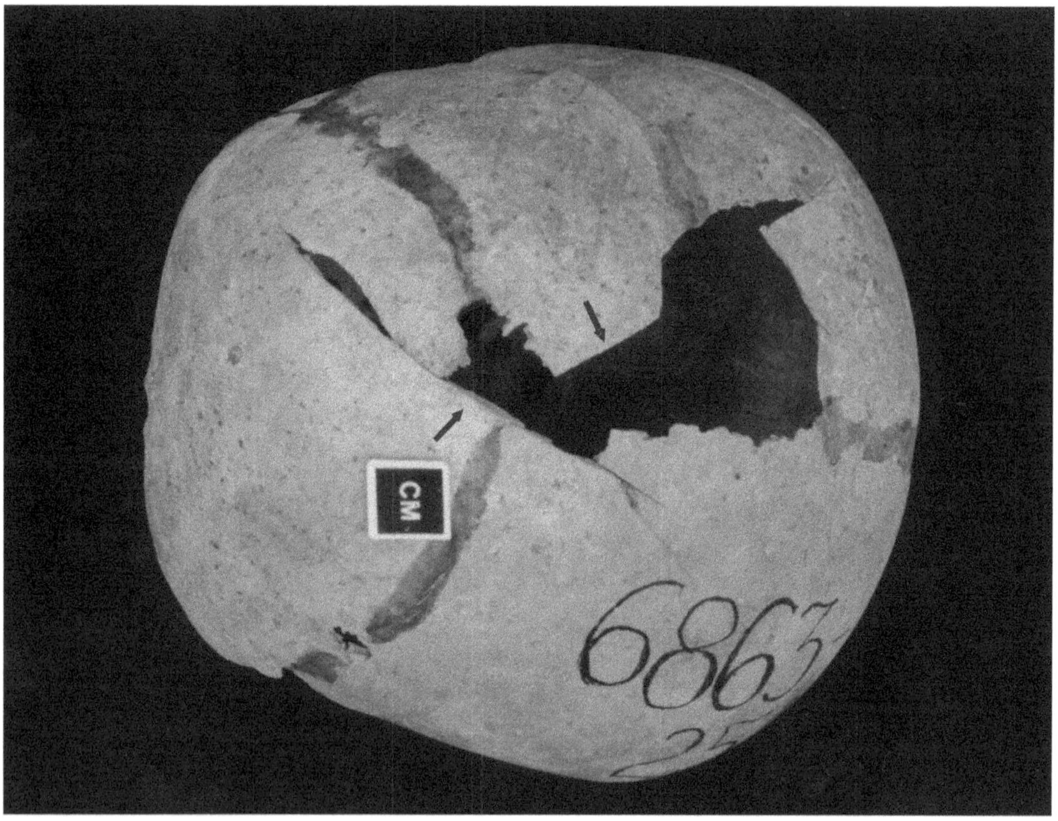

Plate 44: Hunno-Sarmatian period - Posterior view of the two sharp sword cuts in the left parietal and the left parietal-occipital of Skeleton XXXI. 59 (i) a 1-1.5 year old infant.

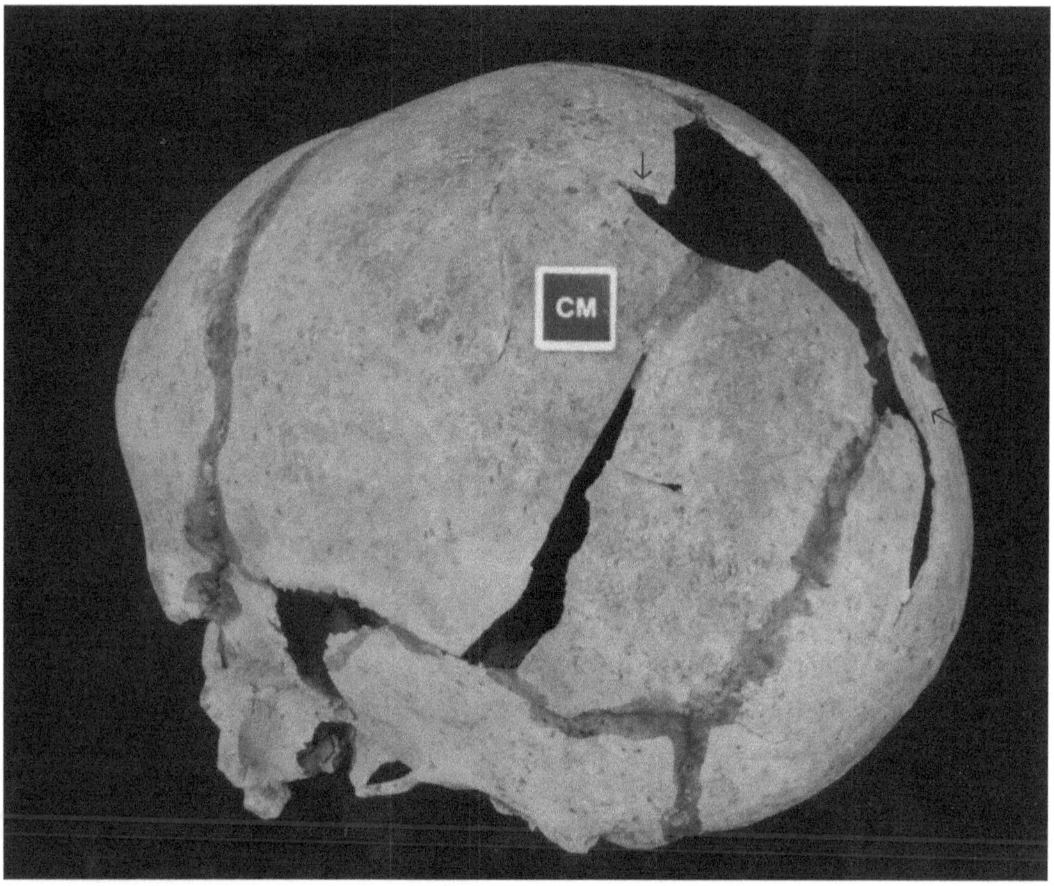

Plate 45: Hunno-Sarmatian period - Lateral view of the two sharp sword cuts in the left parietal and the left parietal-occipital of Skeleton XXXI. 59 (i) a 1-1.5 year old infant.

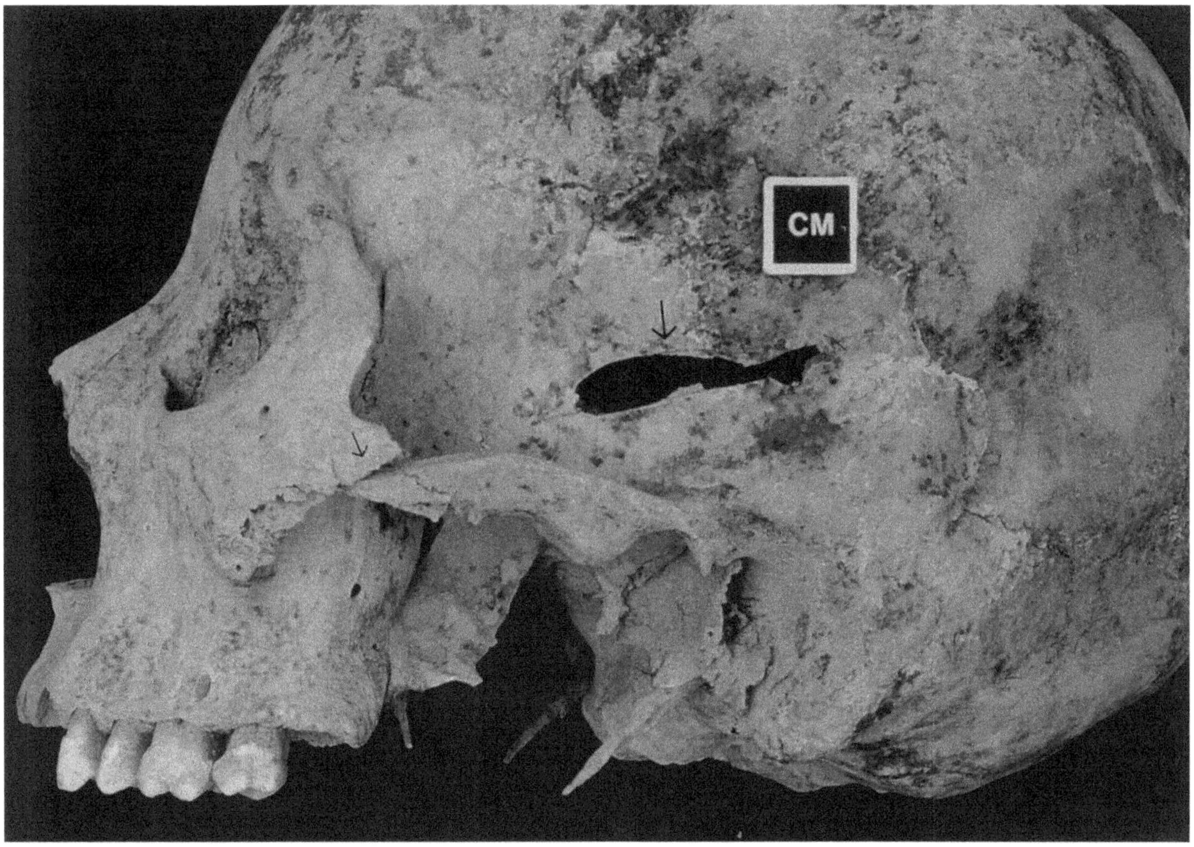

Plate 46: Hunno-Sarmatian period - Sword cuts in the left temporal and zygomatic of Skeleton XXXI. 115, an adult female.

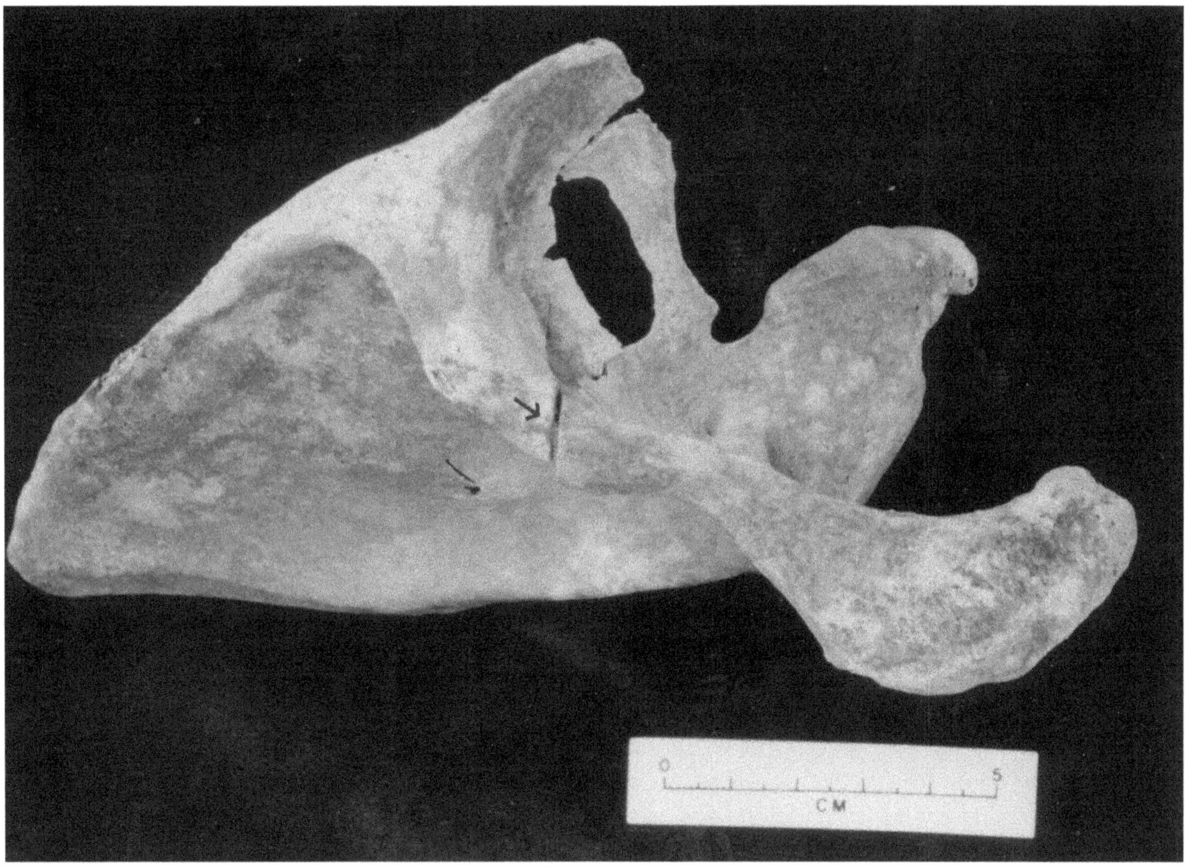

Plate 47: Hunno-Sarmatian period - Sword trauma in the left scapula of Skeleton XXXI. 115, an adult female.

Plate 48: Hunno-Sarmatian period - Sword trauma in the left humerus of Skeleton XXXI. 115, an adult female.

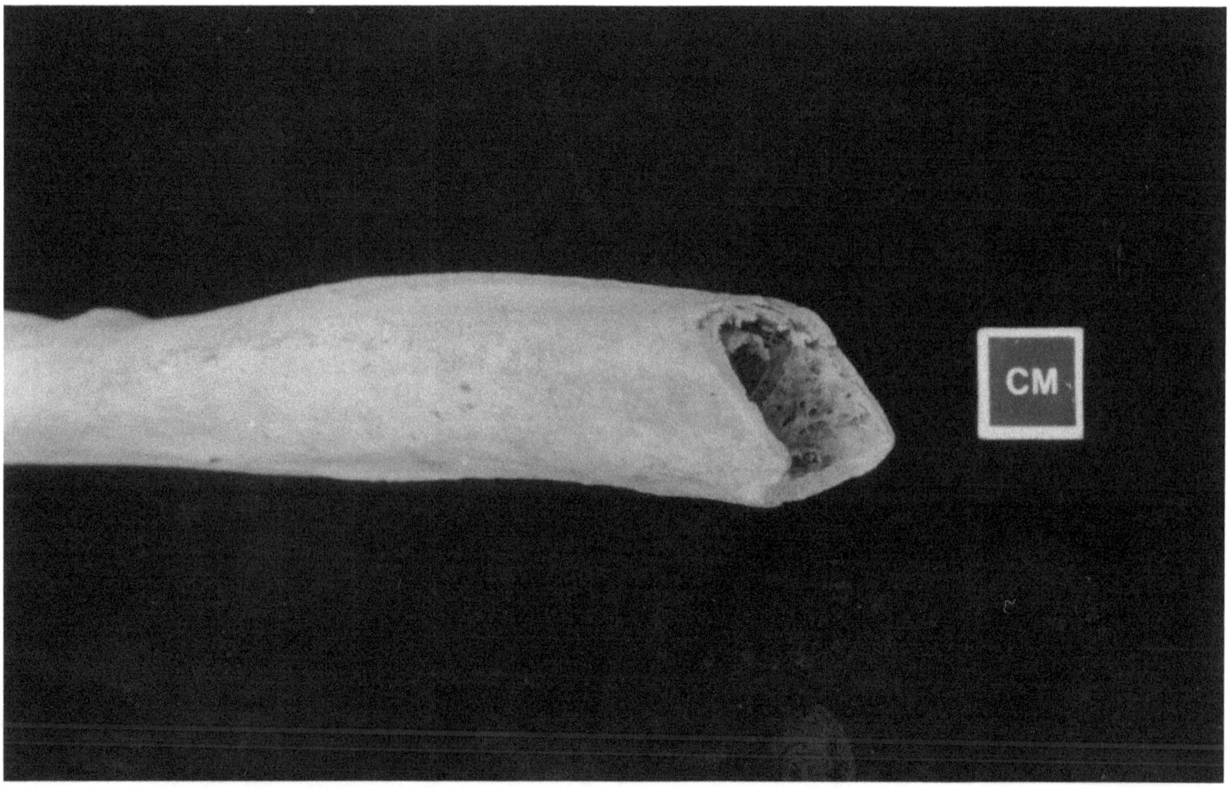

Plate 49: Hunno-Sarmatian period - Sword trauma at the midpoint of the right humerus in Skeleton XXXI. 176.

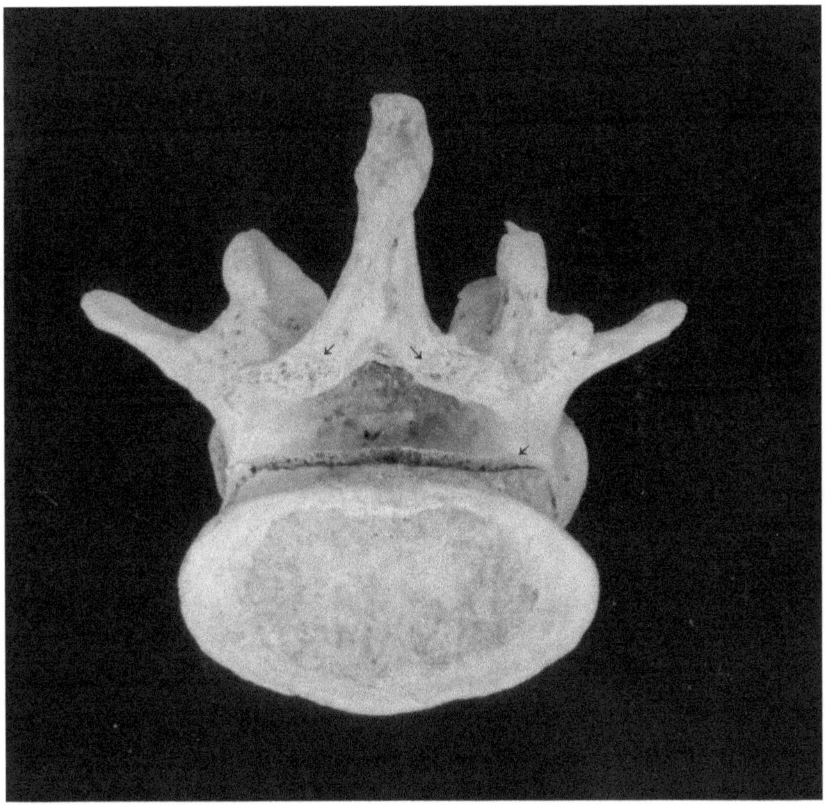

Plate 50: Hunno-Sarmatian period - Sword trauma running transversely across the posterior surface of the body of the fourth lumbar vertebrae in Skeleton XXXI. 176.

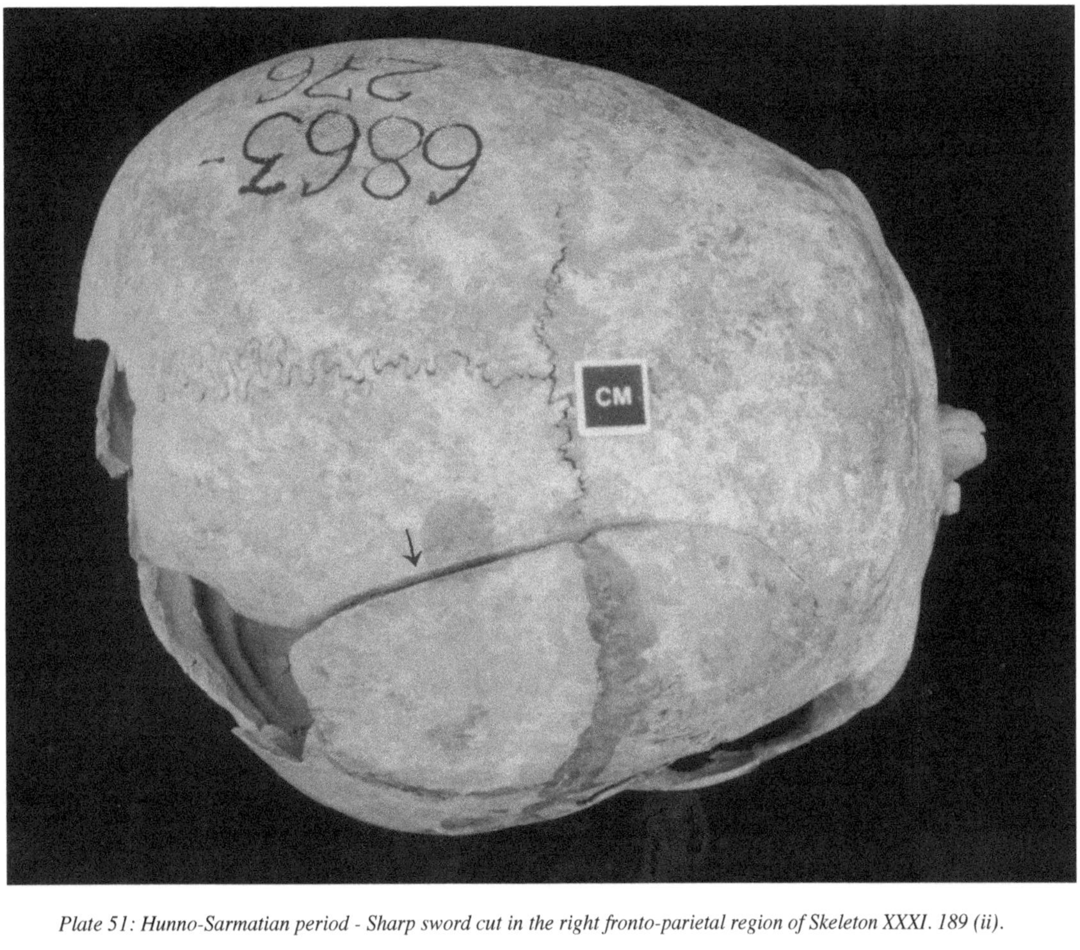

Plate 51: Hunno-Sarmatian period - Sharp sword cut in the right fronto-parietal region of Skeleton XXXI. 189 (ii).

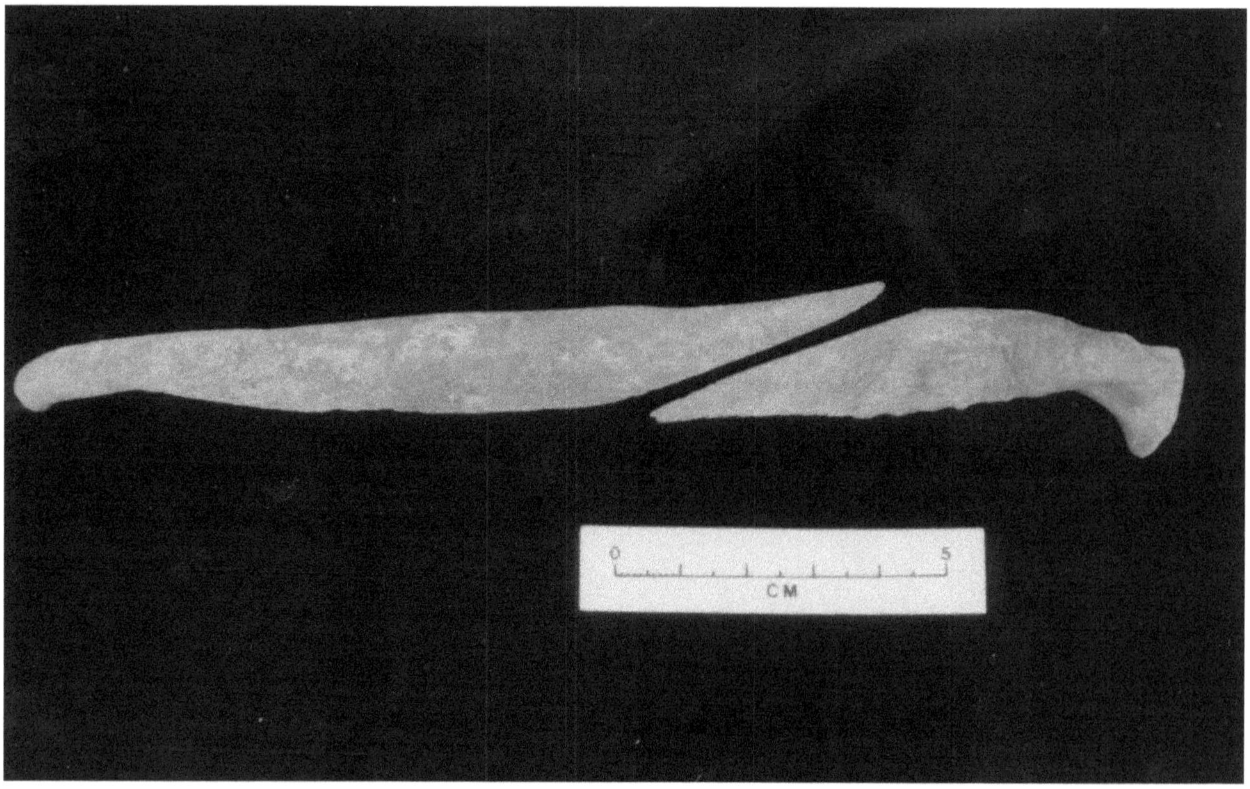

Plate 52: Hunno-Sarmatian period - Sharp sword cut in the left tenth rib of Skeleton XXXI. 189 (ii).

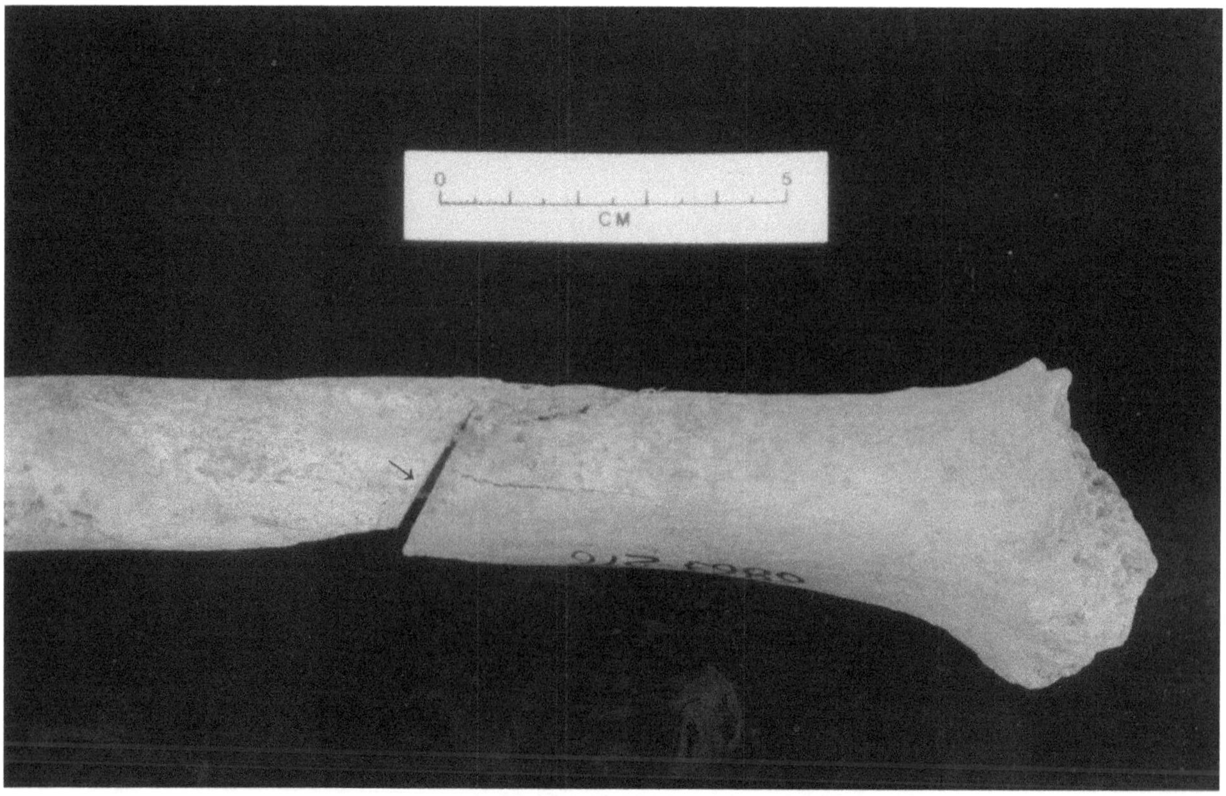

Plate 53: Hunno-Sarmatian period - Sharp sword cut in the left humerus of Skeleton XXXI. 189 (ii).

trauma. In Skeleton XXXI. 115 the lesions are indicative of free-moving combat, and are similar in distribution to the injuries apparent in the male individuals. It is therefore probable that this individual was also a warrior. It is more difficult to interpret the injuries apparent in Skeleton XXXI. 185, the second female, since she only displayed a single sword cut on her right femur. In addition, she had been beheaded (see Section 6.3.1c). It is possible that the injuries had been sustained during combat although it is also possible that the woman had been deliberately and methodically beheaded for punitive reasons or as part of a rite of human sacrifice.

A summary of the prevalence of the injuries in each observable bone in the population group is provided in Table 32 and Figure 5.

Bone	No. trauma	No. observable	% trauma
Frontal	3	156	1.9
Parietal	5	312	1.6
Occipital	1	154	0.6
Temporal	1	305	0.3
Zygomatic	1	299	0.3
Mandible	1	143	0.7
Clavicle	1	320	0.3
Scapula	4	315	1.3
Humerus	3	362	0.8
Radius	2	345	0.6
Femur	4	365	1.1
Rib	2	933	0.2
TV	7	1657	0.4
LV	4	1062	0.4
Sacrum	1	126	0.8

Table 32: Hunno-Sarmatian period individuals - Prevalence of blade trauma in the adult and subadult cranial bones. A single injury which involved two bones is included in the count for both bones. If a bone displayed more than one injury it is only counted once for the purposes of determining the prevalence (TV = thoracic vertebra, LV = lumbar vertebra).

It is evident from Table 32 and Figure 5 that the bones of the upper body generally showed a greater predilection for blade trauma than those of the lower body. This is indicative that the combat was generally undertaken with the opponents fighting on an equal level (eg. on the ground surface or on horseback).

Skeleton XXXI. 108 also displayed fractured nasal bones (see Section 5.3.1b (i)) and a night-stick fracture of the left ulna (see Section 5.3.1a (ii)). In Skeleton XXXI. 157 an ossified haematoma was apparent on the right femur (see Section 5.3.1c), and in Skeleton XXXI. 189 (ii) the left nasal bone had been fractured (see Section 5.3.1b (i)). It is probable that Skeleton XXXI. 108 and Skeleton XXXI. 189 (ii) had been subject to repeated incidents of interpersonal violence, and these two individuals may have been recognised warriors within their society. It was stated in the archive report that an arrowhead had been embedded in one of the vertebra of Skeleton XXXI. 83 (see Section 6.3.1b), while Skeleton XXXI. 176 also displayed knife cuts on the anterior surface of the body of CV2 (see Section 6.3.1d). This would indicate that individuals in the Hunno-Sarmatian period society could be attacked by aggressors using a variety of weapons. In the latter example it is possible that the individual had been stabbed in the throat during the course of combat. Alternatively, once the victim had been extensively weakened by sword injuries his attacker may have methodically slit his throat.

Three individuals displayed signs of possible ritual activity associated with warfare. In Skeleton XXXI. 87. Sk. 1 and Skeleton XXXI. 87. Sk. 2 there were clear signs of scalping (see Section 6.3.1e). As outlined above, the remains of Skeleton XXXI. 189 (ii) displayed evidence of a particularly violent episode of combat. In addition to evidence of sword cuts, the cranium had been smashed and subjected to deliberate and repeated cuts, presumably to enable the extraction of the brain (Murphy forthcoming a) (see Section 6.3.1f).

6.3.1b Arrowhead trauma

Four Hunno-Sarmatian period individuals displayed evidence of trauma inflicted by arrowheads. A summary of the arrowhead injuries observed in the remains of the individuals is presented in Table 33.

It is evident from Table 33 that males, females and subadult individuals displayed evidence of trauma caused by arrow-

Context	Age	Sex	No. injuries	Location	Morphology	Healing	Identification
XXXI. 1	10.5-12.5	-	1	R ilium	trilobate	+	definite
XXXI. 3	25-35	M	2	L scapula	-	+	definite
				CV2*	-	-	probable
XXXI. 64	25-35	F	1	Frontal	-	+	possible
XXXI. 83	17-25	M	1	Vertebra*	-	-	probable

Table 33: Hunno-Sarmatian period individuals - Details of arrowhead injuries in the population (= details obtained from archive report, + = healing present, o = healing not present, - = not applicable/ uncertain, R = right, L = left, CV = cervical vertebra).*

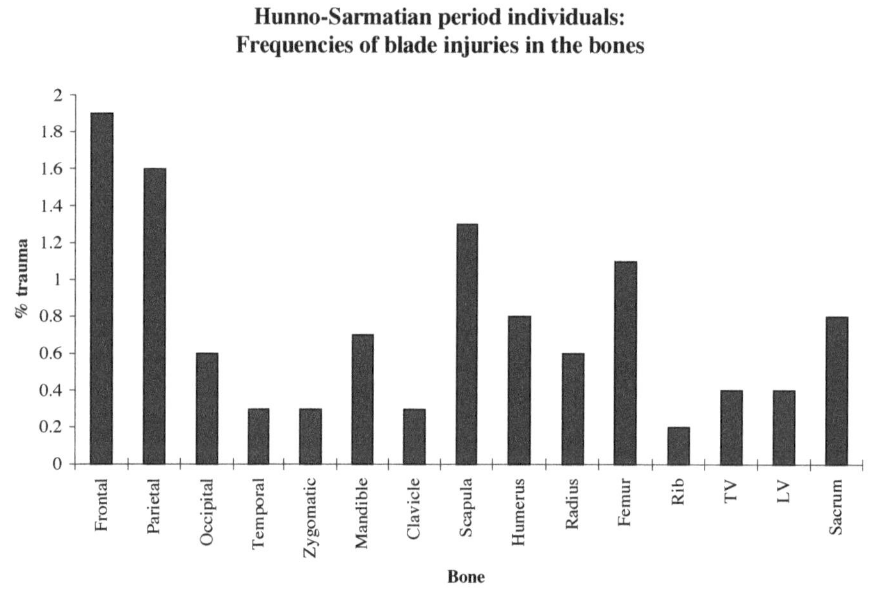

Figure 5: Hunno-Sarmatian period individuals - Frequencies of blade injuries in the bones of the populations.

heads. It was only possible to determine the morphology of the arrowhead which had caused the trauma in Skeleton XXXI. 1, where the injury was clearly made by a trilobate arrow (Plates 54 and 55). In this individual the morphology of the arrowhead was apparent on the posterior surface of the right ilium, indicating that the arrow had been shot by an individual positioned behind the juvenile. The morphology of the arrowhead was also visible on the anterior surface of the ilium, although it was smaller and less distinct, indicating that this perforation represented the exit wound. It is probable that the arrowhead had not gone straight through the bone and that it had lodged in the bone with only the tip protruding through to the posterior surface of the ilium. In Skeleton XXXI. 3 the arrowhead perforation was present at the midpoint of the blade of the left scapula (Plate 56). It would appear that the arrow entered the scapula on the posterior surface, indicating that the arrow was shot from behind the individual. In addition, the angle of the perforation indicates that the arrowhead had probably struck the individual in a downward motion.

All three of the arrowhead injuries which were available for analysis displayed signs of healing. This may indicate that, in general, the arrow was not as effective a weapon for inflicting fatal wounds as the sword (see Section 6.3.1a). In Skeleton XXXI. 1 plaques of woven bone surrounded the arrowhead injury on the anterior surface of the ilium indicating that the individual had survived the weapon trauma for at least a short period of time. Plaques of woven bone were also apparent on the anterior surface of the ilium which were not directly adjacent to the injury, and on the anterior surface of the right ala of the sacrum. These plaques of bone may indicate that the injury had become infected, and that the infection was haematogenously spreading to other parts of the right pelvic/abdominal area. In Skeleton XXXI. 3 a plaque of grey woven bone surrounded the arrowhead injury on the posterior surface of the left scapula suggesting that the individual had survived for a short period of time after this injury. It was stated in the archive report that an iron arrow point had also been protruding from the second cervical vertebra during the archaeological excavation. It is not possible to assess the severity of the vertebral lesion, however, since the cervical vertebra was not available for analysis. It is possible that the injury caused by this second arrow was sustained during the same incident responsible for the arrow wound in the left scapula. Alternatively, it is possible that the individual had been shot in the neck in a further attack shortly after he had sustained the injury which had caused the trauma to the left scapula.

In Skeleton XXXI. 64 a circular depression, which measured 6 mm by 6 mm and had a depth of 2-3 mm, was apparent on the superior surface of the frontal bone immediately to the right of its midpoint. The margins of the lesion were smooth and it did not perforate the inner table of the vault. It is possible that the lesion had been caused by a rounded arrowpoint. The morphology of the lesion indicated that the injury had undergone substantial healing, and it is probable that the individual had survived for a considerable period of time after sustaining the injury.

It was noted in the archive report that an arrowhead had been embedded in one of the vertebra of Skeleton XXXI. 83. Unfortunately, the vertebra was not available for analysis. Evidence of sword and possible knife injuries were also present in the remains of the individual, however, indicating that a variety of weapons had been used during one or more episodes of violence (see Section 6.3.1a). Skeleton XXXI. 3 also displayed a fractured left nasal bone (see Section 5.3.1b (i)) and a night-stick fracture to the left ulna (see Section 5.3.1a (ii). It is probable that Skeleton XXXI. 3 had been subject to repeated incidents of interpersonal violence, and that he was

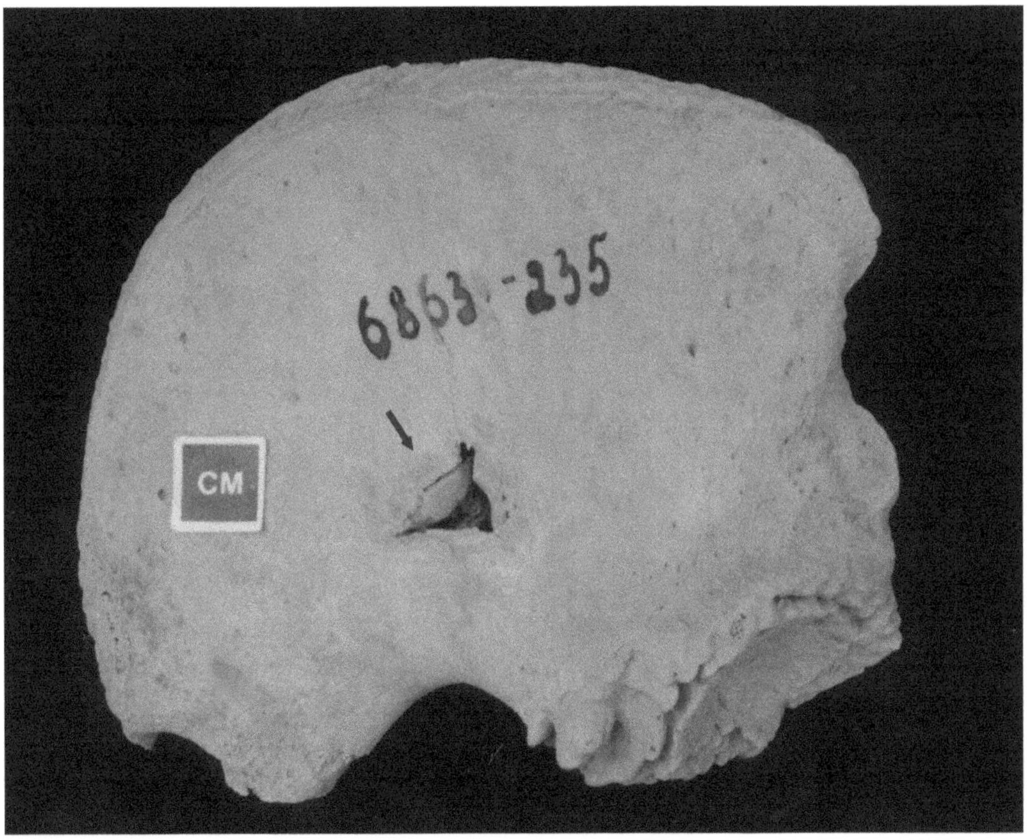

Plate 54: Hunno-Sarmatian period - Injury caused by a trilobate arrowhead on the posterior surface of the right ilium of Skeleton XXXI. 1, a 10.5-12.5 year old juvenile.

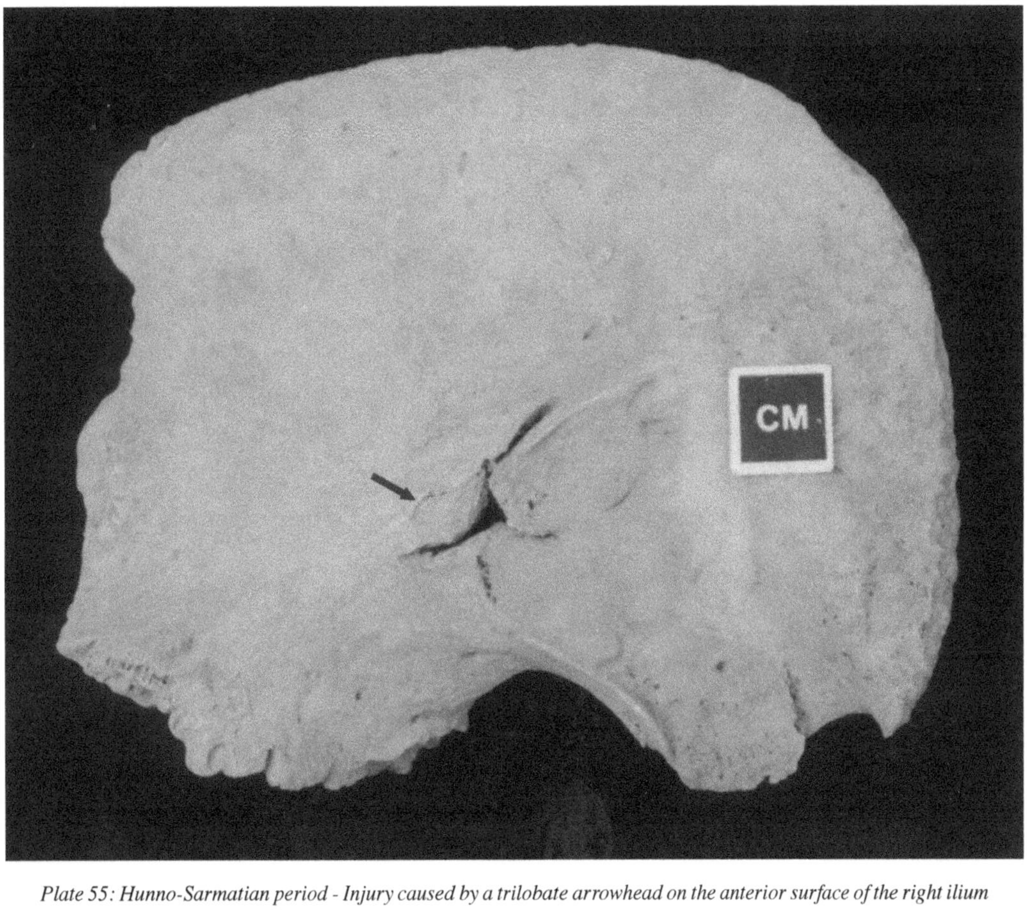

Plate 55: Hunno-Sarmatian period - Injury caused by a trilobate arrowhead on the anterior surface of the right ilium of Skeleton XXXI. 1, a 10.5-12.5 year old juvenile.

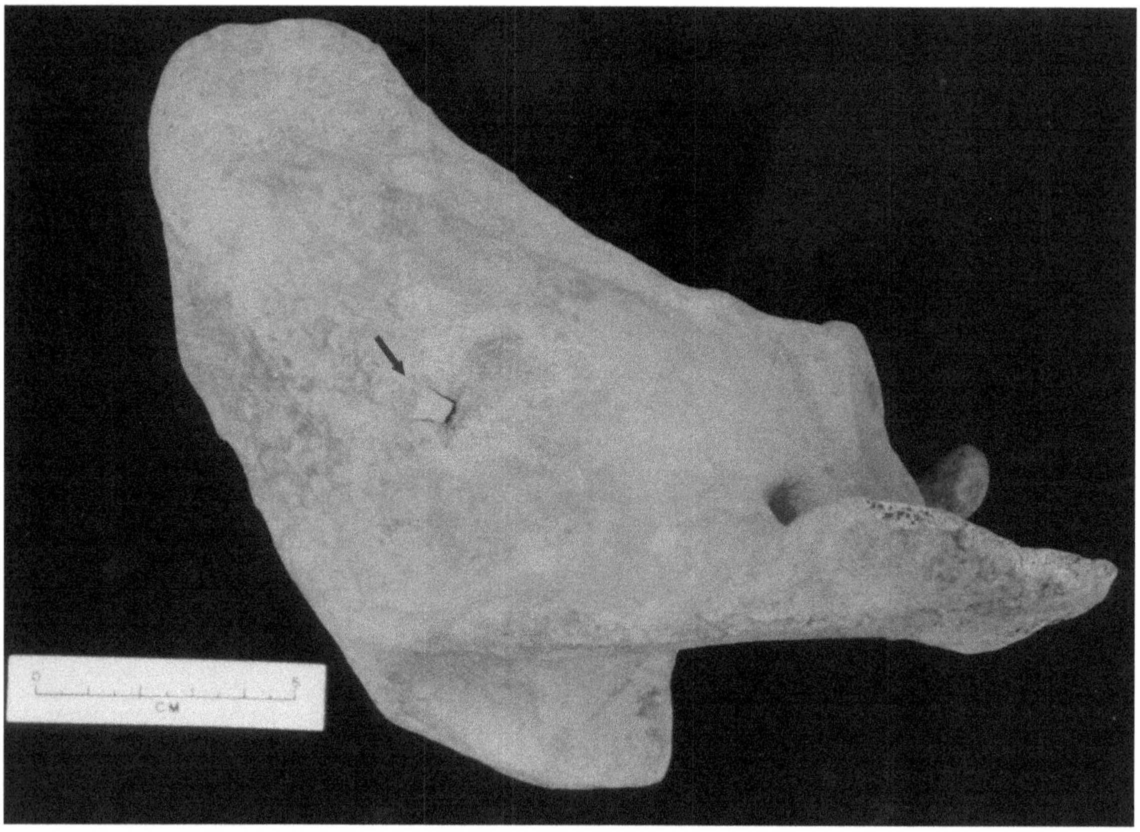

Plate 56: Hunno-Sarmatian period - Arrowhead injury on the posterior surface of the left scapula in Skeleton XXXI. 3.

a recognised warrior in society.

A summary of the prevalence of the arrowhead injuries in each observable bone in the population group is provided in Table 34.

Bone	No. trauma	No. observable	% trauma
Frontal	1	156	0.6
Scapula	1	315	0.3
Vertebra	2	3279	0.1
Ilium	1	358	0.3

Table 34: Hunno-Sarmatian period individuals - Prevalence of arrowhead trauma in the adult and subadult remains.

6.3.1c Decapitation

Five Hunno-Sarmatian period individuals may have been decapitated, providing an approximate frequency of 4% (5/126) of cervical spines with evidence for decapitation. In Skeleton XXXI. 185, a 35-45 year old female, the trauma was definitely indicative of decapitation. A clean, near-transverse sword cut had completely removed the dens of the axis. The posterior margin of the left superior articular facet and the superior surface of the spinous process had also been removed by the blow, exposing the underlying spongy bone. The lateral margin of the right inferior articular facet and the most supero-lateral point of the right transverse process of CV1 had also been removed as a result of the same blow. The weapon used to decapitate the woman must have been extremely sharp, since the evidence indicates that a single blow had been struck to the cervical vertebrae. The angle of the sword chop at the base of the dens indicates that the blow had probably been struck from left to right, with the aggressor positioned posterior to the victim. The skull had been buried with the cadaver, and evidently the head had not been carried away as a trophy. Perhaps the soft-tissue of the right side of the neck had not been completely severed and no further effort was made to remove the head. Alternatively, the individual who had carried out the decapitation may have been prevented from totally removing the woman's head. Other sword trauma was also present on the remains of the individual (see Section 6.3.1a), and it is possible that she was beheaded in the course of combat rather than for punitive or ritual purposes.

The cranium of Skeleton XXXI. 189 (ii), a 17-25 year old male, had also definitely been detached from the remainder of the body in the peri- or post-mortem period. The posterior aspect of the cranium had been bludgeoned, and the bone in the immediate vicinity of the foramen magnum had been smashed open. A number of sharp cutmarks extended from the posterior margin of the damaged foramen magnum (see Section 6.3.1f). In addition, thirteen sword cuts with no evidence of healing were apparent in the remains of the individual (Murphy forthcoming a).

It is stated in the archive report that the skull of Skeleton XXXI. 28. Sk. 2, a 45+ year old male, was not present in the tomb. In addition, the first and second cervical vertebrae of the individual were not present. It was also recorded in the archive report that the skull of Skeleton XXXI. 47, a 25-35 year old male, did not accompany the skeleton in the tomb. The first four cervical vertebrae of the individual were not available for analysis, while the remainder of the vertebrae were present. The archive report also states that the skull of Skeleton XXXI. 61, a 17-25 year old female, was not situated in the tomb. The first and second cervical vertebrae of the individual, however, were present. In all three cases it is highly probable that the individuals had been decapitated prior to burial.

6.3.1d Evidence for deliberate trauma to the throat

Two Hunno-Sarmatian period individuals displayed lesions which may be indicative that they had their throats cut. The frequency of trauma to the cervical vertebrae indicative of throat cutting was 1.6% (2/126). In Skeleton XXXI. 44, an 11-12 year old juvenile, two small transverse cutmarks were present at the anterior margin of the right inferior articular facet of CV1, while transverse cutmarks were also present at the midpoint of the anterior surface of CV2. The cutmarks on CV2 had bevelled margins indicating that they had been made infero-superiorly. A transverse cutmark was also apparent on the anterior surface of the left transverse foramen of CV3. No further traumatic lesions indicative of assault were apparent in the remains and, given the age of the individual, it is feasible to suggest that their throat had been cut during a ritual ceremony.

In Skeleton XXXI. 176, a 25-35 year old male, several tiny cutmarks were present on the anterior surface of CV2, which had probably been made using a very sharp knife. The remains of the individual also displayed six sword cuts with no signs of healing (see Section 6.3.1a). It is possible that the individual had been stabbed in the throat in the course of combat. Alternatively, once the victim had been extensively weakened as a result of his sword injuries, the attacker (or attackers) may have slit his throat. It is probable that the injuries identified in the remains of Skeleton XXXI. 176 were related to an episode of warfare.

6.3.1e Scalping

Two Hunno-Sarmatian period individuals displayed definite evidence of having been scalped (Murphy *et al* 2002). The frequency of scalping in the Hunno-Sarmatian period adult and subadult crania was 1.4% (2/137). Both individuals were buried in the same cist, and it is therefore probable that they had been scalped concomitantly. The cutmarks apparent in Skeleton XXXI. 87. Sk. 1, a 17-25 year old possible male individual (Plates 57 and 58), and in Skeleton XXXI. 87. Sk. 2, a 35-45 year old male (Plate 59), were practically identical in morphology and location. This may indicate that a single individual had carried out both procedures. In both cases numerous small sharp cutmarks were apparent on the surface of the cranial vaults. The cutmarks appeared to follow the circumference of the skull, suggesting that the objective of the scalper had been to carefully remove a circular area of skin from the superior and posterior aspects of the cranium. The cutmarks on the occipital and the posterior regions of the parietals ran in a horizontal direction, parallel to one another. The superior margins of the cutmarks were generally bevelled, suggesting that the person who had made the cutmarks had been positioned superior to the skull, and that the skin had been peeled off in an anterior direction. The cutmarks on the frontal and the anterior right parietal also had bevelled superior margins, which indicates that they had been made in the opposite direction.

Evidence of sword injuries was also present in the remains of both individuals (see Section 6.3.1a). In Skeleton XXXI. 87. Sk. 1 a sword cut was present on the neural arches of TV11 and TV12, indicating that the attacker had been positioned behind the individual. In Skeleton XXXI. 87. Sk. 2 a sword cut was apparent on the right femur. It is probable, therefore, that both of the individuals had been involved in an episode of violence, possibly armed combat, and perhaps following their defeat they had been deliberately and methodically scalped. No signs of healing were associated with the scalping cutmarks on either individuals. It is, therefore, probable that the procedure may have occurred just prior to death or in the peri-mortem period. Given the evidence for weapon trauma on the remains of both individuals it is highly probable that the victor had retained the scalps as war trophies.

6.3.1f Extraction of the brain

Skeleton XXXI. 189 (ii), a 17-25 year old male, displayed lesions which may indicate that his brain was deliberately extracted from his cranium (Plates 60 and 61) (Murphy forthcoming a). The frequency for this type of weapon trauma in the Hunno-Sarmatian period individuals was 0.7% (1/137). Thirteen unhealed sword cuts were also apparent in the remains of the individual, suggesting that he had died in a particularly frenzied attack (see Section 6.3.1a). The posterior aspect of the cranium appeared to have been bludgeoned with a blunt, rounded implement, which had been struck at least four times. One of the blows had struck the lambda, and blows had also been dealt to the areas immediate to the right and to the left of this wound. At least one further blow had been dealt to the area inferior to these three well defined areas of trauma.

The inferior aspect of the cranium in the region of the foramen magnum had been deliberately smashed. The resultant traumatic lesion was oval in shape and extended from the basilar part of the occipital to the right and left styloid processes, ending at the posterior margin of the foramen magnum.

Seven very sharp cutmarks were apparent at the area of the occipital which extended transversely from the posterior margin of the foramen magnum. The cut marks each measured approximately 30-35 mm, and ran supero-inferiorly. The margins of the cutmarks were bevelled, and they appeared to have been made by a right handed person in a supero-inferior motion. The regularity of the cutmarks indicates that they had been made deliberately and methodically. It is probable

Plate 57: Hunno-Sarmatian period - Cutmarks indicative of scalping on the frontal bone and right parietal of Skeleton XXXI. 87. Sk. 1.

Plate 58: Hunno-Sarmatian period - Cutmarks indicative of scalping on the occipital and left parietal of Skeleton XXXI. 87. Sk. 1.

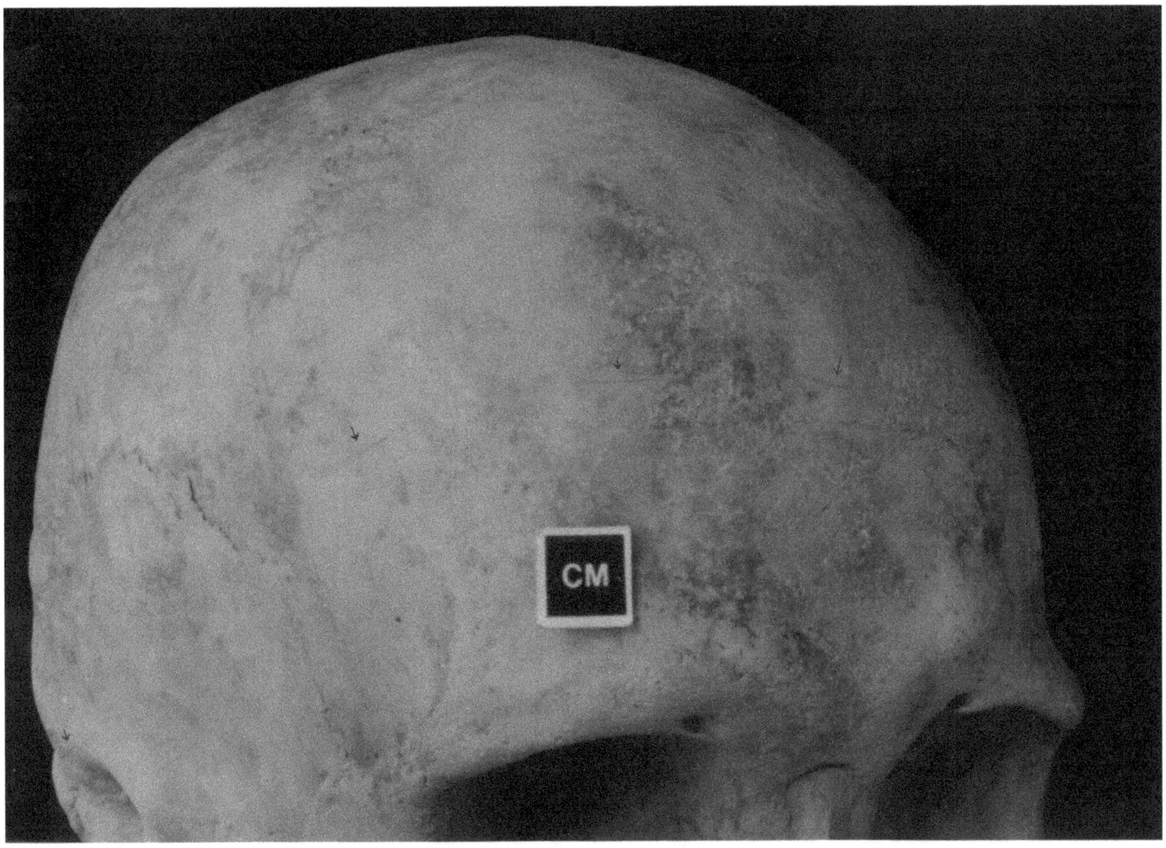

Plate 59: Hunno-Sarmatian period - Cutmarks indicative of scalping on the frontal bone of Skeleton XXXI. 87. Sk. 2.

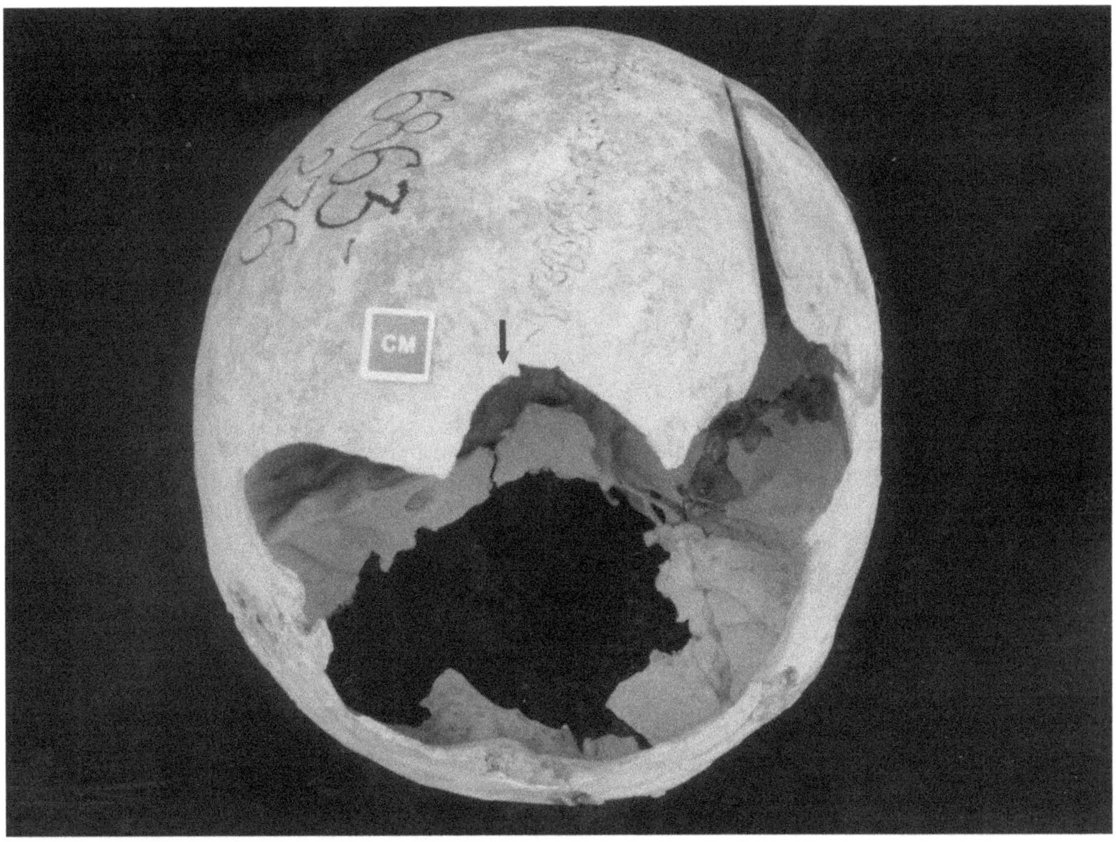

Plate 60: Hunno-Sarmatian period - Evidence of trauma inflicted by a blunt instrument to the occipital and posterior parietal region of Skeleton XXXI. 189 (ii).

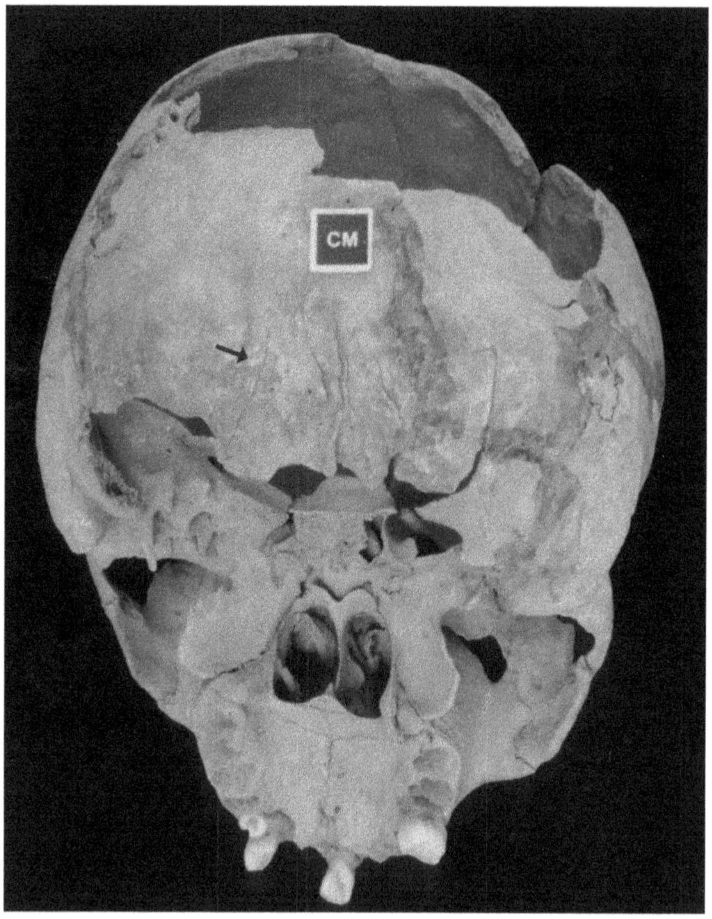

Plate 61: Hunno-Sarmatian period - Evidence of trauma inflicted by a blunt instrument to the occipital and posterior parietal region of Skeleton XXXI. 189 (ii).

that the blunt trauma and the cutmarks at the basioccipital area were made peri-mortem.

The most likely objective of this activity was the extraction of the brain from the victim's cranium. The crudeness of the perimortem trauma and the appearance of numerous examples of sword cuts on the skeleton would suggest that the extraction of the brain was an aggressive post-combat activity. The brain may have been extracted as part of the rituals associated with warfare, although a human brain would not have made a particularly effective war trophy, since it would decompose relatively quickly unless it was chemically treated with, for example, lime. If a trophy was required one would expect the aggressors to have taken the entire head or scalp (see Section 6.3.1e). Consequently, alternative motives for the extraction of the brain must be considered. One possible explanation is the practice of ritual cannibalism which involves the consumption of specific body parts for the purpose of acquiring the special qualities of the deceased (Flinn *et al* 1976, 308). The Zulus, for example, believed that if they ate the forehead and eyebrows of their enemies they would acquire the ability to look unflinchingly in the face of adversity (Masters 1993, 277).

6.4 Summary

The majority of instances of weapon trauma evident in the remains of the Scythian period individuals had been inflicted with a pointed battle axe, with injuries visible on 2.5% of adult and subadult crania, and located on 1.3% of frontals, 1% of parietals, 0.2% of occipitals, 0.2% of temporals and on 0.1% of zygomatics. Seventy-five percent of individuals with pointed axe injuries were male, 17% were subadults and 8% were female. Pointed axe injuries were also present on the remains of a 4-5 year old child and a 7-10 year old juvenile. Since the majority of the injuries were located on either the parietals (40%) or the frontal bone (35%), this may indicate that the opponents were facing each other during the episodes of violence. The majority of injuries on the parietals were situated on the left side, possibly indicating face-to-face combat with a right handed assailant, and 80% displayed no signs of healing. Three Scythian period adult males displayed evidence of blade trauma. The injuries were visible on 0.2% of humeri, 0.2% of femora, 0.8% of sacra, 0.1% of parietals and 0.1% of zygomatics. None of the examples of sword trauma displayed signs of healing.

A single Scythian period child displayed lesions indicative of decapitation, providing a frequency of 1% for the trauma in the observable cervical spines. The child also displayed evidence that his or her throat had been cut. It is possible, therefore, that both the beheading and throat cutting had been undertaken for ritual purposes rather than during combat. Evidence for throat cutting occurred in the Scythian period cervical spines with a frequency of 1%. A single Scythian period adult male displayed evidence of having been scalped, and fractured nasal bones were also apparent in his cranium. Two Scythian period individuals displayed evidence for trepanation, providing a prevalence of 0.4% for the population group. In one of the individuals the perforation was situated on the left side of the frontal bone and displayed signs of healing. In the other individual, however, the perforation was located on the bregma and appeared to have been undertaken posthumously. It is probable that the trepanation in the living individual had been undertaken for magico-therapeutic motives. The trepanation of the dead individual may have been for ritual or trophy purposes, where the head would have been mounted or hung for display. A single Scythian period individual displayed possible evidence for the amputation of his left forearm. The distal humerus displayed signs of healing indicating that the individual had survived the procedure. It is not possible to ascertain whether the procedure had been undertaken deliberately for medical or punitive motives, or if the forearm had been amputated in a domestic accident or during warfare.

The majority of examples of weapon trauma apparent in the remains of the Hunno-Sarmatian period individuals were blade injuries, with 12 skeletons displaying these injuries. The sword trauma occurred with greatest frequency on the frontal bones (1.9%), followed in frequency by the parietals (1.6%), the scapula (1.3%) and the femora (1.1%). It is evident that the bones of the upper body displayed a greater predilection for blade trauma than those of the lower body. Seventy-five percent of the individuals who displayed sword trauma were male, with 17% female and 8% subadults. Sword cuts were also recorded on the cranium of a 1-1.5 year old infant. In ninety-two percent of affected individuals the trauma was not healed, and 58% displayed multiple sword injuries.

Four Hunno-Sarmatian period individuals displayed evidence of arrowhead trauma. The affected individuals included two males, a female and a 10.5-12.5 year old juvenile. Evidence of healing was observed in the three arrowhead injuries which were available for analysis. Arrowhead injuries had a frequency of 0.6% in the frontal bone, 0.3% in the scapula, 0.3% in the ilium and 0.1% in the vertebrae.

Five Hunno-Sarmatian period individuals appear to have been decapitated. In three of the cases it was noted in the archive report that the crania of the individuals (two males, one female) were not present in the tombs where they were buried. The remains of the decapitated female available for analysis showed clear evidence of decapitation on the cervical vertebrae, while the cranium of the adult male displayed modification of the foramen magnum which could not have been undertaken while the head was still attached to the body. An approximate frequency of 4% for decapitation was calculated on the basis of the observable cervical spines. Two Hunno-Sarmatian period individuals displayed evidence for deliberate trauma to the throat. One of the individuals was a subadult, and the other individual was an adult male. Evidence for throat cutting occurred in the Hunno-Sarmatian period cervical spines with a frequency of 1.6%. Two Hunno-Sarmatian period individuals displayed clear signs of having been scalped. Both individuals were adult males, both displayed evidence of blade trauma and both had been buried in the same cist. It is probable, therefore, that the individuals had been scalped so that the victors could obtain war trophies. Scalping was observed in the complete adult and subadult crania with a frequency of 1.4%. The remains of a single Hunno-Sarmatian period individual displayed evidence that his brain had been deliberately extraction from his cranium. The individual also displayed numerous examples of weapon trauma and it is probable that the brain had been extracted as part of a ritual associated with warfare. Extraction of the brain occurred in the Hunno-Sarmatian period population with a frequency of 0.7%.

The majority of instances of weapon trauma in the remains of the Scythian period individuals had been inflicted with a pointed battle axe, whereas none of the examples of weapon trauma recorded among the Hunno-Sarmatian period individuals had been made with such implements. Conversely, the majority of examples of weapon trauma apparent in the remains of the Hunno-Sarmatian period individuals were blade injuries, whereas these occurred relatively infrequently among the Scythian period population. Arrowhead trauma was occasionally present in the remains of Hunno-Sarmatian period individuals, but no such injuries were recorded for the Scythian period. In both population groups the majority of individuals displaying weapon trauma were male, although female and subadult individuals also displayed examples of weapon trauma, which may indicate that warfare was not an exclusively male activity. It must also be considered, however, that the majority of individuals of both sexes and all ages were simply the victims of warfare rather than actual warriors.

Evidence for decapitation was displayed on 1% of cervical spines of Scythian date and approximately 3.2% of cervical spines from Hunno-Sarmatian contexts. The Scythian period individual was a subadult, whereas all four Hunno-Sarmatian period individuals were adults. In addition, the Scythian period child also displayed evidence that his or her throat had been cut, and evidence for throat cutting occurred with a frequency of 1% in the Scythian period cervical spines. Two Hunno-Sarmatian period individuals also displayed evidence that their throats had been cut, providing a frequency of the trauma of 1.6% of the cervical spines. One of the Hunno-Sarmatian period individuals was a subadult. It is possible that the subadults from both periods who had their throats cut may have been killed for ritual purposes. Two Hunno-Sarmatian period individuals (1.4%) had been scalped, while a single Scythian period individual displayed evidence for this practice. In addition, a single Hunno-Sarmatian period individual (0.7%) displayed evidence which indicated that his brain had been extracted, but this practice was not apparent among the Scythian period remains.

None of the Hunno-Sarmatian period individuals displayed evidence of surgical procedures, while two Scythian period individuals displayed evidence of trepanation. One of the Scythian period individuals had probably been trepanned posthumously for ritual purposes, whereas the procedure appeared to have been undertaken on the other individual for magico-therapeutic motives. In addition, a single Scythian period individual displayed possible evidence for the amputation of his forearm.

7. Discussion

7.1 Introduction

The objective of the current chapter is to consider the results obtained during the palaeopathological study, and to compare the findings of the skeletal analysis with information derived from the material culture at Aymyrlyg, the data contained in historical sources pertaining to the Scythian and Hsiung-nu worlds, and the results of ethnographical research on Siberian populations. Apart from the research undertaken by Rokhlin (1965) very little work has been carried out on the palaeopathological characteristics of archaeological populations from Russia and, in particular, south Siberia. Consequently, it has been necessary to utilise data obtained from studies of archaeological skeletal populations from different geographical locations and various time periods for comparative purposes.

A wide variety of traumatic lesions and injuries were observed among both the Scythian and Hunno-Sarmatian period populations. Although the individual frequencies for each type of lesion and bone affected were generally low, when the levels of trauma apparent in the remains are examined cumulatively they attest to the active and warfaring existence of these peoples. Throughout this discussion the term 'fracture' will be applied only to injuries with no clear evidence of weapon trauma, while injuries which were clearly inflicted by a weapon will be referred to as 'weapon injuries'.

7.2 Fractures of the appendicular skeleton with no clear association with interpersonal violence

Fractures were present in 0.9% of all Scythian period and 0.8% of all Hunno-Sarmatian period adult long bones (see Section 5.4). In both groups the ulnae displayed the greatest frequencies of fractures. With the exception of night-stick fractures of the ulnae and possibly boxer's fractures of the fifth metacarpals, the majority of fractures of the appendicular skeleton were characteristic of those obtained as a result of accidents, particularly those caused as a result of falls. Further evidence of injuries having been obtained as a consequence of severe falls presented themselves in the form of a Scythian period male with a dislocated glenohumeral joint and a number of fractures of the appendicular skeleton. In the Hunno-Sarmatian period population compression fractures of the vertebrae occurred with a frequency of 0.7%. It is probable, again, that the majority of these fractures were sustained as the result of heavy falls. The majority of individuals with fractures of the appendicular skeleton were males in both the Scythian (69%) and Hunno-Sarmatian period (64%) populations. This finding indicates that the males of both societies probably engaged in a physical existence which resulted in them being more prone to accidents, particularly falling injuries, than the majority of females.

This finding is particularly interesting in light of Herodotus' and the Pseudo-Hippocratic text's descriptions of male and female activities within Scythian period society. We know that both the Scythian and Hunno-Sarmatian period populations of Tuva were semi-nomads and that they relied heavily upon their horses and oxen for transport across the steppe-lands. The Pseudo-Hippocratic text relates how the Scythian females stayed in the ox-drawn wagons during the seasonal migrations, while the males rode on horseback followed by their herds (Chadwick and Mann 1978, 163). In the account of the Amazons that is presented in the same text the differentiation between males and females is further implied since the Amazons engaged in activities such as horse riding which the Scythian females did not undertake (*ibid.*, 162-163). Herodotus stated that the Scythian females 'stayed at home in their wagons occupied with feminine tasks' (Sélincourt and Burn 1972, 307-308). Perhaps males of both periods were engaged in horse riding to a greater extent than females. Modern studies have indicated that the majority of horse riding injuries are caused by falls (Williams 1988, 140). It is impossible, however, to ascertain whether a fall in antiquity had occurred as a consequence of horse riding or as a result of some other accident. Given the semi-nomadic lifestyles of both populations, however, horse riding is a plausible explanation for the occurrence of at least some of the falling injuries apparent in the remains of these individuals. Although a smaller proportion of females than males displayed fractures of the appendicular skeleton, it must be stated, however, that the occurrence of a number of females with injuries indicative of falling suggests that males were not exclusively engaged in activities which would have resulted in accidental trauma.

7.3 Other examples of trauma with no clear association with interpersonal violence

Other traumatic injuries which attested to the active physical lifestyles of both the Scythian and Hunno-Sarmatian period populations included the occurrence of spondylolysis, clay-shoveller's fractures, ossified haematomas, myositis ossificans traumatica, enthesophytes and osteochondritis dissecans (see Section 5.4). The majority of these lesions occurred in very small frequencies in both populations. Spondylolysis, however, occurred with a frequency of 13% among Scythian period individuals and 5% among Hunno-Sarmatian period individuals. In both populations females were more frequently affected than males. Although genetic predisposition can lead to spondylolysis, it is also thought to occur as a result of repeated stresses and strains on the back. Up to 50% of early modern and modern Inuit may display spondylolysis, a finding which has been explained as a consequence of their difficult and strenuous physical existence (Merbs and Wilson 1960, 160). The occurrence of a greater frequency of spondylolysis in the females of both Aymyrlyg population groups suggests that, contrary to the assertions of the historical accounts relating to the activities of female Scythians, the women did not spend all their time sitting around in wagons but that they were also engaged in heavy physical labour (see Section

2.3.10f). The greater overall frequency of spondylolysis in the Scythian period individuals may indicate that the Scythian lifestyle had involved a greater level of arduous physical tasks than that of the Hunno-Sarmatian period.

A single Scythian and a single Hunno-Sarmatian period individual displayed clay-shoveller's fractures. The affected Scythian period individual was male, while the Hunno-Sarmatian period individual was female. The clay-shoveller's fracture is commonly found in individuals who undertake heavy physical exertion which involves lifting heavy weights (Pathria 1995, 2861). The Hunno-Sarmatian period female with the clay-shoveller's fracture also displayed spondylolysis. The occurrence of these lesions again attests to the heavy physical lifestyle of individuals of both sexes in the two population groups.

7.4 Fractures which may have been related to interpersonal violence

Fractures which had probably been attained as a result of interpersonal violence were observed in individuals from both the Scythian and Hunno-Sarmatian periods. Depressed fractures of the cranial vaults were observed in 1.7% of the Scythian period population's crania, whereas these injuries were not apparent in any of the Hunno-Sarmatian period individuals. Twenty-four percent of the affected individuals were female, 38% were male and 38% of individuals with depressed fractures were subadults. The majority of fractures were present on the left side of the frontals or parietals and this is of note since the head is often the main target for attack in interpersonal and intergroup violence (Roberts and Manchester 1995, 79). The left front-parietal region of the cranium is the site at which an injury would commonly be afflicted in hand-to-hand combat when faced by a right handed opponent (Wakely 1997, 26). It is probable, therefore, that the majority of depressed fractures evident in the Scythian period crania had been deliberately inflicted, possibly by blunt implements.

Fractures of the facial area and the mandible were observed in the remains of individuals from both periods. A greater variety of these forms of injuries were evident in the Scythian period individuals, although the most common facial injury in both groups were fractured nasal bones. These occurred with a frequency of 3.4% among the Scythian period population, and 6% among the Hunno-Sarmatian period population. Males more commonly displayed fractured nasal bones than females in both population groups. The occurrence of depressed cranial fractures and a greater variety of facial injuries present among the Scythian period population suggests that interpersonal face-to-face violence may have been more common during this period relative to the Hunno-Sarmatian period. The greater frequency of fractured nasal bones in the Hunno-Sarmatian period individuals, however, suggests that this is unlikely to be the case. It might be suggested that the differences observed between the groups was due to the use of blunt implements during interpersonal conflict among Scythian period individuals, while the Hunno-Sarmatian period individuals only employed their fists in aggressive activities.

The occurrence of head fractures, night-stick fractures of the ulnae and boxer's fractures of the 5th metacarpals all attests to the occurrence of interpersonal and intergroup conflict in both the Scythian and Hunno-Sarmatian period populations. In the majority of cases the affected individuals were adult males, although interpersonal injuries were not exclusively restricted to these individuals and a number of subadults and adult females also displayed these types of injuries. Indeed, one Scythian period woman displayed multiple fractures - including a night-stick fracture of the left ulna, fractured right metacarpals and a fractured rib - which may indicate that she had procured these injuries during a violent incident.

Studies of other archaeological populations have indicated that, in general, the fracture levels present in archaeological groups are much lower than those apparent among modern populations. It has been suggested that the technological advances of the modern world, particularly the existence of motor vehicles, has resulted in modern populations sustaining such high levels of fractures (Roberts and Manchester 1995, 75). Angel (1974) undertook a survey of the fracture prevalences in Greek and Turkish skeletal populations dating from the 7th millennium BC to the 2nd century AD. In all periods males more frequently displayed fractures than females. The overall prevalences of fractures ranged from 1% to 3.6% (*ibid.*, 19). Night-stick fractures of the ulnae were common in all periods, and fractures of the nasal bones occurred with an overall frequency of 0.2% in females and 5.6% in males (*ibid.*, 11, 15). A survey of six British Medieval sites indicated that fracture prevalences ranged from 0.3% to 6.1%. The 6.1% frequency observed in one of these samples was notably higher than the prevalences observed in the other studies, and it has been suggested that the results reflect a small sample size (Roberts and Manchester 1995, 75). The frequencies of fractures evident in the Scythian and Hunno-Sarmatian period populations from Aymyrlyg, therefore, are well within the expected range for an archaeological population.

7.5 Weapon trauma

Further evidence for the occurrence of interpersonal and intergroup violence was evident among both the Scythian and Hunno-Sarmatian period skeletal remains in the form of a wide variety of weapon injuries. Evidence for the use of different weapons in the two cultures was clearly evident in the morphology of these injuries. The majority of Scythian period weapon injuries had been made with pointed battle axes (*chekany*), whereas the majority of examples of Hunno-Sarmatian period weapon trauma indicated the use of large bladed objects, probably swords. None of the Hunno-Sarmatian period individuals displayed lesions caused by a battle axe. Sword trauma occurred with an overall frequency of 0.1-0.8% in the Scythian period remains, whereas these lesions were evident with a frequency of 0.2-1.9% among the Hunno-Sarmatian period individuals. Arrowhead injuries were not observed in the remains of the Scythian period individuals, while they occurred with a frequency of 0.1-0.6% among the Hunno-Sarmatian period individuals.

It is interesting to note that one of the male individuals recovered from Kurgan 2 at Pazyryk in the High Altai displayed at least two pointed axe injuries on his cranium (Rudenko

1970, 219). In addition, the seven horses recovered from the same kurgan had each been killed with a single blow from a pointed axe which had been applied to the superior aspects of their crania (Jettmar 1951, 175). The remains of other horses of the Pazyryk Culture which had been killed by single blows from pointed battle axes were also recovered during the 1993 excavations at Ukok Plateau in south Siberia (Polosmak 1994, 97). In addition, a number of skulls recovered from the Chowhougou cemetery in western China displayed inuries which has been made using perforated battle axes (Murphy and Davis-Kimball 2001). The tribes buried at all of these sites would have been contemporary neighbours of the Uyuk Culture in Tuva. The occurrence of pointed axe injuries among the dead of these separate cultures may attest to the wide-spread use of the pointed battle axe amongst the tribes of the Scythian World. Artistic depictions of warriors and battle scenes from the Scythian World are common, and a particularly clear depiction of a bearded warrior with a pointed battle axe occurs on a bronze vessel recovered from Kurgan 3 of the Tchastye Kurgans, situated near the town of Voronezh in south Russia. The vase has been dated to the 4th century BC (Fig. 6).

Figure 6: Depiction of a bearded warrior with a pointed battle axe on a bronze vessel of 4th century BC date from Kurgan 3 of the Tchastye Kurgans near Voronezh in South Russia (drawn by Libby Mulqueeny after Schiltz 1994, 390).

All of the Scythian period examples of battle axe trauma occurred on the bones of the crania, and an overall frequency of 2.5% was calculated for these injuries. Seventy-five percent of the individuals with axe trauma were male, 17% were subadults and 8% were female. All three of the Scythian period individuals with sword trauma were male. Seventy-five percent of the Hunno-Sarmatian period individuals with sword trauma were male, 17% were female and 8% were subadults. The four Hunno-Sarmatian period individuals with arrowhead trauma included adult males and females as well as subadults. Although males were predominantly involved in warfare the distribution of weapon trauma in individuals of both periods indicates that females and subadults were not exempt from injury. The occurrences of battle axe trauma in two Scythian period children, who were both less than ten years of age, and clear sword cuts on the cranium of a 1-1.5 year old Hunno-Sarmatian period infant may indicate that not all of the individuals with weapon trauma were killed in battle but that they may have been killed when their camps or wagons were attacked by aggressors. There is, however, no justification for assuming that females and older subadults would not have deliberately participated in the practice of warfare.

Both the archaeological and the historical evidence attests to the existence of female warriors in these semi-nomadic tribes, and a number of burials of definite female warriors have been identified in the funerary record. The majority of burials of armed females have been recovered amongst those of the ordinary classes of the population, rather than in tombs of the upper strata of society. In a number of cemeteries burials of female warriors represented as many as 37% of the dead. In the majority of cases the women were only buried with arrows, as was the case for the males from the ordinary level of society, although some female burials have been found to contain a greater range of weaponry including arrows, spears and, on occasion, swords (Melyukova 1990, 106). The tales of the Amazonian female warriors recounted in both Book IV of Herodotus' *Histories* and the Pseudo-Hippocratic text, and the works of other Classical authors evidently were based on some kernel of truth, hence their repetition, and it would seem probable that there were indeed female warriors amongst the tribes of the steppes. The existence of historical warrior queens like the Scythian Queen Tomyris and the Sacian Queen Zarinaea would also suggest a historical reality for the existence of such women (Gera 1997, 6-9, 89). There is also evidence that Hsiung-nu females were involved with warfare (see Section 2.4.9f). An account of the 1st century BC battle at the Talas fortress by Chinese troops records that Hsiung-nu women fought alongside their male counterparts and a group of Roman mercenaries in defence of the citadel. According to the account the women were the last to leave their posts (Ishjamts 1994, 164).

When we look at the evidence for the participation of subadults in warfare the Chinese sources indicate that Hsiung-nu boys (no mention is made of girls) were trained in the art of warfare from an early age. Young boys initially learned how to ride sheep and shoot small animals with bows and arrows. At a slightly older age they were trained to shoot foxes and hares for food. Consequently, all of the young men of a tribe were familiar with the use of bows and arrows, and were capable of forming an armed cavalry during periods of social unrest (Watson 1961, 155).

It was not possible to identify the burials of warriors on the basis of an examination of the grave goods associated with each of the individuals, since evidence for episodes of grave robbing at the Scythian period tombs seem to indicate that the main objectives of the plunderers had been to remove weaponry (Mandelshtam 1983, 28). The corrosion of iron artefacts in the Hunno-Sarmatian period tombs resulted in these objects being unidentifiable in many cases. In addition, the archive reports for several years of the excavations were unavailable for consultation. Consequently, the grave accoutrements buried with a large proportion of the individuals, particularly the Hunno-Sarmatian period group, could not be established. The only further possible evidence that the individuals whose remains displayed weapon trauma were in fact warriors were the burial of a bone arrowhead with Skeleton XXXI. 42 (an adult male with sword trauma), the association of a number of knives and an iron spearhead with the two adult males from Context XXXI. 87 (both of whom displayed weapon trauma and evidence of scalping) and the apparent deliberate deposition of iron arrowheads with Skeleton XXXI. 83, an adult male with evidence of arrowhead trauma. In addition, knives were recovered from a number of the burials of individuals with weapon trauma, although these may have been used for utilitarian purposes rather than for the purposes of warfare. A number of the individuals with evidence of weapon trauma were clearly not associated with grave goods indicative of warfare, which may indicate that these individuals were not warriors.

Despite not being able to examine the specific grave accoutrements for each individual burial, a cumulative examination of the weaponry present in the graves of the Scythian and Hunno-Sarmatian period individuals is instructive. The weaponry recovered from the Scythian period tombs consisted of bronze and iron daggers, bronze and iron pointed axes, and bronze, bone and wooden arrowheads (Mandelshtam 1983, 28). Consequently, the occurrence of injuries caused by pointed axes is not to be unexpected among the skeletal remains of the Scythian period individuals, although it is perhaps curious that none of the Scythian period individuals displayed injuries caused by arrowheads. This may possibly be explained, however, by arrowheads being embedded in body flesh rather than bone, thereby making the injuries invisible among the skeletal remains. The majority of weapons recovered from the Hunno-Sarmatian period burials consisted of arrowheads, bow brackets of bone, spearheads and darts (Mandelshtam and Stambulnik 1992, 200). Given the frequency of sword injuries among the remains of the Hunno-Sarmatian period individuals, it is curious that more swords were not recovered from the burials. Swords appear to have been rarely recovered at Bay-Dag II, the other main Early Hunno-Sarmatian period Tuvinian cemetery, and the same was true for Late Hunno-Sarmatian period burials in Tuva. Perhaps swords were considered to be too valuable to bury with the dead. The wide array of arrowheads and darts present in the tombs makes it not unexpected that injuries from these weapons were observed among the skeletal remains. It is possible, however, that the frequency of arrowhead injuries among the Hunno-Sarmatian period population is under-represented due to the general lack of damage caused to skeletons by these weapons.

7.6 Scalping

Two Hunno-Sarmatian period adult males displayed clear evidence of scalping (Murphy *et al* 2002). Both individuals were buried in the same tomb, and both displayed evidence of sword cuts. A spearhead and a variety of knives were recovered from the burial, possibly indicative of the warrior status of the two men. The scalping procedure had been undertaken in an identical manner in both individual. A series of small sharp cutmarks were evident, following the circumference of the cranial vault. Throughout the world and during various time periods scalping has been undertaken as a means of retrieving human trophies which are then used as indicators of success and bravery in warfare. Alternatively, however, it has also been suggested that the practise may occur for therapeutic reasons or for magico-religious purposes (Owsley 1994, 337-338). The occurrence of scalping cutmarks on the skulls of two adult male individuals, buried with a range of weapons, whose skeletons also display evidence of weapon trauma would, however, suggest that both men were warriors killed in combat, and that their scalps had been removed for trophy purposes. A Scythian period 25-35 year old male from an uncertain archaeological context also displayed clear evidence for scalping. Although it is unfortunate that the context for this individual was ambiguous, it is still an extremely important addition to the corpus of early cases of Old World scalping.

Herodotus discussed the practice of scalping among the tribes of the Scythian World in great detail, and his description is generally accepted as the earliest historical reference to the practice (Reese 1940, 7). The heads of vanquished enemies were carefully processed with the skin being striped from the head by making a circular cut around the ears. The skull was then shaken out (Sélincourt and Burn 1972, 291). This description corresponds perfectly to the pattern of cutmarks evident in the remains of the two scalped Hunno-Sarmatian period individuals from Aymyrlyg. Herodotus went on to relate how the flesh was scraped from the skin using an ox rib, cleaned and worked by the warrior until supple, and then used as a form of handkerchief. The handkerchiefs were hung on the bridle of the warrior's horse as symbols of battle prowess (*ibid.*, 291). In antiquity the Greeks considered scalping to be so typical of the Scythians that they invented a special verb to denote the process - *aposkythizein*. The written sources indicate that to lose the hair from one's head was considered a disgrace (Rolle 1989, 82). Here at Aymyrlyg we have clear evidence that the scalping practised among the Scythian World of Herodotus continued to be practised into the Hunno-Sarmatian period.

The male individual with battle axe injuries from Kurgan 2 at Pazyryk in the High Altai also displayed evidence for scalping. The skin above the forehead appeared to have been cut through from ear to ear and then torn backwards exposing the skull as far as the neck (Rudenko 1970, 221). Again, this description tallies with the account of Herodotus. The majority of evidence for scalping originates from the New World (Roberts and Manchester 1995, 85), and a large corpus of osteological cases have been investigated. This evidence indicates that scalping was practised in the Americas as early as the 5th or 6th centuries AD (Owsley 1994, 337). While we do have a number of documentary references to scalping in the Classical sources, Aymyrlyg's Scythian period case and two Hunno-Sarmatian period examples, in addition to the example from Pazyryk of Scythian period date, are among the most definitive Old World cases known to the author to have been described in the palaeopathological record (Murphy *et al* 2002).

7.7 Decapitation and throat cutting

A Scythian child and five Hunno-Sarmatian period individuals appear to have been decapitated. The decapitated Hunno-Sarmatian period individuals consisted of three males and two females. One of the Hunno-Sarmatian period females and one of the males also displayed evidence for sword trauma. Clear cut marks indicative of decapitation were apparent on the cervical vertebrae of the female, while the basicranium of the male displayed extensive peri-mortem modifications. The crania and cervical vertebrae of the other three individuals were not available for analysis although it was stated in the archive reports that the skulls of these individuals were not present with the skeletons in the tomb. The neural arches of the third and fourth cervical vertebrae in the Scythian period child displayed clear evidence of medio-lateral chopping. In addition, cut marks visible on the anterior surface of the first and cervical vertebrae of the individual may indicate that the throat of the child had been cut prior to decapitation. Various explanations have been proposed to explain the occurrence of decapitated individuals in archaeological contexts. It is generally accepted that these individuals would have been decapitated for punitive reasons, as a human sacrifice, or for the purposes of obtaining the head of a vanquished enemy as a trophy in warfare (Bush and Stirland 1987).

The historical sources provide enlightenment as to the motives behind Scythian and Hunno-Sarmatian period decapitations. Herodotus recounted that it was a Scythian custom for the heads of all the enemies who had been killed in battle to be brought to the king, and a warrior only received a share of the spoils of war on presentation of a decapitated head. In addition, the skullcaps of the worst enemies of the Scythians were covered with rawhide - and in some cases the endocranial surface was gilded - for use as a drinking vessel. The skulls of warring kinsfolk were supposedly treated in the same manner (Sélincourt and Burn 1972, 291). The *Shih chi* suggests that a similar policy was practised among the Hsiung-nu. Following battle, Hsiung-nu warriors who had decapitated their enemies or taken prisoners were presented with a cup of wine and allowed to keep any of the spoils they had captured (Watson 1961, 165). In one Chinese source the story is related of how a treaty of alliance was made between the Han and Hsiung-nu rulers. It is stated that the dignitaries drank dedicated wine from the skull of a Yüeh-chih lord who had been killed by the *Shan-yü* Laoshan (Rudenko 1970, 291). Since the clear examples of weapon trauma amongst the individuals of both Scythian and Hunno-Sarmatian period date at Aymyrlyg attest to the population's involvement in combat it is feasible that the six individuals had been decapitated during episodes of warfare similar to those referred to in the historical tales. An artistic depiction of Scythian period decapitation is to be found in relief on a cylindrical gold plate which probably once formed the top of a headress, and was recovered from the Kurdzhips Kurgan in the Kuban region. The image on this c. 4th century BC plate seems to depict a warrior handing over a decapitated head to a royal figure (Rolle 1989, 82).

An alternative explanation to account for the decapitation and throat cutting of the Scythian period child has, however, been presented (see Section 6.4) and it is possible that the individual had been ritually killed. Two Hunno-Sarmatian period individuals also displayed evidence for deliberate trauma to the throat. One of the individuals was a juvenile, while the other was an adult male. The adult male also displayed evidence of sword trauma and it is possible that he was stabbed in the throat during the course of combat. As such, it is possible that the Scythian and Hunno-Sarmatian period subadults sustained the traumatic lesions to their throats as a consequence of their involvement in an episode of warfare. Alternatively, however, they may have been ritually killed. Herodotus recounted that the Scythians worshipped the god Ares on a square platform mounted on a huge mound of brushwood. At the summit of the mound was positioned an ancient iron sword which represented the god. Every year prisoners of war, horses and cattle were sacrificed to this sword, the humans having their throats cut (Sélincourt and Burn 1972, 290-291). The Hsiung-nu incident discussed above in relation to the drinking of sacred wine from the skull cap of a vanquished enemy leader also involved moistening the tip of a valuable sword with wine. This account has been interpreted as evidence that a sword cult also existed among the Hsiung-nu (Rudenko 1970, 291). The accounts of throat-cutting as part of the Scythian sword cult, and the possible occurrence of a sword cult amongst the Hsiung-nu, makes it a possibility that the two subadults had been sacrificed as part of the ritual activities associated with some such sword cult.

7.8 Deliberate extraction of the brain

A single Hunno-Sarmatian period adult male displayed evidence for the deliberate extraction of his brain (Murphy forthcoming a). Thirteen sword cuts with no evidence of healing were apparent in the remains of the individual, suggesting that he had died in a particularly frenzied attack. The posterior aspect of his cranium had been bludgeoned with a blunt, rounded implement and it had been struck at least four times. In addition, the inferior aspect of the cranium in the region of the foramen magnum had been deliberately smashed, and seven very sharp cut marks were apparent at the area of the occipital which extended transversely from the posterior margin of the foramen magnum. The regularity of the cutmarks indicated that they had been deliberately and methodically executed. It is probable that the blunt trauma and the cutmarks at the basioccipital area were made peri-mortem.

The most likely objective of this activity was the extraction of the brain from the victim's cranium. The crudeness of the peri-mortem trauma and the appearance of numerous examples of sword cuts on the skeleton would suggest that the extraction of the brain was an aggressive post-combat activity, and that it was part of a war ritual. One might suspect that a human brain would not have made a particularly effective war trophy, however, since it would decompose relatively quickly, and it might be expected that the entire head or scalp would have made a more durable trophy. It is possible, however, that the brain of the Aymyrlyg individual had been processed in some way to aid in its preservation. The practice of calcification of the brains of vanquished warriors is documented in the early historic Irish tales of the *Ulster Cycle* (Stokes 1887, 63; Meyer 1906, 5). It is recorded that the Ulster warriors would extract the brains of all enemies killed in single combat, treat them with lime so that they formed hard balls, and then display the 'balls' as trophies (Meyer 1906, 5). Since no record of such a practice is documented for either of the populations of the Scythian or Hunno-Sarmatian worlds, however, alternative motives for the extraction of the individual's brain must also be considered. One possible explanation is the practice of ritual cannibalism which involves the consumption of specific body parts for the purpose of acquiring the special qualities of the deceased (Flinn *et al* 1976, 308). The Zulus, for example, believed that if they ate the forehead and eyebrows of their enemies they would acquire the ability to look unflinchingly in the face of adversity (Masters 1993, 277). In addition, the brain may have been extracted for mutilation purposes since certain Native American tribes believed that a mutilated person would not get to the afterlife. Historic documentation of the warfare practices of Plains Indians, for example, states that as part of the extensive mutilation of corpses the '...brains (were) taken out and placed on rocks with other members of the body....' (Quaife 1950 as quoted in Bullock 1991, 10).

Olsen and Shipman (1994, 383) described cases of foramen magnum opening in three individuals from North American Woodland period sites of approximately AD 400. All of the crania had been extensively defleshed, and it was postulated that the foramen magnum had been opened to facilitate the removal of the brain during the preparation of the body for secondary burial. While battle trauma was apparent on the remains of the individual from Aymyrlyg, there was no evidence for defleshing or disarticulation. It is therefore more likely that the brain had been extracted during post-combat mutilation and ritual activity.

7.9 Trepanation

Non-weapon related but deliberate human modification of skeletal elements was also present among the Scythian period population from Aymyrlyg. Trepanations have been reported from elsewhere in the former Soviet Union, and Rokhlin (1965, 290) described the occurrence of healed trepanations in a Mesolithic and two Neolithic crania from the cemetery of Vasilyevka 3 in Ukraine. At Aymyrlyg we have further evidence of this practise to augment the number of known cases (Murphy 2003).

Two Scythian period individuals, one female and the other an adolescent, displayed evidence for trepanation. In the case of the female individual the operation is thought to have been performed ante-mortem, and healing was apparent surrounding the opening in the skull. The morphology of the perforation suggested that the operation had been undertaken using the scraping technique. It is considered, however, that the adolescent individual had been operated on post-mortem. An oval perforation was apparent at the bregma, and this had a smooth, glossy bevelled edge. Shallow scrape marks were present over the entire surface of the cranial vault.

Lisowski (1967, 659) has provided three possible motives to explain why trepanation occurred ante-mortem during antiquity. It may have been carried out either for genuine therapeutic reasons (such as to alleviate pain from head injuries), to rid the individual of evil spirits (probably headaches and epilepsy) or for magico-ritual reasons. In addition, there is substantial evidence that certain individuals among a number of the societies of the Scythian World practiced post-mortem trepanation as part of the mummification process (Mednikova 2000). The crania of the two individuals buried in Kurgan 2 at Pazyryk in the High Altai, for example, had both been trepanned. The perforated area measured approximately 40 mm by 55 mm in the cranium of the male. The trepanation facilitated the removal of the brain and the subsequent filling of the cranium with soil, pine needles and larch cones. The plate of bone that had been removed during the procedure was then replaced and the skin secured to the skull with a twist of black horsehair. The individuals recovered from Kurgan 5 at Pazyryk had also been trepanned (Rudenko 1970, 280-281). In addition, mummies from the Oglakty cemetery in the Minusinsk Hollow, which approximately date to the Scythian period, also displayed evidence of post-mortem trepanation. Again it was considered that the trepanation had been undertaken to facilitate the removal of the brain during the mummification process (Tallgren 1936, 73). Photographs of the trepanned skulls indicate that the perforations were large, measuring approximately 90 mm by 45 mm.

It is probable that the trepanation evident in the Scythian period female had been undertaken for one of Lisowski's (1967) three reasons since it displayed evidence for healing. Only the skull of the individual was preserved, however, and the only palaeopathological lesions present were related to the woman's dentition so it is not possible to ascertain the exact motive behind the operation. When we look at the evidence for the second Scythian period trepanation we find subtle differences from the previous case. The post-mortem trepanation perforation evident in the remains of the Scythian period adolescent had polished margins which may be indicative of wear (Owsley *et al* 1994, 368), while the scrape marks were distributed randomly over the cranial vault and were fairly coarse in morphology. This would suggest that the cranium had not been scalped since in scalping the incisions are generally made in a deliberate manner and one that follows the curvature of the cranium (Hamperl 1967, 630; Murphy *et al* 2002). The cutmarks apparent on the cranium of the adolescent were more reminiscent of defleshing, which is characterised by abundant cutmarks with a wide distribution including the face, basicranium and scalp region (Olsen and Shipman 1994, 381).

These differences may mean that the three motivational factors outlined by Lisowski are not strictly appropriate to this second case study. Given this, an alternative reason must be sought. Ethnographic studies in Africa have indicated that a common motive behind posthumous trepanation operations was to obtain roundels of human skull, generally circular in shape, which would be perforated, polished and worn on necklaces. It was considered that such roundels had religico-magical properties and they were regarded as amulets or charms (Lisowski 1967, 659). It is possible, therefore, that the individual from Aymyrlyg had been trepanned post-mortem for ritual purposes which involved the removal of a roundel of bone at the bregma. The possibility that the Aymyrlyg individual had been trepanned post-mortem as part of a mummification process also needs to be considered. The cranial perforations evident in the individuals from Pazyryk and Oglakty, however, were all much larger than that apparent in the individual from Aymyrlyg. It is unlikely, therefore, that the perforation in the cranium from Aymyrlyg would have been large enough to facilitate the extraction of the brain during mummification. In addition, a number of other Aymyrlyg individuals displayed cutmarks indicative of defleshing and disarticulation for secondary burial processes, but none of these individuals displayed perforations in their crania (Murphy 2000; Murphy and Mallory 2000; 2001). This finding may indicate that the population group buried at Aymyrlyg did not remove the brain by means of an artificial perforation in the cranium as part of their secondary burial processes.

An additional theory to account for the removal of circular sections of bone from a skull after death can be found in the work of Gillman (1876), who reported on a series of artificially perforated crania from Moundbuilder sites in Michigan, North America. The majority of these skulls were not associated with skeletal remains, and the skull from Aymyrlyg also appears to have been buried without clear association with a skeleton. The perforations on the Moundbuilder crania were all considered to have been made post-mortem and were invariably positioned on the superior aspect of the cranium. The openings were normally situated at or near the bregma, were circular, and measured approximately 10-25 mm in diameter (*ibid.*, 243-244). Gillman did not refer to any evidence for decapitation in these crania, and there was no evidence to suggest the deliberate detachment of the Aymyrlyg cranium from the remainder of the body. It is possible, however, that in all cases the skulls were detached from the bodies by cutting through the neck and that the signs would have been evident on the cervical vertebrae rather than the base of the cranium. Gillman (1876, 238-239) recounted ethnographic studies among the Dyaks of Borneo which revealed that each village had a central house in which the inhabitants kept the heads of dead prisoners. The head was suspended by means of a string which passed through an artificial perforation in the superior aspect of the cranium. The descriptions of the perforations evident in the crania from Michigan correlate very closely with that apparent in the cranium from Aymyrlyg. It is possible, therefore, that whoever had perforated the crania of the individual from Aymyrlyg was not intent on removing a roundel of bone but wished to display the cranium, possibly by suspension. The theory that the perforation had been used to enable the display of the cranium from Aymyrlyg is considered the most plausible explanation. Consequently, it is probable that the posthumous perforation had been made in the cranium of the Aymyrlyg individual for either ritual or warfare reasons, or as a combination of the two.

7.10 Amputation

A single adult male of Scythian period date displayed possible evidence for amputation of the left forearm which appeared to have been amputated at the distal end of the humerus. The trochlea and capitulum of the humerus were not present and only the superior aspect of the medial epicondyle of the distal humerus was apparent. In addition, the skeleton was in a good state of preservation and none of the bones of the left forearm were present. Evidence of amputation in the palaeopathological record is rare, probably due to the difficulty of identifying this form of trauma, although Rokhlin (1965, 290) recorded evidence for the amputation of the little finger of a Mesolithic female from the Murzak-Kobah cave in Crimea.

Amputation of limbs is known to have occurred in the past for a variety of reasons. The amputation could have been voluntary and permitted by the person, perhaps as part of a surgical procedure, but amputation may also have been involuntary, with the individual's limb removed in the course of punitive action against a criminal or captive (Ortner and Putschar 1981, 94). In addition, accidental amputation could occur during battle, as a result of a blade injury (Mays 1996, 110), or as the result of an accident in the course of everyday life (Roberts and Manchester 1995, 89).

In the discussion of the Scythian practice of sacrificing animals and prisoners to the god Ares, Herodotus recounted that part of the ceremony involved severing the right hands and arms of the slaughtered prisoners and tossing them into the air (Sélincourt and Burn 1972, 290-291). Perhaps the forearm of the Scythian period person from Aymyrlyg had been deliberately severed as part of a ritual sacrifice to the god Ares. The individual had, however, been buried with a number of grave goods, including four arrowheads, which would indicate that this was the grave of a warrior, and it is more plausible that his arm had been amputated during the course of battle.

8. Conclusions

The objective of this monograph has been to provide a detailed account of the archaeological record discovered at Aymyrlyg and the evidence for trauma identified among the human skeletal remains recovered at this Iron Age cemetery. In addition, it is hoped that the text has provided the reader with an overview of the Scythian and Hunno-Sarmatian period in the region through the marriage of information contained in the relevant historical texts with the archaeological evidence.

The cemetery at Aymyrlyg was generally unknown to Western academics, even though Drs. Mandelshtam and Stambulnik undertook a major programme of research at the site over a period of at least 15 years. It is hoped that a final presentation of the excavation reports will be forthcoming from Russian archaeologists in the not too distant future since the excavations are of the utmost importance to furthering our understanding of life and death in Iron Age Siberia. The people buried in the cemetery were not 'Royal Scythians' as was found to be the case in the frozen kurgans at Pazyryk. The Aymyrlyg excavations enable us to gain real insight into the lives and lifestyle of the 'ordinary' members of these semi-nomadic societies. The purpose of this monograph has been to present the archaeological evidence, set the site into its chronological context, highlight how the archaeological record correlates with the historical sources and how palaeopathology is an important means of deriving information from the skeletal remains on trauma, which in turn helps elucidate information on the lives of past societies. South Siberia and Mongolia are amongst the regions of the Old World with the most ancient traditions of pastoralism (Vainshtein 1980, 51). The analysis of the skeletal remains from Aymyrlyg has therefore provided a rare opportunity for the examination of a substantial corpus of skeletal remains of Iron Age semi-nomadic pastoralists from the vast Eurasian steppe-lands. This programme of research has represented one of the first palaeopathological studies of an archaeological population from south Siberia to have been undertaken and, as such, it has made a major contribution to our understanding of life and death in Iron Age Central Asia.

The corpus of skeletons from Aymyrlyg are considered to have originated from two distinct groups, those from the Scythian period (or Uyuk Culture) and those from the Hunno-Sarmatian period (or Shurmak Culture). This division was established by Russian academics on the basis of differences in funerary monuments and material culture between the two groups, combined with information contained in Chinese and Classical texts on the inhabitants of the region. Russian archaeologists consider the Uyuk Culture in Iron Age Tuva to have belonged to the Scythian World. This culture has been approximately dated to between the 8th and the 2nd century BC and the majority of the Scythian period burials at Aymyrlyg were considered to date from the 3rd to 2nd centuries BC. Historical texts indicate that there was a spread of influence of the Hsiung-nu Empire into Tuva during the late 3rd and early 2nd century BC, and that this led to cultural and/or population changes at this time (Mandelshtam 1983, 331; 1992, 183, 185). Russian archaeologists have postulated that the Shurmak Culture developed during the 3rd to the 1st century BC as a result of a complicated and prolonged process that occurred in a climate of rapid political change when new population groups moved into Tuva. These Hunno-Sarmatian period people are considered to be related to the Hsiung-nu of the historical texts (Stambulnik 1983a, 34; Mandelshtam and Stambulnik 1992, 196). It has further been suggested that the Hunno-Sarmatian period population at Aymyrlyg belongs to the earliest phase of this cultural change, when the Scythian period culture was ending and the new culture was gaining ground. As such, it is believed that the Hunno-Sarmatian period burials incorporate elements of both old and new cultures (Stambulnik 1983a, 39).

During the Scythian period at Aymyrlyg people were either buried in communal log house tombs or individual stone cists. The majority of bodies were buried on their left sides, with their legs in a flexed position. Numerous grave goods were associated with the skeletons including two components of the Scythian Triad - weaponry, and artefacts decorated with the Animal Style of artwork. Other objects recovered from the graves included tools, household objects, foodstuffs, clothing and adornments, some horse accoutrements, personal items and ritual objects. When we move to the Hunno-Sarmatian period there is a clear change in the burial rituals associated with the dead. Burial was predominantly made on an individual basis in a composite wooden coffin, log or wooden block, with the bodies mostly laid in an extended supine position. The remains of weapons, tools, household objects, foodstuffs, clothing and adornments, personal objects and pieces of art were recovered from the tombs.

Archaeological evidence recovered from Scythian period funerary monuments in Tuva indicates that the economy of the highland-steppe peoples was based upon a semi-nomadic form of pastoralism which was combined with land-cultivation, hunting and gathering. The Scythian period tribes of the mountain-steppe regions of Tuva are thought to have made seasonal migrations (Vainshtein 1980, 51). This form of economy would have involved regular repeated seasonal movements between the mountains and the steppes (termed 'vertical shifts') within the borders of a relatively defined territory (Mandelshtam 1992, 193). A number of the Scythian period bodies from Aymyrlyg displayed cutmarks which were indicative that they had been disarticulated, and in some cases defleshed, after death. This finding may be indicative of secondary burial, which was related to the seasonal migrations that these semi-nomadic peoples practiced (Murphy 2000b; Murphy and Mallory 2000; 2001). The Hunno-Sarmatian period Shurmak Culture in Tuva is considered to have had a similar semi-nomadic pastoralist economy to that of the preceding Scythian period Uyuk Culture but with a greater degree of mobility (Vainshtein 1980, 52). The distribution of large tribal burial grounds in Tuva from both periods indicates that cyclic migration with fixed routes and set winter campsites existed amongst these populations (*ibid.*, 96). Presumably, herds would have been pastured in the mountains during the summer and in the more low-lying land during the winter (Bokovenko 1995, 255).

The study of the skeletal population from Aymyrlyg has provided great insights into the lifestyle of both Scythian and Hunno-Sarmatian period populations. A wide variety of traumatic lesions and injuries, probably caused during interpersonal violence and warfare, were observed among both populations. Night-stick fractures of the ulnae and boxer's fractures of the metacarpals, both commonly attained during incidents of interpersonal violence, were included among the corpus of fractures of the appendicular skeletons. Depressed fractures of the cranial vaults, also probably indicative of interpersonal violence, were observed among Scythian period people of all age groups, although it is interesting to note that such injuries were not apparent among any of the Hunno-Sarmatian period population. In addition, a wider variety of facial fractures were evident among the Scythian period population, although fractured nasal bones were particularly common among the males of both population groups. The discrepancy in distribution of cranial injuries between the two populations may be indicative that different fighting methods were practised during minor scuffles. Blunt implements may have been used during the Scythian period while it would seem that fist-fights were more likely used to settle disagreements during the Hunno-Sarmatian period.

A wide variety of weapon injuries were apparent among the skeletal remains of both populations, although clear cultural differences were also evident, with battle axe injuries predominating during the Scythian period and sword trauma prevalent during the Hunno-Sarmatian period. Arrowhead injuries were also observed solely among the Hunno-Sarmatian period population. In both populations, however, the majority of cases of weapon trauma were present among males, although a proportion of these injuries were also recorded among females and subadults. When we consider the historical references to women having been involved in warfare as warriors there is no reason why the Aymyrlyg females might not have obtained their cases of weapon trauma during pitched battles, where they would have joined the men and young adults in defence of the clan and its possessions. Two Scythian period and one Hunno-Sarmatian period children, who had ages that were less than ten years old, displayed evidence for weapon trauma. It would seem probable that these are the only individuals in either population group who can be discounted from having been involved in warfare due to their young age-at-death.

A number of cases of trauma identified in the palaeopathological record for Aymyrlyg find echoes in the available historical texts. One Scythian period and two Hunno-Sarmatian period males displayed definite evidence for scalping (Murphy *et al* 2002). The morphology of the lesions in all cases validates Herodotus' description of this practise among Scythian tribes. The occurrence of weapon trauma on the two Hunno-Sarmatian period individuals may indicate that their scalps had been taken as war trophies. Several cases of decapitation were observed among male, female and subadults of both populations. In the majority of these cases it is probable that the practice had occurred for the purposes of obtaining war trophies, as is recounted in both Herodotus' *Histories* and the ancient Chinese Chronicles. One Scythian period subadult, a Hunno-Sarmatian period subadult and a Hunno-Sarmatian period adult all displayed cutmarks indicative that they had had their throats slashed. While this may have occurred during warfare, it is equally possible that they were sacrificial victims, perhaps to one of the sword cults described in the historical texts.

Further examples of traumatic lesions include a single Hunno-Sarmatian period adult male who displayed evidence for the deliberate extraction of his brain in addition to numerous sword injuries. It is probable that the brain had been extracted as part of a war ritual (Murphy forthcoming a). Two Scythian period cases of trepanation were also identified. The first had been undertaken ante-mortem and displayed a degree of healing, while the other had been carried out after death (Murphy 2003). In the former case the motives may have been therapeutic, while it is possible that the latter individual had been trepanned in order to display the skull for ritual or prestige purposes. A single Scythian period adult male revealed possible evidence for the amputation of his left forearm. Whatever the cause of the amputation, the individual had survived for a considerable period of time following its occurrence.

The presence of numerous individuals with weapon trauma and the wide diversity of these traumatic lesions attests to the warfaring nature of both semi-nomadic populations and adds further support to the depiction of their lifestyle as outlined in the historical accounts. In addition, the wide array of traumatic lesions which may not necessarily have been related to warfaring practice is an indication of the active physical lifestyle of these people.

A range of academic papers has already been published on aspects of the completed thesis. These include a case study on a possible case of neurofibromatosis (Murphy *et al* 1998), and an overview of the evidence for disability based on the more severe developmental defects apparent among the population groups (Murphy 2000a). Papers have also been written on the evidence for secondary burial practice (Murphy 2000b; Murphy and Mallory 2000; 2001), scalping (Murphy *et al* 2002), trepanation (Murphy 2003), and a case-study of an individual who appeared to have died as a consequence of a particularly frenzied attack (Murphy forthcoming a). An overview of the findings of the analysis of the Scythian period population has been published in Russian (Murphy 2001), and a Russian version of the results for the Hunno-Sarmatian period group is forthcoming (Murphy forthcoming b). Aspects of the palaeopathological analysis remains to be published, including the evidence for neoplastic, joint, metabolic, infectious and dental disease. The current author is in the process of developing further research projects which will involve stable isotope analysis of the remains from Aymyrlyg for dietary analysis, and statistical analysis of the burials for the purposes of mortuary analysis. In addition, it is intended to subject samples of the skeletons for DNA analysis, both to gain a greater understanding of some of the more unusual palaeopathological conditions and as a means of determining the genetic relationships of individuals both within and between the two population groups.

Bioarchaeology is a widely recognised discipline throughout the world and it is recognised as one of the best ways of gaining a true understanding of ancient people (e.g. Larsen 1997). This method has proved to be of great value for interpreting the remains from Aymyrlyg. The research included in the current text and the author's doctoral thesis represent one of the first major palaeopathological studies to have been undertaken on Siberian archaeological populations and, as such, it is hoped that it will be an important benchmark for further palaeopathological research, both in Russia and throughout the world.

9. References

Abetekov, A. and H. Yusupov, 1994: 'Ancient Iranian Nomads in Western Central Asia', pp. 23-33 in J. Harmatta (ed.), *History of Civilisations of Central Asia, Vol. III, the Development of Sedentary and Nomadic Civilisations: 700 BC to AD 250*. Paris: UNESCO.

Ackernecht, E. H., 1967: 'Primitive Surgery', pp. 635-650 in D. Brothwell and A. T. Sandison (eds.), *Diseases in Antiquity*. Springfield: Charles C. Thomas.

Adams, J. C., 1978: *Outline of Fractures*. London: Churchill Livingstone (7th edition).

Adams, J. C. and D. L. Hamblen, 1990: *Outline of Orthopaedics*. London: Churchill Livingstone (11th edition).

Alekseev, V. P. and I. I. Gohkman, 1986: 'Antropologicheskii Sostav I Proishozhdeniye Naseleniya, Ostabibshevo Mogilnik Kokel', pp. 87-107 in *Problemi Antropologii Drevnevo I Sovremennovo Naseleniya Sovetskoi Azii*. Novosibirsk: Nauka (Anthropological study of the populations from the Kokel burial ground).

Angel, J. L., 1974: 'Patterns of Fractures from Neolithic to Modern Times', *Anthropologiai Közlemények* 18, 9-18.

Antonini, C. S., 1994: 'On Nomadism in Central Asia between the Saka and the Xiongnu: the Archaeological evidence', pp. 287-309 in B. Genito (ed.), *The Archaeology of the Steppes: Methods and Strategies*. Naples: Instituto Universitario Orientale.

Apley, A. G. and L. Solomon, 1988: *Concise System of Orthopaedics and Fractures*. Oxford, Butterworth-Heinemann Ltd.

Askarov, A., V. Volkov and N. Ser-Odjav, 1992: 'Pastoral and Nomadic Tribes at the Beginning of the First Millennium BC', pp. 459-472 in A. H. Dani and V. M. Masson (eds.), *History of Civilisations of Central Asia, Vol. 1, The Dawn of Civilisation: Earliest times to 700 BC*. Paris: UNESCO.

Axtell, J. and W. C. Sturtevant, 1980: 'The Unkindest Cut, or Who Invented Scalping?', *William and Mary Quarterly* 37, 451-472.

Baker, J. and D. Brothwell, 1980: *Animal Diseases in Archaeology*. London: Academic Press.

Bendann, E., 1969: *Death Customs: An Analytical Study of Burial Rites*. London: Dawsons of Pall Mall (reprint of 1930 1st edition).

Bogdanova, V. I and A. B. Radzjun, 1991: 'Paleoantropologicheskiye Materiali Gunno-Sarmatskova vremeni iz Tsentralnoy Tuvi', pp. 55-74 in *Noviye Kollektsii I Issledovaniya Po Antropologii I Arheologii*. Spornik Muzeya Antropologii I Etnografii XLIV (Human Skeletal Remains of the Hunno-Sarmatian Age from Tuva).

Bökönyi, S., 1994: 'The Role of the Horse in the Exploitation of Steppes', pp. 17-30 in B. Genito (ed.), *The Archaeology of the Steppes: Methods and Strategies*. Naples: Instituto Universitario Orientale.

Bokovenko, N. A., 1994a: 'Tomb of Saka Princes Discovered in the Sayans, Siberia', pp. 48-53 in A. G. Kozintsev, V. M. Masson, N. F. Solovyova and V. Yu. Zuyev (eds.), *New Archaeological Discoveries in Asiatic Russia and Central Asia*. Saint Petersburg: Institute of the History of Material Culture.

Bokovenko, N. A., 1994b: 'Le Kourgane "Royal" D'Arjan et son Temps', *Les Dossiers D'Archeologie* 194, 30-37.

Bokovenko, N. A., 1995a: 'History of Studies and the Main Problems in the Archaeology of Southern Siberia During the Scythian Period', pp. 255-261 in J. Davis-Kimball, V. A. Bashilov and L. T. Yablonsky (eds.), *Nomads of the Eurasian Steppes in the Early Iron Age*. Berkeley: Zinat Press.

Bokovenko, N. A., 1995b: 'Tuva During the Scythian Period', pp. 265-281 in J. Davis-Kimball, V. A. Bashilov and L. T. Yablonsky (eds.), *Nomads of the Eurasian Steppes in the Early Iron Age*. Berkeley: Zinat Press.

Brooks, S. and J. M. Suchey, 1990: 'Skeletal age determination based on the os pubis: a comparison of the Acsádi-Nemeskéri and Suchey-Brooks methods', *Human Evolution* 5, 227-238.

Brothwell. D. R., 1981: *Digging up Bones*. New York: British Museum and Cornell University Press (3rd edition).

Buikstra, J. and D. H. Ubelaker (eds.), 1994: *Standards for Data Collection from Human Skeletal Remains*. Arkansas: Arkansas Archeological Survey Research Series No. 44.

Bullock, P. Y., 1991: 'A Reappraisal of Anasazi Cannibalism', *The Kiva* 57, 5-16.

Bush, H. and A. Stirland, 1987: *Decapitated Burials in Roman Britain*, Unpublished paper presented at the 14th

Annual Meeting of the Paleopathology Association, New York.

Chadwick, J. and W. N. Mann, 1978: 'Medicine', pp. 61-276 in G. E. R. Lloyd (ed.), *Hippocratic Writings*. London: Penguin Classics (revised edition).

Chamberlain, A., 1994: *Human Remains*. London: British Museum Press.

Clanton, T. D. and J. C. DeLee, 1982: 'Osteochondritis Dissecans - History, Pathophysiology and Current Treatment Concepts', *Clinical Orthopaedics and Related Research* 167, 50-64.

Clenova, N. L., 1994: 'On the Degree of Similarity between Material Culture Components within the "Scythian World"', pp. 499-540 in B. Genito (ed.), *The Archaeology of the Steppes: Methods and Strategies*. Naples: Instituto Universitario Orientale.

Culebras, A., 1993: 'Trephination of the skull in pre-Columbian cultures', *Journal of the History of the Neurosciences* 2, 11-12.

Czaplicka, M. A., 1914: *Aboriginal Siberia: A Study in Social Anthropology*. Oxford: Clarendon Press.

Davidova, A. V., 1996: *Ivolginskii Arkheologicheskii Kompleks: Tom 2, Ivolginskii Mogilnik*. St. Petersburg, Centre for Oriental Studies (The Ivolga Archaeological Complex: Part 2, The Ivolga Cemetery).

Ewing, T. E., 1981: 'The Forgotten Frontier: South Siberia (Tuva) in Chinese and Russian History, 1600-1920', *Central Asiatic Journal* 25, 174-212.

Ferembach, D., I. Schwidetzky and M. Stloukal, 1980: 'Recommendations for Age and Sex Diagnoses of Skeletons', *Journal of Human Evolution* 9, 517-549.

Flinn, L., C. G. Turner, and A. Brew, 1976: 'Additional evidence for cannibalism in the Southwest: The case of LA 4528', *American Antiquity* 41, 308-318.

Gavrilyuk, N. A., 1987: 'Pishsha Stephih Skifov', *Sovetskaya Arkheologiya* 1, 21-34 (The Diet of the Steppe Scythians).

Gera, D., 1997: *Warrior Women, The Anonymous Tractatus De Mulieribus*. New York: E. J. Brill.

Gilbert, B. M. and T. W. McKern, 1973: 'A method for ageing the female os pubis', *American Journal of Physical Anthropology* 38, 31-38.

Gillman, H., 1876: 'Certain Characteristics Pertaining to Ancient Man in Michigan', pp. 234-245 in *Annual Report of the Board of Regents of the Smithsonian Institution for 1875*. Washington DC: Smithsonian.

Grach, A. D., 1971: 'Noviye danniye o drevnei istorii Tuvy', *Uch. zap. YuvNIIYaLI T. 15* (New Data on the Ancient History of Tuva).

Gryaznov, M. P., 1983: 'Nachalnaya faza razvitiya skifo-sibirskih kultur', *Arkheologiya Yuzhnoi Sibiri*, Kemerovo (The Initial Phase of the Development of the Scytho-Siberian Cultures).

Hamperl, H., 1967: 'The Osteological Consequences of Scalping', pp. 630-634 in D. Brothwell and A. T. Sandison (eds.), *Diseases in Antiquity*. Springfield: Charles C. Thomas.

Harmatta, J. (ed.), 1994: 'Conclusion', pp. 485-492 in *History of Civilisations of Central Asia, Vol. III, the Development of Sedentary and Nomadic Civilisations: 700 BC to AD 250*. Paris: UNESCO.

Hildinger, E., 1997: *Warriors of the Steppe: A Military History of Central Asia, 500 BC to 1700 AD*. Staplehurst: Spellmount.

Hillson, S., 1986: *Teeth* (Cambridge Manuals in Archaeology). Cambridge: Cambridge University Press.

Hillson, S., 1996: *Dental Anthropology*. Cambridge: Cambridge University Press.

Iscan, M. Y., S. R. Loth and R. K. Wright, 1984: 'Age Estimation from the Rib by Phase Analysis: White Males', *Journal of Forensic Sciences* 29, 1094-1104.

Iscan, M. Y., S. R. Loth and R. K. Wright, 1985: 'Age Estimation from the Rib by Phase Analysis: White Females', *Journal of Forensic Sciences* 30, 853-863.

Ishjamts, N., 1994: 'Nomads in Eastern Central Asia', pp. 151-169 in J. Harmatta (ed.), *History of Civilisations of Central Asia, Vol. III, the Development of Sedentary and Nomadic Civilisations: 700 BC to AD 250*. Paris: UNESCO.

Janaway, R. C., 1987: 'The preservation of organic materials in association with metal artefacts deposited in inhumation graves', pp. 127-148 in A. Boddington, A. N. Garland and R. C. Janaway (eds.), *Death, decay and reconstruction*, Manchester: Manchester University Press.

Jettmar, K., 1951: *The Altai Before the Turks*. Stockholm: The Museum of Far Eastern Antiquities Bulletin 23.

Johnson, D. L., 1969: *The Nature of Nomadism*. Chicago: The University of Chicago Department of Geography Research Paper No. 118.

Khazanov, A. M., 1975: *Sotsialnaya Istoriya Skifov*. Nauka: Moscow (The Social History of the Scythians).

Khazanov, A. M., 1978: 'Characteristic Features of Nomadic Communities in the Eurasian Steppes', pp. 119-

126 in W. Weissleder (ed.), *The Nomadic Alternative*. Paris: Mouton Publishers.

Khazanov, A. M., 1984: *Nomads and the Outside World*. Cambridge: Cambridge University Press.

Klepinger, L. L., D. Katz, M. S. Micozzi and L. Carroll, 1992: 'Evaluation of Cast Methods for Estimating Age from the *Os Pubis*', *Journal of Forensic Sciences* 37, 763-770.

Krogman, W. M. and M. Y. Iscan, 1986: *The Human Skeleton in Forensic Medicine*. Springfield: Charles C. Thomas (2nd edition).

Kyzlasov, L. R., 1958: 'Etapy drevnei istorii Tuva', Vestnik MGU, *Seriya istorko-filologicheskaya 4* (Stages of the Ancient History of Tuva).

Larsen, C. S., 1997: *Bioarchaeology: Interpreting behaviour from the human skeleton*. Cambridge: Cambridge University Press.

Lindholm, T. S., K. Österman and E. Vankka, 1980: 'Osteochondritis Dissecans of Elbow, Ankle and Hip: A Comparison Survey', *Clinical Orthopaedics and Related Research* 148, 245-253.

Lisowski, F. P., 1967: 'Prehistoric and Early Historic Trepanation', pp. 651-672 in D. Brothwell and A. T. Sandison (eds.), *Diseases in Antiquity*. Springfield: Charles C. Thomas.

Litvinsky, B. A. and Z. Guang-da, 1996: 'Historical Introduction', pp. 19-33 in B. A. Litvinsky, Z. Guang-da and R. Shabani Samghabadi (eds.), *History of Civilisations of Central Asia, Vol. 3, The Crossroads of Civilisations: AD 250 to 750*. Paris: UNESCO.

Lovejoy, C. O., R. S. Meindl, T. R. Pryzbeck and R. P. Mensforth, 1985: 'Chronological Metamorphosis of the Auricular Surface of the Ilium: A New Method for the Determination of Adult Skeletal Age at Death', *American Journal of Physical Anthropology* 68, 15-28.

Maenchen-Helfen, J. O., 1973: *The World of the Huns*. London: University of California Press.

Mandelshtam, A. M., 1969: 'Issledovaniya Mogilnika Aymyrlyg I Gorodishsha Bazhin-Alak', pp. 189-191 in *Archeologicheskie Otkritiya 1968*. Moscow: Nauka (Research at the Aymyrlyg burial ground and at the fort of Bazhin-Alak).

Mandelshtam, A. M., 1970: 'Raskopki na Mogilnikye Aymyrlyg I Gorodishshye Bazhin-Alak', pp. 183-184 in *Archeologicheskie Otkritiya 1969*. Moscow: Nauka (Excavations at the Aymyrlyg burial ground and the fort of Bazhin-Alak).

Mandelshtam, A. M., 1971: 'O Rezultatah Rabot Na Mogilnikye Aymyrlyg', p. 217 in *Archeologicheskie Otkritiya 1970*. Moscow: Nauka (On the results of the excavations at the Aymyrlyg burial ground).

Mandelshtam, A. M., 1972:' Issledovaniya Na Mogilnikye Aymyrlyg', pp. 280-281 in *Archeologicheskie Otkritiya 1971*. Moscow: Nauka (Research at the Aymyrlyg burial ground).

Mandelshtam, A. M., 1973: 'Raskopki Mogilnika Aymyrlyg', p. 228 in *Archeologicheskie Otkritiya 1972*. Moscow: Nauka (Excavations at the Aymyrlyg burial ground).

Mandelshtam, A. M., 1974: 'Raskopki na Mogilnikye Aymyrlyg', p. 211 in *Archeologicheskie Otkritiya 1973*. Moscow: Nauka (Excavations at the Aymyrlyg burial ground).

Mandelshtam, A. M., 1975: 'Raskopki na Mogilnikye Aymyrlyg', pp. 219-220 in *Archeologicheskie Otkritiya 1974*. Moscow: Nauka (Excavations at the Aymyrlyg burial ground).

Mandelshtam, A. M., 1976: 'Mogilnik Aymyrlyg', pp. 259-260 in *Archeologicheskie Otkritiya 1975*. Moscow: Nauka (The Aymyrlyg burial ground).

Mandelshtam, A. M., 1977: 'Raboto Btorova Otryada Sayano-Tuvinskoi Exspeditsii', pp. 217-218 in *Archeologicheskie Otkritiya 1976*. Moscow: Nauka (Excavations undertaken by the Second Squadron of the Sayano-Tuvinskoy Expedition).

Mandelshtam, A. M., 1983: 'Issledovaniye na mogilnom polye Aymyrlyg: Nekotoriye itovi i perspektii', pp. 25-33 in *Drevniye Kulturi Euraziiskih Stepei*. Leningrad: Institute for the History of Material Culture (Research at the Aymyrlyg burial ground: some results and perspectives).

Mandelshtam, A. M., 1992: 'Ranniye kochevniki Skifskova perioda na territorii Tuvi', pp. 178-196 in M. G. Moshkova (ed.), *Stepnaya Polosa Aziatskoi Chasti SSSR v Skifo-Sarmatskoye Vremya* (Archeologiya SSSR). Moskva: Nauka (Early nomads of the Scythian period in the territory of Tuva).

Mandelshtam, A. M. and E. U. Stambulnik, 1992: 'Gunno-Sarmatskii period na territorii Tuvi', pp. 196-205 in M. G. Moshkova (ed.), *Stepnaya Polosa Aziatskoi Chasti SSSR v Skifo-Sarmatskoye Vremya* (Archeologiya SSSR). Moskva: Nauka (The Hunno-Sarmatian period in the territory of Tuva).

Mandelshtam, A. M., E. U. Stambulnik and T. A. Sharovskaya, 1978: 'Raskopki na Mogilnikah Baidag II I Aymyrlyg', pp. 253-254 in *Archeologicheskie Otkritiya 1977*. Moscow: Nauka (Excavations at the Baidag II and Aymyrlyg burial grounds).

Mannai-Ool, M. H., 1970: *Tuva v Skifskoye Vremya*

(Uyukskaya Kultura). Nauka: Moscow (Tuva in Scythian Times (Uyukskaya Culture)).

Markov, G. E., 1978: 'Problems of Social Change among the Asiatic nomads', pp. 305-311 in W. Weissleder (ed.), *The Nomadic Alternative*. Paris: Mouton Publishers.

Masters, B., 1993: *The Shrine of Jeffrey Dahmer*. London: Hodder and Stoughton Ltd.

Mays, S., 1996: 'Healed Limb Amputations in Human Osteoarchaeology and their Causes: A Case Study from Ipswich, UK', *International Journal of Osteoarchaeology* 6, 101-113.

McKern, T. W., 1970: 'Estimation of Skeletal Age: From Puberty to About 30 Years of Age', pp. 41-56 in T. D. Stewart (ed.), *Personal Identification in Mass Disasters*. Washington: Smithsonian Institution.

McKern, T. W. and T. D. Stewart, 1957: *Skeletal Age Changes in Young American Males, Analyzed from the Standpoint of Identification* (Technical Report EP-45). Natick, Massachusetts: Headquarters, Quartermaster Research and Development Command.

Mednikova, M., 2000: 'Post-mortem Trepanations in Central Asia: Types and Trends', pp. 269-278 in J. Davis-Kimball, E. M. Murphy, L. Yablonsky and L. Koryakova (eds.), *Kurgans, Ritual Sites, and Settlements: The Eurasian Bronze and Iron Age* (Papers from the European Association of Archaeologists 4th and 5th Annual Meetings, BAR International Series 890). Oxford: BAR Publishing.

Meindl, R. S. and C. O. Lovejoy, 1985: 'Ectocranial Suture Closure: A Revised Method for the Determination of Skeletal Age at Death Based on the Lateral-Anterior Sutures', *American Journal of Physical Anthropology* 68, 57-66.

Melyukova, A. I., 1990: 'The Scythians and Sarmatians', pp. 97-117 in D. Sinor (ed.), *The Cambridge History of Early Inner Asia*. Cambridge: Cambridge University Press.

Merbs, C. F. and W. H. Wilson, 1960: 'Anomalies and Pathologies of the Sadlermuit Eskimo Vertebral Column', pp. 154-180 in *National Museum of Canada Bulletin 180, Contribution to Anthropology Part 1*.

Merbs, C. F., 1989: 'Trauma', pp. 161-189 in M. Y. Iscan and K. A. R. Kennedy (eds.), *Reconstruction of Life from the Skeleton*. New York: Alan Liss Inc.

Meyer, K., 1906: *The Death-tales of the Ulster Heroes* (Royal Irish Academy Todd Lecture Series 14). Dublin: Figgis and Co.

Moorrees, C. F. A., E. A. Fanning and E. E. Hunt Jr., 1963: 'Formation and Resorption of Three Deciduous Teeth in Children', *American Journal of Physical Anthropology* 21, 205-213.

Moskova, M. G., 1994: 'On the Nature of the Similarity and Difference in the Nomadic Cultures of the Eurasian Steppes of the 1st Millennium BC', pp. 231-241 in B. Genito (ed.), *The Archaeology of the Steppes: Methods and Strategies*. Naples: Instituto Universitario Orientale.

Murphy, E. M., 1998: *An osteological and palaeopathological study of the Scythian and Hunno-Sarmatian period populations from the cemetery complex of Aymyrlyg, Tuva, South Siberia*. Unpublished PhD thesis School of Archaeology and Palaeoecology, Queen's University Belfast.

Murphy, E. M., 2000a: 'Developmental defects and disability: the evidence from the Iron Age semi-nomadic peoples of Aymyrlyg, south Siberia', pp. 60-80 in J. Hubert (ed.), *Madness, Disability and Social Exclusion* (One World Archaeology 40). London: Routledge.

Murphy, E. M., 2000b: 'Mummification and Body Processing - Evidence from the Iron Age in Southern Siberia', pp. 279-292 in J. Davis-Kimball, E. Murphy, L. Koryakova and L. Yablonsky (eds.), *Kurgans, Ritual Sites, and Settlements: The Eurasian Bronze and Iron Age* (Papers from the European Association of Archaeologists 4th and 5th Annual Meetings, BAR International Series 890). Oxford: BAR Publishing.

Murphy, E. M., 2001: 'Obsor rezultatov Palaeopatologicheskovo Analiza Pogrebenii Skifskovo Perioda na Mogilnike Aymyrlyg (Tuva)', *Arheologicheskiye Vesti* 8, 125-150 (An overview of the results obtained during the Palaeopathological analysis of the Scythian period burials from the cemetery of Aymyrlyg, Tuva, South Siberia).

Murphy, E. M, 2003: 'Trepanations and Perforated Crania from Iron Age South Siberia: an exercise in differential diagnosis', in R. Arnott, S. Finger and C. U. M. Smith (eds.), *Trepanation: History, Discovery, Theory*. Netherlands: Swets and Zeitlinger, 209-221.

Murphy, E. M., forthcoming a: 'Battle frenzy and brain extraction: A case from Iron Age South Siberia'.

Murphy, E. M., forthcoming b: 'An overview of the results obtained during the Palaeopathological analysis of the Hunno-Sarmatian period burials from the cemetery of Aymyrlyg, Tuva, South Siberia', *Arheologicheskiye Vesti*.

Murphy, E. M. and J. Davis-Kimball, 2001 (published 2002): 'Weapon trauma on Bronze Age human skeletal remains from the Chowhougou cemeteries', *Journal of Palaeopathology* 13, 27-34.

Murphy, E. M., U. M. Donnelly and G. E. Rose, 1998: 'Possible Neurofibromatosis in an individual of the Scythian period from Tuva, South Siberia', *International Journal of Osteoarchaeology* 8, 424-430.

Murphy, E. M., I. I. Gokhman, Y. K. Chistov and L. L.

Barkova, 2002: 'A Review of the evidence for Old World Scalping: New Cases from the cemetery of Aymyrlyg, South Siberia', *American Journal of Archaeology* 106, 1-10.

Murphy, E. M. and J. P. Mallory, 2000: 'Herodotus and the Cannibals', *Antiquity* 74, 388-394.

Murphy, E. M. and J. P. Mallory, 2001: 'Herodotus and the "Cannibals"', *Discovering Archaeology* 3 (1), 16.

Murray, K. A. and T. Murray, 1991: 'A test of the auricular surface ageing technique', *Journal of Forensic Science* 36, 1162-1169.

Okladnikov, A. P., 1970: *Yakutia, before its incorporation into the Russian State*. London: McGill-Queen's Press.

Olsen, S. L. and P. Shipman, 1994: 'Cutmarks and Perimortem Treatment of Skeletal Remains on the Northern Plains', pp. 377-387 in D. W. Owsley and R. L. Jantz (eds.), *Skeletal biology of the Great Plains. Migration, warfare, health and subsistence*. Washington: Smithsonian Institution Press.

Ortner, D. J. and W. G. J. Putschar, 1981: *Identification of Pathological Conditions in Human Skeletal Remains* (Smithsonian Contributions to Anthropology 28). London: Smithsonian Institution Press.

Owsley, D., 1994: 'Warfare in coalescent traditional populations of the Northern Plains', pp. 333-343 in D. W. Owsley and R. L. Jantz (eds.), *Skeletal biology of the Great Plains. Migration, warfare, health and subsistence*. Washington: Smithsonian Institution Press.

Owsley, D., R. W. Mann and T. G. Baugh, 1994: 'Culturally Modified Human Bones from the Edwards I Site', pp. 363-375 in D. W. Owsley and R. L. Jantz (eds.), *Skeletal biology of the Great Plains. Migration, warfare, health and subsistence*. Washington: Smithsonian Institution Press.

Pathria, M., 1995: 'Physical Injury: Spine', pp. 2825-2898 in D. Resnick (ed.), *Diagnosis of Bone and Joint Disorders* (Volume 5). London: W. B. Saunders (3rd edition).

Pavlov, H., 1995: 'Physical Injury: Sports Related Abnormalities', pp. 3229-3263 in D. Resnick (ed.), *Diagnosis of Bone and Joint Disorders* (Volume 5). London: W. B. Saunders (3rd edition).

Phillips, E. D., 1965: *The Royal Hordes: Nomad Peoples of the Steppes*. London: Thames and Hudson.

Phillips, E. D., 1972: 'The Scythian domination in Western Asia: its record in history, scripture and archaeology', *World Archaeology* 4, 129-138.

Polosmak, N., 1994: 'A Mummy Unearthed from the Pastures of Heaven', *National Geographic* 186, 80-103.

Raevskii, D. S., 1982: 'Recent Soviet Work - Introduction', *Soviet Anthropology and Archaeology* 21, 5-32.

Reese, H. H., 1940: 'The History of Scalping and its Clinical Aspects', *Yearbook of Neurology, Psychiatry and Endocriminology*, 3-19.

Resnick, D. and G. Niwayama, 1995: 'Enostosis, Hyperostosis, and Periostitis', pp. 4396-4466 in D. Resnick (ed.), *Diagnosis of Bone and Joint Disorders* (Volume 6). London: W. B. Saunders (3rd edition).

Resnick, D. and T. G. Goergen, 1995: 'Physical Injury: Extraspinal Sites', pp. 2693-2824 in D. Resnick (ed.), *Diagnosis of Bone and Joint Disorders* (Volume 5). London: W. B. Saunders (3rd edition).

Resnick, D., T. G. Goergen and G. Niwayama, 1995: 'Physical Injury: Concepts and Terminology', pp. 2561-2692 in D. Resnick (ed.), *Diagnosis of Bone and Joint Disorders* (Volume 5). London: W. B. Saunders (3rd edition).

Roberts, C., 1991: 'Trauma and treatment in the British Isles in the Historic Period: A design for multidisciplinary research', pp. 225-240 in D. J. Ortner and A. C. Aufderheide (eds.), *Human Paleopathology: Current Syntheses and Future Options*. London: Smithsonian Institution Press.

Roberts, C. A., 1996a: 'Forensic Anthropology 1: The contribution of biological anthropology to forensic contexts', pp. 101-121 in J. Hunter, C. Roberts and A. Martin (eds.), *Studies in Crime: An Introduction to Forensic Archaeology*. London: Routledge.

Roberts, C. A., 1996b: 'Forensic Anthropology 2: Positive identification of the individual; cause and manner of death', pp. 122-138 in J. Hunter, C. Roberts and A. Martin (eds.), *Studies in Crime: An Introduction to Forensic Archaeology*. London: Routledge.

Roberts, C. and K. Manchester, 1995: *The Archaeology of Disease*. Stroud: Alan Sutton Publishing Limited (2nd edition).

Rokhlin, D. G., 1965: *Bolezni Drevnih Loodei (Kosti Loodei Razlichnih Epoh- Normalniye I Patologicheski Izmenenniye)*. Moscow: Nauka (Diseases of Ancient People (Bones of the people of various epochs normal and pathological changed)).

Rolle, R., 1989: *The World of the Scythians*. London: B. T. Batsford Ltd.

Romanes, G. J., 1986: *Cunningham's Manual of Practical Anatomy, Volume One - Upper and lower limbs*. Oxford: Oxford University Press (15th edition).

Rudenko, S. I., 1970: *Frozen Tombs of Siberia: the Pazyryk Burials of Iron Age Horsemen*. London: J. M. Dent & Sons.

Saunders, S. R., 1992: 'Subadult Skeletons and Growth Related Studies', pp. 1-20 in S. R. Saunders and M. A. Katzenberg (eds.), *Skeletal biology of past peoples*. New York: Wiley-Liss Inc.

Schiltz, V., 1994: *Les Scythes et les nomades des steppes*. Verona: Gallimard.

Schmorl, G. and H. Junghanns, 1971: *The Human Spine in Health and Disease*. London: Grune & Stratton (2nd edition).

Sélincourt, A., de and A. R. Burn, 1972: *Herodotus: The Histories*. Middlesex: Penguin Books Ltd. (revised edition).

Semenov, Vl. A., 1996: 'Issledovaniye Rannih Pamyatnikov Skifskoi Kulturi V Tuve', pp. 25-29 in *Novie Arheologicheskiye Otkritiya I Izoocheniye Kulturnoi Transformatsii*. St. Petersburg: Institute for the History of Material Culture (Investigations of Early Scythian Sites in Tuva).

Smith, B. H., 1991: 'Standards of Human Tooth Formation and Dental Age Assessment', pp. 143-168 in M. A. Kelley and C. S. Larsen (eds.), *Advances in Dental Anthropology*. New York: Wiley-Liss Inc.

Stambulnik, E. U., 1978: *Otchet o polevih rabotah 2-go otryada Sayano-Tuvinskoy ekspeditsii na mogilnike Aymyrlyg 1977 godoo*. Unpublished archive report held in the Institute for the History of Material Culture, St. Petersburg (Report of the field work undertaken in 1977 by the 2nd team of the Sayano-Tuvinskyaya Archaeological Expedition at the Aymyrlyg burial ground).

Stambulnik, E. U., 1980: *Otchet o polevih rabotah VI otryada Sayano-Tuvinskoy archaeologicheskoi ekspeditsii LONA AN SSSR v 1979 goddo na mogilnike Aymyrlyg*. Unpublished archive report held in the Institute for the History of Material Culture, St. Petersburg (Report of the field work undertaken in 1979 by the 6th team of the Sayano-Tuvinskyaya Archaeological Expedition at the Aymyrlyg burial ground).

Stambulnik, E. U., 1981: *Otchet o polevih issledovaniyah 6-go otryada Sayano-Tuvinskoy Ekspeditsii LONA AN SSSR v 1980 godoo* Unpublished archive report held in the Institute for the History of Material Culture, St. Petersburg. (Report of the field work undertaken in 1980 by the 6th team of the Sayano-Tuvinskyaya Archaeological Expedition).

Stambulnik, E. U., 1983a: 'Noviye Pamyatniki Gunno-Sarmatskova Vremeni B Tyvye: Nekotoriye Itogi Rabot', pp. 34-41 in *Drevniye Kulturi Euraziiskih Stepei*. Leningrad: Institute for the History of Material Culture (New Monuments of the Hunno-Sarmatian period in Tuva: Some results of the work).

Stambulnik, E. U., 1983b: *Otchet o polevih rabotah VI otryada Sayano-Tuvinskoy archeologicheskoy ekspeditsii LOIA AN SSSR v 1981 godoo*. Unpublished archive report held in the Institute for the History of Material Culture, St. Petersburg (Report of the field work undertaken in 1981 by the 6th team of the Sayano-Tuvinskyaya Archaeological Expedition).

Stokes, W., 1887: 'The Siege of Howth', *Revue Celtique* 8, 47-64.

Stuart-Macadam, P., 1991: 'Anaemia in Roman Britain: Poundbury camp', pp. 101-113 in H. Bush and M. Zvelebil (eds.), *Health in Past Societies* (BAR International Series 567) Oxford: BAR Publishing.

Suchey, J. M., D. V. Wisely and D. Katz, 1986: 'Evaluation of the Todd and McKern-Stewart methods of ageing the male *Os pubis*', pp. 33-67 in K. Reichs (ed.), *Forensic Osteology: Advances in the identification of human remains*. Springfield: Charles C. Thomas.

Sulimirski, T., 1970a: *Prehistoric Russia*. London: John Baker Ltd.

Sulimirski, T., 1970b: *The Sarmatians*. London: Thames and Hudson.

Talbot Rice, T., 1966: 'Introduction', pp. 9-83 in M. I. Artamonov (ed.), *Treasures from Scythian Tombs*. London: Thames and Hudson.

Tallgren, A. M., 1936: 'The South Siberian Cemetery of Oglakty from the Han Period', *Eurasia Septentrionalis Antiqua* 11, 69-90.

Taylor, T., 1994: 'Thracians, Scythians and Dacians 800 BC - AD 300', pp. 373-410 in B. Cunliffe (ed.), *The Oxford Illustrated Prehistory of Europe*. London: BCA.

Todd, T. W., 1920: 'Age changes in the pubic bone I: The male white pubic', *American Journal of Physical Anthropology* 3, 285-334.

Trotter, M., 1970: 'Estimation of Stature from Intact Long Limb Bones', pp. 71-83 in T. D. Stewart (ed.), *Personal Identification in Mass Disasters*. Washington: Smithsonian Institution.

Trotter, M. and G. C. Gleser, 1952: 'Estimation of Stature from Long Bones of American Whites and Negroes', *American Journal of Physical Anthropology* 10, 463-514.

Trotter, M. and G. C. Gleser, 1958: 'A Re-Evaluation of Estimation of Stature Taken During Life and of Long Bones after Death', *American Journal of Physical Anthropology* 16, 79-123.

Ubelaker, D. H., 1989a: *Human Skeletal Remains* (Manuals on Archaeology 2). Washington: Smithsonian Institution (2nd edition).

Ubelaker, D. H., 1989b: 'The Estimation of Age at Death from Immature Human Bone', pp. 55-70 in M. Y. Iscan (ed.), *Age Markers In The Human Skeleton*. Springfield: Charles C. Thomas.

Vainshtein, S. I., 1958: 'Nekotorie itogi rabot arkheologicheskoi ekspeditsii TuvNIIYaLI v 1956-1957', *Uch. zap. YuvNIIYYaLi T. 6* (Some Results of the Work of the Archaeological Expedition of YuvNIIyaLI in 1956-1957).

Vainshtein, S. I., 1978: 'The Problem of the Origin and Formation of the Economic-Cultural Type of Pastoral Nomads in the Moderate Belt of Eurasia', pp. 127-133 in W. Weissleder (ed.), *The Nomadic Alternative*. Paris: Mouton Publishers.

Vainshtein, S. I., 1980: *Nomads of South Siberia: The Pastoral Economies of Tuva*. Cambridge: Cambridge University Press.

Vernadsky, G., 1943: *Ancient Russia*. London: Oxford University Press.

Wakely, J., 1997: 'Identification and analysis of violent and non-violent head injuries in the osteoarchaeological material', pp. 24-46 in J. Carman (ed.), *Material Harm, archaeological studies of war and violence*. Glasgow: Cruithne Press.

Watson, B., 1961: *Records of the Grand Historian of China Translated from the Shih Chi of Ssu-Ma Ch'ien*. London: Columbia University Press.

Wells, C., 1974: 'Osteochondritis Dissecans in Ancient British Skeletal Material', *Medical History* 18, 365-369.

Williams, J. P. R, 1988: *Barron's Sports Injuries Handbook*. New York: Barron's.

Woodward, A. H. and A. J. Bianco, 1975: 'Osteochondritis Dissecans of the Elbow', *Clinical Orthopaedics and Related Research* 110, 35-41.

Yü, Ying-Shih, 1990: 'The Hsiung-nu', pp. 118-149 in D. Sinor (ed.), *The Cambridge History of Early Inner Asia*. Cambridge: Cambridge University Press.

Zadneprovskiy, Y. A., 1994: 'The Nomads of Northern Central Asia after the Invasion of Alexander', pp. 457-472 in J. Harmatta (ed.), *History of Civilisations of Central Asia, Vol. III, the Development of Sedentary and Nomadic Civilisations: 700 BC to AD 250*. Paris: UNESCO.

Appendices

Appendix 1. Preservation of the skeletal elements

Scythian period individuals

Preservation of skeletons

The skeletal remains of the Scythian period individuals displayed a wide range of preservation, from those individuals which were in a very poor state of preservation to those which were in an excellent state of preservation. The remains of several individuals displayed soft tissue preservation. In many instances the tissue had been preserved as a result of its being positioned in close proximity to bronze artefacts. The action of the copper ions present in the bronze objects was probably directly responsible for the preservation of this soft-tissue material (Janaway 1987, 135). In addition, the remains of venous-arterial material, which contained well preserved red blood cells, were recovered from a rib sample. The rib had displayed no external signs of soft-tissue preservation and it is probable that soft-tissue remains are to be found within the bones of other individuals from Aymyrlyg.

Adults

The relative bone survival for the 436 adults of Scythian period date was examined and the data is included in Table 35. If survival and recovery had been complete there would have 436 examples of the various bones from each side of the body. The only exception was for the hand and foot phalanges which were grouped together into proximal (R and L1), intermediate (R and L2) and distal (R and L3) groups. Consequently, it was expected that there would have been 2180 proximal phalanges, 1744 intermediate phalanges and 2180 distal phalanges. It was evident that certain skeletal elements were grossly under-represented, such as the coccyx for which only 0.9% of the number expected was present. The survival of different skeletal elements ranged from 0.05% for the left 2nd (intermediate) foot phalanges to 80.7% of the right sides of the frontal bones. The minimum number of individuals ranged from one for the left intermediate foot phalanges to 352 for the right side of the frontal bones. The prevalence rates of the various palaeopathological conditions are based on the values included in Table 35.

A graphical portrayal of the mean relative survival of the main skeletal elements, presented in Figure 7, clearly illustrates the relative preservation of the various skeletal elements. It is evident that complete crania and mandibles were most often represented, followed by the long bones. Smaller bones such as the patellae, carpals and tarsals were most poorly preserved. In addition, more delicate bones such as the ribs were markedly under-represented. It is probable that the larger, more robust, bones were in a better state of preservation and had a greater chance of collection by the archaeologists during the excavation. It is known that many of the skeletal elements were not curated due to lack of a reliable context (Prof. Ilyia Gokhman 1996, pers. comm.). It is also likely, therefore, that the relative survival of the skeletal elements was influenced by the preferential selection of elements with reliable contexts by the anthropologists. Given the history of the assemblage and the length of time since the skeletons were excavated it is not possible to make inferences regarding cultural practices based on the selection of particular skeletal elements for burial.

Subadults

The relative bone survival for the 171 subadults of Scythian period date was examined and the data are included in Table 36. If survival and recovery had been complete there would have 171 examples of the various bones from each side of the body. As was the case for the adult individuals, the only exception was for the hand and foot phalanges which were grouped together into proximal (R and L1), intermediate (R and L2) and distal (R and L3) groups. It was expected, therefore, that there would have been 855 proximal phalanges, 684 intermediate phalanges and 855 distal phalanges. It was evident that certain skeletal elements were grossly under-represented, such as the right patella, for which only 1.2% of the number expected was present, or that the elements were not represented at all, for example the sternal body and the coccyx. The survival of different elements ranged from 0% (coccyx, sternal body, carpals, several tarsals and some of the foot phalanges) to 70.8% of the left sides of the frontal bones. The minimum number of individuals ranged from 0 to 121 (left side of the frontal bones). The prevalence rates of the various palaeopathological conditions were based on the values included in Table 36.

A graphical portrayal of the mean relative survival of the main skeletal elements for the Scythian period subadults is presented in Figure 8. It is evident that complete mandibles were most often represented, followed by complete crania and long bones. As was the case for the adult individuals, smaller bones such as the patellae, carpals and tarsals were most poorly preserved. The proportion of ribs present was notably higher for subadults (16.2%) relative to the adult (5.4%) individuals. None of the subadult individuals had completely preserved sacra while 29.8% of the adult individuals were associated with complete sacra. As was the case for the adult remains, it is probable that the larger, more robust bones were in a better state of preservation and had a greater chance of collection by the archaeologists during the excavation. It is likely that the relative survival of the skeletal remains was influenced by the preferential selection of elements with reliable contexts by the anthropologists during curation.

Hunno-Sarmatian individuals

Preservation of skeletons

The skeletal remains of the Hunno-Sarmatian period individuals displayed a wide range of preservation. In some individuals the remains were in a very poor state of preservation, while in others the remains were in an excellent condition. As was the case with the Scythian period individuals, the remains of several individuals displayed soft tissue preservation.

Adults

The relative bone survival for the 148 Hunno-Sarmatian period adults was examined and the data is presented in Table 37. If survival and recovery had been complete there would have been 148 cases of the various bones from each side of the body. The only exception was for the hand and foot phalanges which were grouped together into proximal (R and L1), intermediate (R and L2), and distal (R and L3) groups. Consequently, it was expected that there would have been 740 proximal and distal phalanges and 592 intermediate phalanges. In general, the remains were well preserved with only the smaller bones (such as the patellae, sterna and bones of the hands and feet) being notably under-represented. The survival of different elements ranged from 0.2% for the right 2nd (intermediate) foot phalanges to 93.2% for the left femora, the right distal femora, and the left tibiae. The minimum number of individuals ranged from one for the right 2nd (intermediate) phalanges to 138 for the left femora, the right distal femora, and the left tibiae. The prevalence rates of the various palaeopathological conditions are based on the values included in Table 37.

A graphical representation of the mean relative survival of the main skeletal elements is provided in Figure 9. It is evident from the graph that the various skeletal elements were generally very well represented in the population group. Only the smaller bones, such as the phalanges, were not present in high numbers. It is probable that the poor representation of the smaller bones is partly due to their state of preservation, and to the lack of their systematic recovery by the archaeologists during the excavation. In addition, it is probable that the smaller bones would have had less secure contexts than the larger bones, and that many of them were discarded during the process of selective curation adopted by the anthropologists. It is unlikely that there were any cultural factors dictating the selection of particular skeletal elements for burial in this population group.

Subadults

The relative survival of the various skeletal elements of the fifty-four subadult individuals of Hunno-Sarmatian period date was examined and the data is included in Table 38. If survival and recovery had been complete there would have been fifty-four examples of the various bones from each side of the body. The only exception was for the hand and foot phalanges, for which there would have been 270 proximal and distal phalanges and 216 intermediate phalanges. In general the skeletal elements were very well represented, with the exception of the smaller bones of the hands and feet. The survival of the various elements ranged from 0% for the coccyx, the left 2nd cuneiform, the right 3rd cuneiform, and several of the foot phalanges, to 87% for the right humerus. The minimum number of individuals ranged from zero (coccyx, the left 2nd cuneiform, the right 3rd cuneiform, and several of the foot phalanges) to forty-seven for the right humerus. The prevalence rates of the various palaeopathological lesions are based on the values included in Table 38.

A graphical representation of the mean relative survival of the main skeletal elements is provided in Figure 10. It is evident from the graph that the larger bones were generally very well represented, while the bones of the hands and feet were particularly poorly represented. The proportion of ribs present was notably higher for subadults (72%) relative to the adult individuals (13.5%). As was the case for the adult individuals, it is probable that the larger, more robust bones were in a better state of preservation and had a greater chance of collection by the archaeologists during the excavation. It is also probable that the relative survival of the skeletal elements was influenced by the preferential selection of elements with reliable archaeological contexts by the anthropologists during curation.

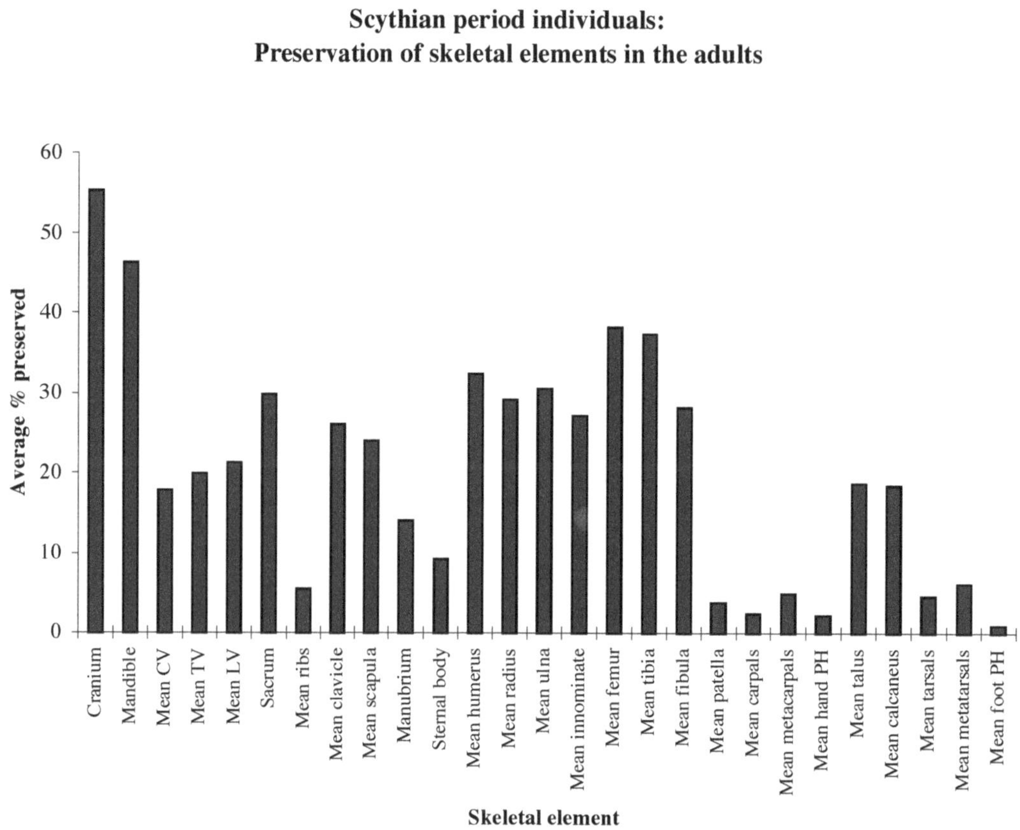

Figure 7: Scythian period individuals - Mean preservation of the main skeletal elements in the adults.

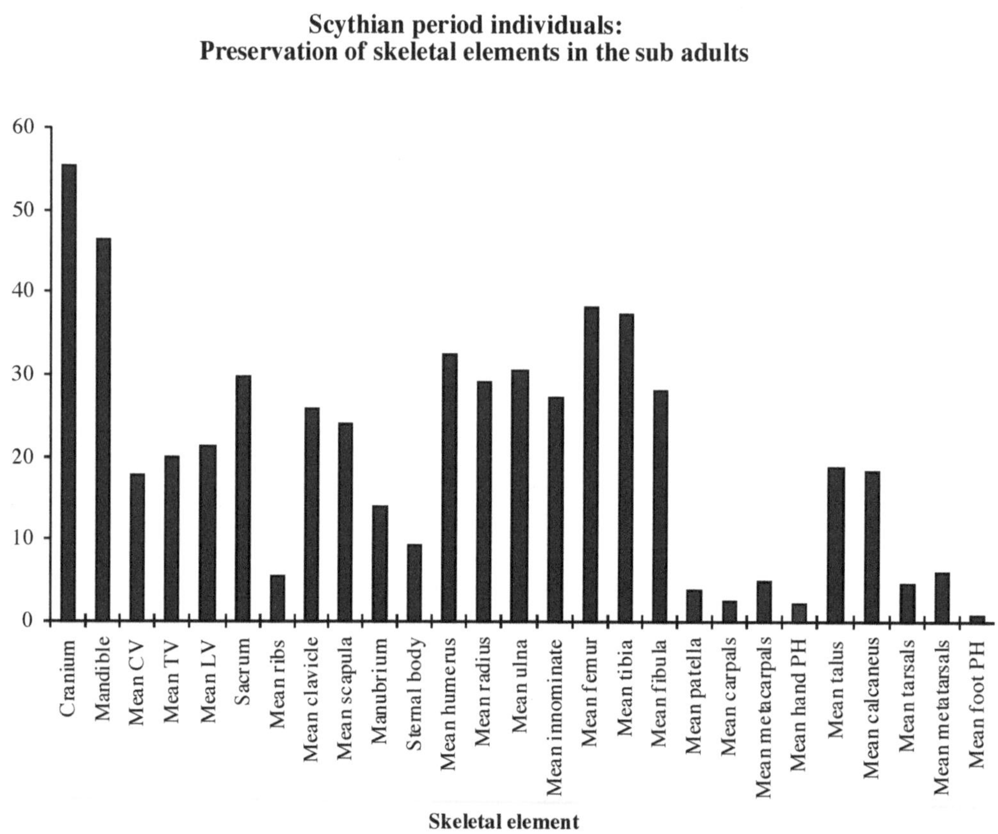

Figure 8: Scythian period individuals - Mean preservation of the main skeletal elements in the subadults.

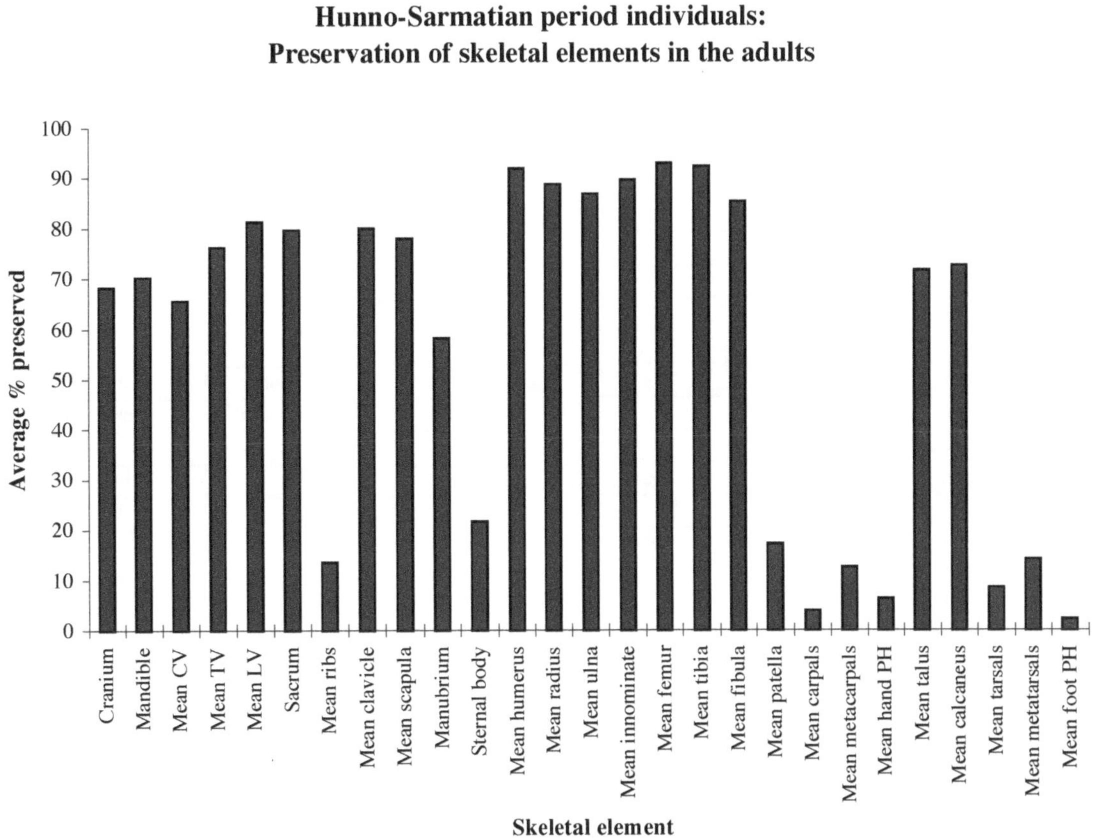

Figure 9: Hunno-Sarmatian period individuals - Mean preservation of the main skeletal elements in the adults.

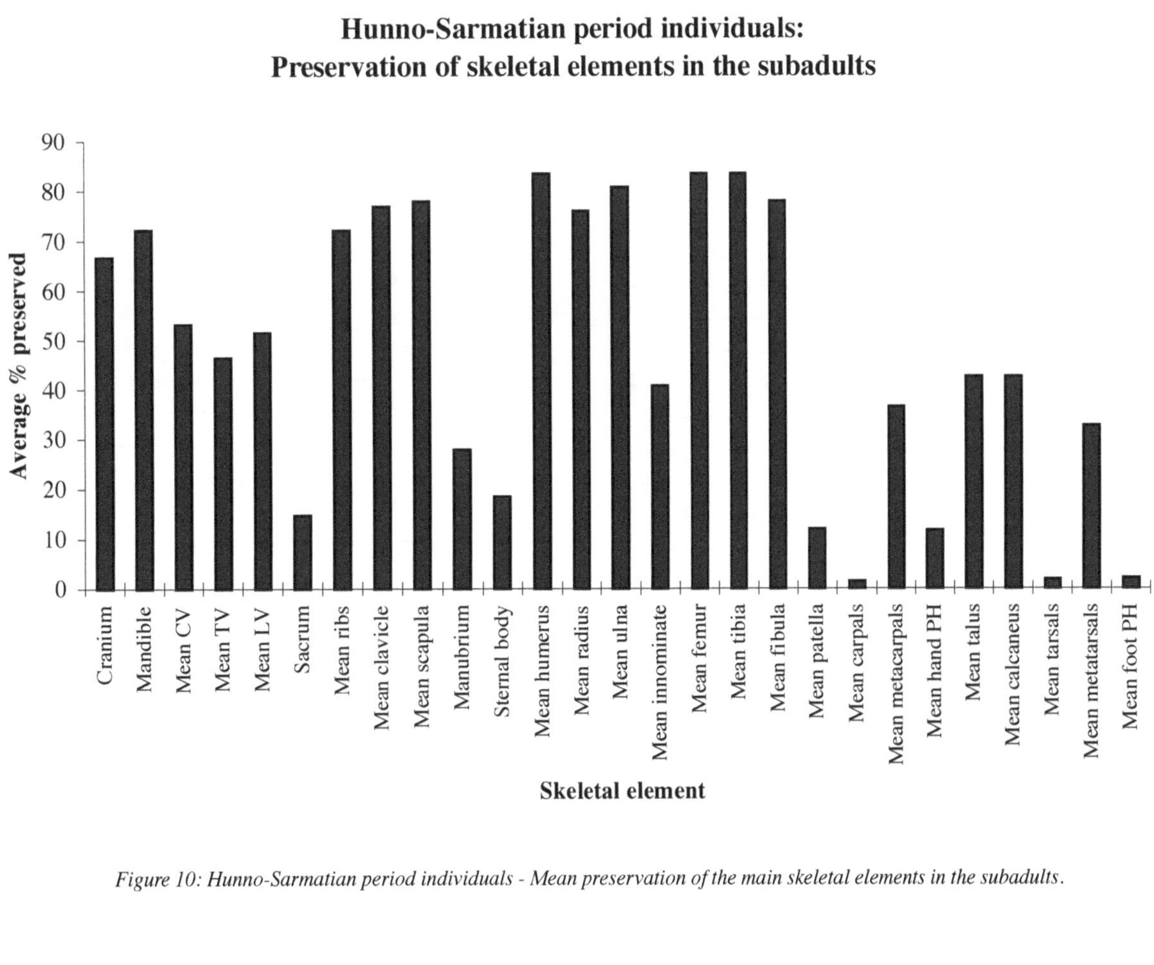

Figure 10: Hunno-Sarmatian period individuals - Mean preservation of the main skeletal elements in the subadults.

Minimum number of bones values

Skeletal element	Side	No. Recovered	% of expected
Skull			
Frontal	L	350	80.3
	R	352	80.7
Parietal	L	345	79.1
	R	350	80.3
Occipital	L	339	77.8
	R	338	77.5
Occipital condyle	L	313	71.8
	R	312	71.6
Petrous temporal	L	318	72.9
	R	321	73.6
Mastoid	L	321	73.6
	R	326	74.8
Zygomatic	L	313	71.8
	R	329	75.4
Maxilla	L	332	76.1
	R	331	75.9
Nasal bones	L	338	77.5
	R	337	77.3
Complete (90%+) cranium		241	55.3
Mandible			
Head	L	230	52.8
	R	233	53.4
Ramus	L	244	56
	R	246	56.4
Body	L	245	56.2
	R	251	57.6
Complete		202	46.3
Post-cranial skeleton			
Vertebrae	CV1	83	19
	CV2	82	18.8
	CV3	66	15
	CV4	74	17
	CV5	74	17
	CV6	76	17.4
	CV7	88	20.2
	TV1	84	19.3
	TV2	79	18.1
	TV3	85	19.5
	TV4	82	18.8
	TV5	91	20.9
	TV6	90	20.6
	TV7	89	20.4
	TV8	90	20.6
	TV9	89	20.4

Skeletal element	Side	No. Recovered	% of expected
	TV10	86	19.7
	TV11	87	20
	TV12	90	20.6
	LV1	88	20.2
	LV2	96	22.2
	LV3	96	22.2
	LV4	90	20.6
	LV5	90	20.6
	S1	140	32.1
	S2	139	31.9
	S3	137	31.4
	S4	133	30.5
	S5	133	30.5
	coccyx	4	0.9
Sacrum	complete	130	29.8
Rib heads	L 1	17	3.9
	L 2	24	5.5
	L 3	26	6
	L 4	28	6.4
	L 5	27	6.2
	L 6	23	5.3
	L 7	30	6.9
	L 8	31	7.1
	L 9	27	6.2
	L 10	23	5.3
	L 11	16	3.7
	L 12	14	3.2
	R 1	15	3.4
	R 2	24	5.5
	R 3	27	6.2
	R 4	26	6
	R 5	27	6.2
	R 6	25	5.7
	R 7	23	5.3
	R 8	26	6
	R 9	24	5.5
	R 10	28	6.4
	R 11	22	5
	R 12	18	4.1
Clavicle	L sternal	107	24.5
	L acromial	107	24.5
	L complete	107	24.5
	R sternal	120	27.5
	R acromial	120	27.5
	R complete	120	27.5
Scapula	L body	105	24.1

Table 35: Proportion of bones recovered from the 436 adult individuals from the Scythian period.

Skeletal element	Side	No. Recovered	% of expected
	L glenoid	106	24.3
	L acromion	103	23.6
	L coracoid	106	24.3
	L complete	102	23.4
	R body	110	25.2
	R glenoid	111	25.4
	R acromion	109	25
	R coracoid	109	25
	R complete	107	24.5
Sternum	manubrium	61	14
	body	40	9.2
Humerus	L proximal	141	32.3
	L distal	149	34.2
	L complete	137	31.4
	R proximal	149	34.2
	R distal	155	35.6
	R complete	146	33.5
Radius	L proximal	129	29.6
	L distal	128	29.4
	L complete	126	28.9
	R proximal	136	31.2
	R distal	132	30.3
	R complete	129	29.6
Ulna	L proximal	132	30.3
	L distal	130	29.8
	L complete	130	29.8
	R proximal	141	32.3
	R distal	138	31.7
	R complete	137	31.4
Femur	L proximal	169	38.9
	L distal	162	37.2
	L complete	163	37.4
	R proximal	175	40.1
	R distal	173	39.7
	R complete	170	39
Tibia	L proximal	164	37.6
	L distal	165	37.8
	L complete	162	37.2
	R proximal	165	37.8
	R distal	165	37.8
	R complete	164	37.6
Fibula	L proximal	124	28.4

Skeletal element	Side	No. Recovered	% of expected
	L distal	129	29.6
	L complete	121	27.8
	R proximal	126	28.9
	R distal	131	30
	R complete	125	28.7
Patella	L	20	4.6
	R	13	3
Carpals	L scaphoid	15	3.4
	L lunate	12	2.8
	L triquetral	4	0.9
	L pisiform	6	1.4
	L trapezium	13	3
	L trapezoid	8	1.8
	L capitate	17	3.9
	L hamate	12	2.8
	R scaphoid	16	3.7
	R lunate	12	2.8
	R triquetral	5	1.1
	R pisiform	5	1.1
	R trapezium	8	1.8
	R trapezoid	9	2.1
	R capitate	16	3.7
	R hamate	12	2.8
Metacarpals	L 1	18	4.1
	L 2	24	5.5
	L 3	22	5
	L 4	22	5
	L 5	23	5.3
	R 1	15	3.4
	R 2	23	5.3
	R 3	26	6
	R 4	21	4.8
	R 5	22	5
Hand PH	L 1	90	4.1
	L 2	39	2.2
	L 3	23	1
	R 1	82	3.8
	R 2	34	1.9
	R 3	11	0.5
Tarsals	L calcaneus	82	18.8
	L talus	78	17.9
	L 1st cuneiform	27	6.2
	L 2nd cuneiform	17	3.9
	L 3rd cuneiform	18	4.1

Table 35: Proportion of bones recovered from the 436 adult individuals from the Scythian period.

Skeletal element	Side	No. Recovered	% of expected
	L cuboid	26	6
	L navicular	24	5.5
	R calcaneus	79	18.1
	R talus	85	19.5
	R 1st cuneiform	22	5
	R 2nd cuneiform	14	3.2
	R 3rd cuneiform	16	3.7
	R cuboid	22	5
	R navicular	24	5.5
Metatarsals	L 1	31	7.1
	L 2	25	5.7
	L 3	24	5.5
	L 4	23	5.3
	L 5	24	5.5
	R 1	33	7.6
	R 2	30	6.9
	R 3	30	6.9
	R 4	24	5.5
	R 5	23	5.3
Foot PH	L 1	51	2.3
	L 2	1	0.05
	L 3	4	0.2
	R 1	43	2.5
	R 2	5	0.3
	R 3	5	0.3
Innominate			
acetabulum	L	147	33.7
	R	154	35.3
Pubic symphysis	L	134	30.7
	R	145	33.2
Ischial tuberosity	L	125	28.7
	R	140	32.1
Sciatic notch	L	147	33.7
	R	156	35.8
Auricular surface	L	147	33.7
	R	154	35.3
Complete	L	114	26.1
	R	124	28.4
	R 2	34	1.9
	R 3	11	0.5

Table 35: Proportion of bones recovered from the 436 adult individuals from the Scythian period.

Skeletal element	Side	No. Recovered	% expected
Skull			
Frontal	L	121	70.8
	R	119	69.6
Parietal	L	119	69.6
	R	115	67.2
Occipital	L	114	66.7
	R	116	67.8
Occipital condyle	L	112	65.5
	R	109	63.7
Petrous temporal	L	104	60.8
	R	105	61.4
Mastoid	L	100	58.5
	R	104	60.8
Zygomatic	L	108	63.2
	R	110	64.3
Maxilla	L	118	69
	R	116	67.8
Nasal bones	L	95	55.6
	R	95	55.6
Complete cranium		73	42.7
Mandible			
Head	L	104	60.8
	R	111	64.9
Ramus	L	106	62
	R	112	65.5
Body	L	111	64.9
	R	114	66.7
Complete		100	58.5
Post-cranial skeleton			
Vertebrae	CV1	20	11.7
	CV2	23	13.4
	CV3	21	12.3
	CV4	23	13.4
	CV5	15	8.8
	CV6	19	11.1
	CV7	17	9.9
	TV1	17	9.9
	TV2	11	6.4
	TV3	14	8.2
	TV4	17	9.9
	TV5	19	11.1
	TV6	16	9.4
	TV7	15	8.8
	TV8	15	8.8
	TV9	17	9.9
	TV10	15	8.8

Table 36: Proportion of bones recovered from the 171 subadult individuals from the Scythian period.

Skeletal element	Side	No. Recovered	% of expected
	TV11	20	11.7
	TV12	21	12.3
	LV1	30	17.5
	LV2	24	14
	LV3	28	16.4
	LV4	26	15.2
	LV5	30	17.5
	S1	26	15.2
	S2	17	9.9
	S3	11	6.4
	S4	6	3.5
	S5	1	0.6
	coccyx	0	0
Sacrum	complete	0	0
Rib heads	L 1	30	17.5
	L 2	26	15.2
	L 3	29	17
	L 4	33	19.3
	L 5	39	22.8
	L 6	32	18.7
	L 7	29	17
	L 8	25	14.6
	L 9	27	15.8
	L 10	25	14.6
	L 11	19	11.1
	L 12	18	10.5
	R 1	27	15.8
	R 2	25	14.6
	R 3	31	18.1
	R 4	33	19.3
	R 5	32	18.7
	R 6	29	17
	R 7	30	17.5
	R 8	27	15.8
	R 9	25	14.6
	R 10	24	14
	R 11	24	14
	R 12	25	14.6
Clavicle	L sternal	40	23.4
	L acromial	40	23.4
	L complete	40	23.4
	R sternal	45	26.3
	R acromial	45	26.3
	R complete	45	26.3
Scapula	L body	45	26.3
	L glenoid	46	26.9

Skeletal element	Side	No. Recovered	% of expected
	L acromion	45	26.3
	L coracoid	45	26.3
	L complete	45	26.3
	R body	47	27.5
	R glenoid	47	27.5
	R acromion	47	27.5
	R coracoid	47	27.5
	R complete	47	27.5
Sternum	manubrium	5	2.9
	body	0	0
Humerus	L proximal	59	34.5
	L distal	59	34.5
	L complete	59	34.5
	R proximal	56	32.7
	R distal	56	32.7
	R complete	56	32.7
Radius	L proximal	38	22.2
	L distal	37	21.6
	L complete	37	21.6
	R proximal	34	19.9
	R distal	34	19.9
	R complete	33	19.3
Ulna	L proximal	41	24
	L distal	41	24
	L complete	41	24
	R proximal	41	24
	R distal	37	21.6
	R complete	37	21.6
Femur	L proximal	65	38
	L distal	64	37.4
	L complete	64	37.4
	R proximal	63	36.8
	R distal	62	36.2
	R complete	62	36.2
Tibia	L proximal	55	32.2
	L distal	54	31.6
	L complete	54	31.6
	R proximal	56	32.7
	R distal	55	32.2
	R complete	55	32.2
Fibula	L proximal	39	22.8
	L distal	41	24
	L complete	39	22.8

Table 36: Proportion of bones recovered from the 171 subadult individuals from the Scythian period.

Skeletal element	Side	No. Recovered	% of expected
	R proximal	40	23.4
	R distal	40	23.4
	R complete	39	22.8
Patella	L	3	1.8
	R	2	1.2
Carpals	L	0	0
	R	0	0
Metacarpals	L 1	3	1.8
	L 2	9	5.3
	L 3	6	3.5
	L 4	6	3.5
	L 5	4	2.3
	R 1	8	4.7
	R 2	8	4.7
	R 3	7	4.1
	R 4	7	4.1
	R 5	4	2.3
Hand PH	L 1	17	9.9
	L 2	2	1.2
	L 3	1	0.6
	R 1	16	9.4
	R 2	3	1.8
	R 3	1	0.6
Tarsals	L calcaneus	19	11.1
	L talus	19	11.1
	L 1st cuneiform	1	0.6
	L 2nd cuneiform	0	0
	L 3rd cuneiform	0	0
	L cuboid	0	0
	L navicular	0	0
	R calcaneus	22	12.9
	R talus	22	12.9
	R 1st cuneiform	0	0
	R 2nd cuneiform	0	0
	R 3rd cuneiform	0	0
	R cuboid	0	0
	R navicular	0	0
Metatarsals	L 1	6	3.5
	L 2	9	5.3
	L 3	9	5.3
	L 4	5	2.9
	L 5	5	2.9
	R 1	8	4.7

Skeletal element	Side	No. Recovered	% of expected
	R 2	6	3.5
	R 3	8	4.7
	R 4	5	2.9
	R 5	4	2.3
Foot PH	L 1	2	1.2
	L 2	0	0
	L 3	0	0
	R 1	2	1.2
	R 2	0	0
	R 3	0	0
Innominate			
Ilium	L	53	30.9
	R	55	32.2
Ischium	L	28	16.4
	R	31	18.1
Pubis	L	34	19.9
	R	38	22.2
Complete	L	20	11.7
	R	22	12.9

Table 36: Proportion of bones recovered from the 171 subadult individuals from the Scythian period.

Skeletal element	Side	No. Recovered	% of expected
Skull			
Frontal	L	110	74.3
	R	109	73.6
Parietal	L	109	73.6
	R	109	73.6
Occipital	L	109	73.6
	R	109	73.6
Occipital condyle	L	107	72.3
	R	107	72.3
Petrous temporal	L	108	73
	R	108	73
Mastoid	L	108	73
	R	108	73
Zygomatic	L	108	73
	R	107	72.3
Maxilla	L	107	72.3
	R	108	73
Nasal bones	L	108	73
	R	107	72.3
Complete (90%+) cranium		101	68.2
Mandible			
Head	L	105	70.9
	R	104	70.3
Ramus	L	106	71.6
	R	105	70.9
Body	L	106	71.6
	R	105	70.9
Complete		104	70.3
Post-cranial skeleton			
Vertebrae	CV1	96	64.9
	CV2	103	69.6
	CV3	95	64.2
	CV4	92	62.2
	CV5	94	63.5
	CV6	98	66.2
	CV7	102	68.9
	TV1	112	75.7
	TV2	111	75
	TV3	112	75.7
	TV4	111	75
	TV5	112	75.7
	TV6	113	76.4
	TV7	114	76.4
	TV8	114	76.4
	TV9	115	77.7
	TV10	112	75.7
	TV11	114	77
	TV12	116	78.4
	LV1	117	79
	LV2	120	81.1
	LV3	120	81.1
	LV4	122	82.4
	LV5	123	83.1
	S1	128	86.5
	S2	127	85.8
	S3	126	85.1
	S4	125	84.4
	S5	122	82.4
	coccyx	12	8.1
Sacrum	complete	118	79.7
Rib heads	L 1	17	11.5
	L 2	21	14.2
	L 3	22	14.9
	L 4	20	13.5
	L 5	21	14.2
	L 6	22	14.9
	L 7	21	14.2
	L 8	21	14.2
	L 9	22	14.9
	L 10	20	13.5
	L 11	19	12.8
	L 12	16	10.8
	R 1	14	9.4
	R 2	20	13.5
	R 3	21	14.2
	R 4	21	14.2
	R 5	22	14.9
	R 6	23	15.5
	R 7	22	14.9
	R 8	22	14.9
	R 9	22	14.9
	R 10	19	12.8
	R 11	19	12.8
	R 12	14	9.4
Clavicle	L sternal	120	81.1
	L acromial	120	81.1
	L complete	120	81.1
	R sternal	117	79
	R acromial	117	79
	R complete	117	79
Scapula	L body	115	77.7
	L glenoid	117	79

Table 37: Proportion of bones recovered from the 148 adult individuals from the Hunno-Sarmatian period.

	L acromion	117	79
	L coracoid	117	79
	L complete	115	77.7
	R body	116	78.4
	R glenoid	118	79.7
	R acromion	118	79.7
	R coracoid	118	79.7
	R complete	116	78.4
Sternum	manubrium	86	58.1
	body	32	21.6
Humerus	L proximal	137	92.6
	L distal	136	91.9
	L complete	136	91.9
	R proximal	136	91.9
	R distal	136	91.9
	R complete	136	91.9
Radius	L proximal	131	88.5
	L distal	130	87.8
	L complete	130	87.8
	R proximal	134	90.5
	R distal	133	89.9
	R complete	133	89.9
Ulna	L proximal	129	87.2
	L distal	128	86.5
	L complete	128	86.5
	R proximal	131	88.5
	R distal	129	87.2
	R complete	129	87.2
Femur	L proximal	138	93.2
	L distal	138	93.2
	L complete	138	93.2
	R proximal	137	92.6
	R distal	138	93.2
	R complete	137	92.6
Tibia	L proximal	138	93.2
	L distal	138	93.2
	L complete	138	93.2
	R proximal	135	91.2
	R distal	136	91.9
	R complete	135	91.2
Fibula	L proximal	128	86.5
	L distal	129	87.2
	L complete	128	86.5

Skeletal element	**Side**	**No. Recovered**	**% of expected**
	R proximal	125	84.4
	R distal	127	85.8
	R complete	125	84.4
Patella	L	29	19.6
	R	22	14.9
Carpals	L scaphoid	5	3.4
	L lunate	3	2
	L triquetral	3	2
	L pisiform	6	4
	L trapezium	7	4.7
	L trapezoid	6	4
	L capitate	8	5.4
	L hamate	8	5.4
	R scaphoid	12	8.1
	R lunate	3	2
	R triquetral	4	2.7
	R pisiform	2	1.4
	R trapezium	5	3.4
	R trapezoid	7	4.7
	R capitate	9	6.1
	R hamate	5	3.4
Metacarpals	L 1	19	12.8
	L 2	18	12.2
	L 3	17	11.5
	L 4	17	11.5
	L 5	19	12.8
	R 1	17	11.5
	R 2	21	14.2
	R 3	22	14.9
	R 4	20	13.5
	R 5	15	10.1
Hand PH	L 1	63	8.5
	L 2	28	4.7
	L 3	15	2
	R 1	70	9.4
	R 2	30	5.1
	R 3	14	1.9
Tarsals	L calcaneus	106	71.6
	L talus	103	69.6
	L 1st cuneiform	17	11.5
	L 2nd cuneiform	8	5.4
	L 3rd cuneiform	13	8.8
	L cuboid	16	10.8
	L navicular	16	10.8

Table 37: Proportion of bones recovered from the 148 adult individuals from the Hunno-Sarmatian period.

	R calcaneus	109	73.6
	R talus	109	73.6
	R 1st cuneiform	14	9.4
	R 2nd cuneiform	7	4.7
	R 3rd cuneiform	10	6.8
	R cuboid	16	10.8
	R navicular	16	10.8
Metatarsals	L 1	21	14.2
	L 2	23	15.5
	L 3	20	13.5
	L 4	21	14.2
	L 5	20	13.5
	R 1	22	14.9
	R 2	22	14.9
	R 3	21	14.2
	R 4	20	13.5
	R 5	19	12.8
Foot PH	L 1	41	5.5
	L 2	2	0.3
	L 3	6	0.8
	R 1	43	5.8
	R 2	1	0.2
	R 3	6	0.8
Innominate			
acetabulum	L	134	90.5
	R	133	89.9
Pubic symphysis	L	131	88.5
	R	131	88.5
Ischial tuberosity	L	131	88.5
	R	133	89.9
Sciatic notch	L	134	90.5
	R	135	91.2
Auricular surface	L	134	90.5
	R	135	91.2
Complete	L	132	89.2
	R	133	89.9

Table 37: Proportion of bones recovered from the 148 adult individuals from the Hunno-Sarmatian period.

Skeletal element	Side	No. Recovered	% of expected
Skull			
Frontal	L	46	85.2
	R	47	87
Parietal	L	47	87
	R	47	87
Occipital	L	45	83.3
	R	45	83.3
Occipital condyle	L	44	81.5
	R	44	81.5
Petrous temporal	L	46	85.2
	R	43	79.6
Mastoid	L	46	85.2
	R	43	79.6
Zygomatic	L	44	81.5
	R	40	74.1
Maxilla	L	45	83.3
	R	43	79.6
Nasal bones	L	45	83.3
	R	45	83.3
Complete cranium		36	66.7
Mandible			
Head	L	43	79.6
	R	39	72.2
Ramus	L	43	79.6
	R	41	75.9
Body	L	43	79.6
	R	41	75.9
Complete		39	72.2
Post-cranial skeleton			
Vertebrae	CV1	29	53.7
	CV2	30	55.5
	CV3	29	53.7
	CV4	28	51.8
	CV5	29	53.7
	CV6	29	53.7
	CV7	27	50
	TV1	25	46.3
	TV2	26	48.1
	TV3	26	48.1
	TV4	24	44.4
	TV5	25	46.3
	TV6	25	46.3
	TV7	25	46.3
	TV8	24	44.4
	TV9	24	44.4
	TV10	25	46.3

Table 38: Proportion of bones recovered from the 54 subadult individuals from the Hunno-Sarmatian period.

Skeletal element	Side	No. Recovered	% of expected
	TV11	25	46.3
	TV12	27	50
	LV1	26	48.1
	LV2	29	53.7
	LV3	27	50
	LV4	27	50
	LV5	30	55.6
	S1	25	46.3
	S2	20	37
	S3	18	33.3
	S4	14	25.9
	S5	10	18.5
	coccyx	0	0
Sacrum	complete	8	14.8
Rib heads	L 1	34	63
	L 2	40	74.1
	L 3	41	75.9
	L 4	39	72.2
	L 5	41	75.9
	L 6	38	70.4
	L 7	40	74.1
	L 8	37	68.5
	L 9	38	70.4
	L 10	39	72.2
	L 11	32	59.2
	L 12	33	61.1
	R 1	33	61.1
	R 2	37	68.5
	R 3	40	74.1
	R 4	45	83.3
	R 5	48	88.9
	R 6	45	83.3
	R 7	45	83.3
	R 8	40	74.1
	R 9	42	77.8
	R 10	37	68.5
	R 11	37	68.5
	R 12	32	59.2
Clavicle	L sternal	39	72.2
	L acromial	40	74.1
	L complete	39	72.2
	R sternal	44	81.5
	R acromial	44	81.5
	R complete	44	81.5
Scapula	L body	42	77.8
	L glenoid	42	77.8

Skeletal element	Side	No. Recovered	% of expected
	L acromion	42	77.8
	L coracoid	42	77.8
	L complete	42	77.8
	R body	42	77.8
	R glenoid	42	77.8
	R acromion	42	77.8
	R coracoid	42	77.8
	R complete	42	77.8
Sternum	manubrium	15	27.8
	body	10	18.5
Humerus	L proximal	43	79.6
	L distal	44	81.5
	L complete	43	79.6
	R proximal	47	87
	R distal	47	87
	R complete	47	87
Radius	L proximal	41	75.9
	L distal	41	75.9
	L complete	41	75.9
	R proximal	41	75.9
	R distal	41	75.9
	R complete	41	75.9
Ulna	L proximal	44	81.5
	L distal	44	81.5
	L complete	44	81.5
	R proximal	44	81.5
	R distal	43	79.6
	R complete	43	79.6
Femur	L proximal	44	81.5
	L distal	44	81.5
	L complete	44	81.5
	R proximal	47	87
	R distal	46	85.2
	R complete	46	85.2
Tibia	L proximal	44	81.5
	L distal	45	83.3
	L complete	44	81.5
	R proximal	46	85.2
	R distal	46	85.2
	R complete	46	85.2
Fibula	L proximal	41	75.9
	L distal	41	75.9
	L complete	41	75.9

Table 38: Proportion of bones recovered from the 54 subadult individuals from the Hunno-Sarmatian period.

Skeletal element	Side	No. Recovered	% of expected
	R proximal	43	79.6
	R distal	43	79.6
	R complete	43	79.6
Patella	L	9	16.7
	R	4	7.4
Carpals	All	12	1.4
Metacarpals	L 1	13	24.1
	L 2	20	37
	L 3	18	33.3
	L 4	15	27.8
	L 5	21	38.9
	R 1	22	40.7
	R 2	25	46.3
	R 3	26	48.1
	R 4	21	38.9
	R 5	16	29.6
Hand PH	L 1	60	22.2
	L 2	15	6.9
	L 3	5	1.8
	R 1	76	28.1
	R 2	22	10.2
	R 3	1	0.4
Tarsals	L calcaneus	22	40.7
	L talus	23	42.6
	L 1st cuneiform	1	1.8
	L 2nd cuneiform	0	0
	L 3rd cuneiform	1	1.8
	L cuboid	2	3.7
	L navicular	0	0
	R calcaneus	24	44.4
	R talus	23	42.6
	R 1st cuneiform	1	1.8
	R 2nd cuneiform	1	1.8
	R 3rd cuneiform	0	0
	R cuboid	2	3.7
	R navicular	0	0
Metatarsals	L 1	20	37
	L 2	15	27.8
	L 3	17	31.5
	L 4	14	25.9
	L 5	16	29.6
	R 1	23	42.6
	R 2	20	37

Skeletal element	Side	No. Recovered	% of expected
	R 3	21	38.9
	R 4	17	31.5
	R 5	13	24.1
Foot PH	L 1	12	4.4
	L 2	0	0
	L 3	1	0.4
	R 1	19	7
	R 2	0	0
	R 3	0	0
Innominate			
Ilium	L	45	83.3
	R	44	81.5
Ischium	L	31	57.4
	R	35	64.8
Pubis	L	25	46.3
	R	24	44.4
Complete	L	23	42.6
	R	21	38.9

Table 38: Proportion of bones recovered from the 54 subadult individuals from the Hunno-Sarmatian period.

Appendix 2. Gazetteer of individuals with trauma

Introduction

The gazetteer included in Appendix 2 provides details on all of the individuals from the cemetery complex at Aymyrlyg that displayed evidence of trauma in their skeletal remains. The skeletons are currently curated in the Research Division of the Peter the Great Museum of Anthropology and Ethnography, St. Petersburg, Russia, and were subject to osteological and palaeopathological examination by the current author during September to November 1995, March to September 1996, and July 1997.

The first line of each gazetteer entry for each individual contains their archaeological context number assigned to them by the archaeologists during excavation. The context number consists of the group letter or number, followed by the number of the burial [e.g. VI. 6]. If more than one individual was recovered from a burial the number of the skeleton is also provided [e.g. VI. 6. Sk. 2]. In those cases where individuals have the same burial number, or the context number is unclear, a small Roman Numeral in brackets accompanies the burial number to enable differentiation to be made between individuals [e.g. VI. 6. Sk. 2 (i)]. Where possible, the age-at-death, sex and stature for each adult individual are included on the second line. Only the context number and the age-at-death are included for subadult individuals. Information on the photographs included in the text of the monograph relating to each individual is included in the next field.

The year when the skeleton was excavated by the archaeologists, and the number assigned during curation in the museum are then provided. This is followed by three sections of information on the burial record associated with each skeleton: The first section provides information on the tomb type, while the second section notes the orientation and preservation of the skeleton when it was excavated. Information on the grave goods associated with the skeleton is then provided in the third section.

A visual inventory (not to scale) is included to show which parts of each individual skeleton were available for examination. Bones, which were present, are shaded in black, while bones that were in a fragmentary or eroded condition are coloured grey. This is followed by details on the state of preservation of the remains. The condition of preservation for individuals with post-cranial remains is categorised as follows:

Excellent	90-100% bones present.
Very good	70-90% bones present.
Good	50-70% bones present.
Fair	40-50% bones present.
Poor	20-40% bones present.
Very poor	< 20%

The physical condition of the bones was also taken into consideration, and where remains were fragmentary or had been subject to surface erosion this was also noted.

The next subdivision includes descriptive information on any examples of cranial palaeopathology present in the skeleton. In those cases where cribra orbitalia was observed this is detailed using the classification system of Stuart Macadam (1991, 109).

Descriptive details of post-cranial palaeopathology are provided in the next section of the entry. If an individual displayed evidence of surgery, weapon trauma or decapitation the descriptive information is provided in a separate category at this point in the text. The next two sections contain details of cranial and post-cranial non-metric traits apparent in the remains, while a section entitled 'Other observations' provides information on cranial deformation, evidence of post-mortem defleshing or disarticulation, and details of any green or orange discoloration apparent on the bones caused by their association with iron or bronze artefacts in the tomb.

The dental formula for each individual is then provided, based on the Zsigmondy System of dental recording. The symbols used to indicate the condition of the teeth are as follows:

/	Tooth lost PM	**A**	Abscess	**E**	Erupting
X	Tooth lost AM	**C**	Caries	**PE**	Partially erupted
NP	Tooth not present	**R**	Root only	**B**	Broken PM
-	Jaw not present	**PU**	Secondary dentine / pulp cavity exposed	**U**	Unerupted

Separate subdivisions then provide details of bite, molar attrition patterns, calculus deposition, periodontal disease, caries, abscesses, ante-mortem tooth loss, dental enamel hypoplasia and the presence of any dental anomalies. The molar attrition patterns, calculus deposition and periodontal disease grading systems of Brothwell (1981) were followed in this section of the entry.

Abbreviations used in Appendix 2

Metrical information:

mm	millimetres
cm	centimetres

Names of bones:

CV	cervical vertebra
TV	thoracic vertebra
LV	lumbar vertebra
S	sacral vertebra
MC	metacarpal
MT	metatarsal
PH	phalanx

Side of bone:

R	right
L	left

Disease:

DJD	degenerative joint disease

Characteristics of individual:

Mat.	mature
Poss.	possible
Prob.	probable
Sk.	skeleton

Gazetteer

1. Scythian period individuals

Context B. 7. Sk. 1

Adult 25-35 yrs: Female: 153.5 cm, 5'2"

Year of excavation: ?1974

Anthropological number: 6862-5

Archaeological context:

Log house tomb - The information for this context was not included in the archive reports.

Visual inventory:

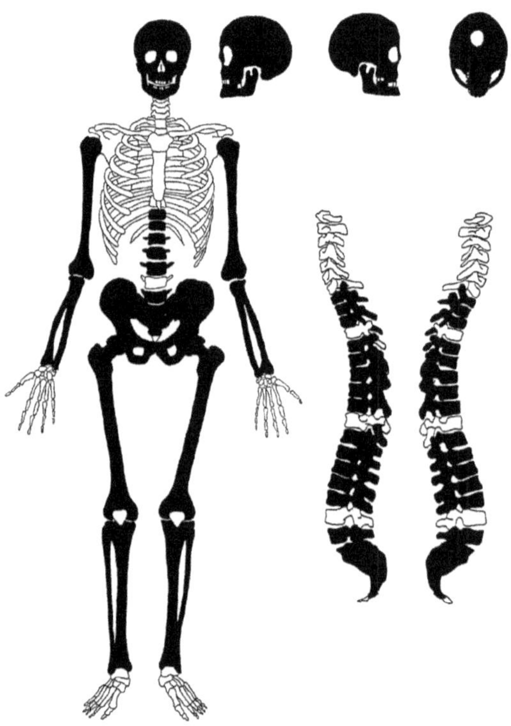

Preservation: The remains were in a very good state of preservation with 70-80% of bones present.

Cranial pathology: An ivory osteoma, which measured approximately 9 mm by 7.5 mm, was present on the L coronal suture just lateral to the sagittal suture.

Post-cranial pathology: A bony exostosis was apparent on the medial surface of the proximal third of the R tibia immediately inferior to the groove for the semimembranosus. The exostosis had a supero-inferior measurement of 15 mm, and probably represented an enthesophyte of the tibial collateral ligament. Vertebral osteophytosis was apparent in TV1, TV5-TV12, LV3 and LV5.

Cranial non-metric traits: R and L highest nuchal lines, R parietal foramen, ossicle at bregma, R and L mastoid foramina extrasutural, R and L posterior condylar canals open, R and L incomplete foramina ovale, R open foramen spinosum, R and L accessory supraorbital foramina, R and L anterior ethmoid frontal foramina extrasutural, R and L accessory infraorbital foramina.

Post-cranial non-metric traits: R and L Poirier's fossae, R and L hypotrochanteric fossae, R and L lateral tibial squatting facets, R and L acetabular creases, R and L accessory sacral facets.

Other observations: Areas of slight green discoloration and small fragments of body tissue were apparent on the most superior surfaces of the frontal and parietals, the posterior areas of the parietals, the anterior and posterior surfaces of the R pubic bone and on the region of the R ilium just superior to the acetabulum. Marked green discoloration was apparent on the proximal third of the R femur and a large fragment of body tissue was attached to the greater trochanter.

Dentition:

~~8~~	~~7~~	6	5	4	~~3~~	~~2~~	~~1~~	~~1~~	2	~~3~~	~~4~~	~~5~~	6	7	8
8	7	6	~~5~~	~~4~~	~~3~~	2	~~1~~	~~1~~	2	~~3~~	4	5	6	7	8 NP

Bite: Overbite.

Attrition: Molar wear patterns of Stage 1 for the 3rd molars, Stage 3 for the 2nd molars and Stage 4 for the 1st molars were apparent.

Calculus: Flecks of calculus were apparent on the buccal and lingual surfaces of 6⌉, while slight deposits were present on the lingual surfaces of 2⌉ ⌈2 7.

Enamel hypoplasia: Single hypoplastic grooves were present on 2⌉ ⌈2 indicating that the individual had been subject to at least one incident of physiological stress at the age of 3 - 3.5 years.

Dental anomalies: The left mandibular 3rd molar was genetically absent. Mandibular tori were present.

Context B. 7. disturbed

Adult: Male: 171 cm, 5'7"

Year of excavation: ?1974

Anthropological number: 6862-6

Archaeological context:

Log house tomb - The information for this context was not included in the archive reports.

Visual inventory:

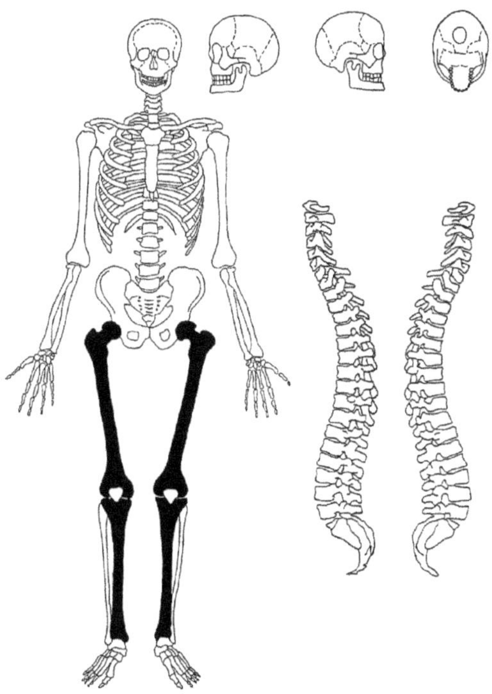

Preservation: The remains were in a very poor state of preservation with 10-20% of bones present.

Post-cranial pathology: The remains were from an extremely robust individual. Enthesopathies were present in the femora (gluteus maximus, adductor longus, piriformis, gluteus minimus, gluteus medius, psoas major, iliacus), and the tibiae (soleus). A bony nodule was apparent at the medial surface of the proximal third of the L tibia in the area immediately inferior to the groove for the semimembranosus. This probably represented an enthesophyte of the tibial collateral ligament. Osteophytes were present in the distal R femur and the proximal tibiae.

Other observations: Two oblique cutmarks were apparent on the superior surface of the proximal condyles of the L tibia. The cutmarks were probably due to post-mortem defleshing/disarticulation.

Context B. 8. Sk. 1

Adult 35-45 yrs: Male: 174 cm, 5'8"

Plate: 32

Year of excavation: ?1974

Anthropological number: 6862-7

Archaeological context:

Log house tomb - The information for this context was not included in the archive reports.

Visual inventory:

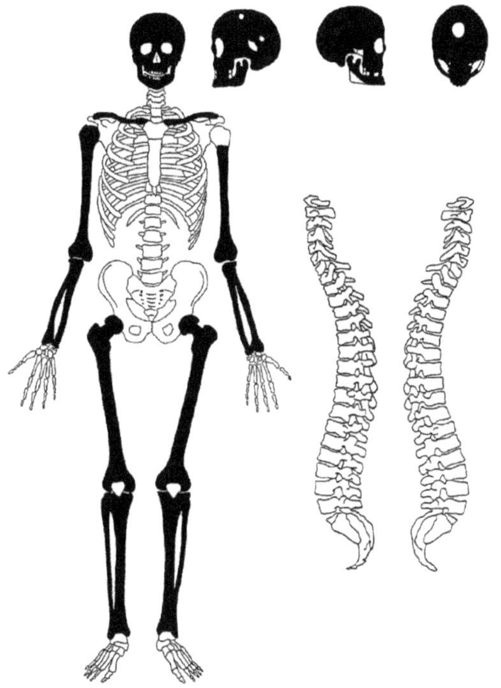

Preservation: The remains were in a fair state of preservation with 40-50% of bones present. Some of the bones were fragmentary.

Post-cranial pathology: A simple, transverse fracture was apparent in the distal third of the shaft of the L ulna (Night-stick fracture). The fracture was extremely well healed with only slight hypertrophy. The fracture line was barely perceptible and the bone fragments were undisplaced. Large osteophytes were present surrounding the distal articular surface, particularly at the antero-medial margin. It is probable that the fracture had caused excess strain to the radio-ulnar joint and that there had been a degree of subluxation at this joint. An ossified haematoma was evident in the postero-lateral surface of the proximal third of the shaft of the L femur. This measured approximately 22 mm supero-inferiorly by 7 mm medio-laterally.

Rotator cuff disease was present in the R humerus. Osteophytes were evident in the L proximal femur and the R distal ulna. Enthesopathies were apparent in the R clavicle (costoclavicular ligament) and the R femur (gluteus minimus).

Weapon trauma: Two injuries made by a pointed battle axe were apparent in the cranium. The first injury was located in the L parietal immediately posterior to the coronal suture, and 25 mm lateral to the sagittal suture. It had completely perforated the vault, and was near-circular in shape measuring 13 mm medio-laterally by 18 mm antero-posteriorly. The margins had bevelled edges but there was no sign of healing. The second injury was situated in the posterior surface of the L parietal, approximately 30 mm superior to the asterion. Again, the injury had completely perforated the vault and measured 15 mm medio-laterally by 17.5 mm antero-posteriorly. The margins were sharp with definite oblique bevelling at the posterior edge. There was no sign of healing.

Cranial non-metric traits: R and L highest nuchal lines, 2 ossicles at the lambda, R parietal foramen, R and L mastoid foramina extrasutural, L incomplete foramen ovale, R and L open foramina spinosum, R and L accessory supraorbital foramina, R and L anterior ethmoid frontal foramina extrasutural, R and L accessory infraorbital foramina.

Post-cranial non-metric traits: R and L plaque, R and L hypotrochanteric fossae, R and L exostoses in trochanteric fossae, R and L lateral tibial squatting facets.

Other observations: Green discoloration was present on the medial aspect of the distal third of the shaft of the L radius and on the posterior and medial surfaces of the midshaft of the L fibula. Cutmarks were visible on the anterior surface of the L distal humerus and on the proximal half and the lateral epicondyle of the R humerus. The cutmarks were probably due to post-mortem defleshing/ disarticulation.

Dentition:

NP															NP
8	7	6	5	4	~~3~~	~~2~~	~~1~~	~~1~~	~~2~~	~~3~~	4	5	6	7	8
~~8~~	~~7~~	6	~~5~~	4	~~3~~	~~2~~	~~1~~	~~1~~	~~2~~	~~3~~	4	5	6	7	~~8~~

Bite: Overbite.

Attrition: Molar wear patterns of Stage 3 for the 2nd molars, Stage 4+ for the mandibular 1st molars and Stage 5 for the maxillary 1st molars were apparent. In general, the maxillary teeth displayed more extensive attrition than the mandibular teeth.

Calculus: Slight deposits of calculus were present on the buccal surfaces of 4⌋ ⌊4 and 4⌉, the lingual surface of ⌈4 and on the buccal and lingual surfaces of 6⌋ ⌊5. Medium deposits of calculus were apparent on the buccal surfaces of ⌊6 7 and on the lingual surface of ⌈5.

Periodontal disease: Periodontal disease of slight severity was apparent.

Dental anomalies: The maxillary 3rd molars were genetically absent.

Context D. 2 (i)

Young adult 17-25 yrs: Male: 166 cm, 5'5"

Year of excavation: 1974

Anthropological number: 6862-29

Archaeological context:

Log house tomb - The remains of at least five individuals were recovered from this tomb. The original location of this individual within the tomb is unknown.

Visual inventory:

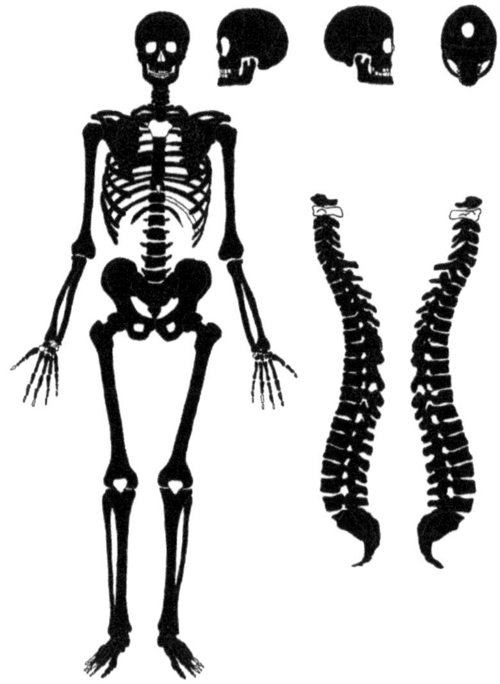

Preservation: The remains were in an excellent state of preservation with 90-100% of bones present.

Post-cranial pathology: Spondylolysis of LV5 was apparent. Osteophytes and porosity were apparent in the distal articular surface of the L MT1. Porosity was evident in several of the rib heads. Slight vertebral osteophytosis was present in TV3-TV12 and LV1-LV5. Schmorl's nodes were visible in LV1-LV4.

Cranial non-metric traits: Ossicle at lambda, L ossicle in lambdoid suture, R and L parietal foramina, L ossicle at pterion, R and L mastoid foramina extrasutural, R and L posterior condylar canals open, R and L accessory supraorbital foramina, R and L anterior ethmoid frontal foramina extrasutural.

Post-cranial non-metric traits: R and L Allan's fossae, R and L Poirier's fossae, R and L hypotrochanteric fossae, R and L third trochanters, R and L lateral tibial squatting facets, R and L acromial articular facets, R and L circumflex sulci, L os trigonum, R and L double anterior calcaneal facets, R and L CV5, CV6, CV7 transverse foramina bipartite.

Dentition:

													CA		
~~8~~	7	6	5	~~4~~	3	2	~~1~~	1	2	~~3~~	4	5	6	7	8
8	~~7~~	~~6~~	5	~~4~~	~~3~~	~~2~~	~~1~~	1	2	~~3~~	~~4~~	5	6	7	8
	A														U

Bite: Overbite.

Attrition: Molar wear patterns of Stage 1 for the 3rd molars, Stages 2+-3- for the 2nd molars, and Stages 4+-5 for the 1st molars were apparent.

Calculus: Slight deposits of calculus were present on the buccal surface of ⌊7, the lingual surfaces of 2⌋ ⌊1 2, the buccal and distal surfaces of ⌊8, the buccal and lingual surfaces of ⌈6, the lingual and distal surfaces of 8⌉ and the buccal, lingual, distal and mesial surfaces of ⌈7. Medium deposits were apparent on the buccal surfaces of 5 3 2⌋ ⌊1 2, the buccal and lingual surfaces of ⌈2 3, and the buccal, mesial and distal surfaces of ⌈5. Heavy deposits were visible on the buccal surfaces of 6⌋ ⌊4 5.

Periodontal disease: Periodontal disease of considerable severity was evident.

Caries: A large caries was present in the occlusal surface of ⌊6.

Abscesses: A medium sized externally draining abscess was associated with 7⌉ and a large externally draining abscess was associated with ⌊6.

Ante-mortem tooth loss: 6⌉ had been lost ante-mortem and the socket was in the early stages of remodelling.

Dental anomalies: The maxillary R lateral incisor and the L central and lateral incisors were shovel shaped. The maxillary L 3rd molar was peg shaped. An enamel pearl was present on the root surface of the buccal aspect of the maxillary L 1st molar.

Context D. 6. Sk. 1

Mature adult 45+ yrs: Female: 152 cm, 5'1"

Year of excavation: 1974

Anthropological number: 6862-38

Archaeological context:

Log house tomb - The information for this context was not included in the archive reports.

Visual inventory:

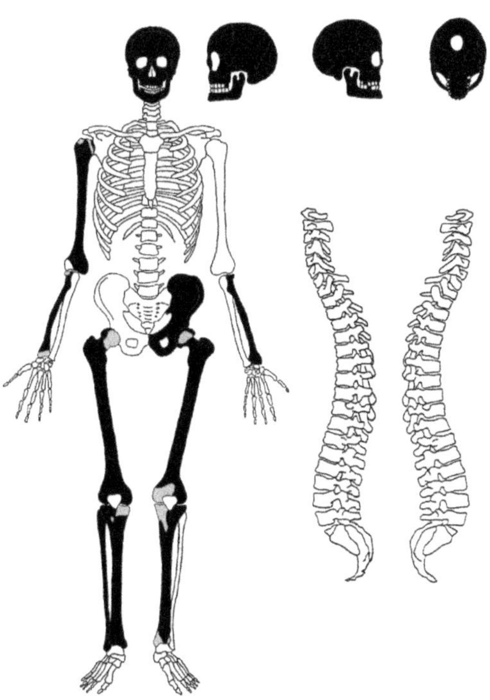

Preservation: The remains were in a poor state of preservation with 30-40% of bones present. Surface erosion was evident on many bones.

Post-cranial pathology: A smooth bony nodule, which measured 14 mm supero-inferiorly by 5 mm medio-laterally, was apparent on the posterior surface of the midshaft of the R tibia. The nodule may have been an ossified haematoma. Enthesopathies were present on radii (biceps), and the R fibula (extensor hallucis longus).

Cranial non-metric traits: 2 large ossicles were present at the lambda, L ossicle in the lambdoid suture, R parietal foramen, L ossicle at asterion, R and L mastoid foramina extrasutural, R and L posterior condylar canals open, L incomplete foramen ovale, R and L open foramina spinosum, R and L accessory supraorbital foramina, L anterior ethmoid frontal extrasutural, R and L accessory infraorbital foramina.

Post-cranial non-metric traits: R and L exostoses in trochanteric fossae, R lateral tibial squatting facet, L acetabular crease.

Other observations: Green discoloration was present on the central area of the superior aspect of the frontal and the lateral surface of the proximal third of the shaft of the L femur.

Dentition:

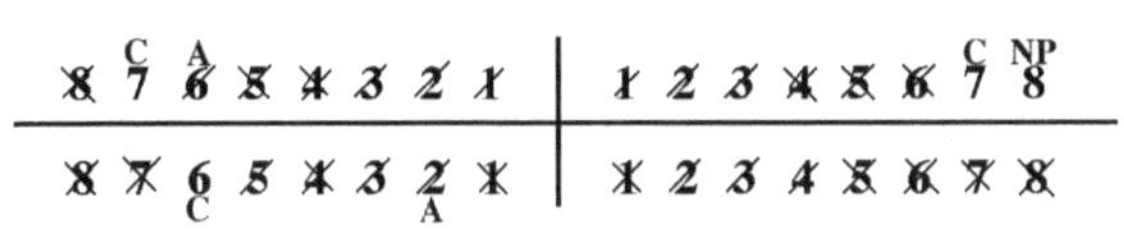

Bite: Overbite.

Attrition: Molar wear patterns of Stages 2+-4 for the 2nd molars and Stage 5 for the 1st molars were apparent.

Periodontal disease: Periodontal disease of considerable severity was evident.

Caries: Medium sized caries were present in the mesial surfaces of 6⌉ and 7⌋, while a large caries was apparent in the buccal surface of ⌊7.

Abscesses: A medium sized externally draining abscess was associated with 2⌉. A medium sized abscess, which was draining into the maxillary sinus, was associated with 6⌋.

Ante-mortem tooth loss: 8 5 4⌋ ⌊4 5 6 and 8 7 4 1⌉ ⌈1 5 6 7 8 had all been lost ante-mortem. The sockets of the teeth which had been lost ante-mortem on the R side of the mouth were all completely remodelled, while those from the L side were in the mid-stages of remodelling.

Dental anomalies: A parastyle was present in the mesiobuccal side of the L maxillary 2nd molar. The L maxillary 3rd molar was genetically absent.

Context D. 7. Sk. 3

Child 2 - 3 years

Plate: 21

Year of excavation: ?1974

Anthropological number: 6863-207

Archaeological context:

Log house tomb - The information for this context was not included in the archive reports.

Visual inventory:

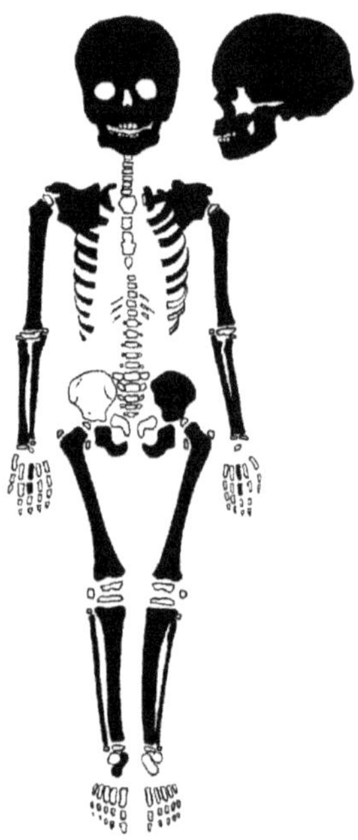

Preservation: The remains were in a very good state of preservation with 70-80% of bones present.

Cranial pathology: A localised area of increased vascular impressions was present at the midpoint of the R parietal. The affected area was associated with the middle meningeal vessels, and measured approximately 60 mm by 60 mm. It is possible that the increased vascularisation was the result of intracranial haemorrhaging, possibly due to direct trauma.

Dentition:

e	d	c	b	~~a~~	~~a~~	~~b~~	c	d	e
e	d	c	~~b~~	~~a~~	~~a~~	~~b~~	c	d	e

Bite: Overbite.

Attrition: Molar wear patterns of Stages 1-2+ were apparent on the 1st and 2nd deciduous molars.

Calculus: Flecks of calculus were present on the buccal surface of ⌈e, the lingual surfaces of d⌋ and ⌈d, and the buccal and lingual surfaces of ⌊d e. Slight deposits were apparent on the buccal surface of d⌋ and the buccal and lingual surfaces of e c b⌋ and e d⌉ ⌈c.

Context D/G

Child 5.5 - 6.5 years

Plate: 37

Year of excavation: unknown

Anthropological number: 6863-202

Archaeological context:

All that is known of this individual is that he or she had originated from either Group D or Group G.

Visual inventory:

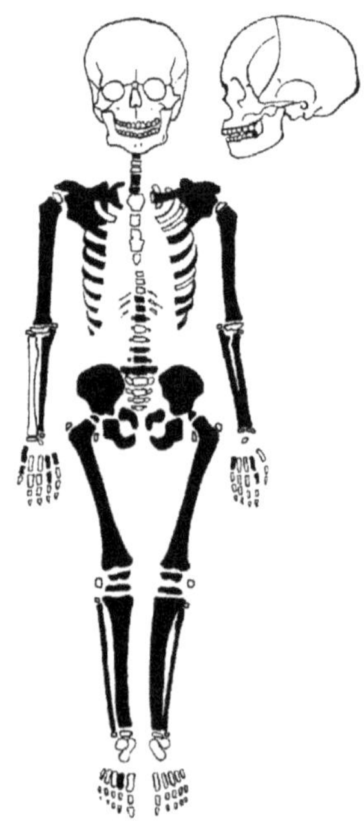

Preservation: The remains were in a very good state of preservation with 70-80% of bones present.

Decapitation: Small transverse cutmarks were present on the anterior surfaces of CV1 and CV2. The superior aspects of each cutmark had a bevelled edge suggesting that they had been made in an upward manner. The cutmarks are similar to those apparent on the cervical vertebrae of adult individuals, which were thought possibly indicative of throat cutting.

Two chop marks were visible on the inferior surface of CV3. The posterior aspect of the R inferior articular facet had been cleanly chopped through. The L side of the inferior surface of the vertebral body displayed an oblique chop mark which extended to the medial margin of the L transverse foramen. Two chop marks were also apparent on the superior surface of CV4. A chop mark was visible on the medial margin of the L superior articular facet and a further chop mark was present on the lateral margin of the R superior articular facet. It is probable that the cutmarks on CV3 and CV4 had been attained in one incident, and that the head and neck had been severed from the remainder of the body by these blows.

The skull of the individual was not present, and it is possible that after having had his or her throat slit, that the individual had been decapitated.

Context E. 9. Sk. 2

Adult 25-35 yrs: Female: 156.3 cm, 5'2"

Year of excavation: ?1974 or 1975

Anthropological number: 6862-56

Archaeological context:

Log house tomb - The information for this context was not included in the archive reports.

Visual inventory:

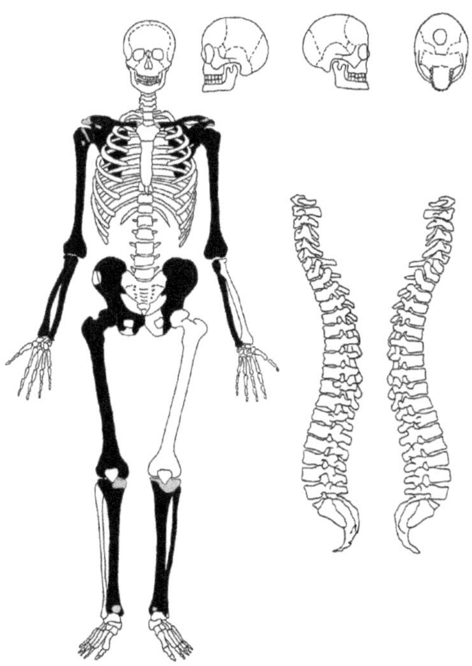

Preservation: The remains were in a fair state of preservation with 40-50% of the bones present. Many of the bones were fragmentary with extensive surface erosion.

Post-cranial pathology: A bony nodule, which measured 8 mm supero-inferiorly by 5 mm medio-laterally, was present at the lateral lip of the intertubercular groove of the R humerus. This may be indicative of an enthesophyte of pectoralis major. Enthesopathies were apparent in the R radius (biceps) and the clavicles (conoid ligament). Although post-mortem damage was apparent the L innominate appeared to have been completely ankylosed to the L side of the sacrum at the sacro-iliac joint.

Post-cranial non-metric traits: R Poirier's fossa, R hypotrochanteric fossa, R exostosis in trochanteric fossa, R third trochanter, R and L lateral tibial squatting facets, L supracondyloid process, L septal aperature, L acromial articular facet, R circumflex sulcus.

Other observations: Green discoloration was present on the anterior superior iliac spine of the R innominate.

Context G. 3. Sk. 2

Adult 25-35 yrs: Female: 158.6 cm, 5'3"

Year of excavation: ?1974

Anthropological number: 6862-16

Archaeological context:

Log house tomb - The information for this context was not included in the archive reports.

Visual inventory:

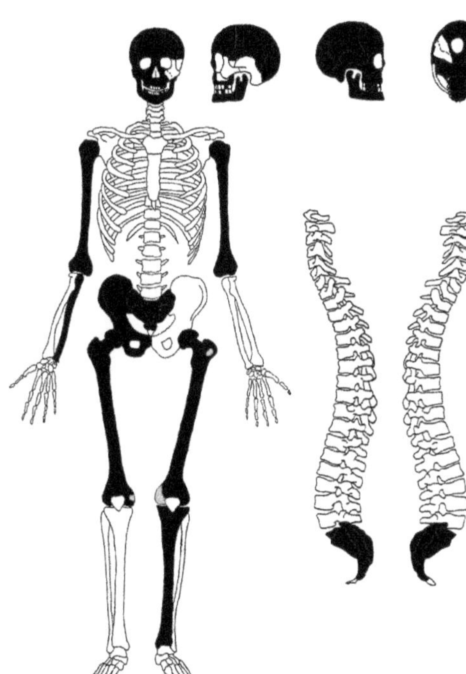

Preservation: The remains were in a fair state of preservation with 40-50% of bones present. Fragmentation and quite extensive surface erosion were apparent on a large proportion of the bones.

Post-cranial pathology: An ossified haematoma was situated on the anterior surface of the midshaft of the R femur. This was represented by a small, smooth oval-shaped area of raised bone, which measured approximately 14 mm supero-inferiorly by 2.5 mm medio-laterally. Enthesopathies were apparent in the femora (gluteus maximus, adductor magnus, adductor brevis, adductor longus).

Cranial non-metric traits: L parietal foramen, R mastoid foramen extrasutural, R posterior condylar canal open, R and L accessory lesser palatine foramina, R absent zygomaticofacial foramen, R bridging of supraorbital notch, R and L accessory supraorbital foramina, R and L anterior ethmoid frontal foramina extrasutural, R accessory infraorbital foramen.

Post-cranial non-metric traits: R and L Allan's fossae, R and L hypotrochanteric fossae, L lateral tibial squatting facet, R acetabular crease.

Dentition:

NP 8	~~7~~	6	5	4	3	~~2~~	~~1~~	~~1~~	2	~~3~~	~~4~~	~~5~~	6	~~7~~	NP 8
~~8~~	7	6	5	~~4~~	~~3~~	~~2~~	~~1~~	~~1~~	~~2~~	3	4	5	6	7	8

Bite: Overbite.

Attrition: Slight attrition was evident with molar wear patterns of Stage 1 for the 3rd molars, Stages 1-2+ for the 2nd molars and Stage 3- for the 1st molars.

Calculus: Slight deposits of calculus were apparent on the lingual surfaces of ⌈3 4 5 and the distal and lingual surfaces of ⌈8.

Periodontal disease: Periodontal disease of slight severity was evident.

Dental anomalies: The maxillary 3rd molars were genetically absent. The tuberculum was prominent and had formed a well defined cusp in the L maxillary lateral incisor.

Context G. 5. Sk. 1

Mature adult 45+ yrs: Female: 159 cm, 5'3"

Year of excavation: 1975

Anthropological number: 6862-18

Archaeological context:

Log house tomb - The remains of four individuals were recovered from the floor of the tomb.

Body position - The individual was represented by scattered arm and leg bones situated near the southwestern wall of the tomb.

Grave goods - A fragment of a bronze awl and a bone arrowhead were recovered from the southwestern corner. In addition, remnants of a leather sheath were discovered near the southwestern wall.

Visual inventory:

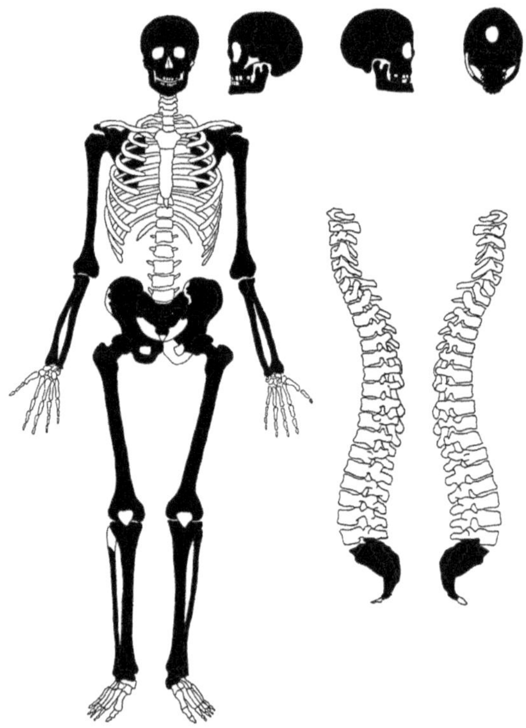

Preservation: The remains were in a very good state of preservation with 70-80% of bones present.

Cranial pathology: A well healed depressed fracture was present on the L frontal, approximately 50 mm superior to the orbit. The lesion was circular and measured approximately 6 mm by 6 mm.

Post-cranial pathology: Enthesopathies were present in the radii (biceps, pronator teres), the femora (psoas major, iliacus, gluteus minimus, piriformis, obturator internis, gemilli and gluteus minimus, vastus lateralis), the tibiae (soleus, interosseous ligament), the L fibula (interosseous ligament) and the L calcaneus (Achilles' tendon). Rotator cuff disease was apparent in the humeri. Osteophytes and porosity were visible in the superior body surface of S1.

Cranial non-metric traits: Ossicle at lambda, L parietal foramen, R and L mastoid foramina extrasutural, R and L posterior condylar canals open, L open foramen spinosum, R and L bridging of supraorbital notches, R and L accessory supraorbital foramina, R anterior ethmoid frontal foramen extrasutural, R accessory infraorbital foramen.

Post-cranial non-metric traits: R and L hypotrochanteric fossae, R and L exostoses in trochanteric fossae, R and L lateral tibial squatting facets, R peroneal tubercle.

Other observations: Green discoloration was apparent on the posterior surface of the L parietal, the occipital, the midshaft of the L femur and the proximal half of the L fibula. Green discoloration and fragments of body tissue were visible on the posterior surface of the R parietal and the proximal half of the R femur.

Dentition:

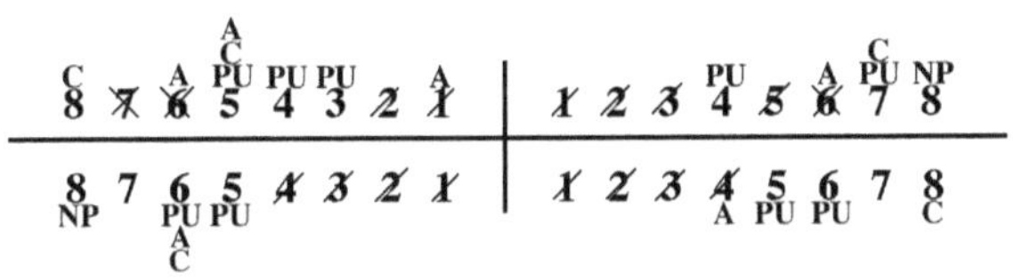

Bite: Overbite.

Attrition: Extensive attrition was apparent on all teeth, with molar wear patterns of Stage 4 for the 3rd molars, Stage 5 for the 2nd molars and Stage 5+ for the 3rd molars. Secondary dentine or pulp cavity exposure was evident in 5 4 3⌋ ⌊4 7 and 6 5⌉ ⌈5 6.

Calculus: Flecks of calculus were apparent on the mesial surface of ⌊7, while slight deposits were present on the buccal surface of ⌈5, the mesial surface of 8⌋ and the mesial and lingual surfaces of 5⌉.

Periodontal disease: Periodontal disease of considerable severity was visible.

Caries: Small caries were present in the occlusal surfaces of 8 5⌋ and ⌈8, while medium sized caries were apparent in the occlusal surface of 6⌉ and the distal surface of ⌊7.

Abscesses: Medium sized externally draining abscesses were associated with 1⌋, 6⌉ and ⌈4. In addition, a single medium sized externally draining abscess was associated with 6 5⌋. A large externally draining abscess was associated with ⌊6.

Ante-mortem tooth loss: 7 6⌋ ⌊6 had been lost ante-mortem.

Dental anomalies: The R mandibular and the L maxillary 3rd molars were genetically absent. A root fragment of the deciduous 2nd molar had been retained in the area between, and just posterior to, ⌊6 and ⌊7.

Context G. 6. Sk. 4 (i)

Mature adult 45+yrs: Prob. female: 156.8 cm, 5'2"

Year of excavation: 1975

Anthropological number: 6862-22

Archaeological context:

NB: 6862-22 and 6863-197 were both labelled as Context G. 6. Sk. 4.

Log house tomb - The remains of four individuals were recovered from the floor of the tomb.

Body position - The disturbed remnants of the skull, arm and leg bones of the individuals were located alongside the northeastern wall of the structure.

Visual inventory:

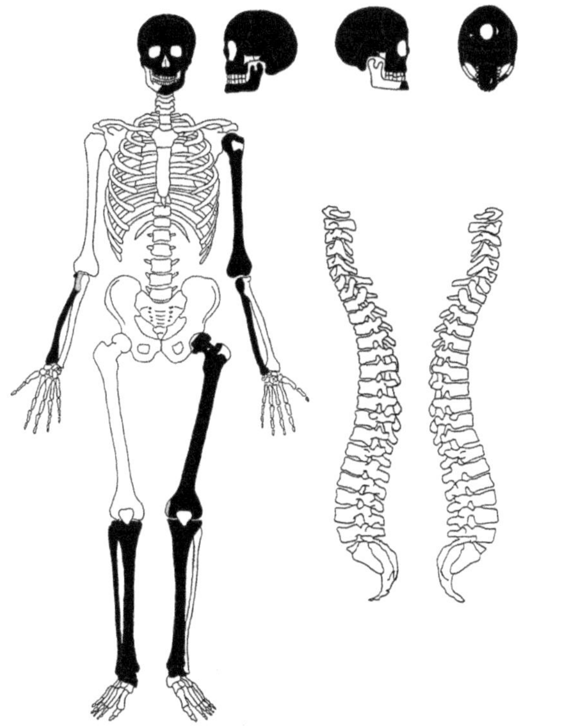

Preservation: The remains were in a poor state of preservation with 20-30% of bones present. Surface erosion was apparent on several bones.

Post-cranial pathology: Large osteophytes were present at the superior margin of the lateral aspect of the most anterior area of the L calcaneus adjacent to the anterior talar articular surface. The lesions may have been indicative of strain of the calcaneocuboid and calcaneonavicular parts of the bifurcate ligament. Enthesopathies were apparent in the L femur (gluteal maximus, pectineus, adductor magnus, adductor longus, adductor brevis), the L tibia (tibialis anterior) and the calcanei (Achilles' tendon).

Cranial non-metric traits: R and L single ossicles in lambdoid suture, R ossicle at parietal notch, R and L mastoid foramina extrasutural, R double anterior condylar canal, R and L open foramina spinosum, R and L accessory supraorbital foramina, R and L anterior ethmoid frontal foramina extrasutural.

Post-cranial non-metric traits: L plaque, L exostosis in trochanteric fossa, R and L lateral tibial squatting facets, R medial talar facet, R and L peroneal tubercles.

Dentition:

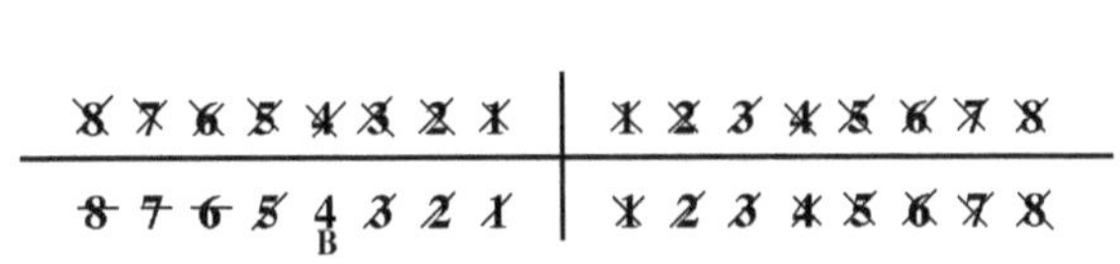

Periodontal disease: Periodontal disease of considerable severity was evident.

Ante-mortem tooth loss: Extensive ante-mortem tooth loss was evident with the maxilla near edentulous. The sockets for all teeth lost ante-mortem had been completely remodelled indicating that the teeth had been lost a significant amount of time prior to death.

Context G. 7. Sk. 5

Adult 25-35 yrs: Male: 157.3 cm, 5'2"

Year of excavation: 1975

Anthropological number: 6862-25

Archaeological context:

Log house tomb - Several groups of disarticulated remains were recovered from the interior of the grave. The original location of this individual within the tomb is unknown.

Visual inventory:

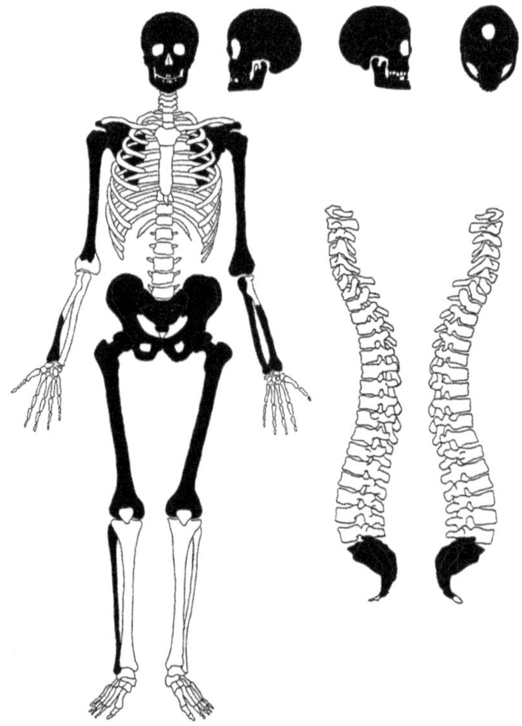

Preservation: The remains were in a poor state of preservation with 40-50% of bones present.

Cranial pathology: There was a simple oblique fracture of the L ramus of the mandible. The fracture had completely healed, but there had been a degree of overlap of approximately 1-2 cm of the fracture fragments resulting in the L side of the mandible being shorter than the R. The L condyle was also positioned approximately 5 mm lower than the R. Consequently, there was secondary subluxation of the L temporomandibular joint. The L condyle displayed a bony exostosis at its lateral margin which may indicate muscle strain, while the articular surface of the L temporal was enlarged relative to the R. It is probable that this overall enlargement represents the development of a false joint which would have enabled freer movement of the subluxated joint. Enthesopathies were apparent at the inferior surfaces of the zygomatic arches (masseter) and the inferior margins of the orbits (levator labii superioris). It is likely that the overall instability of the mandible and its articulation with the cranium had caused strain to these muscles. Porosity was apparent in the superior surface of the frontal and the anterior surfaces of the parietals, which may be indicative of healed porotic hyperostosis.

Weapon trauma: A perforation was visible at the midpoint of the posterior surface of the L parietal. The lesion was circular and measured approximately 4 mm by 4 mm. The surrounding area was depressed and the margins were smooth. It is possible that the perforation represented a healed puncture injury, possibly caused by a pointed battle axe.

Cranial non-metric traits: 3 L and 2 R ossicles in lambdoid suture, R and L mastoid foramina extrasutural, R and L accessory supraorbital foramina, R and L anterior ethmoid frontal foramina extrasutural.

Post-cranial non-metric traits: R and L hypotrochanteric fossae, R and L acromial articular facets.

Other observations: The interior surface of the R side of the mandible displayed green discoloration.

Dentition:

	C	PU											PU		C
~~8~~	7	6	5	4	3	2	~~1~~	~~1~~	2	3	4	5	6	7	8
8	7	6	~~5~~	~~4~~	3	~~2~~	~~1~~	~~1~~	~~2~~	~~3~~	4	~~5~~	6	7	8
													PU	C	C

Bite: Overbite.

Attrition: Extensive attrition was apparent in most teeth with molar wear patterns of Stage 2+ for the 3rd molars, Stage 4 for the 2nd molars and Stage 5 for the 1st molars. Secondary dentine or pulp exposure was evident in 6⌋ ⌊6 and ⌈6.

Calculus: Slight deposits of calculus were present on the buccal surface of 7⌋ and on the distal and lingual surfaces of ⌈8.

Periodontal disease: Slight periodontal disease was visible.

Caries: Small caries were present in the occlusal surface of ⌈7 and the buccal surfaces of ⌊8 and ⌈8, while a medium sized caries was apparent in the occlusal surface of 7⌋.

Enamel hypoplasia: Single hypoplastic lines were present on 3 2⌋ ⌊2 3 4 which indicated that the individual had been subject to at least one incident of physiological stress at the age of 3.5-4 years.

Context I. 3. Sk. 4

Adult 25-35 yrs: Male: 171.5 cm, 5'7"

Year of excavation: 1969

Anthropological number: 6862-64

Archaeological context:

Log house tomb - The very disturbed remains of at least nine individuals were recovered from the floor of the tomb.

Body position - Skeleton 4 and the remains of four other individuals were positioned in the western row. The bodies were lying on their L sides in flexed positions, with the heads facing west-north-west. The upper half of the body of Skeleton 4 was disturbed, but both arms were extended anteriorly. The pelvic bones and the femora were disturbed, while the lower leg bones were partially covered by a plate on which the skull of Skeleton 8 was positioned.

Grave goods - Two plates were lying in close proximity to the skull. Fragments of an iron clasp were situated on one of the plates.

Visual inventory:

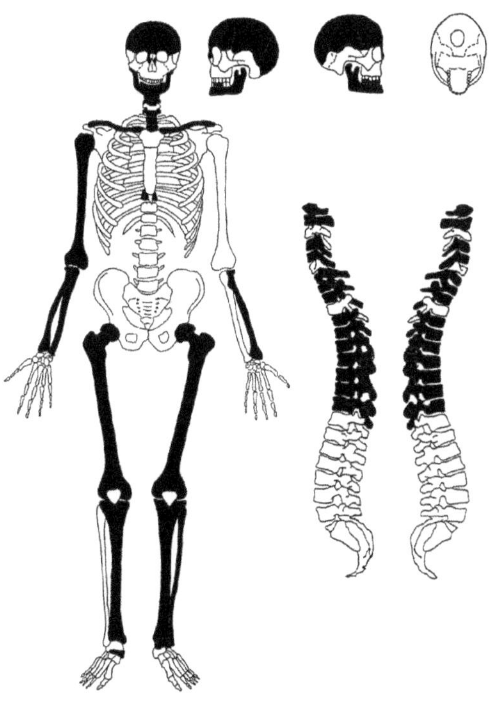

Preservation: The remains were in a fair state of preservation with 40-50% of the bones present.

Post-cranial pathology: A very well healed simple transverse fracture was evident at the distal third of the shaft of the R radius. The distal end of the bone was displaced posteriorly by approximately forty degrees. There was hypertrophy at the point of fracture, with an antero-posterior dimension of 19 mm (LAP=12 mm) and a medio-lateral measurement of 20.2 mm (L ML=16.6 mm). The R radius had been shortened and had a greatest length of 249 mm (L GL=261 mm) which indicated that the parts of the fracture were over-riding by approximately 12 mm.

A very well healed simple transverse fracture was also apparent on the distal third of the R ulna. The distal end was displaced laterally by approximately thirty degrees. There was no hypertrophy or overlapping of the fracture parts.

A further very well healed simple oblique fracture was present at the acromial end of the R clavicle. The parts of the fracture were not malaligned although the R clavicle was shorter than the L (GL=156 mm) and measured 148 mm. This indicated that there was overlap of the fracture parts by approximately 8 mm. The state of healing of the three fractures makes it feasible that they were all obtained in a single incident.

A bony exostosis, which measured 20 mm antero-posteriorly by 15 mm supero-inferiorly was present on the posterior surface of the midshaft of the L femur just lateral to the gluteal ridge. It is probable that this exostosis was an osteochondroma, although differential diagnosis includes an enthesophyte of the vastus lateralis. Osteophytes and porosity were present on the proximal R radius and the proximal R ulna. The lesions were probably secondary to the distal fractures apparent in these two bones, and indicative of strain of the proximal radio-ulnar joint. Marginal osteophytes were present in both femoral heads. Vertebral osteophytosis was apparent in CV1, CV2, CV4, CV5, TV1, TV2, and TV4-TV10. Schmorl's nodes were evident in TV6-TV10. Enthesopathies were apparent in the L radius (biceps), the R humerus (pectoralis major, latissimus dorsai, teres major), the femora (gluteus medius, iliacus, psoas major), the R tibia (interosseous ligament), the R fibula (interosseous ligament) and the L calcaneus (Achilles' tendon).

Cranial non-metric traits: R and L parietal foramina, R and L ossicles at asterion, L bridging of supraorbital notch, R and L accessory supraorbital foramina.

Post-cranial non-metric traits: R and L hypotrochanteric fossae, R and L lateral tibial squatting facets, R double inferior articular talar facet, L double anterior calcaneal facet, L peroneal tubercle, R CV4 transverse foramen bipartite.

Other observations: Green discoloration was present on the L parietal.

Dentition:

~~8~~	~~7~~	~~6~~	~~5~~	~~4~~	~~3~~	~~2~~	~~1~~	~~1~~	~~2~~	~~3~~	~~4~~	~~5~~	~~6~~	~~7~~	~~8~~
8	7	6	5	4	3	2	1	1	2	3̸	4	5	6	7̸	8

Bite: Overbite.

Attrition: Molar wear patterns of Stage 2 for the 3rd molars, Stage 2+ for the 2nd molars and Stage 3 for the 1st molars were evident.

Calculus: Slight deposits of calculus were apparent on the buccal surfaces of 7 6⌉ and the buccal, lingual and distal surfaces of 8⌉. Medium deposits were present on the lingual surfaces of 7 6⌉, the buccal and lingual surfaces of 5 4 3 2 1⌉⌈1 2 4 5 and the buccal, lingual and distal surfaces⌈6.

Periodontal disease: Periodontal disease of slight severity was apparent.

Dental anomalies: The mandibular 1st premolars displayed three cusps, while five cusps were visible on the mandibular 1st molars.

Context I. 15

Young adult 17-25 yrs: Female: 145.4 cm, 4'8"

Year of excavation: 1968

Archaeological context:

Ground pit - A small depressed area was apparent on the ground surface, beneath which was a rectangular pit orientated south-west to north-east. The remains of an adult skeleton were recovered from the pit.

Body position - The individual was partially preserved. She was lying in a prone position with the legs flexed, and the head to the south-west. The skull was disturbed.

Grave goods - The remains of a pot were lying adjacent to the skull and the pelvis.

Visual inventory:

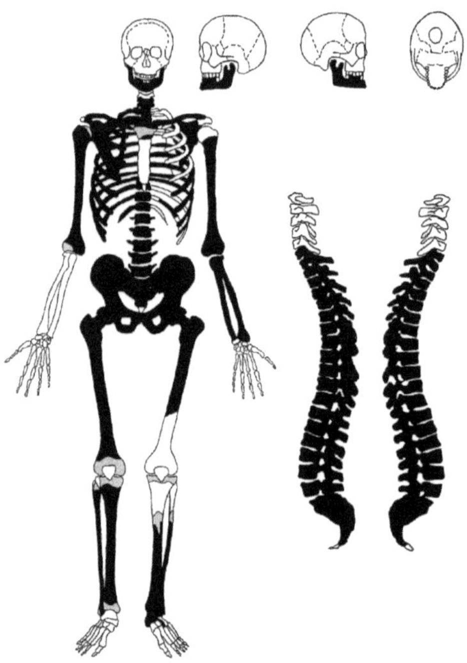

Preservation: The remains were in a fair state of preservation with 50-60% of bones present. The majority of bones displayed extensive surface erosion.

Post-cranial pathology: Spondylolysis was apparent in LV5. Osteophytosis was present in TV9-LV3 and LV5, while Schmorl's nodes were visible in LV2 and LV3.

Post-cranial non-metric traits: R and L septal apertures.

Dentition:

~~8~~	~~7~~	~~6~~	~~5~~	~~4~~	~~3~~	~~2~~	~~1~~	~~1~~	~~2~~	~~3~~	~~4~~	~~5~~	~~6~~	~~7~~	~~8~~
8 U	7	6	5	4̸	3̸	2̸	1̸	1̸	2̸	3	4	5	6	7	8 U

Attrition: Molar wear patterns of Stage 1 for the 2nd molars and Stage 3 for the 1st molars were apparent.

Calculus: Flecks of calculus were present on the distal surface of ⌈7, while slight deposits were apparent on the lingual surfaces of 7 6⌉ ⌈3 5.

Context II. 3 (ii)

Adult: Probable male

Plates: 6, 13

Year of excavation: 1969

Anthropological number: 6862-76

Archaeological context:

Log house tomb - The remains of several individuals were recovered from throughout the tomb. The original location of this individual within the tomb is not known.

Visual inventory:

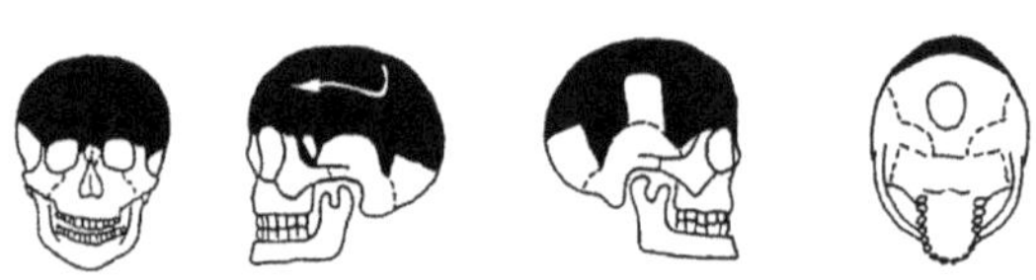

Preservation: Only the cranium was present, which was fragmentary.

Cranial pathology: A simple depressed fracture was evident at the midpoint of the superior margin of the L orbit. A triangular area which measured approximately 20 mm had been depressed so that it was positioned posterior to the orbital margin. The most lateral point of the L orbital margin was thicker than the corresponding area of the R margin. A fracture line was clearly visible at the medial aspect of the fracture, while the line was obliterated at the lateral area of the fracture. It is possible that the injury had been caused by a direct blow with a fairly small implement to the superior orbit.

Cranial non-metric traits: Metopic suture (full set of traits not recorded).

Context II. 4. Sk. 3

Juvenile 11.5 - 12.5 years

Plate: 11

Year of excavation: 1969

Anthropological number: 6863-6

Archaeological context:

Log house tomb - The remains of six fragmentary and one complete skeleton were recovered from the floor of the tomb. The individuals were lying in a western row (Skeletons 1-6) and an eastern row (Skeleton 7).

Body position - Due to the position of the leg bones it appeared that the individuals in the western row were lying on their L sides, with the heads to the west. It is stated in the archive that Skeleton 3 consisted only of the remains of partially disturbed feet bones. Only the cranium was available for analysis.

Grave goods - Two stone slabs were lying near the western wall opposite Skeleton 3.

Visual inventory:

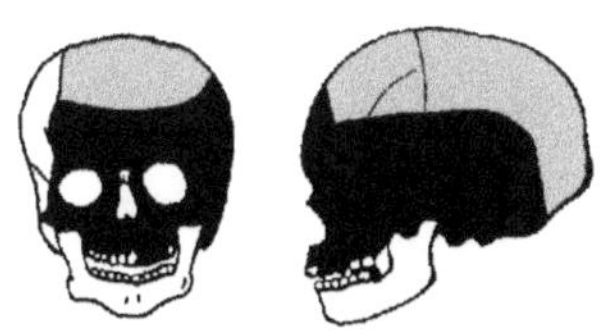

Preservation: Only the cranium was present, which was fragmentary with extensive surface erosion.

Cranial pathology: Cribra orbitalia of Type 2 was present in Areas 1 and 2 of the R orbit and in Areas 1, 2 and 3 of the L orbit. A prominent depressed area was apparent at the midpoint of the frontal bone. The depression ran obliquely and measured 43 mm supero-inferiorly from the R to the L, by 12 mm medio-laterally. It is possible that the depression represented a well healed depressed fracture which had been caused by a blow from a blunt object.

Dentition:

8 (U)	7	6	5̸	4̸	3̸	2̸	1̸	1̸	2̸	3̸	4	5	6	7̸	8 (U)
~~8~~	~~7~~	~~6~~	~~5~~	~~4~~	~~3~~	~~2~~	~~1~~	~~1~~	~~2~~	~~3~~	~~4~~	~~5~~	~~6~~	~~7~~	~~8~~

Attrition: Molar wear patterns of Stage 1 for the 2nd permanent molars and Stage 3 for the 1st permanent molars were apparent.

Calculus: Flecks of calculus were present on the buccal surface of ⌊4.

Context II. 4. Sk. 6

Adolescent 12 - 14 years

Plate: 39

Year of excavation: 1969

Anthropological number: 6863-8

Archaeological context:

Log house tomb - The remains of six fragmentary and one complete skeleton were recovered from the floor of the tomb. The individuals were lying in a western row (Skeletons 1-6) and an eastern row (Skeleton 7).

Body position - Due to the position of the leg bones it appeared that the individuals in the western row were lying on their L sides, with the heads to the west. The remains of Skeleton 6 were of a child and consisted of the skull and the partially disturbed remains of the legs. The skull was lying on a stone slab and was the only element of the skeleton available for analysis.

Grave goods - A flask was situated between the slab associated with the skull and the western wall, while a kabarga tooth pendant was recovered lying adjacent to the bones of the legs.

Visual inventory:

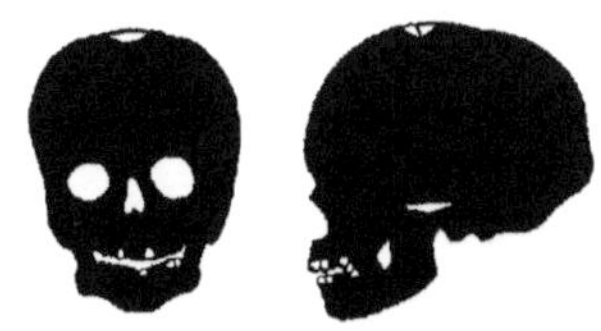

Preservation: Only the skull was present, which was in a very good state of preservation.

Surgery/Post-mortem alteration: An oval perforation which measured 18 mm medio-laterally by 8 mm antero-posteriorly was present on the coronal suture immediately to the R of the bregma. The perforation had a smooth, bevelled margin which had a thickness of 8 mm. The bevelled margin displayed a glossy appearance which may indicate that it was made using a metal implement. It was initially thought that the perforation represented a trepanation operation made using the scraping technique. The regularity of the lesion and the presence of shallow scrape marks over the entire surface of the vault may, however, indicate that the perforation had been made post-mortem. If the perforation was made during a trepanation operation the individual had not survived the procedure since no signs of healing were apparent. In addition, trepanations in the past are considered to have not generally been carried out on an area overlying a suture.

Cranial non-metric traits: Ossicles were present in the lambdoid suture (full set of traits not recorded).

Dentition:

8 (U)	7 (E)	6	e (PU)	4̸	c̸	2	1̸	1̸	2	c̸	4	e (C, PU)	6	7 (E)	8 (U)
8 (U)	7 (U)	6	e (PU)	d̸	3 (E)	2	1̸	1̸	2̸	c̸	d	e (PU)	6	7 (U)	8 (U)

Bite: Overbite.

Attrition: Molar wear patterns of Stage 4+ were apparent for the 1st and 2nd deciduous molars, while a molar wear pattern of Stage 3- was present on the 1st permanent molars. Secondary dentine was exposed in e⌋ ⌊e and e⌉ ⌈e.

Calculus: Flecks of calculus were present on the buccal surfaces of ⌊2 and ⌈d and the lingual surface of 6⌉. Slight deposits were visible on the buccal surfaces of 6⌋ ⌊6, the lingual surfaces of ⌊4 and e⌉ and the buccal and lingual surfaces of 2⌉.

Caries: A medium sized caries was present in the mesial surface of ⌊e.

Enamel hypoplasia: Hypoplastic grooves were present on 6 2⌋ ⌊2 4 which indicated that the individual had been subject to at least two incidents of physiological stress at the ages of 3 - 3.5 and 4 -4.5 years.

Dental anomalies: The maxillary lateral incisors were shovel - shaped.

Context II. 4. Sk. 12

Adult 25-35 yrs: Probable male

Year of excavation: 1969

Anthropological number: 6863-11

Archaeological context:

Log house tomb - The remains of six fragmentary and one complete skeleton were recovered from the floor of the tomb. Details of Skeleton 12 were not recorded in the archive report.

Visual inventory:

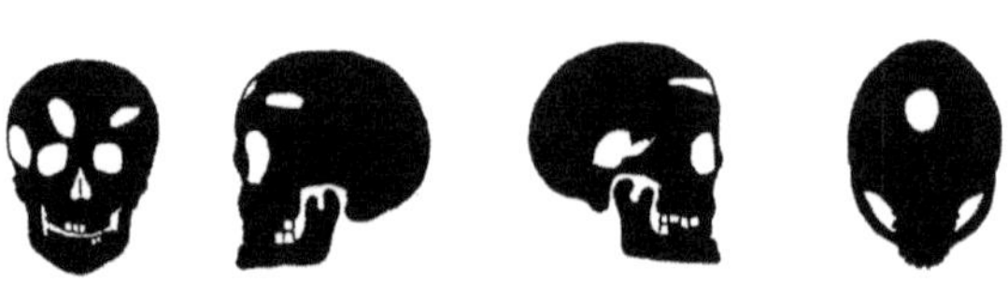

Preservation: Only the skull was present, which was in a very good state of preservation.

Cranial pathology: Cribra orbitalia of Type 1 was present in Areas 1, 2, 3 and 5 of the R orbit, while cribra orbitalia of Type 3 was apparent in Areas 1, 2, 3, 4, 5 and 6 of the L orbit. Porosity was visible in the superior aspects of the frontal, the parietals and the occipital, which may be indicative of healed porotic hyperostosis.

Weapon trauma: Several instances of weapon trauma were apparent on the cranium. An oval perforation, which measured 44 mm antero-posteriorly by 24.5 mm supero-inferiorly, was present on the L side of the frontal bone, between the superior and inferior temporal lines. A fracture line radiated inferiorly from the inferior margin of the perforation.

A large oval shaped perforation, which measured 64 mm antero-posteriorly by approximately 25 mm supero-inferiorly, was present on the R side of the anterior aspect of the frontal bone. A large fracture line extended from the most posterior point of the perforation for approximately 75 mm posteriorly. Another fracture line radiated from the midpoint of the inferior margin of the perforation connecting it with the third example of trauma described below. The area between the two weapon injuries, immediately superior to the glabella, was depressed and it is possible that the fracture lines had weakened this area of the cranium.

A third perforation was present on the R temporal which had obliterated the bone from the sphenoid to the mastoid process. The perforation was irregularly shaped and measured 67 mm antero-posteriorly by 50 mm supero-inferiorly at its largest point.

It is probable that all three injuries had been made by an individual wielding a pointed battle axe. The attacker had dealt at least three blows to the head causing extensive damage to the cranium. It is not surprising that no signs of healing were associated with any of the injuries.

Dentition:

NP 8	7̸	6̸	5	4̸	3̸	A 2̸	A 1̸	1̸	2	3	4	5	6	7	NP 8
8 NP	7	6	5	4	3	2	1̸	1̸	2̸	3	4	5	6	7̸	8 NP

Bite: Overbite.

Attrition: Molar wear patterns of Stages 1-2+ for the 2nd molars and Stage 3 for the 1st molars were apparent.

Calculus: Flecks of calculus were present on the buccal surfaces of ⌊2 3 7 and 6 4 3 2⌉, the lingual surface of ⌈3 and the buccal and lingual surfaces of ⌈6. Slight deposits were visible on the buccal surfaces of 5⌋ ⌊5, the lingual surfaces of 6 5 4 3 2⌉ ⌈4 5 and the buccal, lingual and distal surfaces of 7⌉. Medium deposits were present on the buccal surface of ⌊6.

Periodontal disease: Periodontal disease of slight severity was apparent.

Abscesses: A single large internally draining abscess was associated with 2 1⌋. The abscess was in the process of destroying the palatal surface.

Enamel hypoplasia: Hypoplastic grooves were present on ⌊2 3 4 5 7 and 4 3 2⌉ ⌈3 4 5 which indicated that the individual had been subject to at least five physiological insults at the ages of 2.5-3, 3.5-4, 4-4.5, 4.5-5 and 5-5.5 years.

Dental anomalies: All four 3rd molars were genetically absent.

Context II. 8 (iii)

Adult 25-35 yrs: Probable female

Plate: 33

Year of excavation: 1972

Anthropological number: 6862-97

Archaeological context:

Log house tomb - The remains of several individuals were recovered from the tomb. The original location of this individual within the tomb is unknown.

Visual inventory:

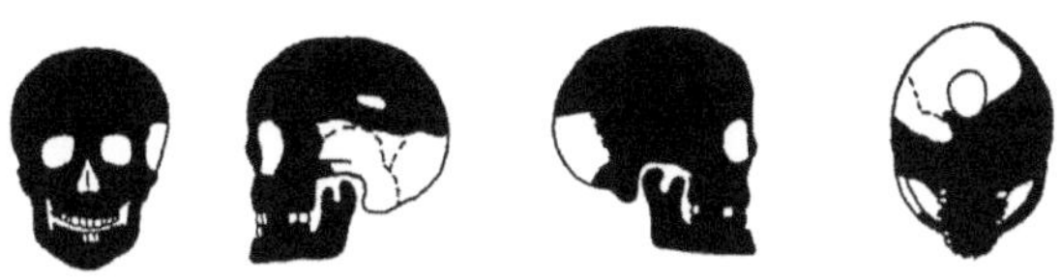

Preservation: Only the skull was present, which was fragmentary.

Weapon trauma/surgery: An irregularly shaped perforation was present on the L parietal just superior to the squamosal suture. The margins of the lesion were smooth and slightly raised, and the superior area of the lesion appeared to have been 'filled in' by new bone. It is probable that the original lesion had been completely circular, and that it would have measured approximately 24 mm by 24 mm. The lesion may be a healed pointed axe injury. If the axe had entered the skull at an angle it is possible that it would have depressed the superior area of the injury and only completely perforated the vault at its inferior aspect. An alternative explanation is that the lesion is indicative of a trepanation operation made using the scraping technique. Whatever the aetiology, there were definite signs of healing which indicated that the lesion was not the direct cause of death.

Dentition:

		PU												A	A
~~8~~	7	6	~~5~~	4	~~3~~	~~2~~	~~1~~	~~1~~	~~2~~	~~3~~	~~4~~	5	6	~~7~~	~~8~~
~~8~~	7	6	5	4	3	2	~~1~~	~~1~~	~~2~~	~~3~~	4	5	6	7	8

Bite: Overbite.

Attrition: Molar wear patterns of Stage 3+ for the 2nd molars and Stage 4+ for the 1st molars were apparent. There was no attrition in the L mandibular 3rd molar which may be related to the ante-mortem loss of the corresponding L maxillary 3rd molar. Secondary dentine was exposed in 6⌋.

Calculus: Slight deposits of calculus were present on the buccal surface of 6⌋, the buccal and lingual surfaces of 7 4⌋ ⌊5 6 and 7 6 5 4 3 2⌉ ⌈4 5 6 7 and the buccal, lingual and distal surfaces of ⌈8.

Periodontal disease: Periodontal disease of slight severity was evident.

Abscesses: A medium sized externally draining abscess was associated with ⌊7, and a large externally draining abscess was related to ⌊8.

Ante-mortem tooth loss: 8⌋ ⌊8 and 8⌉ had been lost ante-mortem.

Context II. 13. south

Adult 25-35 yrs: Male: 154.7 cm, 5'2"

Year of excavation: 1968

Anthropological number: 6862-100

Archaeological context:

Ground pit - The remains of three adults were present at the bottom of the pit, all of which had their heads to the west. The disturbed remains of a child orientated north were lying at the feet of the adults.

Body position - The southern skeleton was lying on his L side in a flexed position. The skull was positioned on a fragment of stone slab.

Grave goods - Fragments of an unidentifiable iron object were situated in close proximity to the stone slab associated with the skull.

Visual inventory:

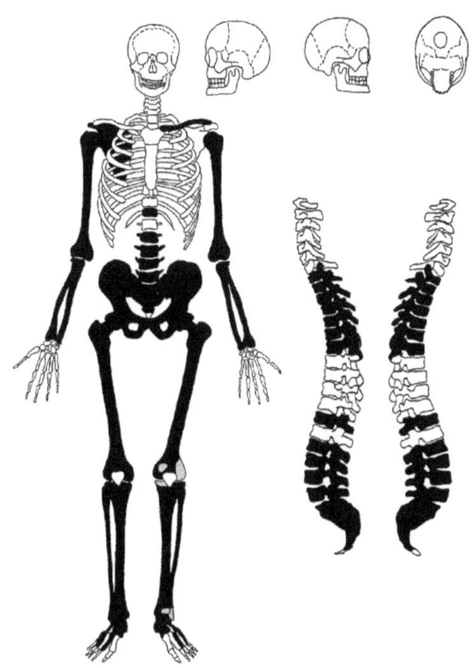

Preservation: The remains were in a good state of preservation with 50-60% of the bones present. Some surface erosion was apparent.

Post-cranial pathology: Sacralization was evident with LV5 completely fused to S1. The neural arch of LV5 was bifurcated, while the neural arches of S1-S4 were cleft. The lesions were probably due to developmental delay during the early formation of the vertebral column.

Archaeology & Trauma

Enthesopathies were apparent in the calcanei (Achilles' tendon). Vertebral osteophytosis was present in TV7, and Schmorl's nodes were visible in TV12.

Weapon trauma: A possible sword cut was present on the supero-posterior surface of the head of the R humerus. The cutmark had sharp edges and measured approximately 23 mm antero-posteriorly. Another possible sword cut, which measured 20 mm antero-posteriorly, was apparent on the lateral aspect of the head of the R femur.

Context III. 3

Young adult 17-25 yrs: Male: 173.8 cm, 5'8"

Year of excavation: 1968

Archaeological context:

Cist - Two parallel rows of slabs were apparent on the ground surface. A ceilingless rectangular cist, orientated south-east to north-west, was situated beneath the slabs. A single skeleton was contained within the cist.

Body position - The individual was lying on his L side in a flexed position, with the head to the north-west. The skull was positioned on a slab, and the bones of the arms were disturbed.

Grave goods - A pot was situated in the northwestern corner of the cist and an iron pin was located in close proximity to the pot. Seven iron pins were recovered from posterior to the skull, while a strip of iron was lying adjacent to the L elbow. Four iron pipes were situated between the L hand and the pelvis. The remains of a girdle made from iron chain, a bronze buckle, and five cone shaped objects were lying on the pelvis. An iron pipe was lying adjacent to the L knee, and another iron pipe was situated at the R foot.

Visual inventory:

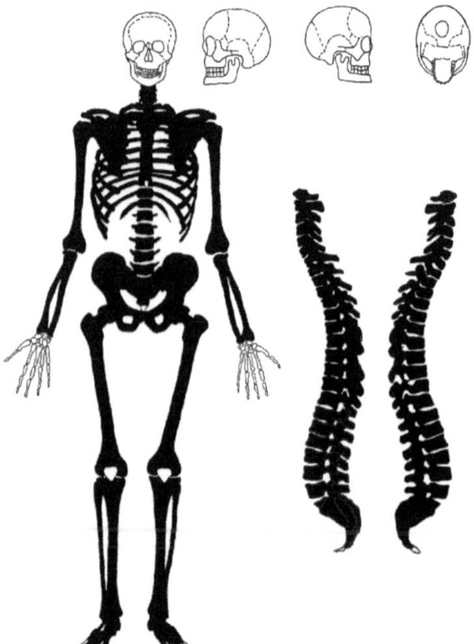

Preservation: The remains were in a very good state of preservation with 70-80% of bones present.

Post-cranial pathology: A bony exostosis was apparent on the distal third of the shaft of the L fibula. It is possible that the exostosis was an enthesophytes of the extensor digitorum longus. Vertebral osteophytosis was present in TV1, TV2, TV6, TV8-TV10, LV3 and LV4. Schmorl's nodes were visible in TV8-TV12.

Context III. 11

Adult 25-35 yrs: Female: 151.1 cm, 5'0"

Year of excavation: 1968

Anthropological number: 6862-116

Archaeological context:

Cist - A pentagonal shaped cist, orientated south-east to north-west, was discovered beneath a group of stone slabs. Slabs were also present at the bottom of the structure.

Body position - An adult skeleton lying on her R side in a crouched position, with the head to the north-west was recovered from the bottom of the cist.

Grave goods - An iron earring was lying on the skull in the region of the external auditory meatus, and a similar earring was recovered from beneath the skull.

Visual inventory:

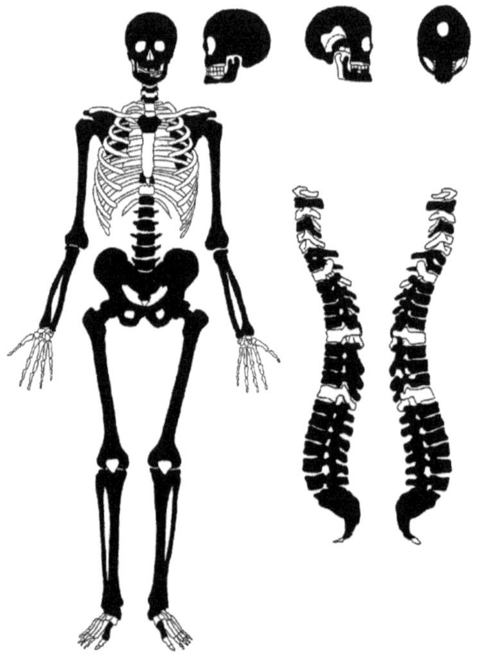

Preservation: The remains were in a very good state of preservation with 80-90% of the bones present.

Cranial pathology: Cribra orbitalia of Type 2 was present in Areas 1, 2 and 4 of both orbits.

Post-cranial pathology: An ossified haematoma, which had a supero-inferior measurement of 30 mm and an antero-posterior measurement of 10 mm, was present on the medial aspect of the midshaft of the R femur. Enthesopathies were apparent in the calcanei (Achilles' tendon). Schmorl's nodes were present in LV1, and vertebral osteophytosis was apparent in LV4.

Cranial non-metric traits: L parietal foramen, R ossicle at parietal notch, R and L mastoid foramina extrasutural, R and L posterior condylar canals open, R double anterior condylar canal, R and L open foramina spinosum, R and L accessory supraorbital foramina, R and L anterior ethmoid frontal foramina extrasutural, R and L accessory infraorbital foramina.

Post-cranial non-metric traits: R and L hypotrochanteric fossae, R and L third trochanters, R and L lateral tibial squatting facets, R and L accessory sacral facets, R and L acromial articular facets, R and L circumflex sulci, R medial talar facet, R and L double anterior calcaneal facets, R and L peroneal tubercles.

Dentition:

		2A			A		PU			PU	PU		A	A	
~~8~~	~~7~~	~~6~~	~~5~~	~~4~~	~~3~~	~~2~~	1	~~1~~	~~2~~	3	4	5	~~6~~	~~7~~	~~8~~
8	7	~~6~~	5	4	3	2	~~1~~	~~1~~	~~2~~	~~3~~	~~4~~	~~5~~	~~6~~	~~7~~	~~8~~
2C	C				PU	PU					A				

Bite: Overbite.

Attrition: Extensive attrition was apparent on all teeth, with molar wear patterns of Stage 4+ for the 2nd molars and Stage 5 for the 1st molars. Secondary dentine or pulp cavity exposure was evident in 1⌋ ⌊3 4 and 3 2⌉.

Calculus: Medium deposits of calculus were present on the distal aspect of 3⌉.

Periodontal disease: Periodontal disease of considerable severity was evident.

Caries: A large caries was present in the distal surface of 7⌉. Multiple caries were present in 8⌉, which displayed a small caries on its lingual aspect and a medium sized caries on its mesial surface.

Abscesses: Small externally draining abscesses were associated with ⌊6 and ⌊7, while large externally draining abscesses were associated with ⌈4 and 3⌋. A small externally draining abscess, and a medium sized internally draining abscess were associated with 6⌋. The internal abscess was draining directly into the maxillary sinus.

Ante-mortem tooth loss: 8 7 6 5 4⌋ ⌊2 6 7 8 and 6 1⌉ ⌈1 5 6 7 8 had been lost ante-mortem. The sockets for the maxillary teeth were in the mid-stages of remodelling, while those for the mandibular teeth were completely remodelled.

Context III. 16 (ii)

Adult 35-45 yrs: Probable male

Year of excavation: 1968

Anthropological number: 6862-103

Archaeological context:

Cist burial - A rectangular cist was orientated south-east to north-west, and was situated beneath stone slabs.

Body position - The individual was lying in a crouched position on his L side, with the head to the north-west. The skull was positioned on a fragment of slab, and was the only skeletal element available for examination.

Grave goods - A pot was situated behind the skull.

Visual inventory:

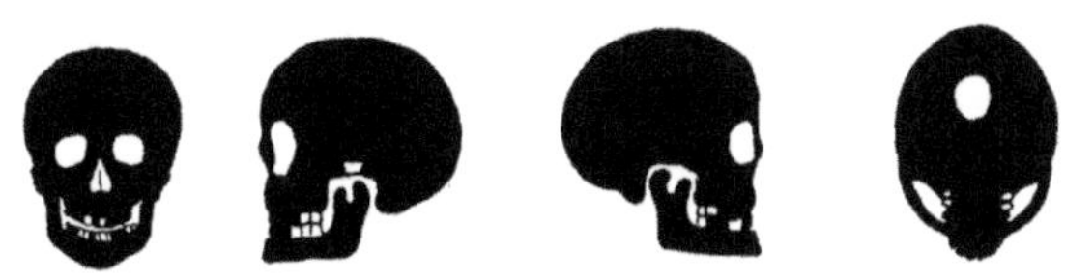

Preservation: Only the skull was present, which was in a very good state of preservation.

Cranial pathology: Well healed fractures were apparent at the anterior aspects of the nasal bones. In both bones the fracture lines were still visible. The fracture was simple and transverse in the L nasal bone, while in the R bone it was simple and oblique. The anterior areas of the fractures were slightly depressed relative to the posterior areas of the bones.

Cranial non-metric traits: Metopic suture (full set of traits not recorded).

Dentition:

		PU	PU	PU								PU	PU		
~~8~~	~~7~~	6	5	4	3	~~2~~	1	1	2	3	4	5	6	~~7~~	~~8~~
8	7	6	~~5~~	4	~~3~~	2	~~1~~	~~1~~	~~2~~	3	4	5	~~6~~	~~7~~	~~8~~
NP								A			PU A		A		

Bite: Overbite.

Attrition: Extensive attrition was apparent on the majority of teeth, with molar wear patterns of Stage 3+ for the 2nd molars and Stages 4+-5 for the 1st molars. Secondary dentine or pulp cavity exposure was visible in 6 5 4⌋ ⌊5 6 and ⌈4.

Calculus: Slight deposits of calculus were present on the buccal surfaces of 6 5 3⌋ ⌊6 and the buccal and lingual surfaces of 4⌋ ⌊3 4 5 and 7 6 4 2⌉ ⌈3 4 5.

Periodontal disease: Periodontal disease of considerable severity was apparent.

Abscesses: Medium sized externally draining abscesses were associated with ⌈1 and ⌈4, and a large externally draining abscess was associated with ⌈6.

Ante-mortem tooth loss: ⌊8 and ⌈8 had been lost ante-mortem.

Dental anomalies: The R mandibular 3rd molar was genetically absent.

Context III. 18

Adult 25-35 yrs: Female: 147.6 cm, 4'9"

Year of excavation: 1968

Anthropological number: 6862-118

Archaeological context:

Cist - A group of horizontal slabs marked the location of a cist which was orientated south-east to north-west. The bottom of the cist was lined with slabs.

Body position - The remains of an adult were situated on the slabs. The individual was lying on her L side in a flexed position, with the head to the north-west.

Grave goods - A pot was situated near the western wall, while fragments of an iron pin were recovered from behind the skull. A dark organic matter was located beneath the pelvic and arm bones. The remains of a leather sheath which contained an bronze knife and fragments of a leather bag were retrieved from the area between the pelvic and arm bones.

Visual inventory:

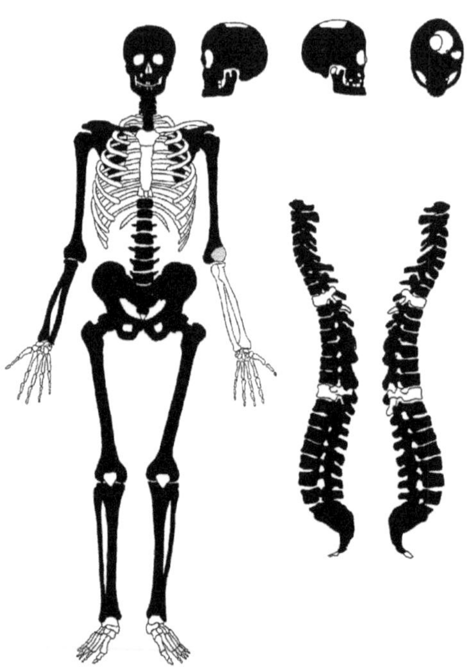

Preservation: The remains were in very good state of preservation with 80-90% of the bones present.

Cranial pathology: Very well healed fractures were present on the anterior thirds of the nasal bones. In both cases the fractures were simple and oblique. A triangular area at the medial aspect of each bone was slightly depressed inferiorly.

Post-cranial pathology: Os acromiale was apparent in the L scapula. Vertebral osteophytosis was present in TV6-TV8. Schmorl's nodes were visible in LV1, LV2 and LV4.

Cranial non-metric traits: R and L ossicles at pterion, L ossicle at parietal notch, R ossicle at asterion, R and L mastoid foramina extrasutural, L posterior condylar canal open, L open foramen spinosum, R and L bridging of supraorbital notches, R accessory supraorbital foramen, R and L anterior ethmoid frontal foramina extrasutural.

Post-cranial non-metric traits: R and L third trochanters, R and L lateral tibial squatting facets, R and L accessory sacral facets, R acromial articular facet, R and L circumflex sulci. R CV7 transverse foramen bipartite.

Dentition:

			A C PU										PU		NP
8	~~7~~	6	5	~~4~~	3	2	~~1~~	1	2	3	4	5	6	7	8
8	7	~~6~~	~~5~~	4	3	~~2~~	1	1	2	3	4	5	6	7	8
NP					PU			PU	PU	PU	PU				NP

Bite: Overbite.

Attrition: Moderate to heavy attrition was apparent in the majority of teeth, with molar wear patterns of Stage 1 for the 3rd molars, Stage 3+ for the 2nd molars and Stage 4+ for the 1st molars. The R maxillary 3rd molar displayed a wear pattern of Stage 1 which was probably related to the genetic absence of the corresponding mandibular tooth. Secondary dentine or pulp cavity exposure was apparent in 5⌋ ⌊6 and 3⌉ ⌈1 2 3 4.

Calculus: Flecks of calculus were visible on the buccal surfaces of ⌊4 and ⌈4 7. Slight deposits were present on the buccal surfaces of 6 3 2⌋ ⌊1 2 3 7 and 4 3 1⌉ ⌈1 2, the lingual surfaces of ⌈4 5 6 and the buccal and lingual surfaces of ⌈3. Medium deposits were apparent on the buccal surfaces of ⌊5 6 and the lingual surface of 7⌉.

Periodontal disease: Periodontal disease of medium severity was evident.

Caries: A large caries was present in the occlusal surface of 5⌋.

Abscesses: A large and very distinct externally draining abscess was associated with 5⌋.

Dental anomalies: The L maxillary and both mandibular 3rd molars were genetically absent.

Context III. 22 (ii)

Young adult 17-25 yrs: Female: 149.3 cm, 5'0"

Year of excavation: 1968

Anthropological number: 6862-122

Archaeological context:

NB: Both 6862-121 and 6862-122 were labelled as Context III. 22.

Cist - A rectangular cist which was orientated south-east to north-west was located beneath a group of stone slabs.

Body position - The skeleton of an adult lying in a prone position with the legs flexed and the head to the north-west, was lying on the bottom of the cist.

Grave goods - A fragment of a stone slab was lying near the skull. A clay pot and a bone borer were situated in the northern corner of the tomb. An iron knife was lying adjacent to the phalanges of the L foot.

Visual inventory:

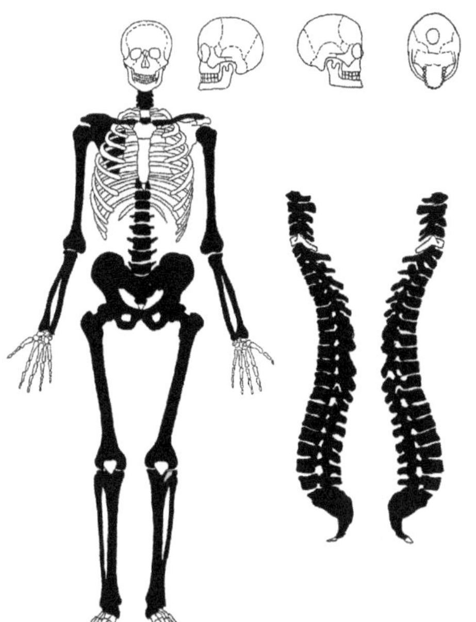

Preservation: The remains were in a very good state of preservation with 80-90% of the bones present.

Post-cranial pathology: Spondylolysis was evident in LV5. Osteophytes and porosity were present in the inferior surfaces of the superior articular facets. The neural arches of S1, S3, S4 and S5 were completely cleft, while the neural arch for S2 was bifurcated. These defects had probably occurred as a result of developmental delay during the early morphogenesis of the spine. Osteophytosis was evident in LV2- LV4, and Schmorl's nodes were visible in LV2 and LV3. Enthesopathies were present in the tibiae (patellar ligament).

Context IV. 4. Sk. 3

Adult 35-45 yrs: Male: 168.6 cm, 5'6"

Year of excavation: 1972

Archaeological context:

Log house tomb - The remains of at least seven individuals were recovered from the floor of the tomb.

Body position - The individual was lying on his L side, with the head to the west. The arms were extended and the legs were flexed. The skull was positioned on a stone slab.

Grave goods - A wooden vessel was situated posterior to the skull. Red paint (?ochre) was apparent on the head of the R femur. A scrap of leather with a bronze button was situated adjacent to the L femur, and the remains of an unidentifiable iron object were lying beneath the leather. A fragment of birch was situated at the feet of the individual.

Visual inventory:

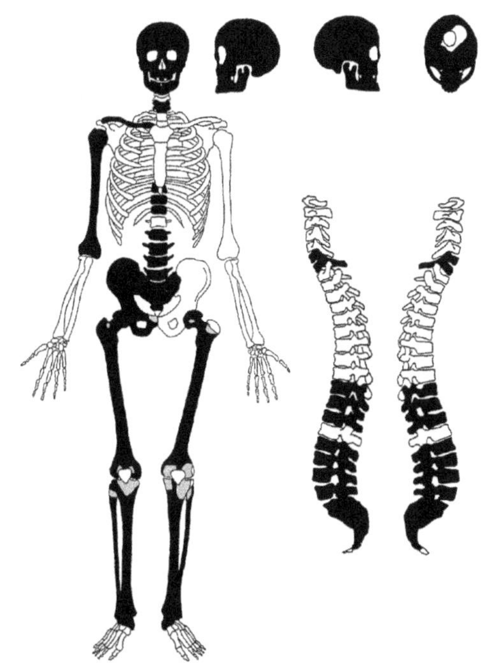

Preservation: The remains were in a fair state of preservation with 40-50% of bones present. The majority of bones were fragmentary with extensive surface erosion.

Post-cranial pathology: A simple oblique, very well healed fracture was apparent on the distal third of the shaft of the L tibia. There was minimal malalignment, almost 100% apposition of the fracture parts and overlap of the fracture sections by approximately 35 mm. The overlap had resulted in hypertrophy at the fracture site. The bone had been subject to post-mortem destruction at its distal end so it was not possible to determine its greatest length. A small sinus was present on the posterior surface of the shaft immediately inferior to the site of the fracture, and it is possible that the fracture had been infected but had healed.

Archaeology & Trauma

A simple oblique, very well healed fracture was also apparent on the proximal half of the shaft of the L fibula. There was overlap of the bone parts by approximately 10 mm, resulting in a greatest length of 347 mm (R GL=353 mm). The inferior part of the fracture had been deflected anteriorly by approximately twenty degrees, and there was only 50% apposition of the fracture parts. Notable hypertrophy was evident at the site of fracture. In addition, prominent vessel impressions and new bone formation in an advanced stage of remodelling was apparent at the fracture site suggesting that this fracture had also been infected. The fractured distal tibia and proximal fibula are an example of a contre coup fracture.

Osteophytosis was present in the R humerus, the R acetabulum, the R femoral head, the L femur, the proximal R tibia and the distal R fibula. DJD was visible at the R acromioclavicular joint. Rotator cuff disease was present in the R humerus. Vertebral osteophytosis was present in CV6, CV7, TV10-TV12 and LV2-S1, while Schmorl's nodes were evident in TV10-TV12 and LV2. Enthesopathies were apparent in the R innominate (adductor magnus), the femora (gluteus medius, gluteus maximus, adductor longus, adductor brevis, adductor magnus), the tibiae (soleus), the R fibula (interosseous ligament) and the calcanei (Achilles' tendon).

Dentition:

~~8~~ 7 6 5 4̸ 3 2 1	1̸ 2̸ 3 4 5 6 7 8̸
8 7 6 5 4 3 2 1̸	1̸ 2 3 4 5 6 7 8

Bite: Overbite.

Attrition: Molar wear patterns of Stage 3+ for the 3rd molars, Stage 4 for the 2nd molars and Stage 4+ for the 1st molars were apparent.

Calculus: Flecks of calculus were present on the buccal surfaces of 1⌋ ⌊4 5 7 and 4⌉, the lingual surfaces of ⌊3 and ⌈1 4 5, the buccal and lingual surfaces of 7 6 5 3⌉ ⌈2 3 6 7 and the buccal, lingual and distal surfaces of 8⌉. Slight deposits were apparent on the buccal surfaces of 5⌋ ⌊6 and ⌈1 8, the lingual surfaces of 3⌋ and 4⌉ and the buccal and lingual surfaces of 7 6⌋ and 2⌉. Medium deposits were present on the lingual and distal surfaces of ⌈8.

Periodontal disease: Periodontal disease of slight severity was apparent.

Context VI. 3. Sk. 3

Adult 35-45 yrs: Male

Year of excavation: 1969

Archaeological context:

Cist - A rectangular group of slabs was apparent on the ground surface, beneath which was a rectangular cist orientated east-south-east to west-north-west. The cist contained the overlying remains of three individuals.

Body position - The individual was situated in the northern area of the cist, and was lying on his L side in a flexed position with the head directed to the north-west. The leg and pelvic bones were displaced.

Grave goods - The remains of a wooden vessel were lying adjacent to the skull, while a fragment of a bronze knife was situated near the knees.

Visual inventory:

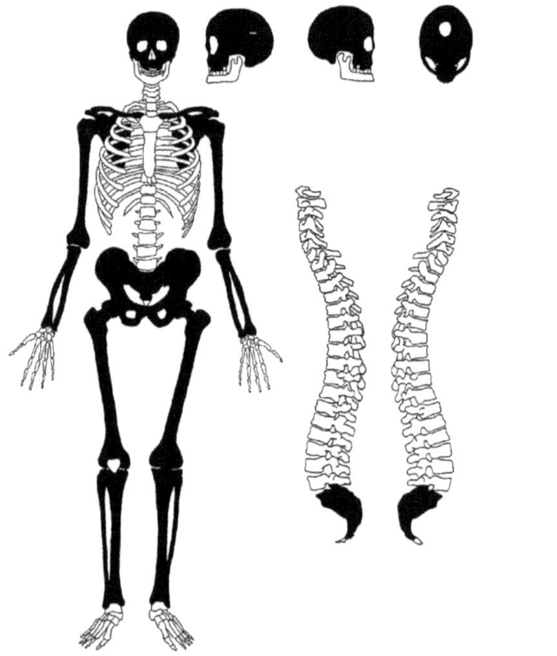

Preservation: The remains were in a good state of preservation with 50-60% of bones present.

Cranial pathology: Porosity was present in the articular tubercles of the temporals, which may be indicative of the early stages of DJD of the temporomandibular joints. The L side was more extensively affected than the R side.

Post-cranial pathology: A bony exostosis, which measured 47.6 mm supero-inferiorly by 8 mm medio-laterally, was present on the lateral supracondylar line of the L femur. The exostosis may have been an enthesophyte of the short head of the biceps. Osteophytosis was apparent in the R distal humerus, the proximal radii and the proximal and distal ulnae. Enthesopathies were present in the radii (biceps) and the fibulae (soleus).

Post-cranial non-metric traits: L vastus notch (full set of traits not recorded).

Dentition:

8 (C) 7̸ (A) 6 (C, A) 5 4 3 2 (B) 1̸	1̸ 2̸ 3 4 5 6 7̸ 8 (NP)
~~8 7 6 5 4 3 2 1~~	~~1 2 3 4 5 6 7 8~~

Bite: Overbite.

Attrition: Molar wear patterns of Stage 3+ for the 3rd molars and Stages 4+-5 for the 1st molars were apparent.

Calculus: Slight deposits of calculus were present on the buccal surface of 3⌋, the lingual surfaces of 4⌋ ⌊3 4 5, the occlusal surface of 8⌋ and the buccal and lingual surfaces of ⌊6. Medium deposits were evident on the buccal surfaces of 4⌋ ⌊3 4 5, the lingual surface of 3⌋, the buccal and lingual surfaces of 5 2⌋, the buccal and distal surfaces of 8⌋ and the buccal, lingual and occlusal surfaces of 6⌋. Heavy deposits were visible on the lingual surface of 8⌋. The presence of calculus on the occlusal surfaces of 8 6⌋ indicated that the individual had not been using the R side of the mouth in mastication, possibly as a result of the caries and abscesses.

Periodontal disease: Periodontal disease of considerable severity was evident.

Caries: A medium sized caries was visible in the buccal surface of 8⌋, while a large caries was apparent in the distal surface of 6⌋.

Abscesses: A medium sized externally draining abscess was associated with 7⌋, and a large externally draining abscess was associated with 6⌋.

Ante-mortem tooth loss: ⌊7 had been lost ante-mortem.

Dental anomalies: The L maxillary 3rd molar was genetically absent.

Context VI. 5. Sk. 6

Adult 25-35 yrs: Probable male

Year of excavation: 1969

Anthropological number: 6862-130

Archaeological context:

Log house tomb - The remains of seven individuals were recovered from the floor of the tomb. The individuals were positioned in a western row (1-6) and an eastern row (7). The individuals in the western row were lying on their L sides in flexed positions, with the heads directed to the north-west, and the remains were very disturbed. The individual in the eastern row was lying on the R side, with the head to the north-east.

Body position - The skull, arm bones and thorax were greatly disturbed.

Grave goods - A stone slab was positioned on top of the knees. The remains of two clay pots were discovered in the northern corner of the tomb.

Visual inventory:

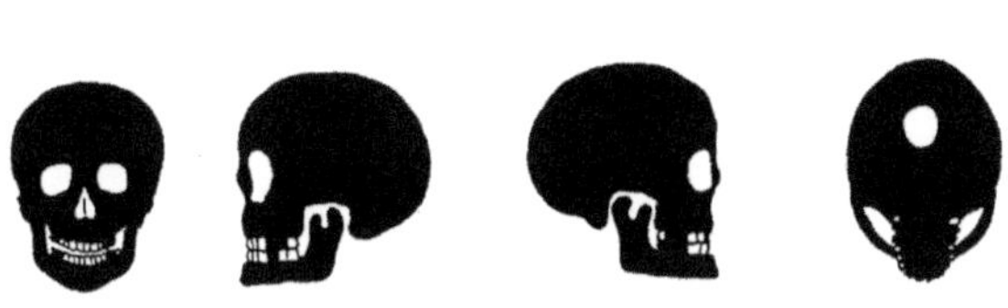

Preservation: Only the skull was present, which was in a very good state of preservation.

Cranial pathology: A very well healed simple oblique fracture was present at the midpoint of the R nasal bone, while a slightly less well healed simple transverse fracture was present at the midpoint of the L nasal bone. The distal half of the fracture on the L nasal bone was slightly depressed. Porosity was present on the superior surface of the frontal, the medial surfaces of the parietals and the superior aspect of the occipital, which may be indicative of healed porotic hyperostosis.

Dentition:

NP 8	7	6	~~5~~	~~4~~	~~3~~	~~2~~	~~1~~	~~1~~	~~2~~	~~3~~	~~4~~	~~5~~	6	7	NP 8
8	7	6	~~5~~	~~4~~	~~3~~	~~2~~	~~1~~	~~1~~	~~2~~	~~3~~	~~4~~	~~5~~	6	~~7~~	8 NP

Bite: Overbite.

Attrition: Molar wear patterns of Stage 2+ for the 3rd molars, Stage 3- for the 2nd molars and Stage 4 for the 1st molars were apparent.

Calculus: Slight deposits of calculus were present on the buccal surfaces of 7⌋ ⌊7, the lingual surface of 6⌉ and the distal surface of 8⌉.

Periodontal disease: Periodontal disease of slight severity was present.

Dental anomalies: The L mandibular 3rd molar and both maxillary 3rd molars were genetically absent.

Context VI. 6. Sk. 4

Adult 25-35 yrs: Male: 169.8 cm, 5'6"

Plates: 1, 40, 41

Year of excavation: 1969

Anthropological number: 6862-134 (skull)

Archaeological context:

Log house tomb - The remains of four individuals were lying on the floor of the tomb. All skeletons had flexed legs and their heads were directed to the north-west.

Body position - The majority of the bones were disturbed, but it was possible to ascertain that the individual was lying on his L side. The skull was positioned on a slab.

Grave goods - A bronze pot, and four tanged arrowheads - three bronze trilobate arrowheads with the remains of shafts, and one wooden trilobate arrowhead - were positioned near the slab on which the skull was situated. A bronze hook, a bronze button and fragments of birch bark were located near the thorax.

Visual inventory:

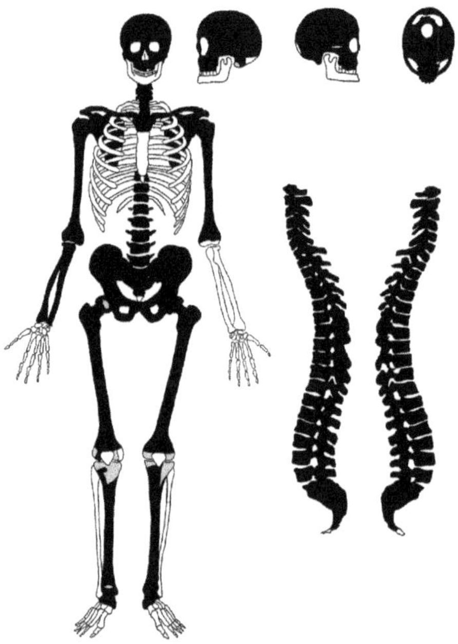

Preservation: The remains were in a very good state of preservation with 60-70% of bones present.

Post-cranial pathology: Vertebral osteophytosis was apparent in TV9-LV5. Osteophytosis was apparent in the L glenoid fossa of the scapula. Rotator cuff disease was visible in the R humerus. Enthesopathies were present in the R humerus (pectoralis major) and the R innominate (reflected head of rectus femoris).

Amputation: The L forearm appeared to have been amputated at the distal end of the humerus. The trochlea and the capitulum were not present and only the superior part of the medial epicondyle had survived. The inferior aspect of the distal end was smooth and curved. A bony nodule was present in the area of the olecranon fossa. The anterior surface of the distal end had a very porotic appearance, and the metaphysis displayed pitted woven bone which was in the advanced stages of remodelling. The presence of new bone formation in the late stages of remodelling may indicate that the amputation was in the advanced stages of healing. The L humerus had a greatest length of 306 mm as opposed to 328 mm for the normal R humerus. The entheses in the L humerus were pronounced and similar to those present in the R humerus. This may suggest that the amputation had occurred quite a short period of time before the individual had died and that he had been actively using his arm until the amputation occurred.

Post-cranial non-metric traits: CV1 L double condylar facet, CV1 L posterior bridging (full set of traits not recorded).

Dentition:

8̸	7	6	5	4	3	2	1	1̸	2̸	3	4	5	6	7	8
~~8~~	~~7~~	~~6~~	~~5~~	~~4~~	~~3~~	~~2~~	~~1~~	~~1~~	~~2~~	~~3~~	~~4~~	~~5~~	~~6~~	~~7~~	~~8~~

Bite: Overbite.

Attrition: Most teeth displayed extensive attrition with molar wear patterns of Stage 3+ for the 3rd molars, Stage 4+ for the 2nd molars and Stage 5 for the 1st molars.

Calculus: Slight deposits of calculus were present on the buccal surfaces of 7 4 3 2 1⌋ ⌊3 6 8 and the buccal and lingual surfaces of 6⌋.

Periodontal disease: Periodontal disease of slight severity was evident.

Context VI. 6

Child 4 - 5 years

Plates: 1, 34

Year of excavation: 1969

Anthropological number: 6863-54

Archaeological context:

Cist - A large mound of stones was present on the ground surface, beneath which was a rectangular cist orientated south-west to north-east. The cist contained the remains of a child.

Body position - The individual was lying on his or her R side, with the head to the north-west. The legs were flexed, and the skull was lying on a stone slab. Only the skull was available for analysis.

Grave goods - A bronze buckle was situated posterior to the skull.

Visual inventory:

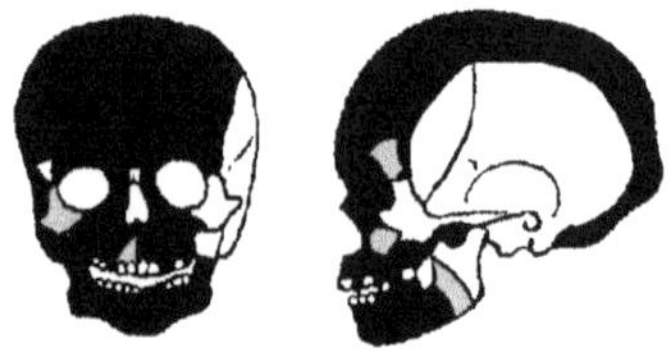

Preservation: Only the skull was present, which was in a poor state of preservation with surface erosion.

Weapon trauma: A circular perforation, which measured 13.5 mm by 13 mm, was present at the midpoint of the L parietal. The supero-anterior margin had a bevelled edge. It is probable that the injury had been made with a pointed battle axe. The angle of the perforation indicates that the attacker had been standing posterior and to the R of the child.

Dentition:

e̸	d	c̸	b̸	a̸	a̸	b	c̸	d	e̸
e̸	d̸	c̸	b̸	a̸	a̸	b̸	c̸	d	e

Bite: Overbite.

Attrition: Molar wear patterns of Stage 3+ for the 1st deciduous molars and Stage 3- for the 2nd deciduous molars were apparent.

Calculus: Slight deposits of calculus were present on the buccal surface of ⌊b.

Context VI. 9. upper (ii)

Juvenile 7 - 10 years

Year of excavation: 1970

Anthropological number: 6863-57

Archaeological context:

Log house tomb - Disarticulated human remains were recovered throughout the tomb. The original location of this individual within the tomb is unknown.

Visual inventory:

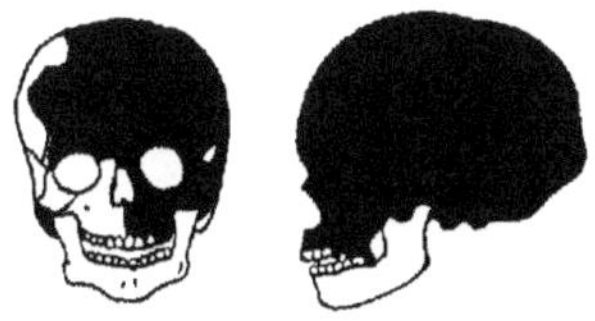

Preservation: Only the cranium was present, which was in a poor state of preservation.

Weapon trauma: A possible pointed battle axe injury was present at the midpoint of the R side of the coronal suture and extending onto the frontal bone. Since the R parietal was not present it was not possible to determine the full extent of the injury.

Dentition:

												PU	PU		U	U
~~8~~	~~7~~	~~6~~	~~e~~	~~d~~	~~c~~	~~2~~	~~1~~	|	1/	2/	c/	d	e	6	7	8
~~8~~	~~7~~	~~6~~	~~e~~	~~d~~	~~c~~	~~2~~	~~1~~	|	~~1~~	~~2~~	~~c~~	~~d~~	~~e~~	~~6~~	~~7~~	~~8~~

Bite: Overbite.

Attrition: Molar wear patterns of Stage 4+ were apparent on the 1st and 2nd deciduous molars. A molar wear pattern of Stage 3- was present for the 1st permanent molars. Secondary dentine was exposed in the deciduous molars.

Calculus: Flecks of calculus were present on the buccal surface of ⌊6, slight deposits were apparent on the buccal surface of ⌊d and medium deposits were visible on the buccal surface of ⌊e.

Context VI. 12. Sk. 3

Adult 25-35 yrs: Female: 158.6 cm, 5'3"

Year of excavation: 1970

Anthropological number: 6862-145

Archaeological context:

Log house tomb - The disturbed remains of three skeletons with their heads to the west were recovered from the floor of the tomb.

Body position - The individual was lying on her R side with the legs flexed. The arms were extended anteriorly with the elbows slightly flexed. The skull was disturbed.

Grave goods - A bone button and fragments of a clay pot were situated near the northern wall.

Visual inventory:

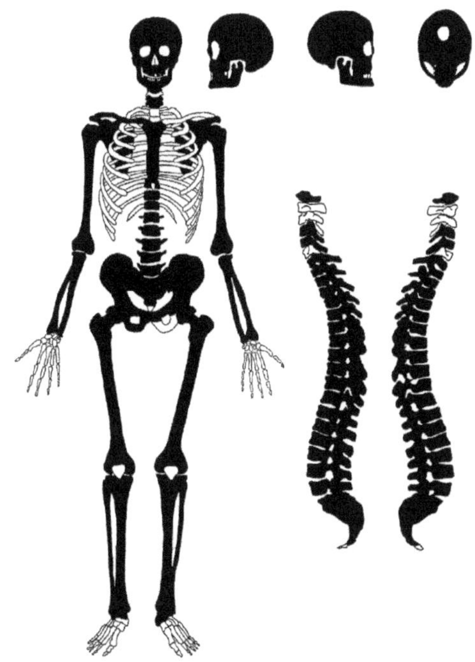

Preservation: The remains were in a very good state of preservation with 80-90% of the bones present.

Post-cranial pathology: Spondylolysis of LV5 was evident with no associated DJD. Vertebral osteophytosis was present in TV6, TV7 and LV5.

Cranial non-metric traits: R and L highest nuchal lines, R and L ossicles in lambdoid suture, R parietal foramen, L ossicle at parietal notch, R and L mastoid foramina extrasutural, R and L posterior condylar canals open, R and L open foramina spinosum, R and L absent zygomaticofacial foramina, R and L anterior ethmoid frontal foramina extrasutural, R and L accessory infraorbital foramina.

Post-cranial non-metric traits: R and L hypotrochanteric fossae, R and L third trochanters, R and L lateral tibial squatting facets, R and L acromial articular facets, R and L circumflex sulci, L medial talar facet, L double inferior articular talar facet, R and L double anterior calcaneal facets, R CV5 transverse foramen bipartite.

Other observations: Tiny cutmarks were evident on the superior surfaces of LV2, LV3 and LV4, the inferior and posterior aspects of the R humeral head, the medial aspect of the iliac blade and the supero-anterior aspect of the L femoral head. The cutmarks are probably indicative of post-mortem defleshing or disarticulation.

Dentition:

	C														
8	7	6	5	4	3	2	1	~~1~~	2	~~3~~	4	5	6	7	~~8~~
8	7	6	~~5~~	~~4~~	~~3~~	~~2~~	~~1~~	1	~~2~~	~~3~~	4	5	6	7	8
2C	C													C	C

Bite: Overbite.

Attrition: Molar wear patterns of Stage 1 for the 3rd molars, Stages 1-2 for the 2nd molars and Stage 3 for the 1st molars were present.

Calculus: Flecks of calculus were present on the buccal surfaces of ⌈4 5, the lingual surface of ⌈8 and the buccal and lingual surfaces of ⌈6 7. Slight deposits were present on the buccal surfaces of 7 6⌋ and ⌈1 8, the lingual surfaces of 7⌉ ⌈4 5, the distal surface of 8⌋, the buccal and lingual surfaces of ⌊5 6 7 and 6⌉ and the buccal, lingual and distal surfaces of 8⌉. Medium deposits were present on the lingual surface of ⌈1.

Periodontal disease: Periodontal disease of slight severity was apparent.

Caries: Small caries were present in the occlusal surfaces of 7⌋, 8⌉, 7⌉ and ⌈7, while two small caries were visible in the occlusal surface of ⌈8.

Dental anomalies: Mandibular tori were apparent.

Context VI. 13. Sk. 8

Adult 25-35 yrs: Male: 169.1 cm, 5'6"

Year of excavation: 1969

Anthropological number: 6862-149

Archaeological context:

Log house tomb - The remains of nine individuals were situated on the floor of the tomb. All skeletons had flexed legs and were positioned in either the western row (4-9) or the eastern row (10-12).

Body position - Skeleton 8 was lying on his L side with the head to the north-west. The arms were extended anteriorly and the skull was positioned on a stone slab.

Grave goods - A fragment of a bead was located near the mandible, and a bronze hook was lying adjacent to the hands. A trapezoid-shaped bronze container and fragments of a thong were situated in close proximity to the R side of the pelvis. Fragments of birch bark and a bronze dagger, under which was an iron knife, were recovered from adjacent to the R femur. A slab was lying across the lower legs. Five pots were situated in the northern corner of the tomb, and a bronze mirror was recovered from inside one of the pots.

Visual inventory:

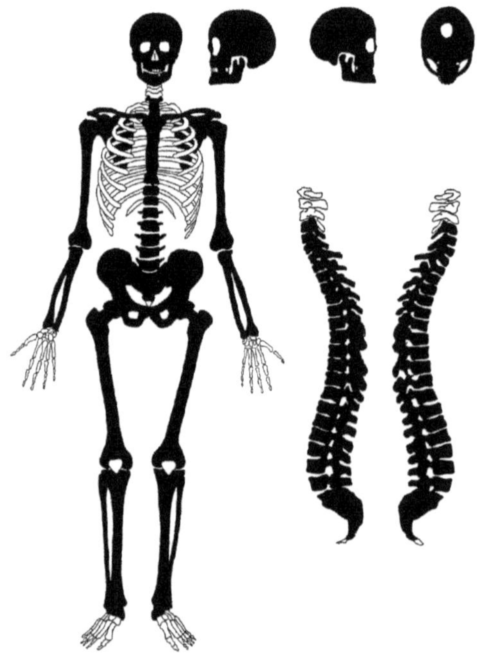

Preservation: The remains were in a very good state of preservation with 80-90% of the bones present.

Cranial pathology: A simple oblique fracture, in the early stages of healing, was present on the R nasal bone. The anterior area of the fracture was depressed and positioned inferiorly to the superior fracture part. Part of the anterior fractured area had been displaced, and it is probable that the fragment would have been present as a detached fragment of bone which was then lost post-mortem. A simple transverse fracture, in the early stages of healing, was present on the L nasal bone. The anterior area was completely malaligned, and was displaced laterally by approximately fifty degrees so that it was positioned superior to the maxilla. A simple transverse fracture was also present on the R side of the margin of the nasal cavity. The fracture line was clearly visible, but there was no malalignment.

Post-cranial pathology: Pitted and striated woven bone was present on the posterior and lateral surfaces of the midshaft of the L femur. A prominent plaque of woven bone was also visible on the lateral surface of the distal half of the femoral shaft. Vertebral osteophytosis was evident in CV6, CV7, TV5-TV11 and LV1-LV4. Schmorl's nodes were visible in LV1 and LV2. Enthesopathies were apparent in the humeri (deltoid) and the calcanei (Achilles' tendon).

Cranial non-metric traits: R and L highest nuchal lines, L parietal foramen, R ossicle at asterion, R and L mastoid foramina extrasutural, R and L posterior condylar canals open, L absent zygomaticofacial foramen, R bridging of supraorbital notch, R and L accessory supraorbital foramina, R and L anterior ethmoid frontal foramina extrasutural, L accessory infraorbital foramen.

Post-cranial non-metric traits: R and L hypotrochanteric fossae, R exostosis in trochanteric fossa, R and L third trochanters, R medial tibial squatting facets, R and L lateral tibial squatting facets, R and L acetabular creases, R and L accessory sacral facets, R and L acromial articular facets, R and L circumflex sulci, R and L CV5, and R and L CV6 transverse foramina bipartite.

Dentition:

8 7 6 5 4 3 2 1	1 2 3 4 5 6 7 8
8 7 6 5 4 3 2 1	1 2 3 4 5 6 7 8

Bite: Overbite.

Attrition: Moderate to extensive attrition was apparent on the majority of teeth, with molar wear patterns of Stages 1-2- for the 3rd molars, Stage 4 for the 2nd molars and Stages 4+-5 for the 1st molars.

Calculus: Slight deposits of calculus were present on the buccal surfaces of 7 6 5⌋ and 6 2⌉ ⌈5 8, the buccal and lingual surfaces of 1⌉ ⌈1 7 and the buccal and mesial surfaces of ⌈6. Medium deposits were visible on the buccal surfaces of 4 3 2 1⌋ ⌊6 7 and 3⌉ ⌈3, the lingual surfaces of 2⌉ ⌈5, the buccal and lingual surfaces of 7 5⌉, the lingual and distal surfaces of ⌈8, the buccal, distal and mesial surfaces of 8⌉ and the buccal, lingual and mesial surfaces of 4⌉. Heavy deposits were observed on the lingual surfaces of 8⌉ ⌈3.

Periodontal disease: Periodontal disease of medium severity was evident.

Ante-mortem tooth loss: 8⌋ had been lost ante-mortem, and the socket was in the early stages of remodelling.

Dental anomalies: An extra root socket was present just buccally to the area between ⌊2 and ⌊3. It is probable that the socket is indicative of retention of a deciduous tooth.

Context VI. 15. Sk. 4

Adult 35-45 yrs: Female

Year of excavation: 1970

Anthropological number: 6862-155

Archaeological context:

Log house tomb - The remains of ten individuals were recovered from the floor of the tomb. Five of the skeletons (2-6) were lying in a row on their L sides with the heads to the north-west.

Body position - The skull was lying on a stone slab. The arms were extended, and the R arm was lying on the remains of Skeleton 5.

Grave goods - A bone button was recovered from between the feet.

Visual inventory:

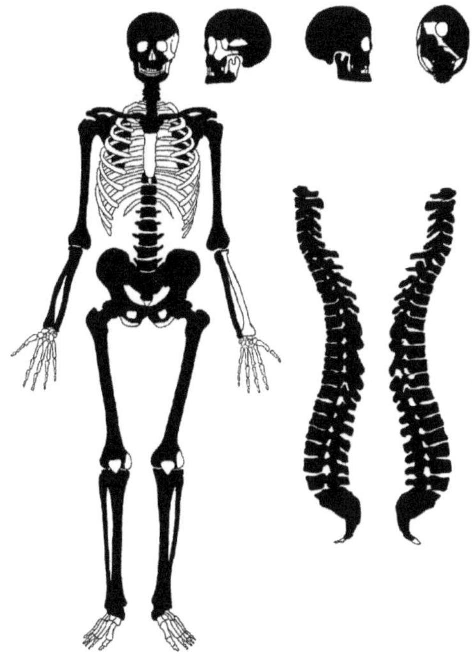

Preservation: The remains were in a very good state of preservations with 80-90% of bones present. The skull was fragmentary.

Post-cranial pathology: Spondylolysis of LV5 was evident. The auricular surfaces of the sacrum and ilia had a very porotic, rugged appearance. The auricular margins on the ilia appeared to have been broken post-mortem, and it is possible that the pelvic bones had been ankylosed at the sacro-iliac joints. Vertebral osteophytosis was evident in CV2, TV8, TV9 and LV5. Schmorl's nodes were present in LV1. Enthesopathies were apparent in the calcanei (Achilles' tendon).

Cranial non-metric traits: R ossicle at pterion, R and L mastoid foramina extrasutural, R accessory supraorbital foramen, R anterior ethmoid frontal extrasutural, R and L accessory infraorbital foramina.

Post-cranial non-metric traits: R and L hypotrochanteric fossae, R and L third trochanters, R and L lateral tibial squatting facets, L septal aperture, R and L acromial articular facets, R and L circumflex sulci, R and L medial talar facets, R and L double inferior articular talar facets, R and L double anterior calcaneal facets, R and L peroneal tubercles, L CV5, R CV6 and L CV7 transverse foramina bipartite.

Dentition:

8 7(C) 6 5 4 3 2 1	1 2 3 4 5 6 7 8
8 7 6 5 4 3 2 1	1 2 3 4 5 6 7 8

Bite: Overbite.

Attrition: Molar wear patterns of Stages 1-2- for the 3rd molars, Stage 3 for the 2nd molars and Stage 4 for the 1st molars were evident.

Calculus: Slight deposits of calculus were present on the lingual surfaces of 7 6 5⌋ ⌊5 and ⌈7 and the mesial surface of 7⌋. Medium deposits were visible on the buccal surfaces of 6 5 3⌋.

Periodontal disease: Periodontal disease of medium severity was apparent.

Caries: A medium sized caries was present in the buccal aspect of 7⌋.

Context VI. 15. Sk. 6

Adult 25-35 yrs: Probable male: 161.2 cm, 5'4"

Plate: 16

Year of excavation: 1970

Anthropological number: 6862-157

Archaeological context:

Log house tomb - The remains of ten individuals were recovered from the floor of the tomb. Five of the skeletons (2-6) were lying in a row on their L sides with the heads to the north-west.

Body position - The skull was lying on a stone slab.

Grave goods - A clay pot was situated behind the slab, on which the skull was positioned. A pot was also located anterior to the skull. A bone object was situated beside the thorax, while another bone object was located on the R side of the pelvis. Fragments of an iron knife were recovered from adjacent to the pelvis.

Visual inventory:

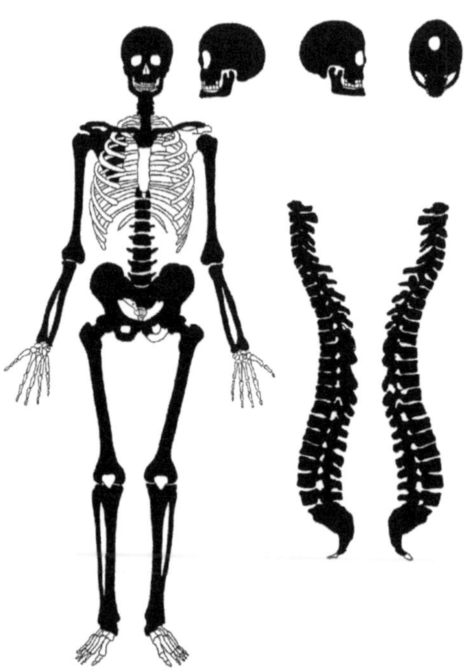

Preservation: The remains were in a very good state of preservation with 70-80% of bones present.

Cranial pathology: Cribra orbitalia of Type 2 was present in Areas 1, 2, 3, and 4 of the R orbit, and cribra orbitalia of Type 1 was present in Areas 1 and 4 of the L orbit.

Post-cranial pathology: A clay-shoveller's fracture was present in TV1, in which the tip of the spinous process had become separated from the remainder of the vertebra. Indentations were apparent at the supero-anterior area of the lateral condyles of both distal femora. The R lesion measured 15 mm medio-laterally by 13 mm supero-inferiorly, while the L lesion measured 9 mm medio-laterally by 7.5 mm supero-inferiorly. The lesions had a similar appearance to tibial squatting facets.

Vertebral osteophytosis was apparent in CV2-CV6, TV6, TV7, TV9-TV12 and LV2-LV5. Schmorl's nodes were present in TV5-LV5. Marginal osteophytes were present at the fibular facet of the L tibia. Enthesopathies were visible on the femora (adductor magnus, adductor longus, adductor brevis), the tibiae (soleus) and the calcanei (Achilles' tendon).

Dentition:

~~8~~	7	6 (PU)	~~5~~	~~4~~	~~3~~	~~2~~	~~1~~	~~1~~	~~2~~	~~3~~	~~4~~	5	6	7	~~8~~
~~8~~	~~7~~	~~6~~	~~5~~	~~4~~	~~3~~	~~2~~	~~1~~	~~1~~	~~2~~	~~3~~	~~4~~	~~5~~	~~6~~	~~7~~	~~8~~

Bite: Overbite.

Attrition: Molar wear patterns of Stage 3+ for the 2nd molars and Stages 4+-5 for the 1st molars were apparent. Secondary dentine was exposed in 6⌋.

Calculus: Slight deposits of calculus were present on the buccal surface of 6⌋, the lingual surface of ⌊5 and the buccal and lingual surfaces of 7⌋ ⌊6 7.

Periodontal disease: Periodontal disease of medium severity was evident.

Ante-mortem tooth loss: 8 7 6 5 4 1⌉ ⌈1 4 5 6 7 8 had been lost ante-mortem.

Context VI. 15 (ii)

Adult 35-45 yrs: Male: 165.1 cm, 5'5"

Year of excavation: 1970

Anthropological number: 6862-160

Archaeological context:

Log house tomb - The remains of ten individuals were recovered from the floor of the tomb. Five of the skeletons (2-6) were lying in a row on their L sides, with the heads to the north-west. The original location of this individual within the tomb is unknown.

Visual inventory:

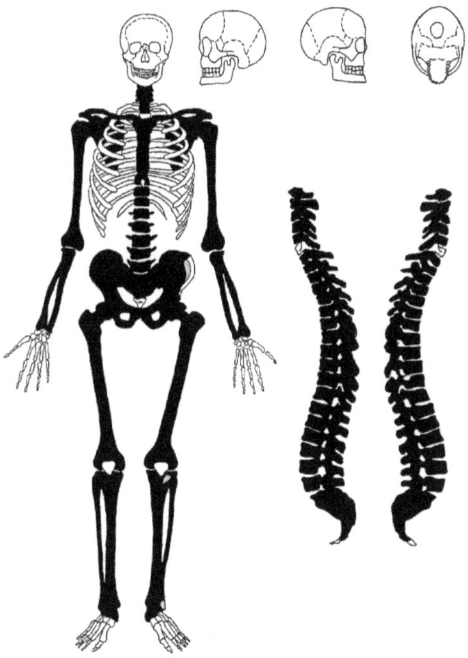

Preservation: The remains were in a good state of preservation with 60-70% of bones present.

Post-cranial pathology: Os acromiale was apparent in the R scapula. Partial sacralization was evident, with the R transverse process of TV5 completely ankylosed to the superior surface of the R ala of the sacrum. Vertebral osteophytosis was visible in TV7 and Schmorl's nodes were present in TV8-LV2.

Weapon trauma: A possible sword cut was present on the sacrum, in which a sharp edged transverse injury had resulted in the complete separation of the distal part of S5 from the remainder of the sacrum. The distal fragment was not present.

Post-cranial non-metric traits: Sternal foramen, large third trochanters (full set of traits not recorded).

Other observations: Green discoloration was evident on the posterior and lateral surfaces of the distal half of the R ulna.

Context VI. 16. Sk. 1

Adult 35-45 yrs: Female: 151.4 cm, 5'0"

Year of excavation: 1971

Anthropological number: 6862-161

Archaeological context:

Log house tomb - The remains of seven individuals were recovered from the floor of the tomb. Each skeleton was lying on its side in a flexed position, with the head to the north-west. Four of the individuals were situated in a lower row, while three were positioned in an upper row.

Body position - The individual was lying on her L side, with the legs markedly flexed. The arm bones were slightly disturbed, and the skull was lying in front of a stone slab.

Grave goods - A wooden jar, a clay pot and an ivory spike were lying adjacent to the stone slab associated with the skull. The remains of two arrowheads and a bronze mirror were discovered in the area posterior to the skull and adjacent to the wall of the tomb. In addition, a broken arrowshaft and fragments of birch were recovered from beneath the skull.

Visual inventory:

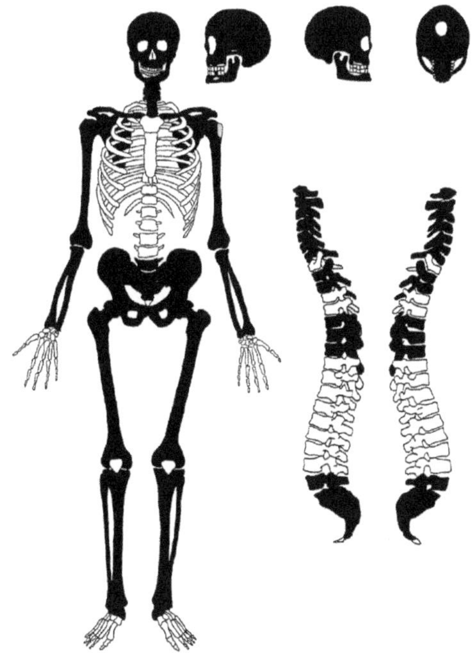

Preservation: The remains were in a very good state of preservation with 70-80% of bones present.

Cranial pathology: Caudal border shifting resulting in partial occipitalization was evident, with a paracondylar bony process situated on the R side of the occipital. The corresponding transverse process of the atlas was enlarged. Cribra orbitalia of Type 2 was present in Areas 1 and 4 of both orbits. Porosity was evident on the articular tubercles of the temporals and on the mandibular condyles, which may be indicative of the early stages of DJD of the temporomandibular joints.

Post-cranial pathology: Partial spondylolysis was evident in LV5, in which the inferior margin of the R superior articular facet was not fused to the lamina. The L side of the vertebra was normal. A simple transverse fracture was present on the distal third of the shaft of the L ulna (Night-stick fracture). The fracture was very well healed with the fracture line barely visible. There was no malalignment of the fracture parts, but some hypertrophy was apparent at the point of fracture, indicating overlap of the fracture parts. The L ulna had a greatest length of 243 mm as opposed to 249 mm in the R ulna, suggesting that there was overlap of the fracture parts by approximately 6 mm.

Osteophytosis was apparent in the ulnar heads, the distal ends of the radii and the acetabulae. Vertebral osteophytosis was present in CV2, CV5-CV7, TV7, TV8 and LV5. Schmorl's nodes were visible in TV8. Enthesopathies were visible on the proximal humeri (latissimus dorsai, pectoralis major, teres major), the radii (biceps), the femora (gluteus maximus, gluteus medius, psoas major, iliacus), the tibiae (soleus) and the calcanei (Achilles' tendon).

Other observations: An area of orange discoloration (rust) was present on the L temporal just superior to the mastoid process.

Dentition:

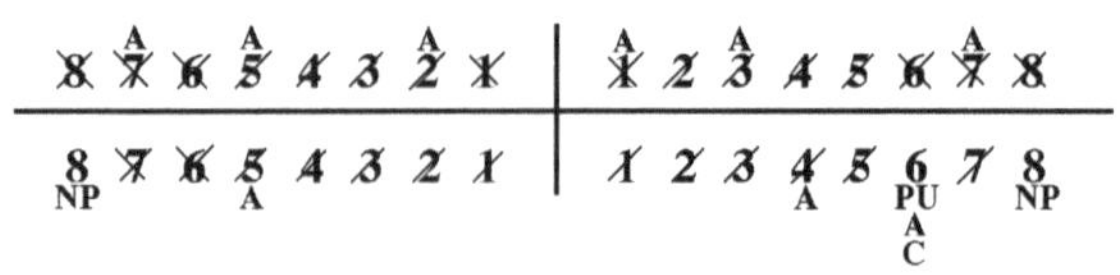

Bite: Overbite.

Attrition: The only tooth present was the L mandibular 1st molar which had a wear pattern of Stage 5. Secondary dentine was exposed.

Periodontal disease: Periodontal disease of considerable severity was apparent.

Caries: A small caries was present in the occlusal surface of ⌈6.

Abscesses: Small externally draining abscesses were associated with 5⌉ and ⌈4, while small internally draining abscesses were associated with ⌊1 and 5⌋. A medium sized externally draining abscess was associated with ⌊3, while medium sized abscesses, which were draining into the maxillary sinuses, were associated with ⌊7 and 2⌋. A large externally draining abscess was associated with ⌈6, and a large abscess, which was draining into the maxillary sinus, was associated with 7⌋.

Ante-mortem tooth loss: 8 7 6 1⌋ ⌊1 6 7 8 and 7 6⌉ had been lost ante-mortem.

Dental anomalies: The mandibular 3rd molars were genetically absent.

Context VI. 16. Sk. 2

Adult 35-45 yrs: Male: 169.2 cm, 5'6"

Year of excavation: 1971

Anthropological number: 6862-162

Archaeological context:

Log house tomb - The remains of seven individuals were recovered from the floor of the tomb. Each skeleton was lying on its side in a flexed position, with the head to the north-west. Four of the individuals were situated in a lower row, while three were positioned in an upper row.

Body position - The individual was lying on his L side. Some of the bones of the skeleton, including the skull, were disturbed. The arms were extended over the sternum, and the skull was lying on a stone slab.

Grave goods - A wooden jar was recovered from behind the slab associated with the skull. Three wooden jars and a clay pot were positioned in a row between the L arm and the wall of the tomb. One of the jars was situated inside another jar.

Visual Inventory:

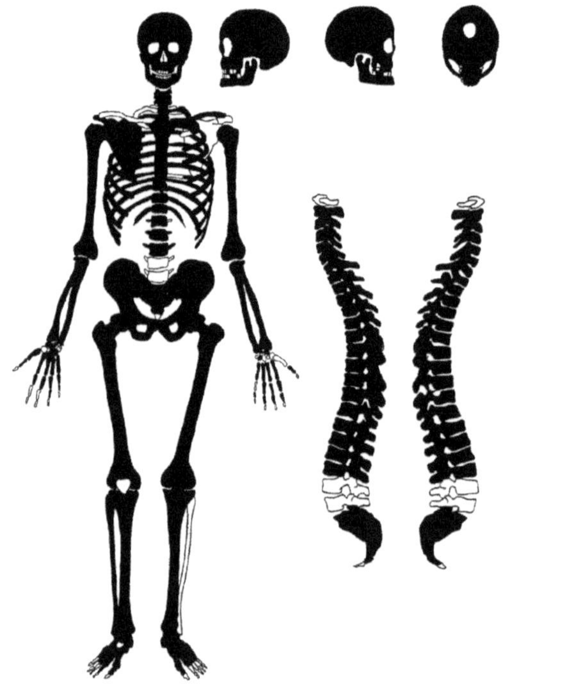

Preservation: The remains were in a very good state of preservation with 80-90% of bones present.

Cranial pathology: Cribra orbitalia of Type 1 was present in Areas 1 and 4 of the R orbit and in Area 1 of the L orbit.

Post-cranial pathology: Simple transverse fractures in the advanced stages of healing were present on the midshafts of the L 7th and 8th ribs. The fracture lines were only barely visible. Osteophytes and porosity were present on the majority of rib heads. Osteophytosis was evident at the distal end of the R 1st proximal foot phalanx, the distal end of the R MC1, the proximal end of the R 1st hand phalanx and the distal radii. Rotator cuff disease was present in the humeri. Extensive DJD was evident in the L hip and both knee joints. Vertebral osteophytosis was visible in CV2 and CV7-LV3. Schmorl's nodes were present in TV7-LV1 and LV3. Enthesopathies were apparent in the humeri (teres major, latissimus dorsai, pectoralis major).

Cranial non-metric traits: R and L highest nuchal lines, 1 L and 2 R ossicles in lambdoid suture, R and L parietal foramina, R and L ossicles at pterion, R and L ossicles at parietal notch, L ossicle at asterion, R and L mastoid foramina extrasutural, R and L posterior condylar canals open, R and L accessory supraorbital foramina, R and L anterior ethmoid frontal foramina extrasutural, R and L accessory infraorbital foramina.

Post-cranial non-metric traits: R plaque, R and L hypotrochanteric fossae, R and L exostoses in trochanteric fossae, R and L lateral tibial squatting facets, R acetabular crease, R and L accessory sacral facets, R acromial articular facet, R circumflex sulcus, L CV4, R and L CV5, R and L CV6 transverse foramina bipartite.

Dentition:

		2A	PU												2A	
Upper	8	7̸	6	5	4	3̸	2̸	1 (X)	1̸	2̸	3̸	4̸	5	6 (X)	7̸	8
Lower	8	7	6	5̸	4̸	3	2	1̸	1̸	2̸	3̸	4̸	5 (X)	6	7	8
	NP		PU			C			A							NP

Bite: Overbite.

Attrition: Moderate to extensive attrition was present on all teeth, with molar wear patterns of 4+-5 for both the 1st and 2nd molars. Pulp was exposed in 6⌋ and 6⌉.

Calculus: Slight deposits of calculus were present on the buccal surfaces of 4⌋ ⌊5 and ⌈6, the buccal and lingual surfaces of 7 2⌉, the buccal, lingual and distal surfaces of 8⌉ and the buccal, lingual, distal and mesial surfaces of 5⌋ ⌊8. Medium deposits were present on the lingual surface of ⌈6, the occlusal surfaces of 8⌋ ⌊8, the buccal and lingual surfaces of 3⌉ and the buccal and distal surfaces of 6⌋. Heavy deposits were visible on the lingual surface of ⌈7. The heavy deposits and the presence of calculus on occlusal surfaces of some teeth indicated that the individual did not use certain teeth in mastication.

Periodontal disease: Periodontal disease of medium severity was apparent.

Caries: A medium sized caries was present on the lingual surface of 3⌉.

Abscesses: A medium sized externally draining abscess was associated with ⌈1. Medium sized internally draining and externally draining abscesses were associated with ⌊7. A large internally draining abscess and a large externally draining abscess associated with 7⌋ had combined to form a single huge abscess fistula.

Ante-mortem tooth loss: 1⌋ ⌊6 and ⌈5 had been lost ante-mortem.

Dental anomalies: The maxillary 3rd molars were positioned very far back in the tooth row, and it is unlikely that the individual would have been able to use them in mastication. The presence of calculus on the occlusal surfaces of the maxillary 3rd molars supports this finding. Both mandibular 3rd molars were genetically absent.

Context VI. 16. Sk. 4

Young adult 17-25 yrs: Male: 173.3 cm, 5'8"

Year of excavation: 1971

Anthropological number: 6862-164

Archaeological context:

Log house tomb - The remains of seven individuals were recovered from the floor of the tomb. Each skeleton was lying on its side in a flexed position, with the head to the north-west. Four of the individuals were positioned in a lower row, while three were situated in an upper row.

Body position - The individual was lying on the L side, with the arms extended in front of the sternum. The skull was positioned on a stone slab.

Grave goods - Four wooden jars were situated posterior to the skull, and a sheep scapula was located beneath one of the jars.

Visual inventory:

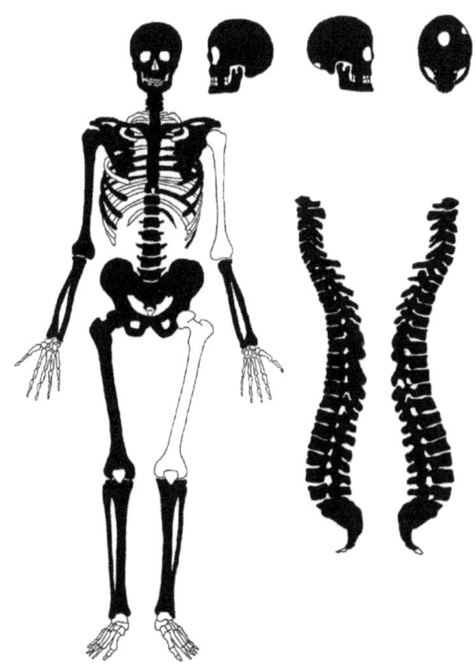

Preservation: The remains were in a very good state of preservation with 70-80% of bones present.

Post-cranial pathology: Spondylolysis of LV5 was present. Vertebral osteophytosis was evident in LV1, and Schmorl's nodes were visible in TV6-LV3.

Cranial non-metric traits: R and L highest nuchal lines, 2 ossicles at lambda, 1 L and 2 R ossicles in lambdoid suture, R parietal foramen, R and L ossicles at pterion, L ossicle at parietal notch, R and L ossicles at asterion, R and L mastoid foramina extrasutural, R posterior condylar canal open, R and L open foramina spinosum, R and L accessory supraorbital foramina, L anterior ethmoid frontal extrasutural, R and L accessory infraorbital foramina. In addition an ossicle was visible in both the R and L squamosal sutures.

Post-cranial non-metric traits: R Allan's fossa, R hypotrochanteric fossa, R and L lateral tibial squatting facets, R acetabular crease, R and L acromial articular facets, R and L circumflex sulci, R and L double atlas facets, R and L CV5, R and L CV6 transverse foramina bipartite.

Dentition:

U							A								U
8	7	6	~~5~~	~~4~~	~~3~~	~~2~~	~~1~~	~~1~~	~~2~~	~~3~~	~~4~~	~~5~~	6	7	8
8	7	6	~~5~~	~~4~~	~~3~~	~~2~~	~~1~~	~~1~~	~~2~~	~~3~~	~~4~~	~~5~~	6	7	8
U															U

Bite: Overbite.

Attrition: Molar wear patterns of Stage 3 for the 2nd molars and Stages 3+-4 for the 1st molars were apparent.

Calculus: Slight deposits of calculus were present on the buccal surface of ⌈6 and the lingual and distal surfaces of ⌈7. Medium deposits were visible on the buccal surfaces of 7⌋ ⌊6 7, the buccal and lingual surfaces of 3⌉ and the buccal, lingual and distal surfaces of 7 6⌉.

Periodontal disease: Periodontal disease of slight severity was evident.

Abscesses: A medium sized externally draining abscess was associated with 1⌋.

Ante-mortem tooth loss: All four 2nd premolars had been lost ante-mortem.

Context VII. 1 (iii)

Adult 35-45 yrs: Probable male

Year of excavation: 1970

Anthropological number: 6862-167

Archaeological context:

Log house tomb - Disarticulated human remains were recovered throughout the tomb. The original location of this individual within the tomb is not known.

Visual inventory:

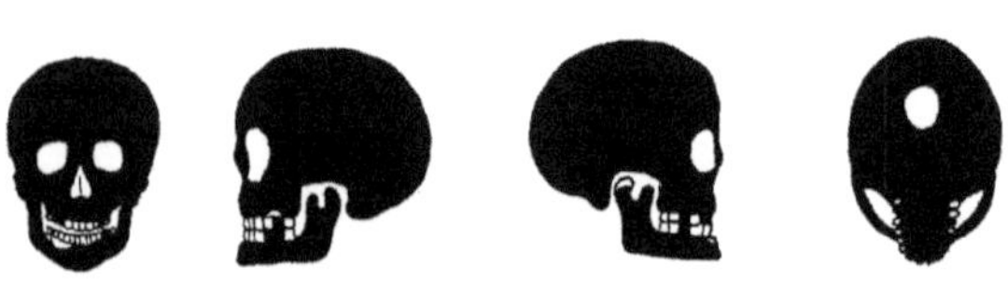

Preservation: Only the skull was present, which was slightly fragmentary.

Cranial pathology: Simple transverse fractures were present at the distal thirds of the nasal bones. In both instances the fracture lines were situated at the medial aspect of the bones, which may indicate that they had been caused by a direct blow to the midpoint of the nose. There was no malalignment.

Dentition:

NP		C PU	A	PU				A					A	A C PU	NP
8	~~7~~	6	~~5~~	4	~~3~~	~~2~~	~~1~~	~~1~~	~~2~~	~~3~~	~~4~~	~~5~~	~~6~~	7	8
8	~~7~~	6	~~5~~	~~4~~	~~3~~	~~2~~	~~1~~	~~1~~	~~2~~	~~3~~	~~4~~	~~5~~	6	7	8
NP		PU		A											NP

Bite: Overbite.

Attrition: Molar wear patterns of Stage 4+ for the 2nd molars and Stages 4+-5 for the 1st molars were apparent. Secondary dentine or pulp cavity exposure was evident in 6 4⌋ ⌊7 and 6⌉.

Calculus: Slight deposits of calculus were present on the buccal surfaces of 6⌋ ⌊7, the lingual surface of 6⌉ and the buccal and lingual surfaces of 4⌋ and ⌈6 7.

Periodontal disease: Periodontal disease of considerable severity was apparent.

Caries: Large caries were present in the occlusal surfaces of 6⌋ and ⌊7.

Abscesses: Medium sized externally draining abscesses were associated with 5⌋, ⌊7 and ⌊1, while large externally draining abscesses were associated with ⌊6 and 4⌉.

Ante-mortem tooth loss: ⌊1 6 and 1⌉ had been lost ante-mortem.

Dental anomalies: All four 3rd molars were genetically absent.

Context VII. 3

Young adult 17-25 yrs: Probable male

Year of excavation: 1970

Anthropological number: 6862-179

Archaeological context:

Log house tomb - The remains of three individuals were recovered from the floor of the tomb. It was not known which individual was represented by 6862-179.

Visual inventory:

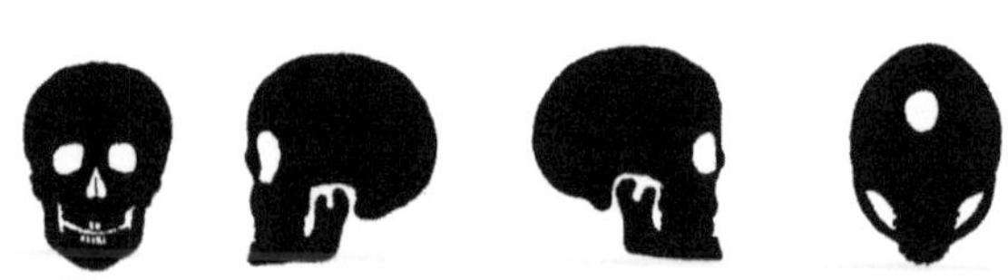

Preservation: Only the skull was present, which was in a very good state of preservation.

Cranial pathology: Cribra orbitalia of Type 2 was present in Areas 1 and 2 of the R orbit, and cribra orbitalia of Type 1 was present in Areas 1 and 2 of the L orbit. A simple oblique fracture was present at the midpoint of the R nasal bone. The fracture was in an advanced state of healing, and the fracture line was barely visible. A simple transverse fracture was evident at the midpoint of the L nasal bone. There was a small gap between the two parts of the fracture, but no malalignment was apparent.

Dentition:

						B									
8	7	6	5	4	3	2	~~1~~	~~1~~	2	3	4	5	6	7	8
8	7	6	5	4	~~3~~	~~2~~	~~1~~	~~1~~	~~2~~	3	4	5	6	7	8
			U									U			

Bite: Overbite.

Attrition: Molar wear patterns of Stage 1 for the 2nd and 3rd molars and Stage 2+ for the 1st molars were apparent.

Calculus: Flecks of calculus were present on the buccal surface of ⌊2, the distal surface of 8⌉ and the buccal and lingual surfaces of 6⌋ and 7⌉. Slight deposits were present on the lingual surfaces of 5 4 3⌋ ⌊7 8 and 6 5⌉ ⌈5, the buccal and lingual surfaces of 7⌋ ⌊3 4 5 and 4⌉ ⌈3 4 6 7, the lingual and distal surfaces of 8⌋ and the buccal, lingual and distal surfaces of ⌈8.

Enamel hypoplasia: Hypoplastic grooves were visible on ⌊2 3 and 4⌉ ⌈3 4 which indicated that the individual had been subject to at least one incident of physiological stress at the age of 3-3.5 years.

Dental anomalies: The maxillary deciduous 2nd molars were retained, while the maxillary permanent 2nd premolars were unerupted. The maxillary lateral incisors were shovel-shaped.

Context VII. 5. Sk. 1

Adult 35-45 yrs: Probable male

Year of excavation: 1970

Anthropological number: 6862-184

Archaeological context:

Log house tomb - The remains of this context were not included in the archive reports.

Visual inventory:

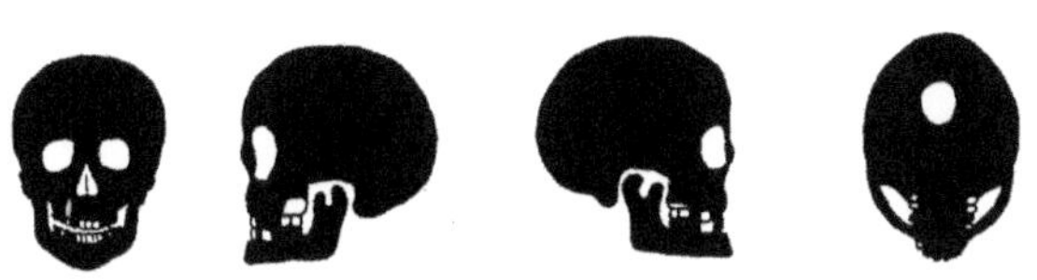

Preservation: Only the skull was present, which was in a very good state of preservation.

Cranial pathology: Cribra orbitalia of Type 1 was present in Areas 1 and 4 of the R orbit. The lesion was not present in the L orbit. A simple transverse fracture was present at the midpoint of the L nasal bone. The fracture was in an advanced state of healing, and the fracture line had been almost completely obliterated. The inferior part of the fracture was displaced supero-laterally. Only the lateral part of the R nasal bone was present. A simple transverse fracture was apparent on the bone. The superior and inferior parts of the R nasal bone were ununited.

Dentition:

		2A											2C		
		C											2A		
NP		PU		PU		PU			A		PU	PU	PU		2A
8	~~7~~	6	~~5~~	4	~~3~~	2	~~1~~	~~1~~	~~2~~	~~3~~	4	5	6	~~7~~	~~8~~
8	7	6	~~5~~	4	3	~~2~~	~~1~~	~~1~~	~~2~~	~~3~~	4	~~5~~	6	7	8
C	C	PU			PU								PU		C

Bite: Overbite.

Attrition: Extensive attrition was evident on the majority of teeth, with molar wear patterns of Stage 3+ for the 3rd molars, Stage 4 for the 2nd molars and Stage 5 for the 1st molars. Secondary dentine or pulp cavity exposure was evident in 6 4 2⌋ ⌊4 5 6 and 6 3⌉ ⌈6.

Calculus: Slight deposits of calculus were present on the buccal surfaces of 6⌋ ⌊6, the buccal and lingual surfaces of 4 2⌋ ⌊4 5 and 7 6 4 3⌉ ⌈4 6 7, the lingual and distal surfaces of ⌈8 and the buccal, lingual and distal surfaces of 8⌉.

Periodontal disease: Periodontal disease of considerable severity was apparent.

Caries: A small caries was present in the mesial surface of ⌈8, and a medium sized caries was visible in the distal surface of 7⌉. Large caries were present in the occlusal surface of 6⌋ and the mesial surface of 8⌉. Both a small and a medium sized caries were apparent in the occlusal surface of ⌊6.

Abscesses: A medium sized externally draining abscess was associated with ⌊2. Medium sized externally draining abscesses and large internally draining abscesses were associated with both 6⌋ and ⌊6. A large internally draining abscess and a large externally draining abscess were both associated with ⌊8.

Ante-mortem tooth loss: 7⌋ ⌊7 8 had been lost ante-mortem.

Dental anomalies: The R maxillary 3rd molar was genetically absent.

Context VII. 5. Sk. 4

Adult 25-35 yrs: Male: 171.4 cm, 5'7"

Year of excavation: 1970

Anthropological number: 6862-185

Archaeological context:

Log house tomb - The remains of this context were not included in the archive report.

Visual inventory:

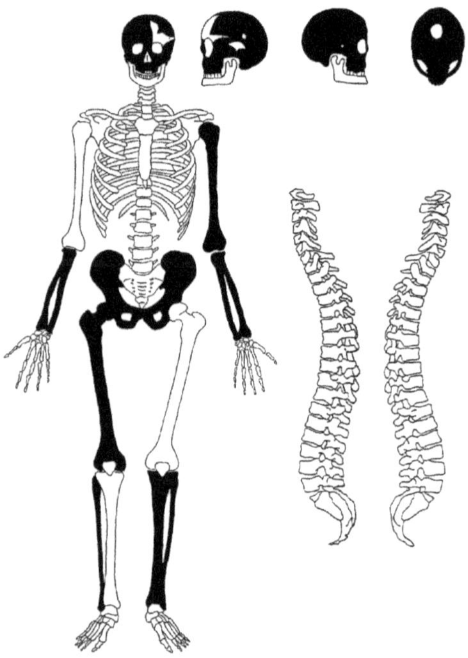

Preservation: The remains were in a poor state of preservation with 30-40% of bones present.

Weapon trauma: The bones of the skull were well preserved, although they were fragmentary. It is possible that the fragmentation was at least partially related to ante-mortem trauma, particularly in the frontal bone. A pointed axe injury was evident in the form of a circular perforation, which measured 11.5 mm medio-laterally by 16.5 mm supero-inferiorly, on the midpoint of the L parietal. The margins of the perforation were chipped and bevelled particularly at the inferior edge. The slope of the superior margin suggests that the perforation had been made by an individual who was positioned at the same level but with the axe raised and to the L side of the victim. Three fracture lines were radiating from the perforation. The fracture line emanating from the inferior margin of the injury may have been at least partially responsible for the fragmentary condition of the L side of the skull.

Other observations: Numerous small sharp cutmarks were present on the L humerus, the R innominate, the R femur and the L tibia, which are probably indicative of post-mortem defleshing or disarticulation. Green discoloration was apparent on the anterior surface of the R ilium and the R pubic bone.

Cranial non-metric traits: R and L highest nuchal lines, L parietal foramen, R and L ossicles at asterion, R and L mastoid foramina extrasutural, L double condylar facet, L double anterior condylar canal, L open foramen spinosum, R accessory supraorbital foramen.

Post-cranial non-metric traits: R Allan's fossa, R hypotrochanteric fossa, R exostosis in trochanteric fossa, R accessory sacral facets.

Dentition:

8 7 6 5 4 3 ~~2~~ ~~1~~	~~1~~ 2 3 4 5 6 7 8
~~8 7 6 5 4 3 2 1~~	~~1 2 3 4 5 6 7 8~~

Bite: Overbite.

Attrition: Molar wear patterns of Stage 1 for the 3rd molars, Stage 2+ for the 2nd molars and Stage 3+ for the 1st molars were apparent.

Calculus: Slight deposits of calculus were present on the buccal surface of ⌊3, the lingual surfaces of 7 3⌋, the buccal and lingual surfaces of 6 4⌋ ⌊5 6 7, the distal and lingual surfaces of 8⌋, the mesial and distal surfaces of ⌊4 and the buccal, lingual, mesial and distal surfaces of 5⌋. Medium deposits were present on the buccal and distal surfaces of ⌊8.

Periodontal disease: Periodontal disease of slight severity was apparent.

Dental anomalies: The L maxillary 2nd premolar was rotated by ninety degrees so that the buccal aspect of the tooth faced mesially. The R maxillary 2nd premolar was rotated by forty-five degrees so that the buccal aspect was facing mesially.

Context VII. 5. Sk. 6

Adult 25-35 yrs: Probable male

Year of excavation: 1970

Anthropological number: 6862-188

Archaeological context:

Log house tomb - The remains of this context were not included in the archive report.

Visual inventory:

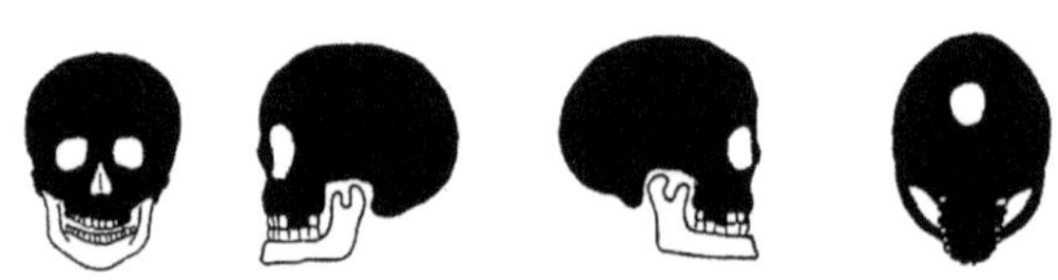

Preservation: Only the cranium was present, which was in a very good state of preservation.

Cranial pathology: Two possible completely healed depressed fractures were present on the L parietal, just lateral to the sagittal suture. The more anterior depression was circular in shape and measured approximately 18.5 mm by 18.5 mm, while the posterior lesion was oval-shaped and measured 37 mm medio-laterally by 12 mm antero-posteriorly. It would appear that the lesions had been made with a relatively small implement. Differential diagnoses includes healed trepanation or healed pointed axe traumatic injuries. Healed porotic hyperostosis was apparent on the superior aspect of the frontal bone.

Dentition:

8̸ 7 6 5̸ 4̸ 3̸ 2̸ 1̸	1̸ 2̸ 3̸ 4̸ 5̸ 6 7 8̸
~~8 7 6 5 4 3 2 1~~	~~1 2 3 4 5 6 7 8~~

Bite: Overbite.

Attrition: Molar wear patterns of Stage 3+ for the 2nd molars and Stage 4+ for the 1st molars were apparent.

Calculus: Slight deposits of calculus were present on the buccal surface of 7⌋, the lingual surface of ⌊6 and the buccal and lingual surfaces of 6⌋ ⌊7.

Periodontal disease: Periodontal disease of slight severity was apparent.

Context VII. 5

Adult 35-45 yrs: Male: 167.8 cm, 5'6"

Year of excavation: 1970

Anthropological number: 6862-187

Archaeological context:

Log house tomb - The remains of this context were not included in the archive report.

Visual inventory:

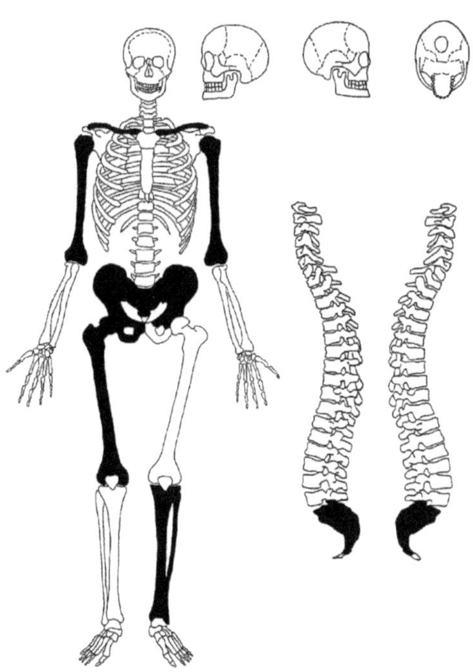

Preservation: The remains were in a poor state of preservation with 20-30% of bones present.

Post-cranial pathology: A simple oblique fracture was present on the distal third of the shaft of the L fibula. There was hypertrophy at the point of fracture, and the fracture parts were overlapped by approximately 30 mm. The parts were malaligned with only 50% apposition. Only a small amount of callus was visible, indicating that the fracture was in an advanced stage of healing. Prominent enthesopathies were apparent on the tibial tuberosity and the fibular notch of the L tibia which may be secondary to the fractured L fibula.

Other observations: Green discoloration and fragments of body tissue were present on the anterior aspect of the R ischial tuberosity.

Context VIII. 4. Sk. 12

Young adult 17-25 yrs: Male: 166.9 cm, 5'6"

Year of excavation: 1970

Anthropological number: 6862-202

Archaeological context:

Log house tomb - The remains of fourteen individuals were recovered from the floor of the tomb. All of the individuals were lying on their R sides, with the legs flexed. Nine skeletons were positioned in a row at the western end, while the other five individuals were located at the eastern end of the tomb.

Body position - The head was directed to the south-west, and the skull was positioned on a stone slab. The arms were extended towards the thighs.

Grave goods - A clay pot and a wooden dish were positioned posterior to the slab associated with the skull. The remains of an iron pin were lying on the slab, and thirty-two beads were situated near the cervical vertebrae.

Visual inventory:

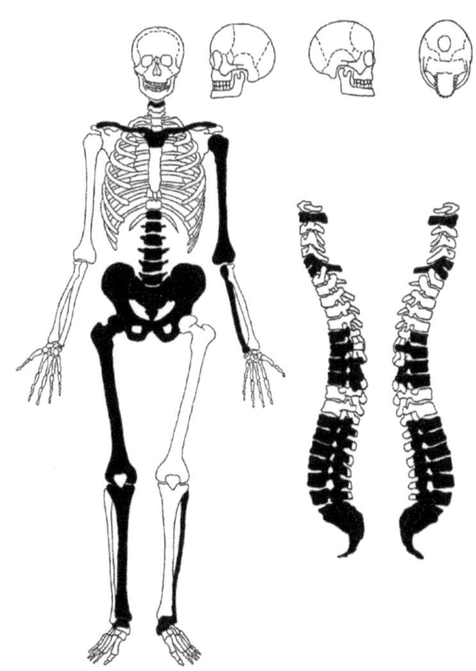

Preservation: The remains were in a poor state of preservation with 30-40% of bones present.

Post-cranial pathology: A small (10 mm) bony exostosis was present just inferior to the medial condyle of the proximal R tibia. It is possible that the exostosis was an enthesophyte of part of the popliteus. The neural arches of S3 and S4 were cleft, probably as a result of developmental delay during the early formation of the vertebral column.

Post-cranial non-metric traits: R Poirier's fossa, R lateral tibial squatting facet, L medial talar extension, L double inferior articular talar facet, L CV6, and R and L CV7 transverse foramina bipartite.

Context VIII. 10 (ii)

Adult 35-45 yrs: Probable male

Plate: 15

Year of excavation: 1971

Anthropological number: 6862-211

Archaeological context:

Log house tomb - Human remains were recovered throughout the tomb. The original location of this individual within the tomb is unknown.

Visual inventory:

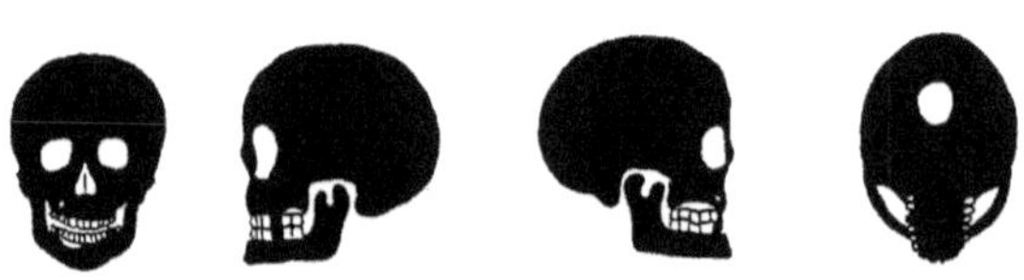

Preservation: Only the skull was present, which was in a very good state of preservation.

Cranial pathology: A simple transverse fracture was present at the midpoint of the R nasal bone. The fracture line originated at the medial aspect of the bone and ran laterally for approximately three-quarters of the width of the bone. The fracture was not well healed, and the two parts of the fracture were ununited. The inferior part was depressed slightly. A simple oblique fracture was present at the midpoint of the L nasal bone. The fracture line began at the medial aspect of the bone, parallel to the fracture line of the R bone and continued until it reached the midpoint of the inferior margin of the bone. The fracture was quite well healed, with the fracture line nearly obliterated. The inferior part of the fracture was slightly depressed.

Dentition:

				A				A					PU		
8	7	6	5	4	3	2	1	1	2	3	4	5	6	7	8
8	7	6	5	4	3	2	1	1	2	3	4	5	6	7	8
				PU		A	2A							A	

Bite: Overbite.

Attrition: Extensive attrition was evident on the teeth, with molar wear patterns of Stage 5 for the 1st molars. Secondary dentine was exposed in ⌊6 and 4⌉.

Calculus: Slight deposits of calculus were present on the buccal surface of ⌊6 and the buccal and lingual surfaces of 4⌉ ⌈6.

Periodontal disease: Periodontal disease of considerable severity was evident.

Abscesses: Medium sized externally draining abscesses were associated with 2⌉, ⌈7 and ⌊1. Both a medium sized externally draining abscess and a medium sized internally draining abscess were associated with 1⌉. A large externally draining abscess was associated with ⌊4.

Ante-mortem tooth loss: 7 6 5 2⌋ ⌊2 5 7 and 6⌉ ⌈7 had all been lost ante-mortem.

Context VIII. 12. Sk. 5

Adult 35-45 yrs: Probable male

Year of excavation: ?1973

Anthropological number: 6862-215

Archaeological context:

Log house tomb - The remains of this context were not recorded in the archive reports. It is possible that the tomb was excavated in 1973, but the archive report for the 1973 excavations was not available for consultation.

Visual inventory:

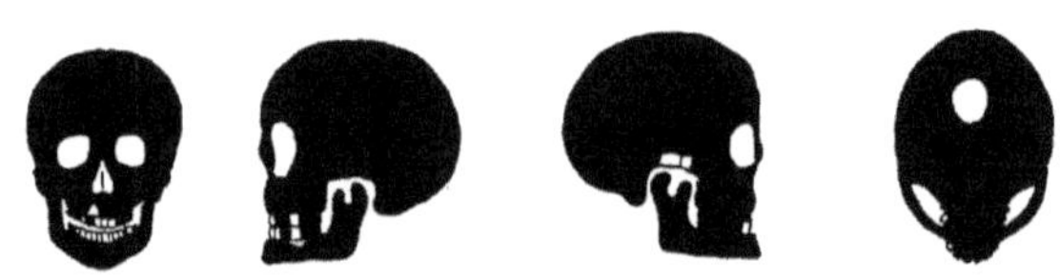

Preservation: Only the skull was present, which was in a very good state of preservation.

Cranial pathology: A possible healed depressed fracture was present on the posterior aspect of the L parietal. The lesion appeared to have been completely healed with smooth margins and a smooth interior surface, and was oval in shape. The medio-lateral measurement was 31.5 mm and the antero-posterior measurement was 17 mm.

Cranial non-metric traits: Metopic suture, ossicles in lambdoid suture (full set of traits not recorded).

Dentition:

C	PU	PU	PU	PU											
8	7	6	5	4	3	2	1	1	2	3	4	5	6	7	8
8	7	6	5	4	3	2	1	1	2	3	4	5	6	7	8
												C	A		

Bite: Overbite.

Attrition: Moderate to extensive attrition was present on the majority of teeth, with molar wear patterns of Stages 3-4+ for the 2nd molars and Stage 5 for the 1st molars. The 3rd molars displayed wear patterns of Stage 1 suggesting that they were not used for mastication. The maxillary teeth displayed more extensive attrition than the mandibular teeth and secondary dentine was exposed in 7 6 5 4⌋.

Calculus: Slight deposits of calculus were present on the buccal surfaces of 6 5 4⌋, the lingual surfaces of ⌊7 and 5⌉, the buccal and lingual surfaces of 8 7 2⌋ ⌊4 5 and 7⌉ ⌈7 and the buccal, lingual and distal surfaces of 8⌉ ⌈8. Medium deposits were visible on the buccal surface of ⌊7, the lingual surface of 6⌉ and the distal surface of 8⌋.

Periodontal disease: Periodontal disease of medium severity was apparent.

Caries: A medium sized caries was present on the mesial aspect of 8⌋, and a medium sized caries was visible on the distal aspect of ⌈5.

Abscesses: A large externally draining abscess was associated with ⌈6.

Context VIII. 12. Sk. 7 (i)

Adult 35-45 yrs: Probable male

Year of excavation: ?1973

Anthropological number: 6862-216

Archaeological context:

NB: 6862-216 and 6862-217 were both labelled as Context VIII. 12. Sk. 7.

Log house tomb - The remains of this context were not recorded in the archive reports. It is possible that the tomb was excavated in 1973, but the archive report for the 1973 excavations was not available for consultation.

Visual inventory:

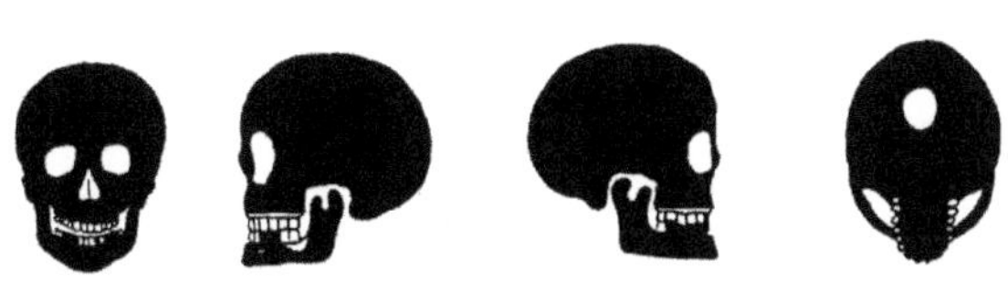

Preservation: Only the skull was present, which was in a very good state of preservation.

Cranial pathology: A very well healed simple transverse fracture was present at the midpoint of the R nasal bone, and a very well healed simple oblique fracture was present at the midpoint of the L nasal bone. In both cases the fracture lines were nearly completely obliterated and there was no malalignment.

Dentition:

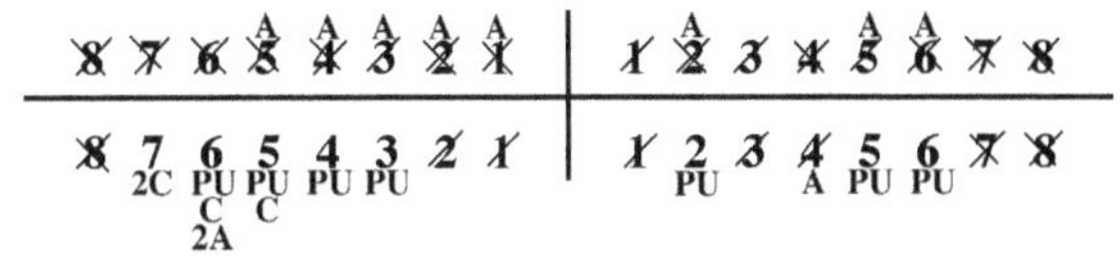

Bite: Overbite.

Attrition: Extensive attrition was evident on all teeth present, with molar wear patterns of Stage 4+ for the 2nd molars and Stage 5 for the 1st molars. Secondary dentine or pulp cavity exposure was evident in 6 5 4 3⌉ ⌈2 5 6.

Calculus: Slight deposits of calculus were present on the buccal and lingual surfaces of 7 6 5 4 3⌉ ⌈5 6.

Periodontal disease: Periodontal disease of considerable severity was apparent.

Caries: Small caries were present in the occlusal surfaces of 5⌉ and 6⌉. A small caries was visible in the buccal surface, and a large caries in the occlusal surface of 7⌉.

Abscesses: Two small externally draining abscesses were associated with 6⌉. Medium sized externally draining abscesses were associated with 1⌋, 2⌋, 3⌋, 4⌋, ⌊2, ⌊5 and ⌈4, and a medium sized internally draining abscess was associated with ⌊6. A large externally draining abscess was associated with 5⌋.

Ante-mortem tooth loss: 8 7 6 5 4 3 2 1⌋ ⌊2 4 6 7 8 and 8⌉ ⌈7 8 had been lost ante-mortem.

Context VIII. 18. Sk. 4

Young adult 17-25 yrs: Female: 153.1 cm, 5'1"

Year of excavation: 1971

Anthropological number: 6862-228

Archaeological context:

Log house tomb - The remains of five individuals were recovered from the floor of the tomb. All of the skeletons were lying on their R sides, with the legs flexed.

Body position - The head was directed to the west, and the skull was positioned on a slab. The bones of the hands were partly disturbed.

Grave goods - The remains of an iron pin were lying on the stone slab associated with the skull, and a wooden jar was situated nearby. A bronze finger ring and a bronze ring for attachment to a belt were situated near the R humerus. The remains of a bronze knife and an awl in a leather scabbard, were recovered from posterior to the spine, adjacent to the northern wall. The remains of a leather needle case with three bronze needles, and two bronze rings attached to a fragment of a belt were situated near the R ulna. The remains of five bronze rings for attachment to belts, a bronze buckle, a bronze finger ring and a bronze pendant were located nearby. A fragment of an ivory plate, a piece of fabric, two beads, a bronze cymbal, four bronze rings on a leather belt, an ivory pendant, a bronze pendant, a boar tusk and bronze rings attached to a belt, and the remains of a wooden object were all recovered from near the R hand phalanges.

Visual inventory:

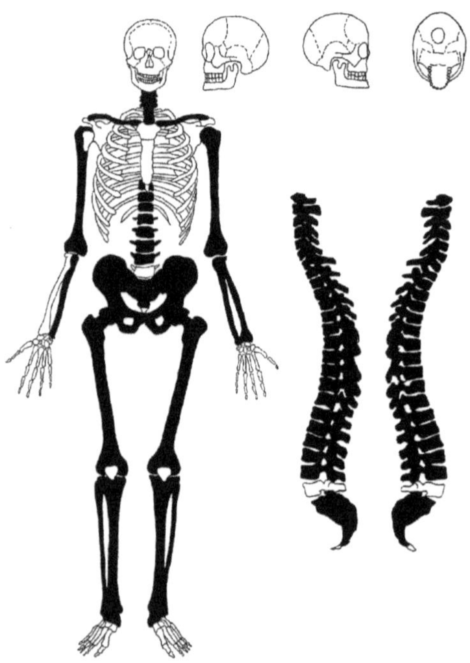

Preservation: The remains were in a good state of preservation with 60-70% of bones present.

Post-cranial pathology: LV5 was not present, but the presence of joint surfaces with porosity on the inferior surfaces of the inferior articular processes of LV4 may indicate that there was spondylolysis of LV5. Vertebral osteophytosis was evident in CV5 and TV5-TV9. Schmorl's nodes were visible in LV1-LV3. Enthesopathies were apparent on the pubic bones (adductor longus).

Context VIII. 21. Sk. 3

Adult 35-45 yrs: Probable female

Plate: 38

Year of excavation: 1970

Anthropological number: 6862-230

Archaeological context:

Log house tomb - Disarticulated human remains were recovered throughout the tomb. The original location of this individual within the tomb is unknown.

Visual inventory:

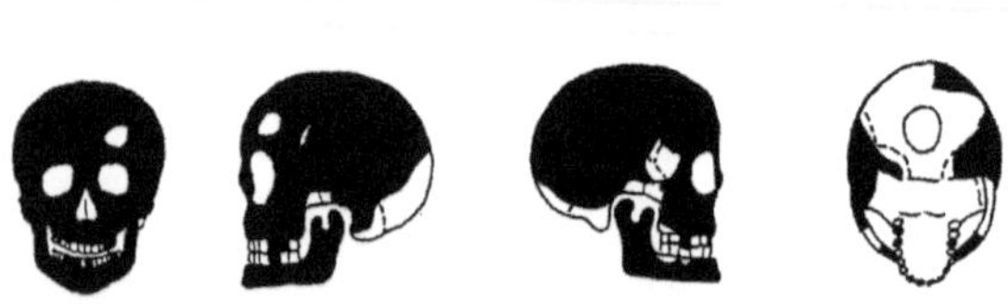

Preservation: Only the skull was present, which was fragmentary.

Cranial pathology: Cribra orbitalia of Type 3 was present in Areas 1, 2 and 4 of both orbits.

Surgery: A possible well healed trepanation perforation was present on the L side of the frontal. The lesion was near circular, measuring 23.5 mm by 24.4 mm, and had very smooth margins. The area surrounding the margins was slightly depressed which may be indicative of healing. The morphology of the lesion would class the perforation as a scraped or gouged trepanation.

Dentition:

~~8~~	~~7~~	6	~~5~~	~~4~~	~~3~~	~~2~~	~~1~~	~~1~~	~~2~~	~~3~~	~~4~~	~~5~~	6	7	~~8~~
~~8~~	~~7~~	~~6~~	~~5~~	~~4~~	3	2	1	~~1~~	2	~~3~~	~~4~~	~~5~~	6	~~7~~	~~8~~
	A										A				A

Bite: Overbite.

Attrition: Molar wear patterns of Stage 4+ for the 2nd molars and Stage 5 for the 1st molars were apparent. The R maxillary 1st molar had a wear pattern of Stage 2+ which may be related to masticatory habits or occupational activities. The corresponding mandibular molar had been lost post-mortem.

Calculus: Slight deposits of calculus were present on the buccal and lingual surfaces of 6⌋ ⌊6 7 and 3 2 1⌉ ⌈2 6.

Periodontal disease: Periodontal disease of medium severity was evident.

Abscesses: A small externally draining abscess was associated with ⌈4, and a small internally draining abscess was associated with ⌈8. A large externally draining abscess was associated with 7⌉.

Ante-mortem tooth loss: 8⌉ had been lost ante-mortem.

Context VIII. 35. Sk. 1

Adult 35-45 yrs: Probable male

Year of excavation: 1972

Anthropological number: 6862-247

Archaeological context:

Log house tomb - The remains of four commingled skeletons were recovered from the floor of the tomb. It was stated in the archive report that it was not possible to separate the remains of each individual.

Grave goods - The remains of the top of an iron pin, which was covered with gold foil, were discovered near the western wall. A stone slab was situated near the southwestern corner of the tomb. The remains of a wooden vessel were situated between the slab and the southern wall. Fragments of arrowshafts, and the remains of wooden vessels were retrieved from the central area of the tomb.

Nine beads were situated at the northwestern corner, and fragments of leather and scraps of fabric were located at the central part of the northern wall. A bronze awl, a bronze arrowpoint, a fragment of leather and a piece of a leather belt were all recovered from the eastern part of the floor of the tomb.

Visual inventory:

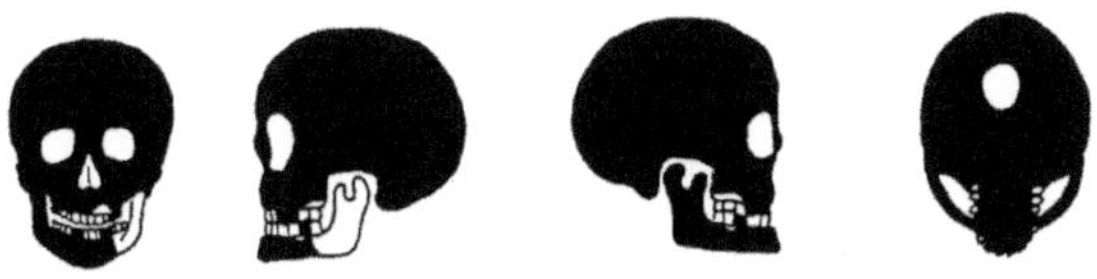

Preservation: Only the skull was present, which was fragmentary with surface erosion.

Cranial pathology: Cribra orbitalia of Type 2 was present in Areas 1 and 4 of the R orbit and cribra orbitalia of Type 4 was present in Areas 1, 2 and 4 of the L orbit. Simple transverse fractures were present at the midpoints of both nasal bones. The fracture lines were still visible indicating that the fractures were not completely healed. There was no malalignment.

Dentition:

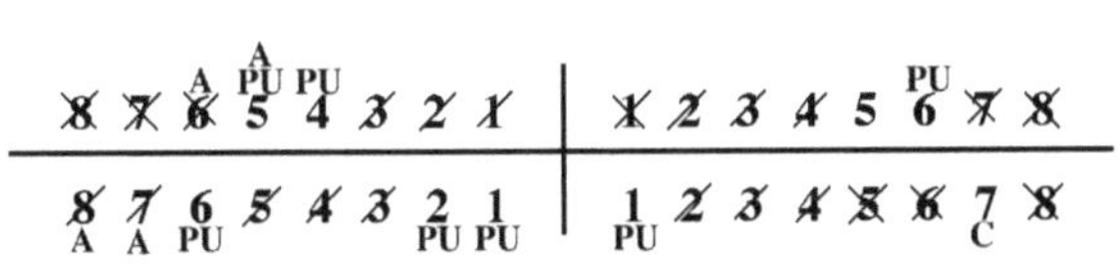

Bite: Overbite.

Attrition: Extensive attrition was evident in the majority of teeth. Molar wear patterns of Stage 4+ for the 2nd molars and Stage 5 for the 1st molars were apparent. Secondary dentine or pulp cavity exposure was evident in 5 4⌋ ⌊6 and 6 2 1⌉ ⌈1.

Calculus: Slight deposits of calculus were present on the buccal surface of ⌊6 and the buccal and lingual surfaces of ⌊5 and 6 2 1⌉ ⌈7.

Periodontal disease: Periodontal disease of considerable severity was apparent.

Caries: A small caries was present in the mesial surface of ⌈7.

Abscesses: A small internally draining abscess was associated with 8⌉, and a medium sized externally draining abscess was associated with 5⌋. Large externally draining abscesses were associated with 7⌉ and 6⌋. A plaque of pitted woven bone surrounded the abscess cavity associated with 6⌋.

Ante-mortem tooth loss: 8 7 6⌋ ⌊1 7 8 and ⌈5 6 8 had been lost ante-mortem.

Dental anomalies: An extra tooth socket was situated between 2⌋ and 3⌋. It is possible that there was retention of deciduous b⌋ or c⌋.

Context VIII. 39 (ii)

Adult: unknown

Plate: 20

Year of excavation: 1970

Archaeological context:

Log house tomb - Disarticulated human remains were recovered from throughout the tomb. The original location of this individual within the tomb is unknown.

Visual inventory:

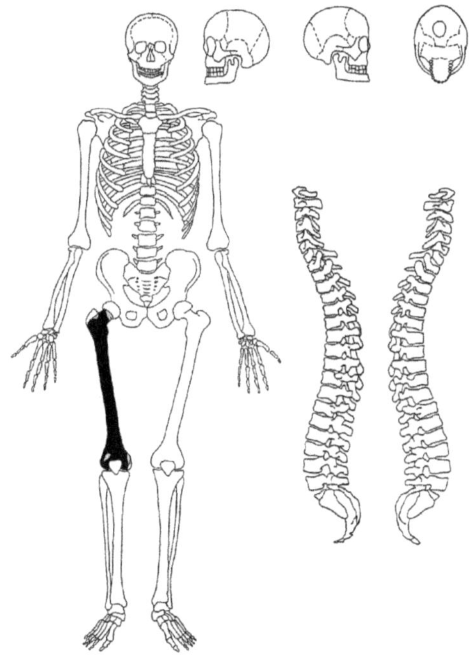

Preservation: Only the R femur was present.

Post-cranial pathology: A spur-like bony process was projecting from the medial supracondylar line of the distal R femur. The process measured 49 mm supero-inferiorly by 3.3 mm medio-laterally by 10 mm antero-posteriorly. It is possible that the exostosis was indicative of myositis ossificans of the adductor magnus.

Context VIII. 54. Sk. 5

Adult 35-45 yrs: Female: 151.1 cm, 5'0"

Year of excavation: 1971

Anthropological number: 6862-287

Archaeological context:

Log house tomb - The very disturbed remains of seven skeletons were recovered from the floor of the tomb. Each individual had the head directed to the west.

Body position - The position of the body was unclear due to the partial disturbal of the bones.

Archaeology & Trauma

Grave goods - A stone slab was located behind the skull, and a clay pot was situated behind the slab.

Visual inventory:

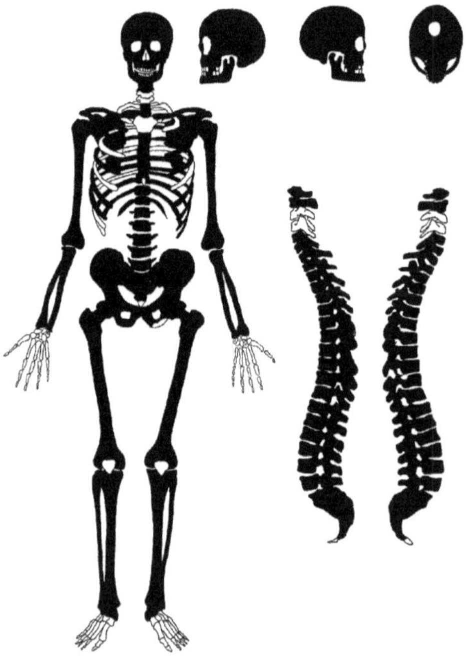

Preservation: The remains were in an excellent state of preservation with 90-100% of bones present.

Cranial pathology: A very well healed simple transverse fracture was present at the medial aspect of the R zygomatic. The fracture line was barely visible, and ran vertically from the inferior margin to the superior margin of the zygomatic. There was 100% apposition and no malalignment.

Post-cranial pathology: A very well healed simple oblique fracture was present at the acromial end of the R clavicle. The fracture line was only barely visible on the anterior surface. There was hypertrophy at the point of fracture, and the greatest length was 134 mm as opposed to 140 mm for the L clavicle. This indicated that there was overlap of the fracture parts by approximately 6 mm. Vertebral osteophytosis was evident in CV2, CV5-CV7 and TV6-TV11. Enthesopathies were apparent on the calcanei (Achilles' tendon).

Cranial non-metric traits: 5 ossicles at lambda, L parietal foramen, R and L ossicles at pterion, R and L mastoid foramina extrasutural, R and L open foramina spinosum, R accessory supraorbital foramen, R and L anterior ethmoid frontal foramina extrasutural, L accessory infraorbital foramen. In addition, 2 ossicles were present in the L and 2 in the R squamosal sutures.

Post-cranial non-metric traits: R and L Poirier's fossae, R and L exostoses in trochanteric fossae, R and L lateral tibial squatting facets, R and L accessory sacral facets, R and L acromial articular facets, R and L circumflex sulci, R and L medial talar facets, R and L double inferior articular talar facets, R and L double anterior calcaneal facets, R and L peroneal tubercles.

Dentition:

8	7	6	5	~~4~~	~~3~~	~~2~~	~~1~~	~~1~~	~~2~~	~~3~~	~~4~~	5	6	7	~~8~~
~~8~~	7	6	~~5~~	~~4~~	~~3~~	~~2~~	~~1~~	~~1~~	~~2~~	~~3~~	~~4~~	5	6	~~7~~	~~8~~

Bite: Overbite.

Attrition: Extensive attrition was present on the majority of teeth, with molar wear patterns of Stage 3+ for the 3rd molars, Stage 4+ for the 2nd molars and Stage 5 for the 1st molars.

Calculus: Slight deposits of calculus were present on the buccal surfaces of 7 5⌋ ⌊6 and ⌈6 and the lingual and mesial surfaces of 6⌉ ⌈5. Medium deposits were present on the buccal surfaces of 6⌋ ⌊7 and the lingual surface of 7⌉.

Periodontal disease: Periodontal disease of medium severity was apparent.

Context IX. 3 (i)

Young adult 17-25 yrs: Male: 174.6 cm, 5'8"

Plate: 35

Year of excavation: 1976

Anthropological number: 6862-292

Archaeological context:

Log house tomb - Human remains were recovered from all levels of the tomb. The original location of this individual within the tomb is unknown.

Visual inventory:

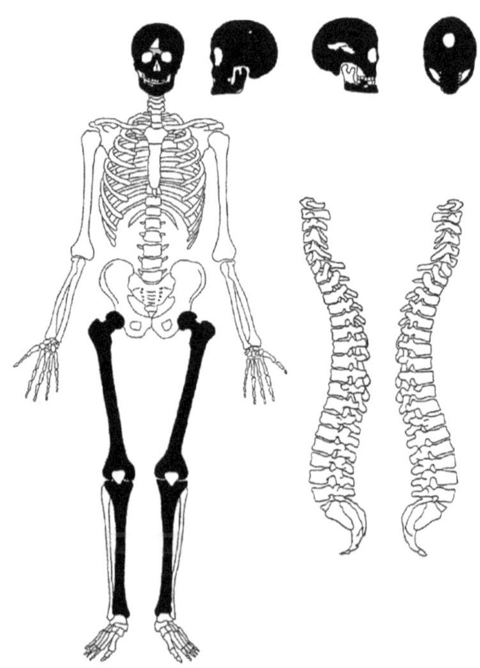

Preservation: The remains were in a very poor state of preservation with 10-20% of bones present.

Weapon trauma: Four instances of weapon trauma were apparent on the cranium. A large oval perforation was present at the midpoint of the glabella. The superior part of the injury completely perforated the vault, and measured 45 mm supero-inferiorly by 31 mm medio-laterally. The inferior part of the lesion only perforated the outer table of the vault into the frontal sinus. The margins of the lesion were sharp, and a portion of the R side of the injury was depressed. It is likely that the lesion was caused by a pointed battle axe which had struck the forehead with a considerable amount of force.

A small diamond shaped perforation was present at the midpoint of the L parietal. The margins of the lesion were sharp, and the most infero-lateral margin had a bevelled edge. The outer table of the supero-medial margin was slightly raised. It is possible that the injury was caused by an arrowhead or a pointed axe which had entered the skull from behind the individual.

A large oval perforation was also present on the R side of the skull. The possible injury extended from the midpoint of the pterion posteriorly for approximately 92.5 mm. The margins had irregular bevelled edges. A large fracture line radiated from the posterior margin of the perforation for approximately 60.5 mm where it united with the supero-lateral margin of the injury described below. It is possible that the injury was caused by repeated blows with a pointed axe or by a larger weapon.

A large oval perforation, which measured approximately 34 mm supero-inferiorly by 22.5 mm medio-laterally, was present immediately to the R of the lambda. The margins of the lesion were sharp. As well as the fracture line described for the injury above, another fracture line was radiating from the inferior margin of the injury to the area just posterior to the L mastoid process. It is possible that the injury had been caused by a pointed axe.

Cranial non-metric traits: R and L highest nuchal lines, R ossicle in lambdoid suture, R and L parietal foramina, L ossicle at asterion, R mastoid foramen extrasutural, L mastoid foramen absent, R and L anterior ethmoid frontal foramina extrasutural, R and L accessory infraorbital foramina.

Post-cranial non-metric traits: R and L Allan's fossae, R and L hypotrochanteric fossae, L third trochanter, R and L lateral tibial squatting facets.

Dentition:

		B														
~~8~~	~~7~~	6	~~5~~	~~4~~	3	~~2~~	~~1~~	|	~~1~~	2	~~3~~	4	5	6	7	~~8~~
~~8~~	7	6	~~5~~	~~4~~	3	2	1	|	1	2	3	4	5	6	7	8
		A			B	B					B					

Bite: Overbite.

Attrition: Molar wear patterns of Stage 1 for the 2nd and 3rd molars, and Stage 3+ for the 1st molars were apparent.

Calculus: Flecks of calculus were present on the lingual surface of 3⌋. Slight deposits were present on the lingual surfaces of ⌊6 and 6 2 1⌉⌈1 2, the buccal and lingual surfaces of ⌊4 and ⌈4 5 6 7, the buccal and distal surfaces of 7⌉ and the buccal, lingual and distal surfaces of ⌈8.

Periodontal disease: Periodontal disease of slight severity was apparent.

Abscesses: A large externally draining abscess was associated with 6⌉. Plaques of periosteal reactive new bone surrounded the abscess sinus.

Enamel hypoplasia: Hypoplastic grooves were present on 3⌋ ⌊2 3 which indicated that the individual had been subject to at least one physiological insult at the age of 2-2.5 years.

Context IX. 19. Sk. 1

Adult 25-35 yrs: Male: 172.9 cm, 5'8"

Year of excavation: 1976

Anthropological number: 6862-311

Archaeological context:

Log house tomb - The remains of two individuals were recovered from the floor of the tomb.

Body position - The leg bones were *in situ*, and the disturbed skull and arm bones were positioned near the southern wall to the west of the leg bones.

Grave goods - A bead and fragments of a leather belt were situated to the west of the arm bones. A bronze rectangular container with a fragment of a belt were located in the central floor area. A bronze container and a scrap of leather associated with a fragment of wood were recovered from near the leg bones.

Visual inventory:

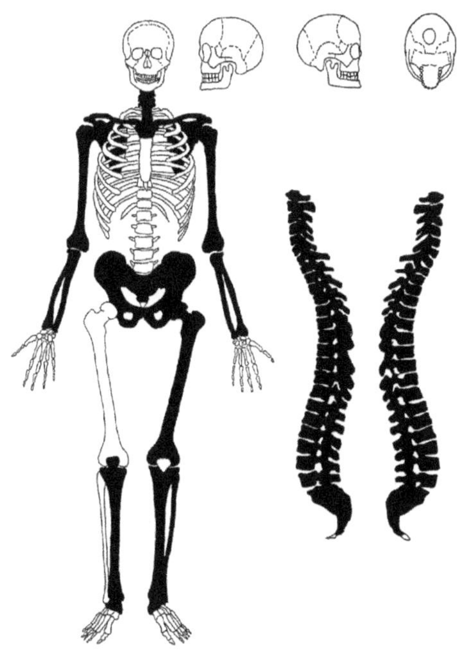

Preservation: The remains were in a good state of preservation with 60-70% of bones present.

Post-cranial pathology: A very well healed simple oblique fracture was present on the distal third of the shaft of the L tibia. There was almost 100% apposition of the fracture parts, and slight overlap of the two fracture parts by approximately 5-10 mm. The distal part of the fracture was displaced posteriorly by approximately five degrees. It is likely that the injury had been splinted in some way since the fracture was so well healed with very little malalignment. It is possible that there was associated subluxation of the distal tibio-fibular articulation. An additional joint surface was present on the L fibula immediately superior to the lateral malleolus. In addition, osteophytes were present on the medial surface of the fibula which may be indicative of stress to the interosseous ligament. Possible secondary lesions to the fracture were osteophytosis on the posterior and medial talar articular surfaces of the L calcaneus and the trochlear surface of the L talus.

Osteophytosis was visible on the manubrium. Vertebral osteophytosis was present in CV2-CV7 and TV4-LV5. Schmorl's nodes were visible in LV4. Enthesopathies were present on the clavicles (costoclavicular ligaments), the humeri (deltoid, pectoralis major, teres major, latissimus dorsai, subscapularis), the R patella (patellar ligament), the R tibia (patellar ligament), the L femur (adductor magnus) and the calcanei (Achilles' tendon).

Context X. 2. Sk. 1

Young adult 17-25 yrs: Probable female

Plate: 12

Year of excavation: 1971

Anthropological number: 6862-315

Archaeological context:

Log house tomb - Disarticulated human remains were recovered throughout the tomb. The original location of this individual within the tomb is unknown.

Visual inventory:

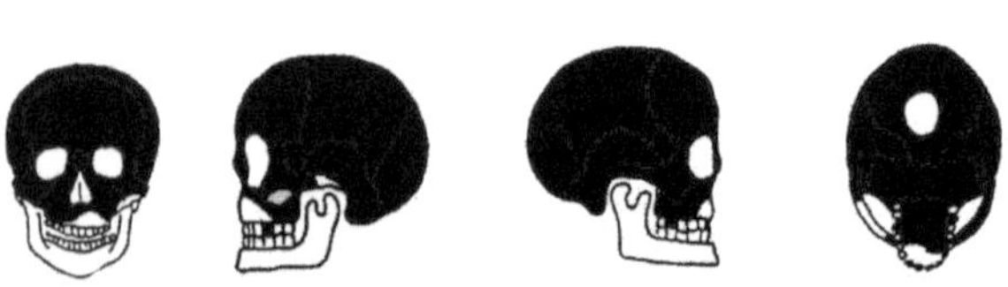

Preservation: Only the cranium was present, which displayed extensive surface erosion.

Cranial pathology: A well healed depressed fracture was present on the posterior surface of the L side of the occipital. The lesion was circular, measuring 20 mm by 20 mm, and had probably been made with a small blunt implement. Cribra orbitalia of Type 1 was present in Areas 1, 2 and 3 of the R orbit, and cribra orbitalia of Type 2 was present in Areas 1, 2 and 3 of the L orbit.

Cranial non-metric traits: 2 R and 2 L ossicles in lambdoid suture, R and L parietal foramina, L ossicle at pterion, R fronto-temporal articulation, R mastoid foramen absent, L posterior condylar canal open, R open foramen spinosum, R and L accessory supraorbital foramina, R and L anterior ethmoid foramina absent.

Dentition:

8	7	6	5	4	3	2	1	1	2	3	4	5	6	7	8
8	7	6	5	4	3	2	1	1	2	3	4	5	6	7	8

Bite: Overbite.

Attrition: Molar wear patterns of Stage 3+ were apparent for the 1st molars.

Context X. 3

Adolescent 14 - 18 years

Year of excavation: unknown

Anthropological number: 6863-147

Archaeological context:

Log house tomb - Details of this context were not recorded in the archive reports.

Visual inventory:

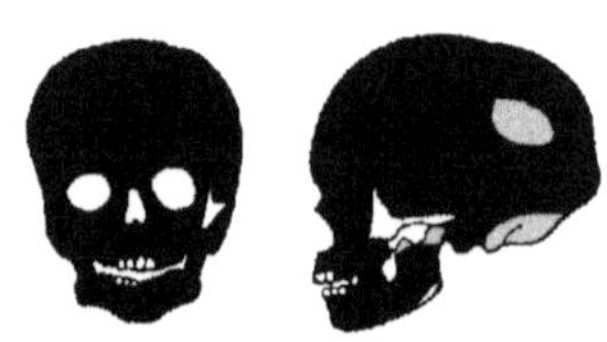

Preservation: Only the skull was present, which was in a good state of preservation.

Cranial pathology: A possible example of a well healed depressed fracture was present at the midpoint of the R parietal. The lesion had smooth margins, and was oval in shape, measuring 15 mm medio-laterally by 5 mm antero-posteriorly.

Dentition:

U 8	7	6	5	4	3	2	1	1	2	3	4	5	6	7	U 8
8 U	7	6	5	4	3	2	1	1	2	3	4	5	6	7	8 U

Bite: Overbite.

Attrition: Molar wear patterns of Stage 1 for the 2nd molars and Stage 2+ for the 1st molars were apparent.

Calculus: Flecks of calculus were present on the buccal surface of ⌊6 and the buccal and lingual surfaces of 6⌋. Slight deposits were visible on the lingual surface of 6⌉.

Enamel hypoplasia: Single hypoplastic grooves were apparent on 6⌋ ⌊6 7 and 7 6⌉ ⌈6 7 which indicated that the individual had been subject to at least two incidents of physiological stress at the ages of 2-2.5 and 5-5.5 years.

Context X. 13. Sk. 6

Adolescent 15-18 years, Probable female

Year of excavation: unknown

Anthropological number: 6863-157

Archaeological context:

Log house tomb - The information for this context was not provided in the archive reports.

Visual inventory:

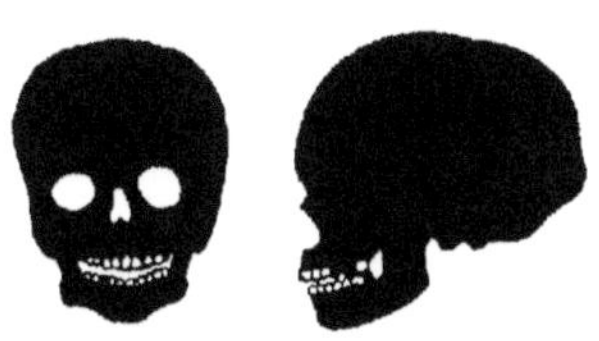

Preservation: Only the skull was present, which was in a very good state of preservation.

Cranial pathology: Possible healed trauma was present at the midpoint of the R parietal. A small transverse depression which ran medio-laterally, and measured 17 mm medio-laterally by 1.5 mm antero-posteriorly, was apparent.

Cranial non-metric traits: Ossicles in lambdoid suture (full set of traits not recorded).

Dentition:

U 8	~~7~~	6	~~5~~	~~4~~	~~3~~	~~2~~	~~1~~	~~1~~	~~2~~	~~3~~	~~4~~	~~5~~	6	7	U 8
8 U	7	~~6~~	~~5~~	~~4~~	~~3~~	~~2~~	~~1~~	~~1~~	~~2~~	~~3~~	~~4~~	~~5~~	~~6~~	~~7~~	8 U

Bite: Overbite.

Attrition: Molar wear patterns of Stage 2+ for the 1st molars and Stage 1 for the 2nd molars were apparent.

Calculus: Flecks of calculus were present on the lingual surface of 6⌋ and the buccal and lingual surfaces of 6⌉ and ⌊6. Slight deposits were visible on the buccal surface of 6⌋.

Context XII. 4 (i)

Adult 25-35 yrs: Male: 175.2 cm, 5'8"

Year of excavation: 1975

Anthropological number: 6862-337

Archaeological context:

Pit and cist - Staining and the presence of disordered slabs were apparent on the ground surface. Cleaning revealed a rectangular cist, orientated south-east to north-west. Disarticulated human remains were scattered inside the cist.

Visual inventory:

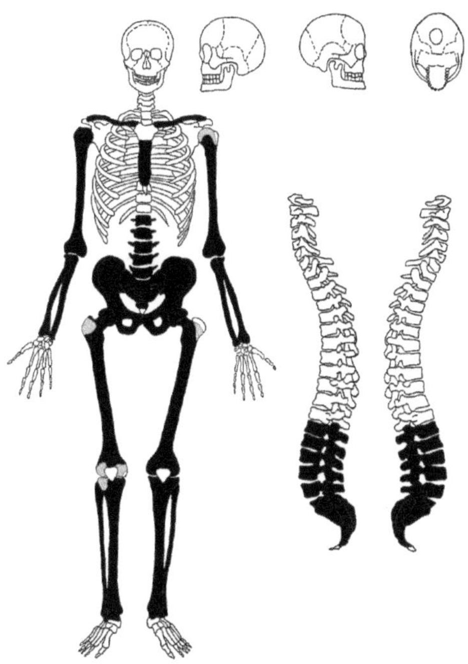

Preservation: The remains were in a fair state of preservation with 40-50% of bones present. All bones were unwashed and displayed extensive surface erosion and root damage.

Post-cranial pathology: LV5 displayed spondylolysis. The detached portion of the vertebra was not present. Partial sacralization was also apparent, with the L transverse process of LV5 articulating with the superior surface of the L ala. The R transverse process also displayed an ala-like process, but this did not articulate with the sacrum. Vertebral osteophytosis was apparent in LV2-LV4, and Schmorl's nodes were present in LV4 and LV5.

Post-cranial non-metric traits: R and L lateral tibial squatting facets, R and L double anterior calcaneal facets, L peroneal tubercle.

Other observations: A layer of organic matter was adhering to most bones, which may be the remains of body tissue. Alternatively, the layer could be composed of soil and/or plant matter.

Context XX. 7. Sk. 4

Young adult 17-25 yrs: Male: 164.4 cm, 5'5"

Plates: 10, 31

Year of excavation: 1975

Anthropological number: 6862-353 (skeleton)

Archaeological context:

Log house tomb - The remains of eight skeletons were recovered from the floor of the tomb. Five of the skeletons were positioned in a row.

Body position - The remains were partially disturbed.

Grave goods - The skull was situated near a stone slab, and fragments of an iron pin and a bone pin were lying on the slab. The remains of a quiver which contained arrowshafts and four arrowheads were positioned on the R femur. A bronze knife in a leather sheath, a bronze template for making relief decoration with the remains of a wooden handle, a bronze dagger in a leather sheath and a bone button were all recovered from adjacent to the R femur.

Visual inventory:

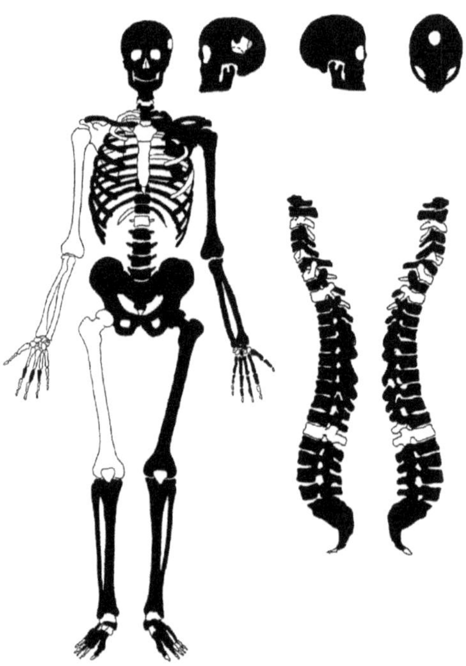

Preservation: The remains were in a very good state of preservation with 70-80% of bones present.

Cranial pathology: Cribra orbitalia of Type 2 was present in Area 2 of the R orbit and Areas 1, 2, 3 of the L orbit. Porosity was apparent on the superior aspect of the frontal, the posterior surfaces of the parietals and the superior aspect of the occipital. The porosity was probably indicative of healed porotic hyperostosis.

Post-cranial pathology: A simple oblique fracture was present on the proximal third of the shaft of the L radius. There was approximately 50% apposition of fracture parts, and they overlapped by 40 mm. The proximal part of the fracture had been displaced posteriorly and medially relative to the distal part, and the fracture was in an advanced stage of healing. A simple oblique fracture was also present on the distal half of the shaft of the L ulna. There was approximately 75% apposition of the fracture parts, and they overlapped by 30 mm. The distal half of the fracture was malaligned laterally by approximately forty-five degrees, and a perforation was present at the midpoint of the fracture line. The fracture was in an advance staged of healing. Both bones would have been considerably shortened as a result of the fractures, and it is extremely probable that they were obtained in a single incident.

Vertebral osteophytosis was evident in TV8, TV9, TV12 and LV5. Schmorl's nodes were present in TV7-TV12 and LV2. Enthesopathies were apparent on the calcanei (Achilles' tendon).

Other observations: Green discoloration was present on the anterior surface of the L femur. A cutmark was visible on TV11, which may be indicative of post-mortem defleshing/disarticulation.

Dentition:

C 8	7	6	5	4	3	2	1	1	2	3	4	5	6	7	8
8	7	6	5	4	3	2	1	1	2	3	4	5	6	7	8

Bite: Overbite.

Attrition: Molar wear patterns of Stage 1 for the 3rd molars, Stage 2+ for the 2nd molars and Stage 3 for the 1st molars were apparent.

Calculus: Flecks of calculus were present on the buccal surfaces of 8 7 3 1⌋ ⌊1 2 3 4 and 5 4⌉ ⌈3 4 5, the lingual surfaces of 8 7⌉ ⌈7 and the buccal and lingual surfaces of 5 4 2⌋ ⌊7 8 and 6 3 2 1⌉ ⌈1 2 6 8. Slight deposits were visible on the buccal surfaces of 6⌋ ⌊5 6 and the lingual surfaces of 5 4⌉ ⌈3 4 5.

Periodontal disease: Periodontal disease of slight severity was apparent.

Caries: A medium sized caries was present on the mesial surface of 8⌋.

Dental anomalies: The maxillary incisors were shovel-shaped.

Context XX. 8. Sk. 2

Mature adult 45+ yrs: Male: 169.3 cm, 5'6"

Plates: 17, 18, 19

Year of excavation: 1975

Anthropological number: 6862-355

Archaeological context:

Stone cist - The cist was trapezium-shaped, and extended from south-west to north-east. Three skeletons were lying at the bottom of the cist. Each of the individuals was lying on the L side with the head directed to the north-west.

Body position - The skull was positioned on a stone slab, and the arms and legs were flexed.

Grave goods - Fragments of a bone object were recovered from between the femora.

Visual inventory:

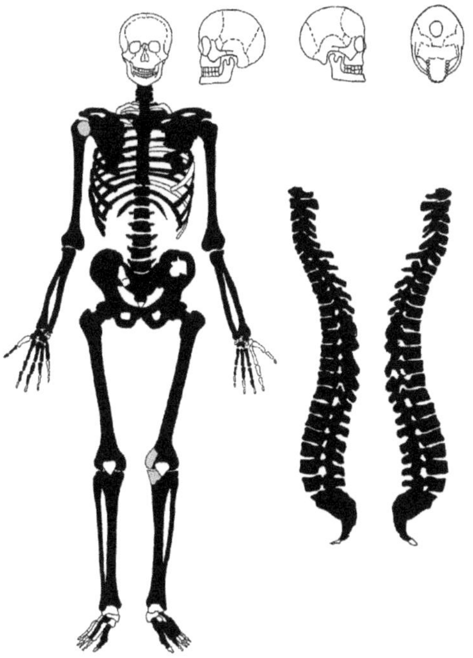

Preservation: The remains were in a very good state of preservation with 80-90% of bones present. Some bones displayed surface erosion.

Post-cranial pathology: The R glenohumeral joint was dislocated with the humerus completely displaced anteriorly. A new joint surface, which displayed osteophytes, porosity and eburnation, was present on the antero-inferior surface of the coracoid process. The R humeral head appeared to have had a compressed fracture, and the head had a flattened appearance with a porotic surface and marginal osteophytes. An area of porosity and eburnation was present on the supero-posterior surface of the anatomical neck. Inferior to this were large osteophytes, which were probably due to strain of the supraspinatus, infraspinatus and teres minor muscles. An enthesopathy of the costoclavicular ligament was present on the R clavicle, and enthesopathies of the deltoid and pectoralis major were apparent on the R humerus. Osteophytosis was present on the proximal and distal ends of the R ulna and the distal end of the R radius. These latter lesions may have been secondary to the dislocation.

A simple transverse fracture was present at the distal end of the shaft of the L radius, immediately superior to the articular surface (Colles' fracture). The fracture was very well healed, and the fracture line was barely visible. There was almost 100% apposition of the fracture parts, with only slight malalignment in which the articular surface was displaced posteriorly by approximately thirty degrees. Overlap of the fracture parts by approximately 8 mm was evident. Possible subluxation of the radio-ulnar joint was apparent with osteophytosis of the ulnar notch of the radius and the ulnar head.

A simple oblique fracture was present at the distal end of the R fibula, immediately superior to the articular surface. The fracture parts displayed almost 100% apposition. There was slight hypertrophy at the fracture site, and slight posterior displacement of the distal part of the fracture by approximately ten degrees. Osteophytes were present at the supero-lateral margin of the distal third, which were probably related to strain of the tibio-fibular interosseous ligament. A possible fracture was also present on the L fibula. A smooth bony nodule was visible on the postero-lateral surface of the midshaft which appeared to be associated with a fracture line which ran supero-inferiorly. If the lesion was due to a fracture then it was very well healed. It is possible that all the examples of trauma were obtained in a single incident, and they may indicate that the individual had been subject to an incident of extreme trauma.

Osteophytosis was visible on the acromial ends of the clavicles, the distal L MC3, the R trapezium, the acetabulae, the R femoral head, the R distal femur, and the R proximal tibia. Osteophytosis was also present on the majority of rib heads, the distal R MT1, the proximal R proximal foot PH1, the distal L MT1, the proximal L foot proximal PH1, and the distal L MT2, MT3, MT4 and MT5. Very extensive vertebral osteophytosis was evident in CV1-CV7 and TV2-LV5. Enthesopathies were apparent on the ischiae (adductor magnus), the femora (psoas major, iliacus, gluteus medius, vastus lateralis, gluteus maximus, adductor magnus) and the tibiae (patellar ligament).

Other observations: A cutmark was present on the posterior surface of the R femoral head, which may be indicative of post-mortem defleshing/disarticulation.

Context XX. 9. Sk. 2

Adult 35-45 yrs: Female: 153.7 cm, 5'1"

Year of excavation: 1975

Anthropological number: 6862-360

Archaeological context:

Log house tomb - The remains of nine individuals were recovered from the floor of the tomb. Five of the skeletons were lying in a row.

Body position - The individual was lying on her L side, with the head to the west. The L arm was extended and the R arm was flexed. The legs were flexed, and the skull was situated on a stone slab.

Grave goods - A clay pot and fragments of an iron pin were situated near the stone slab associated with the skull. The remains of an extensively oxidised bronze earring were recovered from beneath the skull. Three pendants made from boar tusks, and a fragment of an iron object were situated near the L side of the pelvis. A fragment of a Cowrie shell was situated between the femora and the tibiae. Some sheep bones were recovered from the area to the north of the skeleton near the arm bones, and a bronze arrowhead was situated adjacent to the sheep bones.

Visual inventory:

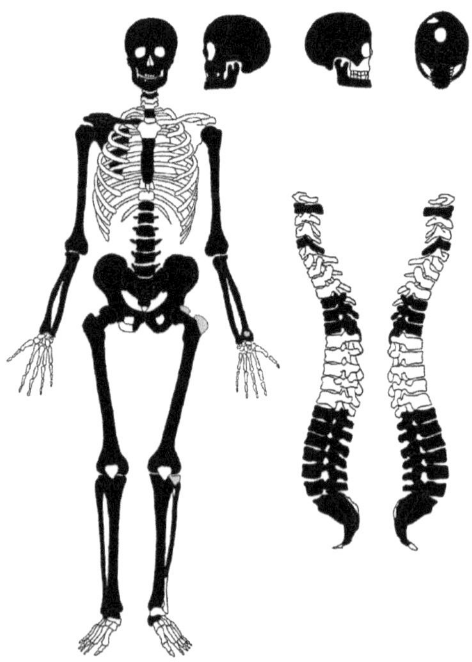

Preservation: The remains were in a very good state of preservation with 70-80% of bones present.

Post-cranial pathology: Spondylolysis of LV5 was apparent. Schmorl's nodes were present in TV4-TV6 and TV12-LV2.

Cranial non-metric traits: Ossicle at lambda, R ossicle in lambdoid suture, R ossicle at pterion, R and L mastoid foramina extrasutural, R and L posterior condylar canals open, R bridging of supraorbital notches, R and L anterior ethmoid frontal foramina extrasutural, R and L accessory infraorbital foramina, R and L CV5 transverse foramina bipartite.

Post-cranial non-metric traits: R and L hypotrochanteric fossae, R and L exostoses in trochanteric fossae, R and L lateral tibial squatting facets, R acromial articular facet, R circumflex sulcus, L medial talar facet, L double inferior articular talar facet, R double anterior calcaneal facet, R peroneal tubercle.

Dentition:

8	7	6	5	~~4~~	~~3~~	~~2~~	~~1~~	~~1~~	~~2~~	~~3~~	~~4~~	5	6	~~7~~	8
~~8~~	7 C	6	~~5~~	4	3	2	1	1	2	3	4	5	~~6~~ C	7	8

Bite: Overbite.

Attrition: Very slight attrition was evident on the teeth for an individual of 35-45 years. Molar wear patterns of Stage 1 for the 3rd molars and Stage 3 for the 1st and 2nd molars were apparent.

Calculus: Slight deposits of calculus were present on the buccal surfaces of 8 5⌋ ⌊4 and ⌈2 4, the distal surfaces of 8⌉ ⌈7, the mesial surfaces of 3 2⌉ and the buccal and lingual surfaces of ⌊6 and 1⌉ ⌈1. Medium deposits were visible on the buccal surfaces of 7 6⌋, the lingual surfaces of 7 4⌉ ⌈3 4 6 7, the lingual and mesial surfaces of ⌊4, the buccal and distal surfaces of ⌊8 and the buccal, lingual and distal surfaces of ⌈8. Heavy deposits were apparent on the lingual surfaces of 3 2⌉ ⌈2 and the mesial surface of ⌈5.

Periodontal disease: Periodontal disease of slight severity was apparent.

Caries: Medium sized caries were visible in the occlusal surfaces of 7⌋ and ⌊6.

Dental anomalies: The mandibular teeth were overcrowded, with the mesial aspect of 3⌉ tilted lingually. The overcrowding had resulted in an extensive build-up of calculus on 2 3⌉. The L mandibular 2nd premolar was not fully erupted. The mandibular incisors and canines were shovel-shaped.

Context XX. 9

Adult 25-35 yrs: Male: 160.3 cm, 5'3"

Year of excavation: 1975

Anthropological number: 6862-358

Archaeological context:

Log house tomb - The remains of nine individuals were recovered from the floor of the tomb. Five of the individuals were lying in a row. It is not known which individual is represented by 6862-358.

Visual inventory:

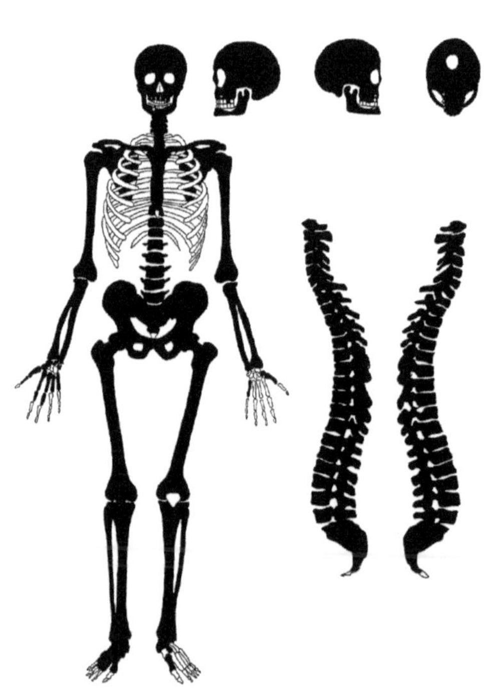

Preservation: The remains were in a very good state of preservation with 80-90% of bones present. Some surface erosion was apparent on the bones.

Cranial pathology: A simple transverse fracture was present at the medial aspect of the most distal end of the R nasal bone. The fractured area was depressed, and the fracture line was clearly visible.

Post-cranial pathology: Porosity and eburnation was apparent on the facet which articulated with the pisiform of the R triquetral. DJD was evident on the distal head of the R MT5. Vertebral osteophytosis was present in TV9 and TV11-LV1. Schmorl's nodes were visible in TV7-LV1 and LV3. Enthesopathies were apparent on the calcanei (Achilles' tendon).

Dentition:

				A	A PU	A	A	A	A	A PU	A	A		C	
8	7	6	5	~~4~~	3	~~2~~	~~1~~	~~1~~	~~2~~	3	~~4~~	~~5~~	~~6~~	7	~~8~~
~~8~~	~~7~~	~~6~~	~~5~~	~~4~~	3	~~2~~	~~1~~	~~1~~	~~2~~	3	~~4~~	~~5~~	~~6~~	~~7~~	~~8~~
				A	A										

Bite: Overbite.

Attrition: Molar wear patterns of Stage 1 for the 3rd molars, Stage 2+ for the 2nd molars and Stage 4 for the 1st molars were apparent. Secondary dentine was exposed in 3⌋ ⌊3.

Calculus: Slight deposits of calculus were present on the buccal surface of 3⌉ and the distal surfaces of 8⌋ ⌊7. Medium deposits were present on the buccal surface of ⌈3.

Periodontal disease: Periodontal disease of considerable severity was apparent.

Caries: A small caries was present in the distal surface of ⌊7.

Abscesses: Small externally draining abscesses were associated with 3⌋ and ⌊2, medium sized externally draining abscesses with 4⌋ and ⌊3, and a large externally draining abscess was associated with ⌊1. Single large externally draining abscesses were associated with 2 1⌋, ⌊4 5 and 4 3⌉.

Ante-mortem tooth loss: ⌊2 6 and 8 7 6 5⌉ ⌈6 7 8 were all lost ante-mortem.

Dental anomalies: The R maxillary 3rd molar was peg-shaped, and the morphology of the socket for the L maxillary molar suggests that it was also peg-shaped.

Context XX. 10. Sk. 2

Adult 25-35 yrs: Male: 170.8 cm, 5'7"

Year of excavation: 1975

Anthropological number: 6862-364

Archaeological context:

Stone cist - The cist was rectangular, and extended from the south-east to the north-west. The remains of four skeletons, three of which were lying in a row, were discovered at the bottom of the cist.

Body position - The individual was lying on his R side, with the head to the north-west. The arms and legs were flexed, and the skull was situated at the edge of a stone slab.

Grave goods - A clay pot was situated near the stone slab associated with the skull.

Visual inventory:

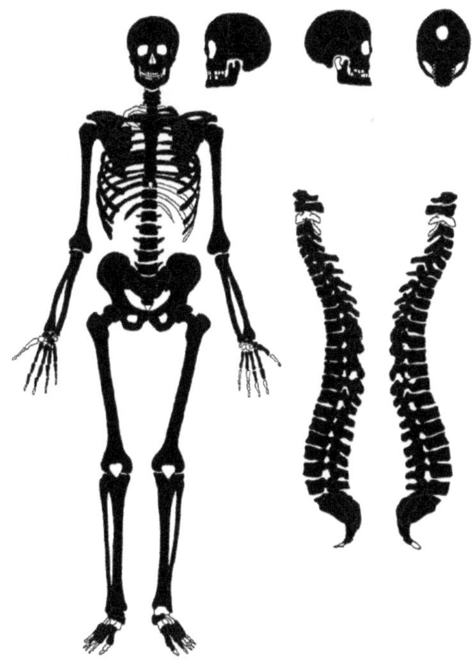

Preservation: The remains were in an excellent state of preservation with 90-100% of bones present. Slight surface erosion was evident on several bones, and the remains were unwashed.

Post-cranial pathology: Plaques of woven bone were visible on the visceral surfaces of the R 2nd - 6th ribs. A well healed, simple oblique fracture was present at the midshaft of the R clavicle. There was almost 100% apposition of the fracture parts, but they were overlapped by approximately 10 mm. Osteophytes and porosity were apparent on the majority of rib heads. Vertebral osteophytosis was present in CV4, TV1, TV2, TV6, TV10, TV11, LV1-LV3 and LV5. Schmorl's nodes were visible in TV5-LV4. Osteophytosis and eburnation were present at the distal end of the L MC1. Enthesopathies were visible on the humeri (deltoid, supraspinatus), the femora (gluteus maximus, adductor magnus), the tibiae (popliteus) and the R calcaneus (Achilles' tendon). An enthesophyte of semimembranosus was present on the L tibia.

Cranial non-metric traits: L highest nuchal line, R ossicle in lambdoid suture, R parietal foramen, R and L mastoid foramina extrasutural, L double anterior condylar canal, R open foramen spinosum, R and L absent zygomaticofacial foramina, L bridging of supraorbital notch, R accessory supraorbital foramen, anterior ethmoid frontal extrasutural.

Post-cranial non-metric traits: R and L Poirier's fossae, R and L third trochanters, R and L lateral tibial squatting facets, R and L acromial articular facets, R circumflex sulcus, R and L vastus notches, R double inferior articular talar facet, R and L CV5, and L CV6 transverse foramina bipartite.

Other observations: A cutmark was present on the anterior surface of the sternum, which may be indicative of post-mortem defleshing/ disarticulation. An area of green discoloration was apparent on the L iliac crest.

Dentition:

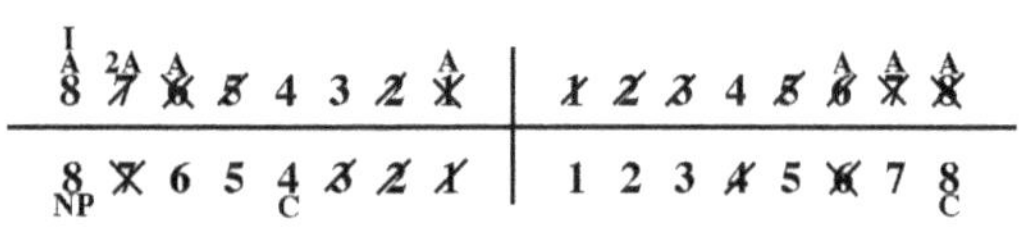

Bite: Overbite.

Attrition: The attrition was extensive for an individual of 25 - 35 yrs, with molar wear patterns of Stage 3 for the 3rd molars, Stage 3+ for the 2nd molars and Stage 5 for the 1st molars.

Calculus: Flecks of calculus were present on the distal surfaces of 8 4⌋, the lingual surface of 4⌉and the buccal and lingual surfaces of 6⌉. Slight deposits were present on the buccal surface of⌈1, and medium deposits were present on the buccal surface of 3⌋.

Periodontal disease: Periodontal disease of considerable severity was evident.

Caries: Small caries were present in the occlusal surfaces of 4⌉and⌈8.

Abscesses: A medium sized externally draining abscess was associated with⌊6, and a large externally draining abscess was associated with 1⌋. A single medium sized internally draining abscess was associated with 8 7⌋, while a single large externally draining abscess was related to 7 6⌋. A single large externally draining abscess was associated with⌊7 8.

Ante-mortem tooth loss: 6 1⌋ ⌊7 8 and 7⌉ ⌈6 had been lost ante-mortem. The socket for ⌈6 had been completely remodelled, while the other sockets were in the early stages of remodelling.

Dental anomalies: The R maxillary 3rd molar was impacted, the R mandibular 3rd molar was genetically absent, and the L mandibular 3rd molar was peg-shaped.

Context XX. 10. Sk. 4

Adult 25-35 yrs: Male: 167.5 cm, 5'6"

Year of excavation: 1975

Anthropological number: 6862-365

Archaeological context:

Stone cist - The cist was rectangular, and extended from the south-east to the north-west. The remains of four skeletons, three of which were lying in a row, were discovered at the bottom of the cist.

Body position - The remains were situated in the southeastern area of the cist. The bones were disturbed, and partly overlay the leg bones of Skeleton 2.

Visual inventory:

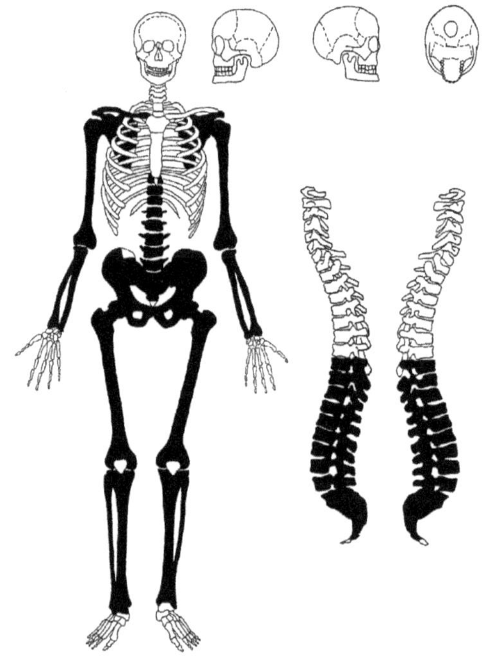

Preservation: The remains were in a very good state of preservation with 70-80% of bones present.

Post-cranial pathology: The bones were generally osteopaenic. A circular lytic lesion with a porotic surface was present in the groove for the attachment of the extensor carpi ulnaris of the R distal ulna. A hook-like lesion, which measured approximately 2 mm by 2 mm, was apparent on the proximal L ulna positioned between the radial notch and the supinator crest. A near circular lytic lesion with smooth margins, which measured approximately 8 mm antero-posteriorly by 9.5 mm supero-inferiorly, was present on the lateral epicondyle of the L distal femur just superior to the groove for the popliteus. The cortex of the articular surfaces of the distal femoral condyles was generally thin and porotic. In addition, areas of periosteal reactive woven bone were present on all surfaces of the distal third of the L femur. The proximal condyles of the L tibia displayed numerous erosive sharp edged marginal lesions. In addition, a circular lesion with sharp margins was present at the midpoint of the medial condyle. In all cases the interior of the lytic lesions was very porotic. The articular facet for the fibula was indistinct, and was represented by a small blunt hook shaped nodule. The proximal articular surface of the L fibula was very porotic, and the facet for articulation with the tibia was concave. The R tibia, fibula and calcaneus all appeared normal. TV12 was ankylosed with the R 12th rib. The sacro-iliac joints appeared normal.

It is possible that this the lesions described above are indicative of a seronegative arthropathy, possibly rheumatoid arthritis. The asymmetrical nature of the lesions would suggest, however, that it is more likely that one of the rheumatoid-like seronegative arthropathies were responsible. It is impossible to give a more specific aetiology for the lesions, since the bones of the hands and feet were not present. Differential diagnosis includes gout.

Complete sacralization of LV5 was evident. Spondylolysis of LV3 was apparent, which had resulted in secondary porosity and osteophytosis of LV2 and LV3. The sacralization and spondylolysis may have resulted in anterior kyphosis of the vertebral column. Osteophytosis was evident on the R glenoid fossa of the scapula, the R humerus and the R femoral head. Vertebral osteophytosis was present in TV9-LV5, and Schmorl's nodes were visible in TV9-LV4. Enthesopathies were apparent on the clavicles (costoclavicular ligament), the R humerus (subscapularis) and the femora (psoas major, iliacus, gluteus maximus, adductor magnus).

Context XXI. 4. Sk. 3

Adult 35-45 yrs: Female: 157.9 cm, 5'3"

Year of excavation: 1975

Anthropological number: 6862-371

Archaeological context:

Log house tomb - The remains of four individuals were recovered from the floor of the tomb. Each individual was lying on the L side, with the head to the west.

Body position - The arms were extended, and the legs were flexed. The skull was situated on a stone slab.

Grave goods - A bronze needle and an iron pin were situated near the skull. A bronze awl and two bronze containers were positioned on the sacrum.

Visual inventory:

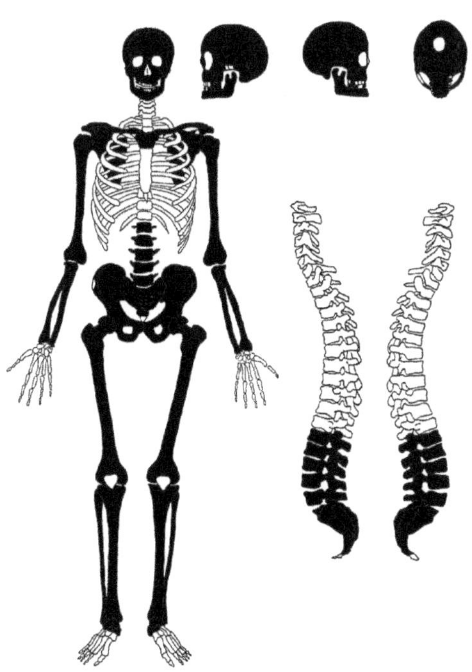

Preservation: The remains were in a good state of preservation with 60-70% of bones present.

Post-cranial pathology: The bones were extremely light which may indicate that the individual had osteoporosis. Spondylolysis of LV4 was evident. Secondary degenerative joint disease with a severity of Grade III was apparent on the superior articular facets of LV5. A very well healed simple transverse fracture was apparent at the distal third of the shaft of the L radius (Colles' fracture). Almost 100% apposition was evident, and the distal part of the fracture was very slightly displaced posteriorly. Although it was not clearly apparent, there was probably a degree of overlap of the fracture parts since the L radius was approximately 10 mm shorter than the R radius. A fracture of the styloid process of the L ulna was also evident and the styloid process was extensively displaced posteriorly. In addition, a possible compression fracture was apparent on the R side of the body of LV1.

Os acromiale was evident in the R scapula. Periosteal reactive woven bone and a possible sinus were present on the intertubercular groove of the R proximal humerus. It is possible that the lesions are indicative of stress or strain to the deltoid, or that there was an infection in this area. Rotator cuff disease was apparent in the humeri. Osteophytosis was evident on the L proximal ulna. Vertebral osteophytosis was present in LV1-LV5, and Schmorl's nodes were visible in LV1-LV3.

Cranial non-metric traits: R and L parietal foramina, R and L mastoid foramina extrasutural, L posterior condylar canal open, R and L bridging of supraorbital notches, R and L accessory supraorbital foramina, R and L anterior ethmoid frontal foramina extrasutural, L accessory infraorbital foramen.

Post-cranial non-metric traits: R and L hypotrochanteric fossae, R exostoses in trochanteric fossae, R and L lateral tibial squatting facets, L septal aperture, L acromial articular facet, R and L circumflex sulci, R and L os trigonums, R and L double inferior articular talar facets, R and L peroneal tubercles.

Other observations: Green discoloration was present on the posterior surface of the R distal radius and the distal end of the R ulna.

Dentition:

~~8~~	7	~~6~~ (A)	~~5~~	~~4~~	~~3~~	~~2~~	~~1~~	~~1~~	~~2~~	~~3~~	~~4~~	~~5~~	6	7	8 (C)
8	7	6	~~5~~	4	3	~~2~~	~~1~~	~~1~~	~~2~~	3	~~4~~	5	6 (C)	7	8

Bite: Overbite.

Attrition: Molar wear patterns of Stage 1 for the 3rd molars, Stage 3+ for the 2nd molars and Stages 4+-5 for the 1st molars were apparent.

Calculus: Slight deposits of calculus were present on the lingual surfaces of ⌊6 and 4 3⌉ ⌈5, the distal surfaces of ⌊7 8 and ⌈8 and the mesial surface of ⌈3. Medium deposits were apparent on the lingual surfaces of ⌈7 8 and the lingual and mesial surfaces of 7 6⌉.

Periodontal disease: Periodontal disease of slight severity was apparent.

Caries: A small caries was present in the occlusal surface of ⌊8, and a medium sized caries was visible in the mesial aspect of ⌈6.

Abscesses: A large internally draining abscess was associated with 6⌋.

Ante-mortem tooth loss: 6⌋ had been lost ante-mortem. The socket was almost completely remodelled.

Enamel hypoplasia: Single hypoplastic grooves were apparent on 4 3⌉ which indicated that the individual had been subject to at least one incident of physiological stress at the age of 3-3.5 years.

Context XXI. 4. Sk. 4

Young adult 17-25 yrs: Male: 169.1 cm, 5'6"

Plate: 22

Year of excavation: 1975

Anthropological number: 6862-372

Archaeological context:

Log house tomb - The remains of four individuals were recovered from the floor of the tomb. Each individual was lying on the L side, with the head to the west.

Body position - The remains were disturbed.

Grave goods - A bronze knife was recovered near the R femur.

Visual inventory:

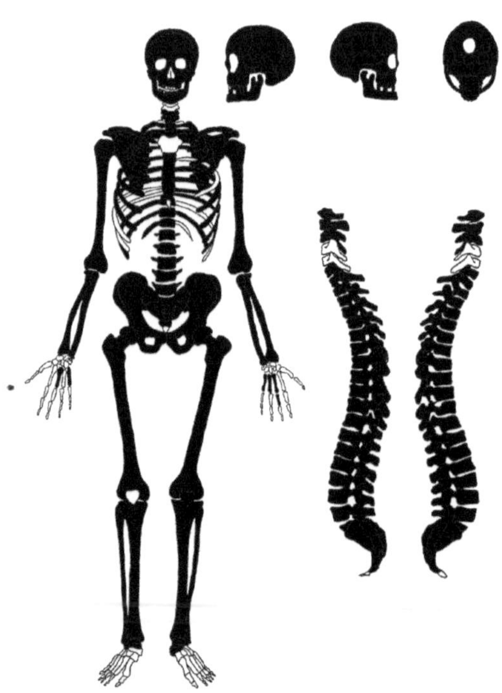

Preservation: The remains were in an excellent state of preservation with 90-100% of bones present.

Cranial pathology: Porosity was evident on the posterior aspect of the parietals and the superior surface of the occipital, which may be indicative of healed porotic hyperostosis.

Post-cranial pathology: The neural arches of S1, S2, S4 and S5 were cleft, which is indicative of developmental delay during the early morphogenesis of the vertebral column. A bony exostosis, which measured approximately 22 mm supero-inferiorly by 12 mm medio-laterally, was present on the medial supracondylar line of the L distal femur. It is possible that the spur was an enthesophyte of the adductor magnus. Osteophytes and porosity were present on the L 6th and 8th rib heads. Schmorl's nodes were visible in TV4-LV3. Enthesopathies were apparent on the femora (adductor magnus, gluteus maximus, psoas, iliacus) and the R navicular (plantar calcaneonavicular ligament).

Weapon trauma: A circular depression, which measured 5 mm by 5 mm, was present on the R frontal, just superior to the orbit. It is possible that the depression represented a healed depressed fracture, which was caused by a pointed object (? pointed axe).

Cranial non-metric traits: 5 L and 3 R ossicles in lambdoid suture, R and L parietal foramina, L ossicle at pterion, R and L mastoid foramina extrasutural, L posterior condylar canal open, L double condylar facet, R open foramen spinosum, R and L accessory supraorbital foramina, R and L anterior ethmoid frontal foramina extrasutural.

Post-cranial non-metric traits: R and L Poirier's fossae, R and L hypotrochanteric fossae, R and L third trochanters, R and L lateral tibial squatting facets, R acetabular crease, R and L acromial articular faces, L medial talar facet, R and L peroneal tubercles, R CV5, and R and L CV6 transverse foramina bipartite.

Other observations: Green discoloration was present on the visceral surface of the L 3rd rib and the greater trochanter of the R femur.

Dentition:

8	7	6	5	4	3	~~2~~	~~1~~	~~1~~	~~2~~	3	4	5	6	7	8
8	~~7~~	6	~~5~~	~~4~~	3	2	1	1	2	3	4	~~5~~	6	7	8

Bite: Overbite.

Attrition: Molar wear patterns of Stage 1 for the 3rd molars, Stage 2+ for the 2nd molars and Stage 3 for the 1st molars were apparent.

Calculus: Flecks of calculus were present on the mesial surfaces of ⌈6 7. Slight deposits were apparent on the buccal surfaces of 3 2 1⌉ ⌈1, the distal surfaces of 6⌋ and 7⌉, the mesial surface of ⌊5 and the buccal and distal surfaces of ⌈4. Medium deposits were present on the buccal surface of 7⌋, the lingual surfaces of 3 2 1⌉ ⌈1, the buccal and lingual surfaces of ⌈2 3 and the mesial and lingual surfaces of ⌈4.

Enamel hypoplasia: Single hypoplastic lines were present on 3⌉⌈4 which indicated that the individual had been subject to at least one incident of physiological stress at the age of 2.5-3 years.

Context XXIII. 4

Adult 25-35 yrs: Male: 168.4 cm, 5'6"

Year of excavation: 1975

Anthropological number: 6862-374

Archaeological context:

Log house tomb - Disarticulated human remains were recovered throughout the tomb. The original location of this individual within the tomb is unknown.

Visual inventory:

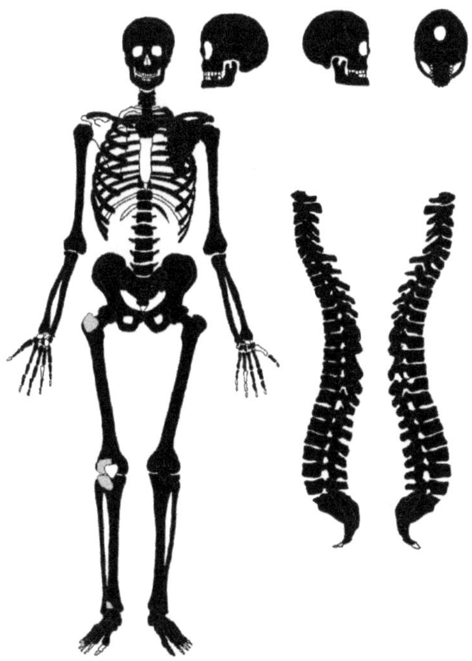

Preservation: The remains were in an excellent state of preservation with 90-100% of bones present. Some bones displayed surface erosion.

Post-cranial pathology: A well healed simple oblique fracture was present at the midshaft of the L MC3. There was no malalignment, but the L MC3 had a greatest length of 69.1 mm making it approximately 5 mm shorter than the corresponding R MC3. This indicates that there was a degree of overlap of the fracture parts. Secondary osteophytes were apparent at the distal end of the bone, and were associated with osteophytes on the capitate. A well healed simple oblique fracture was also apparent on the midshaft of the L MC4. Again, there was no malalignment, but the L MC4 had a greatest length of 57.1 mm making it approximately 4 mm shorter than the R MC4. This indicates that there was a degree of overlap of the fracture parts. Secondary osteophytes were present at the distal end, and were associated with osteophytes on the hamate. Osteophytes were also present on the distal end of the L MC2 and the L trapezoid, which are likely to have been secondary to the fractured MC3 and MC4. The R navicular appeared to have been subject to a compression fracture of the anterior surface. Fracture lines and marginal osteophytes were present on the anterior surface which was hypertrophied. Unusually, all of the other bones of the R foot appeared normal.

A bony tubercle was apparent on the lateral surface of the L tibia, immediately inferior to the lateral proximal condyle, which may have been an enthesophyte of tibialis anterior. Enthesopathies were present on the femora (adductor magnus, gluteus maximus, psoas, iliacus). Porosity was apparent on several of the rib heads. Vertebral osteophytosis was present in TV6, TV8, TV10 and LV5, and Schmorl's nodes were visible in TV6-TV12 and LV2.

Weapon trauma: A near circular lesion, which measured approximately 25 mm medio-laterally by 20 mm supero-inferiorly, was apparent on the R side of the frontal approximately 40 mm superior to the orbital margin. The lesion had a smooth raised margin. Possible aetiologies for the lesion are a healed depressed fracture possibly caused by a pointed axe or an overlying skin ulcer.

Cranial non-metric traits: Ossicle at lambda, R 1 and L 1 ossicles in lambdoid suture, ossicle at bregma, R and L mastoid foramina extrasutural, L double condylar facet, L double anterior condylar canal, R bridging of supraorbital notch, L accessory supraorbital foramen, L anterior ethmoid frontal extrasutural, L accessory infraorbital foramen.

Post-cranial non-metric traits: R and L hypotrochanteric fossae, R exostosis in trochanteric fossa, R and L lateral tibial squatting facets, R and L septal apertures, L acromial articular facet, L circumflex sulcus, L vastus fossa, R and L medial talar facets, R and L double inferior articular talar facets, R and L peroneal tubercles.

Other observations: Venous-arterial material was preserved within the maxilla.

Dentition:

NP 8	7	A ~~6~~	5	~~4~~	3	~~2~~	~~1~~	~~1~~	~~2~~	3	~~4~~	~~5~~	~~6~~	~~7~~	NP 8
8	7	6	~~5~~	~~4~~	~~3~~	~~2~~	~~1~~	~~1~~	2	3	~~4~~	5	6	7	~~8~~

Bite: Overbite.

Attrition: Molar wear patterns of Stage 4 for the 3rd molars, Stage 4+ for the 2nd molars and Stages 4+-5 for the 1st molars were apparent.

Calculus: Slight deposits of calculus were present on the lingual surfaces of 3⌋ and ⌈6 7, the distal surface of 8⌉, the buccal and lingual surfaces of 5⌋, the buccal and mesial surfaces of ⌈2, the mesial and lingual surfaces of ⌈5, the distal and lingual surfaces of 6⌉ and the buccal, mesial and lingual surfaces of ⌊3.

Periodontal disease: Periodontal disease of slight severity was apparent.

Abscesses: A medium sized externally draining abscess was associated with 6⌋.

Ante-mortem tooth loss: ⌊1 6 had been lost ante-mortem. The socket for ⌊1 was in the mid stages of remodelling while the socket for ⌊6 was almost completely remodelled.

Dental anomalies: Both maxillary 3rd molars were genetically absent.

Context XXIII.8

Mature adult 45+ yrs: Female: 156.5 cm, 5'2"

Year of excavation: 1975

Anthropological number: 6862-378

Archaeological context:

Log house tomb - Human remains were recovered throughout the tomb. The original location of this individual within the tomb is unknown.

Visual inventory:

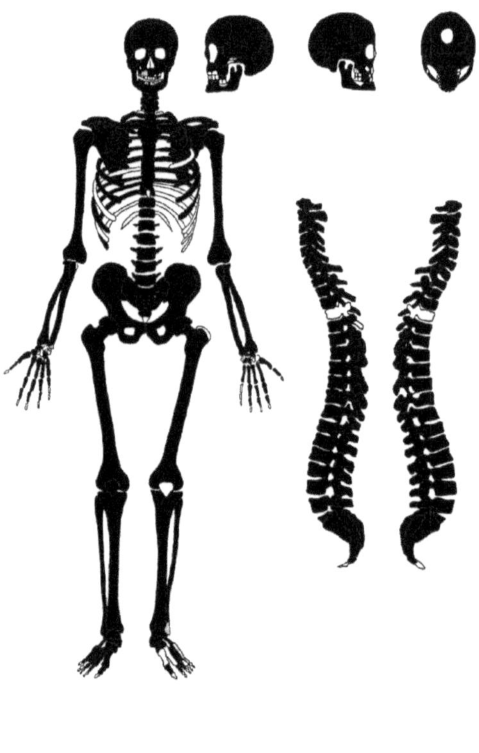

Preservation: The remains were in an excellent state of preservation with 90-100% of bones present.

Cranial pathology: Osteophytes and porosity were apparent on the articular tubercles of the temporals and the mandibular condyles. The changes are probably indicative of DJD of the temporomandibular joints. The L joint would have been slightly more severely affected.

Post-cranial pathology: Several fractures were present in the remains of this individual. A well healed simple transverse fracture was apparent at the midshaft of the R MC4. There was very little malalignment. The R MC4 was 3 mm shorter than the L MC4 indicating that there was slight overlap of the fracture parts. A well healed simple transverse fracture was also apparent at the midshaft of the R MC5. The proximal part of the fracture was displaced anteriorly by approximately forty-five degrees. The R MC5 was approximately 6 mm shorter than the L MC5 indicating that there was a degree of overlap of the fracture parts. Osteophytes, porosity and eburnation were apparent on the R scaphoid and the R trapezium, which are probably secondary to the fractured metacarpals.

A simple transverse fracture was apparent at the midshaft of the R 8th rib. Callus was present at the fracture site indicating that it was still in the process of healing. The anterior part of the fracture was displaced inferiorly by approximately ten degrees. The rib appeared shorter than the adjacent ribs. A well healed simple transverse fracture was apparent at the distal third of the shaft of the L ulna (Night-stick fracture). There was 100% apposition of the fracture parts, but the distal part of the fracture was displaced medially by approximately ten degrees. Hypertrophy was evident at the point of fracture, and the L ulna was approximately 5 mm shorter than the R ulna indicating a degree of overlap of the fracture parts. Secondary osteophytosis was present at the distal articular surface. Os acromiale was evident in the R scapula. Rotator cuff disease was present in the humeri. Osteophytes, porosity, and eburnation were apparent on the distal L MC1 and the proximal end of the L proximal PH1. Osteophytosis was apparent on the majority of rib heads, the sternal and acromial ends of the clavicles, the L glenoid fossa of the scapula, the R distal humerus, the R distal ulna, the R proximal radius, the femoral heads and the R proximal tibia. Vertebral osteophytosis was present in CV1-TV3 and TV5-LV5. Enthesopathies were present on the radii (biceps).

Cranial non-metric traits: R and L highest nuchal lines, 4 ossicles at lambda, 3 R ossicles in lambdoid suture, R parietal foramen, R and L ossicles at parietal notch, R and L mastoid foramina extrasutural, R and L posterior condylar canals open, L double anterior condylar canal, R and L incomplete foramina ovale, R open foramen spinosum, L bridging of supraorbital notch, R and L anterior ethmoid frontal foramina extrasutural, R and L accessory infraorbital foramina.

Post-cranial non-metric traits: R and L Allan's fossae, R and L plaque, R hypotrochanteric fossa, L exostosis in trochanteric fossa, R and L lateral tibial squatting facets, R acromial articular facet, R and L circumflex sulci, R vastus notch, R and L medial talar facets, R double inferior articular talar facet, R double anterior calcaneal facet, R and L peroneal tubercles, L CV5, R and L CV6, and R and L CV7 transverse foramina bipartite.

Dentition:

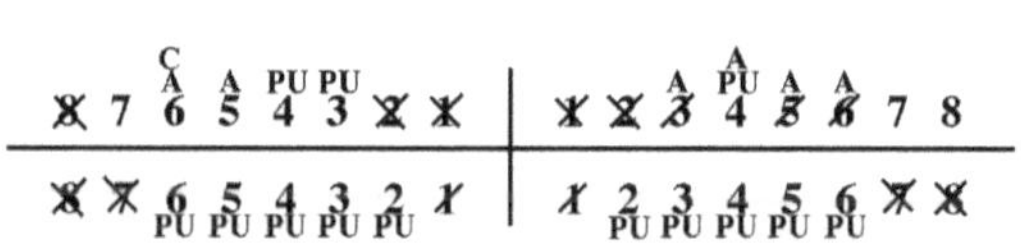

Attrition: Extensive attrition was apparent on the majority of teeth, with molar wear patterns of Stage 3+ for the 3rd molars, Stage 4+ for the 2nd molars and Stage 5 for the 1st molars. Secondary dentine or pulp cavity exposure was evident in 4 3⌋ ⌊4 and 6 5 4 3 2⌉ ⌈2 3 4 5 6.

Periodontal disease: Periodontal disease of considerable severity was apparent.

Caries: A large caries was apparent in the occlusal surface of 6⌋.

Abscesses: A single medium sized externally draining abscess was associated with ⌊3 4. Single large externally draining abscesses were associated with ⌊5 6 and 6 5⌋.

Ante-mortem tooth loss: 8 2 1⌋ ⌊1 2 and 8 7⌉ ⌈7 8 had been lost ante-mortem. The sockets for the maxillary teeth were nearly completely remodelled, while those for the mandibular teeth were in the mid-stages of remodelling.

Dental anomalies: A diastema was apparent between 1⌉ and ⌈1.

Context XXIII. 10. Sk. 1

Adult 25-35 yrs: Male: 158.4 cm, 5'3"

Plate: 14

Year of excavation: ?1974

Anthropological number: 6862-380

Archaeological context:

Log house tomb - Details of this context were not recorded in the archive reports.

Visual inventory:

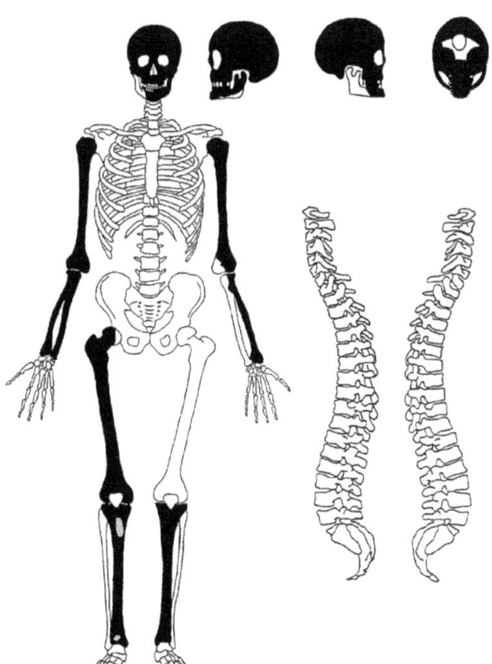

Preservation: The remains were in a poor state of preservation with 30-40% of bones present. Several bones displayed surface erosion and were fragmentary.

Cranial pathology: A well healed simple transverse (supero-inferior) fracture was present on the posterior aspect of the L zygomatic. The associated musculature was affected by the fracture. A prominent bony exostosis at the superior margin of the posterior area of the zygomatic, was indicative of myositis ossificans of the masseter. A well healed simple transverse (supero-inferior) fracture was also apparent at the midpoint of the L zygomatic arch. The area between the fracture line on the zygomatic and the fracture line on the zygomatic arch was depressed, so that there was only approximately 50% apposition of the fracture parts. It is likely that the fracture had been caused by a blow to L side of the cranium. Porosity was apparent on the posterior surface of the parietals, which may be indicative of healed porotic hyperostosis.

Cranial non-metric traits: 1 R and 1 L ossicles in lambdoid suture, R and L parietal foramina, R and L mastoid foramina extrasutural, R and L open foramina spinosum, R and L accessory supraorbital foramina, R and L anterior ethmoid frontal foramina extrasutural, R and L accessory infraorbital foramina.

Post-cranial non-metric traits: R and L lateral tibial squatting facets.

Dentition:

8̸ 7 6̸ 5̸ 4̸ 3 2̸ 1̸	1̸ 2̸ 3 4̸ 5 6 7 8̸
~~8~~ 7 6 5 4̸ 3̸ 2̸ 1̸	1 2 3 4 5 6 7 8

Bite: Overbite.

Attrition: Molar wear patterns of Stage 1 for the 3rd molars, Stages 2-2+ for the 2nd molars and Stage 3+ for the 1st molars were apparent.

Calculus: Slight deposits of calculus were present on the buccal surfaces of 6⌋ ⌊7 and 6 5⌉ ⌈4 5 6 7 8, the lingual surfaces of 7 5 3⌋ and ⌈3, the buccal and lingual surfaces of ⌈1 2 and the buccal, lingual, distal and mesial surfaces of 7⌉. Medium deposits were present on the lingual surfaces of 6⌋ and 6 5⌉ ⌈4 5 6 7, the distal surface of ⌊7, the buccal and distal surfaces of 7⌋ and the distal and lingual surfaces of ⌈8. Subgingival calculus was also present on ⌈8.

Context XXIII. 11. Sk. 3

Adult 25-35 yrs: Male: 168.3 cm, 5'6"

Year of excavation: ?1974

Anthropological number: 6862-385

Archaeological context:

Log house tomb - Details of this context were not recorded in the archive reports.

Visual inventory:

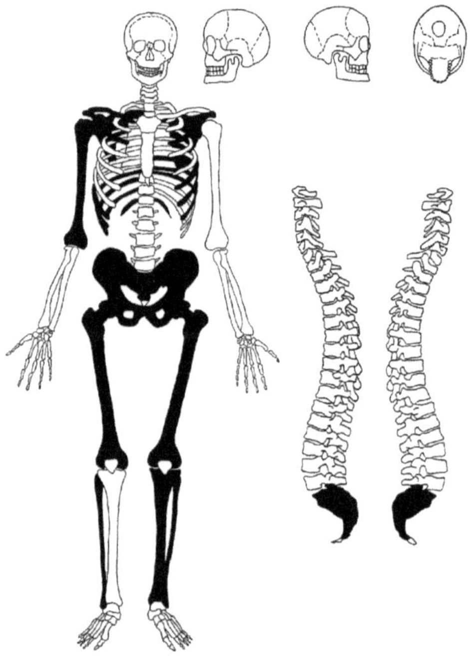

Preservation: The remains were in a fair state of preservation with 50-60% of bones present.

Post-cranial pathology: The neural arch of S1 was cleft, which is indicative of developmental delay during the early morphogenesis of the vertebral column. The popliteal surface of the R femur had a roughened appearance with small bony exostoses. The aetiology of this unusual morphology may have been strain of the gastrocnemius and plantaris muscles. A small bony nodule with a prominent groove was present on the antero-lateral surface of the L tibia, which may have been an enthesophyte of the iliotibial tract. Osteophytosis was present on several rib heads.

Context XXIII. 13 (i)

Adult 25-35 yrs: Probable male

Year of excavation: ?1974

Anthropological number: 6862-387

Archaeological context:

Log house tomb - Details of this context were not recorded in the archive reports.

Visual inventory:

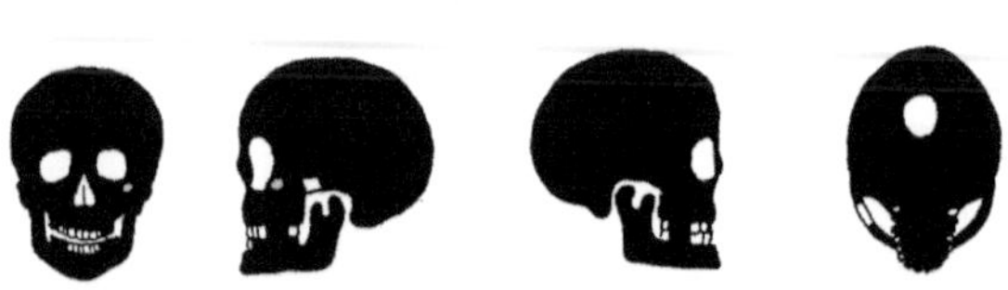

Preservation: Only the skull was present, which was in a very good state of preservation.

Cranial pathology: A simple fracture, which ran supero-inferiorly from the medial aspect of the midpoint of the R nasal bone to its inferior margin, was present. The depressed inferior aspect of the fracture was not visible. It is probable that the injury had been caused by a direct blow with a blunt object. A circular depression, which measured 17 mm by 17 mm, was apparent immediately superior to the L side of the glabella. The lesion is probably indicative of a healed depressed fracture. The mandibular condyles were both hypoplastic, a finding which is indicative of developmental delay of the mandibular anlage, and is a field defect of the 1st branchial arch.

Weapon trauma: A large perforation ran supero-inferiorly through the anterior margin of the L zygomatic. The lesion had smooth margins and a diameter of approximately 10 mm. It is possible that the lesion was due to a perforating injury made by a pointed objected. The zygomatic appeared to have been displaced so that it was lying more medially against the skull than normal. The zygomatic arch was markedly shortened and the posterior aspect of the zygomatic was almost articulating with the posterior aspect of the zygomatic arch. The intermediate aspect of the zygomatic arch itself was not present. It is possible that the zygomatic arch had also been fractured as a result of a blow to the side of the skull. The injuries to the L zygomatic and the zygomatic arch had resulted in orbital asymmetry with the greatest breadth of the L orbit measuring 3 mm longer than that for the R orbit.

Dentition:

NP 8	7	C 6	~~5~~	~~4~~	~~3~~	~~2~~	~~1~~	~~1~~	~~2~~	~~3~~	~~4~~	~~5~~	6	7	NP 8
8 NP	7	6	~~5~~	~~4~~	~~3~~	~~2~~	~~1~~	~~1~~	~~2~~	~~3~~	4	5	6	7	8

Bite: Overbite.

Attrition: Molar wear patterns of Stage 1 for the 3rd molars, Stage 3 for the 2nd molars and Stage 4+ for the 1st molars were apparent.

Calculus: Slight deposits of calculus were present on the buccal surfaces of⌈4 5 6, the lingual surfaces of 6⌋ ⌊6 7 and⌈7, the buccal and lingual surfaces of 7⌋ and 6⌉ and the buccal, lingual and distal surfaces of 7⌉ ⌈8. Medium deposits were apparent on the buccal surfaces of 6⌋ ⌊7, and on the lingual surfaces of⌈4 5 6.

Periodontal disease: Periodontal disease of slight severity was evident.

Caries: A medium sized caries was apparent in the mesial surface of 6⌋.

Dental anomalies: The R mandibular and both maxillary 3rd molars were genetically absent.

Context XXIII. 19. Sk. 1

Young adult 17-25yrs: Female: 152.3cm, 5'1"

Year of excavation: ?1974

Anthropological number: 6862-396

Archaeological context:

Log house tomb - Details of this context were not recorded in the archive reports.

Visual inventory:

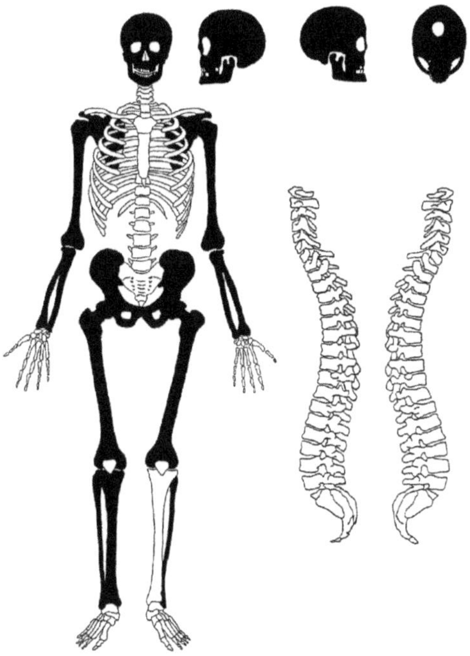

Preservation: The remains were in a good state of preservation with 60-70% of bones present.

Cranial pathology: Cribra orbitalia of Type 1 was present in Area 1 of the R orbit and Areas 1 and 4 of the L orbit.

Post-cranial pathology: A bony nodule was present on the postero-medial aspect of the L femoral neck, which may represent an enthesopathy of part of the capsule attachments of the hip joint.

Cranial non-metric traits: R and L highest nuchal lines, 2 ossicles at the lambda, R and L parietal foramina, L ossicle at pterion, R and L mastoid foramina extrasutural, R and L posterior condylar canals open, R and L open foramina spinosum, R and L anterior ethmoid frontal foramina extrasutural, R and L accessory infraorbital foramina.

Post-cranial non-metrics: L third trochanter, R lateral tibial squatting facet, R and L acetabular creases, R and L accessory sacral facets, R and L acromial articular facets, R and L circumflex sulci. The acetabular creases were very prominent.

Dentition:

8 7 6 5 ~~4~~ ~~3~~ ~~2~~ ~~1~~	~~1~~ ~~2~~ ~~3~~ ~~4~~ 5 6 7 ~~8~~
8 7 6 ~~5~~ ~~4~~ ~~3~~ ~~2~~ ~~1~~	~~1~~ ~~2~~ ~~3~~ 4 ~~5~~ 6 7 8

Bite: Molar wear patterns of Stages 1-2 for the 2nd and 3rd molars and Stage 3 for the 1st molars were apparent.

Calculus: Flecks of calculus were present on the buccal surface of ⌊5. Slight deposits were present on the buccal surfaces of 5⌋ ⌊7, the lingual surface of ⌈4 and the mesial surface of 8⌋. Medium deposits were present on the buccal surfaces of 7 6⌋ ⌊6 and the lingual surfaces of 7 6⌉ ⌈6.

Context XXIII. 19. Sk. 3

Adult 35-45 yrs: Male: 169.4 cm, 5'6"

Year of excavation: ?1974

Anthropological number: 6862-398

Archaeological context:

Log house tomb - Details of this context were not recorded in the archive reports.

Visual inventory:

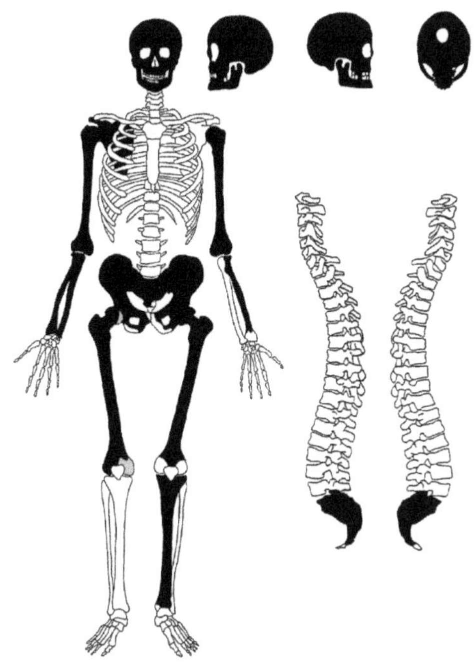

Preservation: The remains were in a fair state of preservation with 40-50% of bones present.

Cranial pathology: A simple transverse fracture was present at the distal end of the L nasal bone. The fracture line was clearly visible, and the two parts of the fracture were well aligned. Extensive cribra orbitalia of Type 3 was present in Areas 1, 2, 3, 4, 5 and 6 of the R orbit and Areas 1, 2, 3, 5 and 6 of the L orbit.

Post-cranial pathology: Rotator cuff disease was evident in the humeri. Osteophytosis was apparent on the distal R ulna and the superior margins of the auricular surfaces. Enthesopathies were present on the radii (biceps).

Cranial non-metric traits: R and L highest nuchal lines, 1 L ossicle in lambdoid suture, R and L parietal foramina, R and L mastoid foramina extrasutural, R and L posterior condylar canals open, R and L double anterior condylar canals, R and L open foramina spinosum, R and L absent zygomaticofacial foramina, R and L accessory supraorbital foramina, R and L anterior ethmoid frontal foramina extrasutural, R and L accessory infraorbital foramina.

Post-cranial non-metric traits: R and L hypotrochanteric fossae, L third trochanter, L lateral tibial squatting facet, R acromial articular facet, R circumflex sulcus.

Dentition:

~~8~~	PU 7	PU 6	~~5~~	~~4~~	~~3~~	~~2~~	~~1~~	~~1~~	~~2~~	A ~~3~~	A ~~4~~	~~5~~	C A 6	PU 7	8
8 NP	~~7~~	6 PU	~~5~~	~~4~~	~~3~~	~~2~~	~~1~~	~~1~~	~~2~~	3	4	5	6	7	8 NP

Bite: Overbite.

Attrition: Molar wear patterns of Stage 2 for the 3rd molars, Stage 4+ for the 2nd molars and Stage 5 for the 1st molars were apparent. Secondary dentine or pulp cavity exposure was evident in 7 6⌋ ⌊7 and 6⌉.

Calculus: Flecks of calculus were present on the distal surface of ⌊8. Slight deposits were present on the buccal surface of ⌊7 and the lingual surfaces of ⌈4 5.

Periodontal disease: Periodontal disease of medium severity was evident.

Caries: A medium sized caries was present on the mesial aspect of ⌊6.

Abscesses: A medium sized externally draining abscess was associated with ⌊6, and a single medium sized externally draining abscess was associated with ⌊3 4. Plaques of woven bone and inflammatory pitting were associated with ⌊6.

Dental anomalies: The mandibular 3rd molars were genetically absent.

Context XXIII

Adult 25-35 yrs: Probable male

Plate: 36

Year of excavation: ?1974

Anthropological number: 6862-400

Archaeological context:

All that was known of this individual was that he had originated from Group XXIII.

Visual inventory:

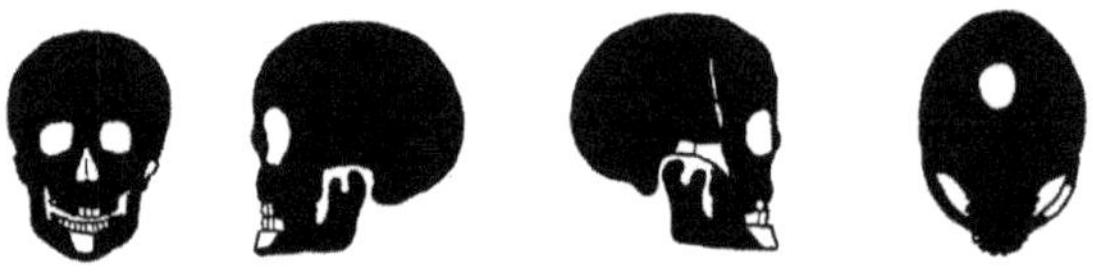

Preservation: Only the skull was present, which was in a very good state of preservation.

Cranial pathology: Cribra orbitalia of Type 1 was present in Area 1 of the R orbit and Areas 1 and 2 of the L orbit.

Weapon trauma: A diagonal sword cut, which measured 59.5 mm, extended supero-inferiorly from the midpoint of the R coronal suture to the greater wing of the sphenoid. A parallel cut mark was also present on the R zygomatic, which appeared to have been cleanly cut in half supero-inferiorly. The two cut marks were definitely attained by a single blow with a sword. It is probable that the attacker was standing in front of, and to the right of the victim. There was no sign of healing and it is fairly definite that the individual died as a result of the weapon trauma.

Other observations: The superior aspect of the occipital had a flattened appearance, which may be indicative of lambdoid cranial deformation.

Dentition:

B 8	B 7	6	5	4	3	~~2~~	~~1~~	~~1~~	~~2~~	~~3~~	~~4~~	5	6	7	8
8 NP	7	6	5 B	~~4~~	~~3~~	~~2~~	~~1~~	~~1~~	~~2~~	~~3~~	~~4~~	~~5~~	6	7	8

Bite: Overbite.

Attrition: Molar wear patterns of Stages 1-2+ for the 3rd molars, Stages 3-3+ for the 2nd molars and Stage 4 for the 1st molars were apparent.

Calculus: Slight deposits of calculus were present on the buccal surfaces of 5⌋ and 6⌉, the lingual surface of 3⌋, the buccal and lingual surfaces of 6 4⌋ ⌊5 6 7 and 8 7⌉ ⌈6 and the buccal, lingual and distal surfaces of ⌊8 and ⌈7.

Periodontal disease: Periodontal disease of medium severity was evident.

Dental anomalies: The R mandibular 3rd molar was genetically absent.

Context XXV. 16. Sk. 4

Young adult 17-25 yrs: Male: 170.7 cm, 5'7"

Year of excavation: 1977

Anthropological number: 6862-407

Archaeological context:

Log house tomb - Disarticulated human remains were recovered throughout the tomb. The original location of this individual within the tomb is unknown.

Visual inventory:

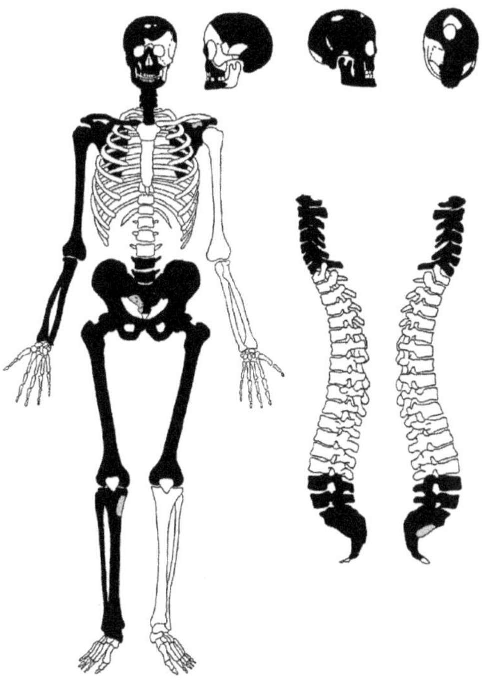

Preservation: The remains were in a fair state of preservation with 40-50% of bones present. Most bones were fragmentary with surface erosion.

Cranial pathology: Healed porotic hyperostosis was present on the superior surfaces of the frontal and the parietals.

Post-cranial pathology: Enthesopathies were apparent on the R tibia (patellar ligament) and the R calcaneus (Achilles' tendon). Schmorl's nodes were present in LV4 and LV5.

Weapon trauma: Three examples of weapon trauma were apparent on the cranium. An oval perforation, which measured approximately 25 mm antero-posteriorly by 12 mm medio-laterally, was present on the superior aspect of the R parietal. The margins of the perforation had a chipped appearance which indicated that the perforation had been caused by a pointed implement which entered the cranium at an angle. It is probable that the perforation was caused by a pointed battle axe. The attacker would have been situated behind the victim and he or she would have wielded the axe above the person's head. The morphology of the lesion indicates that the attacker did not strike a clear blow to the cranium.

A circular perforation, which measured 26 mm antero-posteriorly by 18 mm medio-laterally, was present at the superior surface of the frontal. The margins were slightly eroded but appeared bevelled, and a large fracture line was radiating from the anterior margin of the perforation. The injury was clearly due to a pointed axe, and would have been made with a considerable amount of force.

A circular perforation, which measured 12.5 mm antero-posteriorly by 11.5 mm medio-laterally, was present on the frontal bone just posterior to the perforation described above. The margins were sharp and there was no bevelling. Again, it is probable that the perforation was caused by a pointed axe.

Cranial non-metric traits: R and L highest nuchal lines, 2 ossicles at lambda, 2 R and 1 L ossicles in lambdoid suture, R and L parietal foramina, R and L mastoid foramina extrasutural, R and L posterior condylar canals open, R open foramen spinosum, R absent zygomaticofacial foramen, R accessory supraorbital foramen. In addition, there was a large ossicle in the R squamosal suture.

Post-cranial non-metric traits: R and L hypotrochanteric fossae, R and L third trochanters, R lateral tibial squatting facets, R and L acromial articular facets, R and L circumflex sulci, R medial talar facet, R double inferior articular talar facet, R peroneal tubercle, R CV5, and R and L CV6 transverse foramina bipartite.

Other observations: Green discoloration was evident on the R pubis and the posterior surface of the R femur.

Dentition:

C				B							B				B
8	7	6	5̶	4	3̸	2̸	1̸	1̸	2̸	3̸	4	5	6	7	8
8	7	6	5̸	4̸	3̸	2̸	1̸	1̸	2̸	3̸	4	5̸	6	7̶	8̶
C															

Bite: Overbite.

Attrition: Molar wear patterns of Stage 1 for the 3rd molars, Stage 3- for the 2nd molars and Stage 4 for the 1st molars were apparent.

Calculus: Slight deposits of calculus were present on the buccal surfaces of 6⌋ ⌊5, the lingual surfaces of ⌊6 and 8 7 6⌉ ⌈4 6 and the buccal and lingual surfaces of ⌊7.

Periodontal disease: Periodontal disease of slight severity was evident.

Caries: Small caries were present in the occlusal surfaces of 8⌋ and 8⌉.

2. Hunno-Sarmatian period individuals

Context XXIII. 3. entrance (i)

Adult 25-35 yrs: Female: 162.3 cm, 5'4"

Year of excavation: 1975

Anthropological number: 6862-373

Archaeological context:

NB: 6862-373 and 6863-166 were both labelled as the secondary burial XXIII. 3.

Archaeology & Trauma

Log house tomb - A rectangular ground pit, which was orientated north-east to south-west, was situated above the remains of the log house tomb. The remains of a single individual were recovered from the pit.

Body position - The skeleton was lying in a supine position, with the head to the south-west. The L arm was extended, and the R arm was flexed. The L leg was extended, and the R leg was flexed. The skull and thorax were situated on stone slabs.

Grave goods - Traces of fabric were recovered from below the skull, while sheep bones were situated near the skull.

Visual inventory:

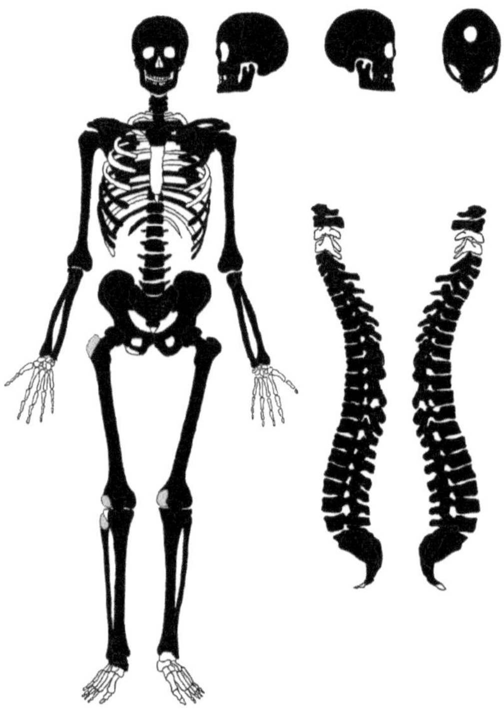

Preservation: The remains were in a very good state of preservation with 80-90% of bones present.

Cranial pathology: An ivory osteoma, which measured 14 mm antero-posteriorly by 10 mm medio-laterally, was present on the L parietal just adjacent to the coronal suture.

Post-cranial pathology: An ossified haematoma, which measured 36 mm supero-inferiorly by 10 mm medio-laterally, was present on the lateral surface of the proximal third of the L fibula. Osteophytes and porosity were apparent on the majority of rib heads. Vertebral osteophytosis was evident in CV1, CV2, CV5, CV6, TV1, TV3-TV12, LV4, LV5 and S1. Schmorl's nodes were visible in TV6, TV8 and TV10-LV3. An enthesopathy was present on the distal R fibula (interosseous ligament).

Cranial non-metric traits: L highest nuchal line, R and L parietal foramina, R and L mastoid foramina extrasutural, L double condylar facet, L double anterior condylar canal, R and L anterior ethmoid frontal foramina extrasutural, R and L accessory infraorbital foramina.

Post-cranial non-metric traits: R and L Poirier's fossae, R and L lateral tibial squatting facets, L septal aperture, R and L accessory sacral facets, R and L acromial articular facets, R and L circumflex sulci, R medial talar facet, R double inferior articular talar facet.

Other observations: The superior aspect of the occipital had a flattened appearance, which may be indicative of the lambdoid form of cranial deformation. Fragments of body tissue were adhering to the R 8th rib and the femora.

Dentition:

8	7	6	5̸	4	3	2	1	1	2	3	4	5	6	7	8 (C)
8 (2C)	7	6	5	4	3̸	2̸	1̸	1	2̸	3̸	4	5	6	7	8

Bite: Overbite.

Attrition: Molar wear patterns of Stages 1-2 for the 3rd molars, Stage 3- for the 2nd molars and Stages 3+-4 for the 1st molars were apparent.

Calculus: Slight deposits of calculus were present on the buccal surfaces of 7⌋ and 6⌉ ⌈1, the lingual surfaces of 6 3 2⌋, the occlusal surfaces of ⌊4 5 7, the occlusal and lingual surfaces of ⌊6, the distal and lingual surfaces of 8⌋ and the buccal and mesial aspects of 4⌋. Medium deposits were apparent on the buccal surfaces of 6 5⌋, the lingual surfaces of ⌊7 and 6⌉ ⌈1, the buccal and lingual surfaces of ⌊5 and 7 5⌉, the mesial and lingual surfaces of 4⌉, the occlusal and lingual surfaces of ⌈4, the buccal, lingual and distal surfaces of ⌊4 and 8⌉ and the buccal, lingual and occlusal surfaces of ⌊8 and ⌈5 7. Heavy deposits were apparent on the buccal surfaces of ⌊6 7, the buccal, lingual and occlusal surfaces of ⌈6 and the buccal, lingual, distal and occlusal surfaces of ⌈8. The occurrence of calculus on the occlusal surfaces of the teeth, and the presence of the heavy deposits of calculus, may indicate that the individual did not chew using the L side of the mouth. This may be related to the large caries situated in the occlusal surface of ⌊8.

Periodontal disease: Periodontal disease of medium severity was apparent.

Caries: A large caries was present in the occlusal surface of ⌊8. Small caries were apparent in the occlusal and the buccal surfaces of 8⌉.

Enamel hypoplasia: It was not possible to examine all the teeth for enamel hypoplasia due to the extensive deposits of calculus. Single hypoplastic grooves were, however, apparent on 3 1⌋ ⌊1 which indicated that the individual had been subject to at least one physiological insult at the age of 2.5-3 years.

Context XXXI. 1

Juvenile 10.5 - 12.5 years: Possible male

Plates: 54, 55

Year of excavation: 1977

Anthropological number: 6863-235

Archaeological context:

Pit and cist - An area of staining was apparent on the ground surface, under which was a rectangular pit with rounded corners. A stone cist, which was orientated south to north and contained the remains of a child, was situated inside the pit.

Body position - The individual was lying in an extended, supine position with the head to the north. The skull, arm and legs bones were disturbed.

Grave goods - An overturned clay pot was situated in the northeastern corner of the cist.

Visual inventory:

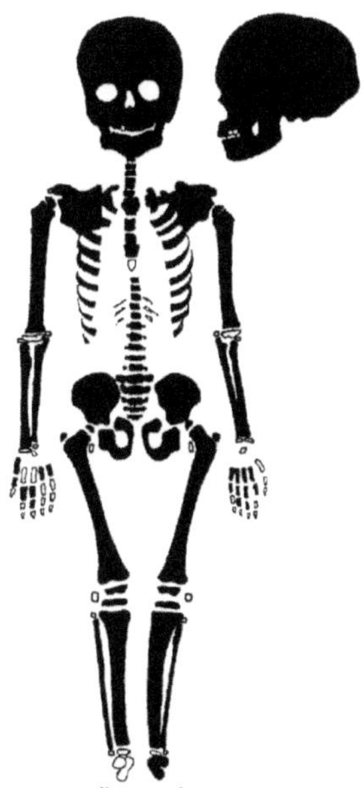

Preservation: The remains were in an excellent state of preservation with 90-100% of bones present.

Weapon trauma: An injury made by a trilobate arrowhead was apparent on the R ilium, approximately 19 mm superior to the sciatic notch. The morphology of the arrowhead was clearly apparent on the posterior surface of the ilium, indicating that the arrow had entered the bone through the posterior surface. The morphology of the arrowhead was also visible on the anterior surface of the ilium, although it was smaller and less distinct indicating that this represented the exit wound. It is probable that the arrowhead had not gone straight through the bone, and it may have been lodged in the bone with only the tip protruding through to the anterior surface.

Plaques of woven bone surrounded the perforation on the posterior surface of the ilium indicating that the individual had survived the weapon trauma for at least a short period of time. Plaques of woven bone were also apparent on the anterior surface of the ilium immediately superior to the sciatic notch and the anterior surface of the R ala of the sacrum. These plaques of bone may indicate that the injury had become infected, and that the infection was haematogenously spreading to other parts of the R pelvic/abdominal area.

Dentition:

U															U
8	7	6	5	4	3	2	1	1	~~2~~	~~3~~	4	5	6	~~7~~	8
8	7	6	5	4	3	2	1	1	2	3	4	5	6	7	8
U			E		B							E			U

Bite: Overbite.

Attrition: Molar wear patterns of Stage 1 for the 2nd molars and Stage 3 for the 1st molars were apparent.

Calculus: Flecks of calculus were present on the buccal surfaces of 5 4⌋ and 4⌉ ⌈5 and the lingual surfaces of 7 6⌉. Slight deposits were visible on the buccal surfaces of 6⌋ ⌊4 5 6, the lingual surfaces of 4⌉ ⌈7 and the buccal and lingual surfaces of 3 2 1⌉ ⌈1 2 3 4.

Enamel hypoplasia: Prominent hypoplastic grooving was present on 4⌋ ⌊6, and 6 5 4 3⌉ ⌈4 5 6 7, and double grooving was visible on 6 5 3 2 1⌋ ⌊1 4 5 and ⌈3 which indicated that the individual had been subject to at least five incidents of physiological stress at the ages of 2-2.5, 2.5-3, 3-3.5, 3.5-4 and 4-4.5 years.

Dental anomalies: The L maxillary central incisor and the R maxillary incisors were shovel-shaped.

Context XXXI. 2

Young adult 17-25 yrs: Male: 170.8 cm, 5'7"

Year of excavation: 1977

Archaeological context:

Cist - The remains of a circular enclosure and a low cairn were present on the ground surface. A rectangular pit, which was orientated north to south, was situated below these features. At the bottom of the pit was a trapezium shaped stone cist which contained a skeleton.

Body position - The individual was lying in an extended, supine position, with the head to the north. The hands were situated on the L side of the pelvis, and the skull was lying on a stone slab.

Grave goods - A clay vessel positioned on a tray was recovered from behind the skull.

Visual inventory:

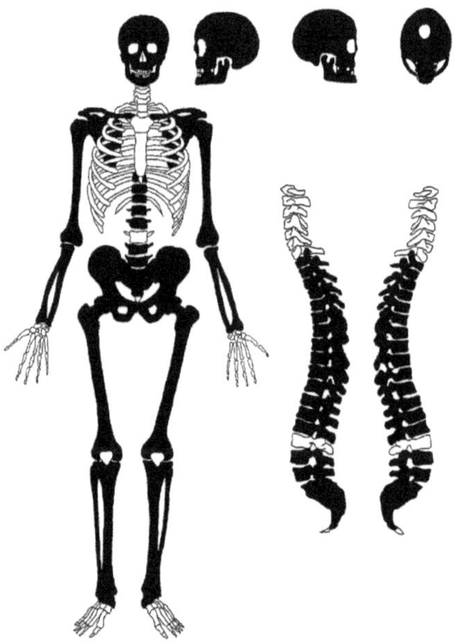

Preservation: The remains were in a very good state of preservation with 70-80% of bones present.

Post-cranial pathology: Early spondylolysis was evident in LV4, where the lamina was partially detached from the pedicle on both sides. Slipped femoral capital epiphysis was present in the L femur. The femoral head was displaced inferiorly and had a flattened morphology with a well defined depression for the ligamentum teres. The joint surface was smooth and intact with no evidence of porosity. The L acetabulum was broader and shallower than normal.

A smooth bony nodule, which measured 32 mm supero-inferiorly by 16 mm antero-posteriorly, was present on the antero-medial aspect of the proximal third of the L tibia. The nodule may be indicative of an ossified haematoma. Vertebral osteophytosis was present in TV6, TV7 and TV9-LV2. Schmorl's nodes were visible in TV6-LV2. Enthesopathies were present on the clavicles (costoclavicular ligaments), the humeri (pectoralis major, latissimus dorsai, teres major), the R tibia (soleus, tibialis anterior) and the L tibia (soleus).

Cranial non-metric traits: R and L highest nuchal lines, 2 L ossicles in lambdoid suture, R and L parietal foramina, metopic suture, R and L ossicles at asterion, R and L mastoid foramina extrasutural, R posterior condylar canal open, R and L double condylar facets, R double anterior condylar canal, R and L open foramina spinosum, R and L bridging of supraorbital notches, R and L accessory supraorbital foramina, R and L anterior ethmoid frontal foramina extrasutural, R and L accessory infraorbital foramina.

Post-cranial non-metric traits: R and L hypotrochanteric fossae, R and L lateral tibial squatting facets, R and L septal apertures, R and L acetabular creases, R and L acromial articular bridges, R and L circumflex sulci, R os trigonum, R and L medial talar facets, R and L double inferior articular talar facets, R and L double anterior calcaneal facets, R and L peroneal tubercles.

Dentition:

~~8~~	7	6	~~5~~	4	~~3~~	~~2~~	~~1~~	~~1~~	~~2~~	~~3~~	~~4~~	~~5~~	6	7	~~8~~
8	7	6	5	~~4~~	3	2	~~1~~	~~1~~	~~2~~	~~3~~	~~4~~	5	6	7	8

Bite: Overbite.

Attrition: Molar wear patterns of Stage 1 for the 2nd and 3rd molars and Stage 3- for the 1st molars were apparent.

Calculus: Flecks of calculus were present on the buccal surfaces of 6⌋ and 3⌉and the buccal and lingual surfaces of 7⌋ ⌊6 7 and 2⌉ ⌈6. Slight deposits were apparent on the lingual surfaces of 7 5 3⌉ ⌈5 and the buccal and distal surfaces of 8⌉.

Context XXXI. 3

Adult 25-35 yrs: Male: 179.4 cm, 6'0"

Plate: 56

Year of excavation: 1977

Archaeological context:

Cist and coffin - The remains of a circular stone enclosure with stone slabs in its central area were present on the ground surface. A trapeziform pit, which was orientated south-south-east to north-north-west, was situated below the centre of the enclosure. A stone cist was situated at the bottom of the pit and contained a wooden board coffin. A skeleton was recovered from the coffin.

Body position - The individual was lying in an extended, supine position, with the head to the north-north-west.

Grave goods - A clay pot positioned on a tray, and partly disturbed bone cases were situated near the skull. An iron arrowpoint was protruding from the second cervical vertebra.

Visual inventory:

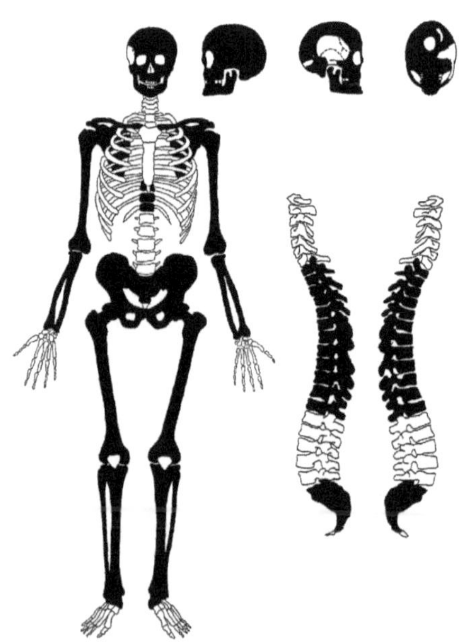

Preservation: The remains were in a very good state of preservation with 70-80% of bones present.

Cranial pathology: A very well healed simple oblique fracture was present at the midpoint of the L nasal bone. There was no malalignment.

Post-cranial pathology: A very well healed simple oblique fracture was present at the distal third of the shaft of the L ulna (Night-stick fracture). No obvious malalignment of the fracture parts was evident, but they did overlap by approximately 5 mm causing the L ulna (GL=279 mm) to be shorter than the R ulna. There was a degree of hypertrophy at the fracture site. An enthesopathy of the interosseous ligament was present on the L radius, which was indicative of strain on the ligament as a result of the fracture. Vertebral osteophytosis was evident in TV5-TV12. Schmorl's nodes were apparent in TV7-TV9, TV11 and TV12. Enthesopathies were apparent on the calcanei (Achilles' tendon).

Weapon trauma: A small perforation, which may have been caused by an arrowhead, was present at the midpoint of the blade of the L scapula. The arrow appeared to have entered the scapula on the posterior surface. The angle of the perforation indicated that the injury had been made from above the individual. A plaque of grey woven bone surrounded the perforation suggesting that the individual lived for a period of time after the injury. It is possible that the iron arrowpoint which was protruding from the second cervical vertebrae was attained in the same incident, or just shortly after the injury which had caused the perforation in the L scapula. Unfortunately, the cervical vertebrae were not available for analysis.

Cranial non-metric traits: R and L highest nuchal lines, 7 L and 6 R ossicles in lambdoid suture, R and L parietal foramina, L ossicle at pterion, R ossicle at asterion, R and L mastoid foramina extrasutural, R posterior condylar canal open, L open foramen spinosum, R and L accessory supraorbital foramina, R and L anterior ethmoid frontal foramina extrasutural, R and L accessory infraorbital foramina.

Post-cranial non-metric traits: R and L medial tibial squatting facets, R and L acetabular creases, R and L acromial articular facets, R and L circumflex sulci, R and L medial talar facets, R and L peroneal tubercles.

Other observations: Orange discoloration (rust) was present on the L sides of TV8 and TV9.

Dentition:

~~8~~	~~7~~	6	5	4	~~3~~	~~2~~	~~1~~	~~1~~	~~2~~	~~3~~	4	5	6	~~7~~	~~8~~
8	7	6	5	4	3	~~2~~	~~1~~	~~1~~	~~2~~	3	4	5	6	7	8

Bite: Overbite.

Attrition: Molar wear patterns of Stage 2+ for the 3rd molars, Stage 3 for the 2nd molars and Stage 3+ for the 1st molars were apparent.

Calculus: Flecks of calculus were present on the lingual surfaces of ⌊6 and 7⌉ ⌈7 8. Slight deposits were apparent on the buccal surface of ⌈3, the lingual surfaces of ⌊5 and 8 6 5⌉ ⌈5 6 and the buccal and lingual surfaces of 4 3⌉ ⌈4.

Periodontal disease: Periodontal disease of slight severity was evident.

Context XXXI. 11. Sk. 1

Adult 35-45 yrs: Female: 153.8 cm, 5'1"

Year of excavation: 1978

Archaeological context:

The archive report for the 1978 excavations was not available for consultation.

Visual inventory:

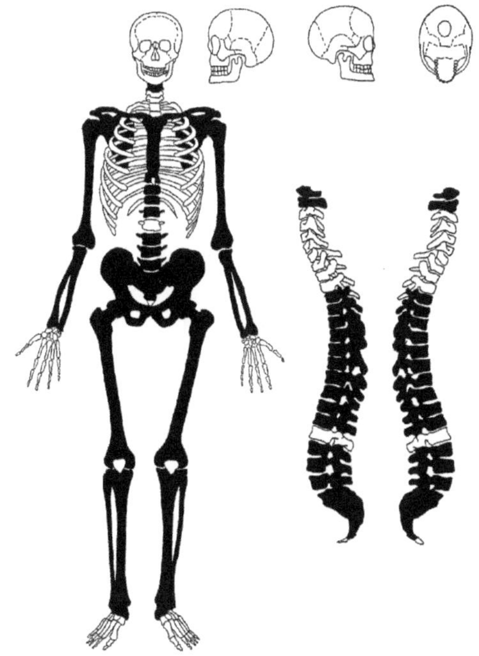

Preservation: The remains were in a good state of preservation with 60-70% of bones present.

Post-cranial pathology: A well healed simple oblique fracture was present on the distal third of the L radius (Colles' fracture). The distal fragment was deflected postero-laterally by approximately forty degrees, and the fracture parts overlapped by 17 mm. There was hypertrophy at the site of the fracture. Secondary marginal osteophytes were present at the distal end of the L radius and the distal end of the L ulna, which may be indicative of dislocation of the inferior radio-ulnar joint. The L styloid process also appeared to have been fractured. Osteophytosis was evident on the acromial ends of the clavicles, the acromions of the scapulae, the humeral heads, the sacro-iliac joints, the distal condyles of the femora and the proximal condyles of the R tibia. Rotator cuff disease was present in the humeri. Vertebral osteophytosis was visible in CV1, CV2, TV5-LV1 and LV3-LV5. A Schmorl's node was present in TV9. Enthesopathies were apparent on the calcanei (Achilles' tendon).

Post-cranial non-metric traits: R and L exostoses in trochanteric fossae, R and L lateral tibial squatting facets, R and L septal apertures, R and L acromial articular facets, R and L circumflex sulci, R and L medial talar facets, R and L double inferior articular talar facets, R and L peroneal tubercles.

Other observations: A small cutmark was present on the superior aspect of the L humeral head, which may be indicative of post-mortem defleshing/disarticulation.

Context XXXI. 13

Adult 35-45 yrs: Male: 170.5 cm, 5'7"

Year of excavation: 1979

Archaeological context:

Ground pit + coffin - A circular stone enclosure was present on the ground surface, inside which was a square pit orientated west to east. A rectangular board coffin, which contained a single skeleton, was situated at the bottom of the pit.

Body position - The individual was lying in an extended, supine position, with the head to the west. The L hand was situated on the L side of the pelvis, and the skull was disturbed.

Grave goods - A cauldron-shaped vessel was positioned near the southwestern corner, and an iron javelin point was located near the northwestern corner. An iron disc-shaped object was recovered near the R humerus, and an iron bar was situated beneath the lower vertebrae and the R side of the pelvis. Two iron knives were positioned between the wall of the coffin and the R arm. An iron oval buckle was situated on the L side of the pelvis. Four poorly preserved iron arrowpoints and a small hook were recovered from beneath the buckle, and the remains of an iron mouthpiece were positioned at the lower edge of the buckle. The remains of an iron oval buckle and an iron knife were situated at the midpoint of the R femur. Sheep bones were recovered from beneath the buckle and the knife.

Visual inventory:

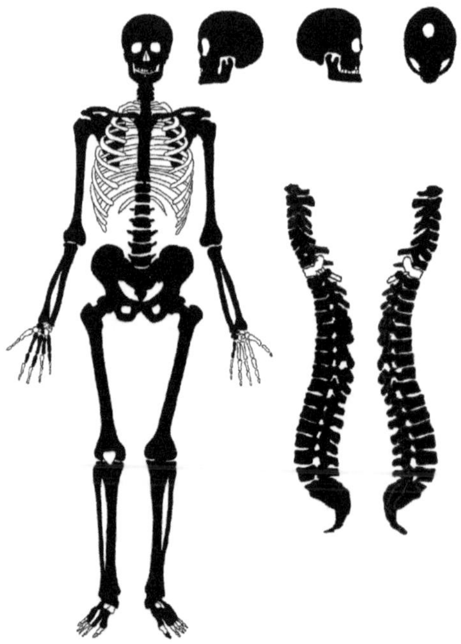

Preservation: The remains were in a very good state of preservation with 80-90% of bones present.

Cranial pathology: Porosity was present on the mandibular condyles and the articular tubercles of the temporals, which may be indicative of the early stages of DJD of the temporomandibular joints. Marginal osteophytes were apparent on the occipital condyles.

Post-cranial pathology: The xiphisternal joint was ankylosed, which may be related to ageing or alternatively could have been due to a developmental defect in which the fibrous lamina separating the xiphoid process from the sternum failed to develop normally. CV2 and CV3 had formed a block vertebrae (Type II Klippel-Feil syndrome) which is a developmental defect due to failure of segmentation of the vertebral sections.

Osteophytosis was evident on the costal and acromial ends of the L clavicle, the scapulae, the proximal ulnae, the distal radii, the acetabulae, the distal femora, the R patella, the proximal tibiae and the distal R MT1 and MT2. Extensive DJD was apparent in the vertebrae, with osteophytosis visible in CV1-TV1 and TV3-S1. TV12 was wedge-shaped, which had resulted in anterior kyphosis of the vertebral column by approximately forty degrees. Enthesopathies were visible on the clavicles (costoclavicular ligament), the humeri (supraspinatus, pectoralis major), the radii (biceps), the femora (gluteus medius, psoas major, iliacus, gluteus maximus, adductor magnus), the tibiae (soleus) and the calcanei (Achilles' tendon).

Cranial non-metric traits: R and L highest nuchal lines, R and L parietal foramina, R and L ossicles at asterion, R and L mastoid foramina extrasutural, R posterior condylar canal open, L double condylar facet, R double anterior condylar canal, R and L accessory supraorbital foramina, R and L anterior ethmoid frontal foramina extrasutural, R and L accessory infraorbital foramina.

Post-cranial non-metric traits: R plaque, R and L hypotrochanteric fossae, L lateral tibial squatting facet, R and L acromial articular facets, R and L circumflex sulci, L vastus notch, R and L medial talar facets, L double atlas facet.

Dentition:

		A PU	PU						PU		PU	PU	A PU		
8	7	6	5	~~4~~	~~3~~	~~2~~	~~1~~	~~1~~	2	~~3~~	4	5	6	7	8
8	7	6	5	4	3	2	~~1~~	1	2	3	4	5	6	7	8
		PU C	PU	PU	PU	PU		PU	PU	PU	PU	PU	PU		

Bite: Overbite.

Attrition: Molar wear patterns of Stage 3+ for the 3rd molars, Stages 4-4+ for the 2nd molars and Stage 5 for the 1st molars were apparent. Secondary dentine or pulp cavity exposure was evident in 6 5⌋ ⌊2 4 5 6 and 6 5 4 3 2⌉ ⌈1 2 3 4 5 6.

Calculus: Flecks of calculus were present on the buccal surfaces of 6 5⌋, the lingual surfaces of ⌊7 8 and 6⌉ ⌈4 5 6, the buccal and lingual surfaces of 4 3 2⌉ ⌈7 and the buccal, lingual and distal surfaces of ⌈8. Slight deposits were visible on the buccal surface of 7⌋, the lingual surfaces of 8 7⌉, the buccal and lingual surfaces of ⌈1

2 3 and the buccal, lingual and distal surfaces of 8⌋.

Periodontal disease: Periodontal disease of medium severity was apparent.

Caries: A small caries was present in the distal surface of 6⌉.

Abscesses: Medium sized externally draining abscesses were associated with 6⌋ and ⌊6.

Context XXXI. 15

Adult 35-45 yrs: Female: 155.2 cm, 5'2"

Year of excavation: 1979

Archaeological context:

Cist - A compact cairn orientated south-east to north-west was apparent on the ground surface. A rectangular pit which contained a stone cist was present beneath the cairn. The remains of thin, horizontally positioned logs and a skeleton were recovered from the cist.

Body position - The individual was lying in an extended, supine position, with the head to the north-west. The R arm was flexed so that the hand was lying on the central area of the vertebral column. The L arm was extended and the L leg was disturbed.

Visual inventory:

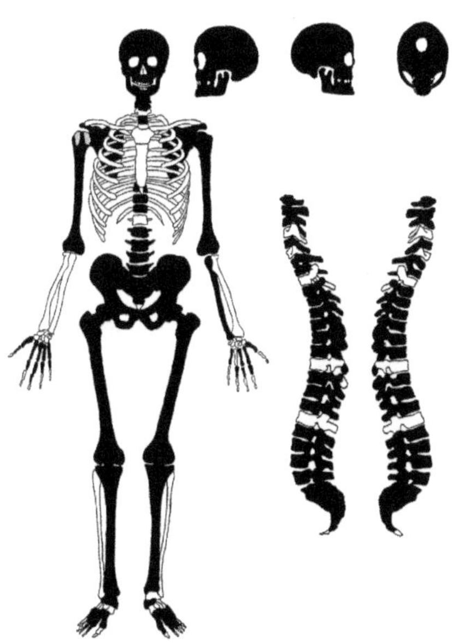

Preservation: The remains were in a good state of preservation with 60-70% of bones present. Surface erosion was apparent on several bones.

Post-cranial pathology: Sacralization of LV5 was apparent with the R transverse process completely fused to the R ala, and the L transverse process articulating with the L ala. A well healed simple oblique fracture was present at the distal third of the shaft of the L ulna (Night-stick fracture). Approximately 80% apposition of the fracture parts was evident, and the inferior part of the fracture was displaced postero-laterally by approximately forty degrees. Hypertrophy was evident at the point of fracture, with the fracture parts overlapping by approximately 24.3 mm.

Extensive DJD was apparent on the R knee with porosity and eburnation of the proximal medial condyle. In addition both condyles had a flattened appearance. The proximal end of the R tibia appeared to have been rotated by approximately forty-five degrees, so that the lateral condyle was positioned more posteriorly, and the medial condyle more anteriorly than normal. Large osteophytes were evident on both tibial condyles, and eburnation and porosity were also present on the medial condyle. It is possible that the R knee had been subject to an incident of trauma in which the joint had become malpositioned, resulting in extensive secondary DJD. Osteophytosis was evident on the glenoid fossae of the scapulae and the humeral heads. Rotator cuff disease was present in the humeri. DJD was evident on the facet for the dens of the axis on the atlas and the dens of the axis. Osteophytosis was apparent in CV3, CV5, CV7, TV1, TV4-TV8, TV10-TV12 and LV2-LV5. Enthesopathies were apparent on the humeri (pectoralis major), the femora (gluteus medius, gluteus maximus), the patellae (patellar ligament) and the calcanei (Achilles' tendon).

Cranial non-metric traits: R and L highest nuchal lines, 1 L and 2 R ossicles in lambdoid suture, R and L parietal foramina, L ossicle at pterion, R ossicle at parietal notch, R ossicle at asterion, R and L mastoid foramina extrasutural, R and L posterior condylar canals open, R double anterior condylar canal, R and L anterior ethmoid frontal foramina extrasutural, R and L accessory infraorbital foramina.

Post-cranial non-metric traits: R and L medial tibial squatting facets, R and L lateral tibial squatting facets, R and L accessory sacral facets, R and L acromial articular facets, R and L circumflex sulci, R medial talar facet, R and L CV5, and L CV7 transverse foramina bipartite.

Dentition:

C	2A	A PU		A								PU	A PU		C
8	~~7~~	6	~~5~~	~~4~~	~~3~~	~~2~~	~~1~~	~~1~~	~~2~~	~~3~~	~~4~~	5	6	~~7~~	8
8	7	6	5	4	~~3~~	~~2~~	~~1~~	~~1~~	~~2~~	3	4	5	6	7	8
		PU											PU		

Bite: Overbite.

Attrition: Extensive attrition was apparent on most teeth, with molar wear patterns of Stage 4 for the 2nd and 3rd molars and Stage 5 for the 1st molars. Secondary dentine or pulp cavity exposure was evident in 6⌋ ⌊5 6 and 6⌉ ⌈6.

Calculus: Flecks of calculus were present on the lingual surface of ⌈6. Slight deposits were apparent on the buccal surfaces of 7 6 5⌉, the lingual surfaces of 4⌉ ⌈3 4, the buccal and lingual surfaces of ⌈5 7 and the buccal, lingual and distal surfaces of ⌈8. Medium deposits were evident on the lingual surfaces of 7 6 5⌉, the buccal and lingual surfaces of ⌊8 and 8⌉ and the buccal, lingual and distal surfaces of 8⌋.

Periodontal disease: Periodontal disease of medium severity was apparent.

Caries: Small caries were present in the mesial surfaces of 8⌋ and ⌊8.

Abscesses: Medium sized externally draining abscesses were associated with 4⌋ and ⌊6, while a large externally draining abscess was associated with 6⌋. Large externally draining and internally draining abscesses were associated with 7⌋. The internal abscess was draining directly into the maxillary sinus.

Context XXXI. 20

Adult 35-45 yrs: Female: 149.9 cm, 5'0"

Year of excavation: 1979

Archaeological context:

Ground pit and stone lined room - The possible remains of a circular stone enclosure and a low cairn were apparent on the ground surface. A rectangular pit orientated south-east to north-west was situated beneath the fence. Steps, horizontal slabs and burnt logs were discovered within the pit. A rectangle (1.9 m by 0.7 m) in the central area of the pit had been deepened and lined with large slabs to form a chamber. The floor of the room had been covered with a reed mat, and a skeleton was situated in the chamber.

Body position - The individual was lying in an extended, supine position, with the head to the north-west. The R arm was extended alongside the body, and the L arm was flexed.

Grave goods - A clay pot was situated in the northeastern corner, near which was a stone spindle whorl and the remains of two iron knives. The remains of a flat wooden plate and a birch casket were recovered near the southwestern wall opposite the skull. The casket contained an unidentifiable iron object. A second birch casket was positioned adjacent to the R lower arm, and a small iron buckle with a moveable lug was recovered from near the midpoint of the L femur.

Visual inventory:

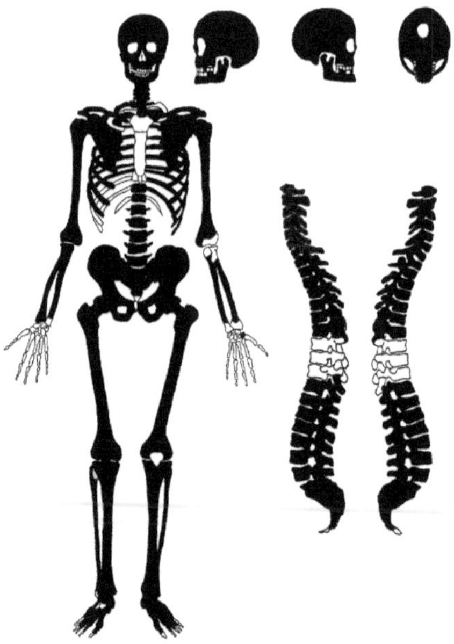

Preservation: The remains were in a very good state of preservation with 70-80% of bones present. Some bones were fragmentary with surface erosion.

Cranial pathology: Cribra orbitalia of Type 4 was present in Areas 1, 2 and 3 of the R orbit. In the L orbit, however, woven bone was apparent. It is possible that the orbital lesions are related to the post-cranial lesions described below. Porosity was present on the L articular tubercle of the temporal, which may be indicative of the early stages of DJD of the L temporomandibular joint.

Post-cranial pathology: Spicules of woven bone were present just medial to the necks of the anterior surfaces, and in the supraspinous fossae of the posterior surfaces of the scapulae. Spicules of woven bone were also apparent on the anterior surfaces of the ilia just superior to the acetabulae, the anterior surfaces of S1 and S2, and the antero-medial surface of the proximal third of the shaft of the L femur. A plaque of woven bone was present on the visceral surface of the neck of the R 4th rib, and the lateral surface of the midshaft of the R humerus. The ribs and the anterior aspects of the vertebral bodies had a generally porous appearance. Areas of small, diffuse lytic lesions were apparent on the medial margins of the scapulae, the posterior surfaces of the ilia, the pubes, and the anterior surface of the sacrum. The L acetabular notch had a trabecular, porous appearance. It was difficult to determine a precise aetiology for these lesions, since they are characteristic of several diseases. It is possible they are indicative of a generalised systemic infection, or a condition which caused haemorrhaging, such as scurvy. A metastatic neoplastic condition must also be considered as a possible aetiology.

A possible fracture was apparent at the midshaft of the L 2nd rib which displayed callus formation. The lesion was, however, obscured by post-mortem damage. Osteophytosis was evident on the majority of rib heads. Extensive rotator cuff disease was apparent in the humeri. Vertebral osteophytosis was present in CV1-CV4, TV1-TV6 and TV12-LV5. Schmorl's nodes were apparent in LV1-LV3. Enthesopathies were visible in the humeri (latissimus dorsai, pectoralis major, teres major, deltoid), the R radius (biceps), the R patella (patellar ligament) and the calcanei (Achilles' tendon).

Cranial non-metric traits: R and L highest nuchal lines, R and L parietal foramina, R ossicle at asterion, R and L mastoid foramina extrasutural, L accessory supraorbital foramen, R and L anterior ethmoid frontal foramina extrasutural, R and L accessory infraorbital foramina.

Post-cranial non-metric traits: R and L exostoses in trochanteric fossae, R lateral tibial squatting facet, R septal aperture, R and L accessory sacral facets, R and L acromial articular facets, R bridging at suprascapular notch, R and L circumflex sulci, R vastus notch, R and L medial talar facets, R and L peroneal tubercles.

Dentition:

NP 8	7̸	6̸	5̸	4	3̸	2̸	1̸	1̸	2̸	3̸	4̸	5̸	6	2A 7̸	NP 8
8̸	7	6	5̸	4	3̸	2̸	1̸	1̸	2̸	3̸	4̸	5̸	6	7	8

Attrition: Molar attrition was light for an individual of 35-45 years, with molar wear patterns of Stage 3+ for the 2nd and 3rd molars and Stages 3+-4 for the 1st molars.

Calculus: Flecks of calculus were apparent on the buccal surfaces of 7⏋ ⎡7, the buccal and lingual surfaces of 4⏌ and 6⏋ and the buccal, lingual and distal surfaces of ⎡8. Slight deposits were present on the lingual surface of ⎡7, while medium deposits were apparent on the buccal and lingual surfaces of 4⏋.

Periodontal disease: Periodontal disease of medium severity was evident.

Abscesses: Both a large externally draining abscess and a large internally draining abscess were associated with ⎣7.

Ante-mortem tooth loss: 7⏌ and 8⏋ had been lost ante-mortem.

Dental anomalies: Both maxillary 3rd molars were genetically absent.

Context XXXI. 24a

Mature adult 45+ yrs: Male: 172.1 cm, 5'7"

Year of excavation: 1979

Archaeological context:

Ground pit and wooden coffin - A cairn and a poorly preserved circular stone enclosure were apparent on the ground surface. A rectangular pit orientated south-east to north-west, which contained a rectangular board coffin, was present inside the enclosure. The burial had been partly destroyed by a stone cist grave of a child (Skeleton XXXI. 24b). A skeleton was recovered from the interior of the coffin.

Body position - The individual was lying in an extended, supine position, with the head to the north-west.

Grave goods - A poorly preserved clay pot was situated between the R side of the sternum and the R side of the pelvis.

Visual inventory:

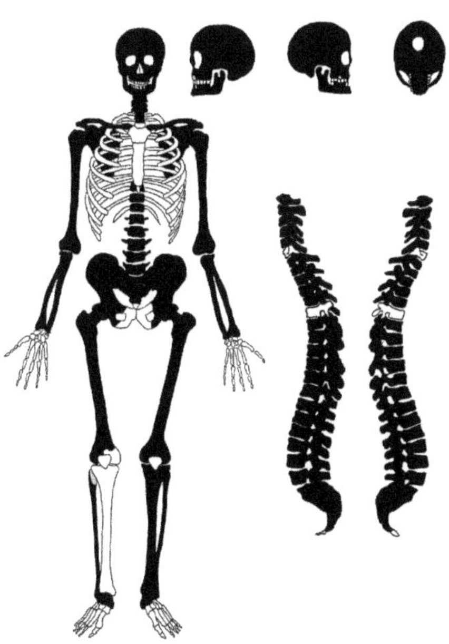

Preservation: The remains were in a good state of preservation with 60-70% of bones present. Many of the bones were fragmentary with surface erosion.

Post-cranial pathology: Osteophytosis was apparent on the acromial ends of the clavicles, the glenoid fossae and the acromial articular facets of the scapulae, the L acetabulum, the proximal and distal femora and the proximal L tibia. Vertebral osteophytosis was evident in CV1-CV5, CV7-TV4 and TV6-LV5. The vertebral body for TV7 was partially compressed, while that for LV1 was completely compressed and had formed a wedge-shaped vertebra. Consequently, there would have been anterior kyphosis of the vertebral column by approximately forty-five degrees. Schmorl's nodes were present in TV7-TV9, TV11-LV1 and LV3.

Enthesopathies were visible on the clavicles (costoclavicular ligament), the radii (interosseous ligament, biceps), the innominates (obturator externus, quadratus lumborum), the femora (gluteus medius, gluteus maximus, psoas major, iliacus, adductor magnus, adductor longus, adductor brevis), the L tibia (patellar ligament), the L fibula (interosseous ligament) and the calcanei (Achilles' tendon).

Cranial non-metric traits: R and L highest nuchal lines, L parietal foramen, R ossicle at parietal notch, R and L mastoid foramina extrasutural, R and L posterior condylar canals open, R and L accessory supraorbital foramina, R and L anterior ethmoid frontal foramina extrasutural, R and L accessory infraorbital foramina.

Post-cranial non-metric traits: L plaque, R hypotrochanteric fossa, L exostosis in trochanteric fossa, R and L acromial articular facets, R and L circumflex sulci, R and L medial talar facets, R and L double inferior articular talar facets, L double anterior calcaneal facet, L CV1, and L CV3 transverse foramina bipartite.

Dentition:

NP		A								A	PU				
8	~~7~~	~~6~~	~~5~~	~~4~~	~~3~~	~~2~~	~~1~~	~~1~~	~~2~~	~~3~~	4	~~5~~	~~6~~	~~7~~	~~8~~
8	7	6	5	~~4~~	~~3~~	~~2~~	~~1~~	~~1~~	~~2~~	3	~~4~~	5	6	7	8
NP	C	PU A	PU A							PU			PU		

Bite: Overbite.

Attrition: Extensive attrition was apparent on the majority of teeth, with molar wear patterns of Stage 4+ for the 2nd molars and Stage 5 for the 1st molars. Secondary dentine or pulp cavity exposure was evident in ⎣4 and 6 5⏋ ⎡3 6. Molar wear patterns of only Stage 2+ was present on ⎡8, which may be related to the ante-mortem loss of the corresponding mandibular molar.

Calculus: Flecks of calculus were present on the lingual surfaces of 6 5⏋ ⎡3 6 and the buccal and lingual surfaces of ⎡5 7 8. Slight deposits were apparent on the distal and lingual surfaces of 7⏋.

Periodontal disease: Periodontal disease of medium severity was evident.

Caries: A medium sized caries was apparent in the occlusal surface of 7⌉.

Abscesses: A small externally draining abscess was associated with 6⌉, while medium sized externally draining abscesses were associated with 5⌉ and 6⌋, and a large externally draining abscess was associated with ⌊3.

Ante-mortem tooth loss: 7⌋ ⌊7 8 had been lost ante-mortem.

Dental anomalies: The R maxillary and the R mandibular 3rd molars were genetically absent.

Context XXXI. 28. Sk. 2

Mature adult 45+ yrs: Male: 155.2 cm, 5'2"

Year of excavation: 1979

Archaeological context:

Pit and wooden coffin - The remains of a circular slab enclosure were apparent on the ground surface. A rectangular pit orientated south-south-east to north-north-west, which contained a log house-shaped coffin, was contained within the enclosure. An arrow with an iron arrowpoint was situated on the lid of the coffin, which contained the remains of an adult and a child.

Body position - The eastern skeleton was that of an adult who was lying on his R side, with the head to the north-north-west. The R arm was extended, and the L arm was flexed so that the hand was positioned on the R side of the pelvis. It was stated in the archive report that the skull of the individual was not present.

Grave goods - A clay pot was situated at the northwestern wall of the coffin. A fragment of an iron ring was recovered near the R scapula, and the remains of an iron object were situated medially to the R ulna. A fragment of fabric was positioned above the L side of the pelvis, and a fragment of an iron buckle was situated just inferior to the fabric. Several wooden sticks were positioned alongside the lower L arm. The remains of an iron ring was located near the superior aspect of the R side of the pelvis, and a fragment of an iron ring was also situated on the L side of the pelvis. The remains of a quiver made of thin leather, with a disc-shaped base made from birch, was situated along the southern part of the northwestern wall. The quiver contained several arrows with extensively oxidised iron points. One of the arrows had been entirely manufactured from wood. An iron hook was located beneath the upper part of the quiver, and the remains of the central part of a wooden bow were positioned on the hook. A fragment of a wooden object was situated between the R carpals and the R femur, and the remains of an iron knife were recovered from beneath the L patella. A fragment of an iron object with a ring-shaped top was situated on top of the L patella. The remains of an iron object with traces of oxidised iron on its surface were positioned near the R foot. The remains of fabric were located under both feet, and sheep vertebrae were situated on top of the feet of the individual.

Visual inventory:

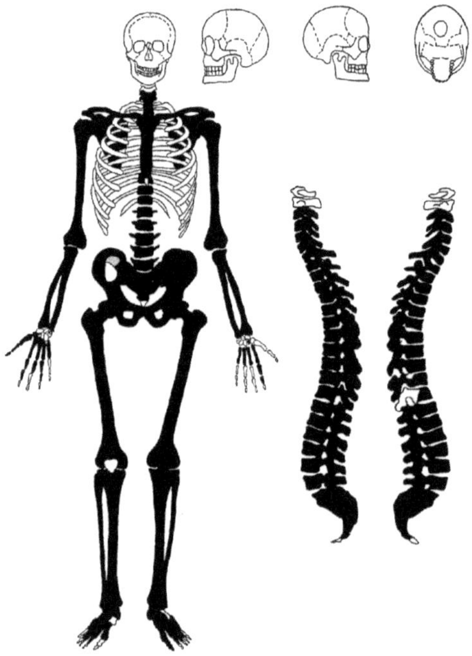

Preservation: The remains were in a very good state of preservation with 70-80% of bones present. Several bones displayed surface erosion.

Post-cranial pathology: Osteophytosis was apparent on the acromial and sternal ends of the clavicles, the costal facets of the manubrium, the glenoid fossae and acromial ends of the scapulae, the R humeral head, the distal L radius, the distal R ulna, the R MC1, the R proximal PH1, the R distal PH1, the acetabulae, the proximal femora, the L patella, the proximal tibiae, the naviculars and the L 1st cuneiform. DJD was present on the distal L ulna and the L ulnar notch of the radius. Rotator cuff disease was visible in the humeri. The L auricular surface of the ilium and the sacrum were ankylosed at the sacro-iliac joint. The ankylosis was limited to the superior aspect of the joint. Vertebral osteophytosis was apparent in CV3-LV5. The vertebral bodies of TV12 and LV1 had been compressed to form wedge-shaped vertebrae. In addition, the body of TV8 was partially compressed. Consequently, there would have been anterior kyphosis of the vertebral column by approximately forty-five degrees at the thoraco-lumbar border.

Enthesopathies were present on the clavicles (costoclavicular ligaments), the humeri (pectoralis major, latissimus dorsai, deltoid), the radii (biceps), the femora (gluteus medius, gluteus maximus, psoas major, iliacus, adductor magnus, adductor longus, adductor brevis), the tibiae (soleus), and the calcanei (Achilles' tendon).

Decapitation: It is stated in the archive that the skull was not present in the tomb. In addition the first and second cervical vertebrae were not present. It is feasible, therefore, that the individual was decapitated prior to burial.

Post-cranial non-metric traits: R and L plaque, R and L third trochanters, R and L lateral tibial squatting facets, R accessory sacral facet, R and L acromial articular facets, R and L circumflex sulci, R vastus notch, R and L medial talar facets, R and L peroneal tubercles, L CV5, R and L CV6, and R CV7 transverse foramina bipartite.

Context XXXI. 30. Sk. 2

Adult 25-35 yrs: Male: 170.8 cm, 5'7"

Plate: 9

Year of excavation: 1979

Archaeological context:

Ground pit, cist and possible wooden coffin - A large cairn and the remains of a circular enclosure were apparent on the ground surface. An oval pit orientated north to south was situated partly inside the circular enclosure and partly underneath the cairn. Layers of slabs and a stone cist, which was lined with the possible remains of coffin boards, were discovered inside the pit. The cist contained two skeletons. Two sheep skulls and leg bones were situated between the feet of the two individuals. Skeleton 1 was situated in the western area of the coffin, while Skeleton 2 was positioned in the eastern part of the coffin.

Body position - The individual was lying in an extended, supine position, with the head to the north. The arms were slightly flexed with the hands positioned at the inferior aspect of the pelvis.

Grave goods - The remains of a clay pot were positioned to the north of the skull, while the remains of a wooden plate were situated to the right of the skull. Two iron knives were recovered from beside the lumbar vertebrae, and two unidentifiable iron objects were situated at the midshaft of the L femur.

Visual inventory:

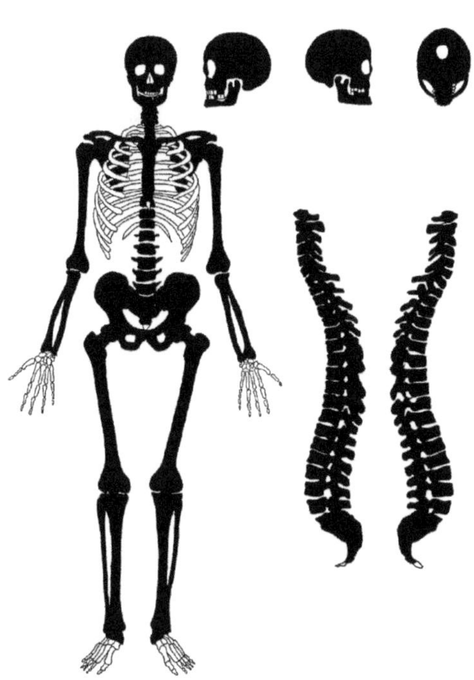

Preservation: The remains were in a very good state of preservation with 70-80% of bones present.

Post-cranial pathology: Sacralization was apparent with LV5 completely fused to the sacrum. A simple oblique fracture was present at the proximal third of the shaft of the R MC5. The fracture was in the early stages of healing and large amounts of callus were present at the fracture site. Hypertrophy was evident at the point of fracture, but there was 100% apposition of the fracture parts and no obvious malalignment was apparent. Vertebral osteophytosis was present in CV1, CV2, CV5, CV6 and TV2-LV5. A Schmorl's node was apparent in LV1. Enthesopathies were visible on the clavicles (costoclavicular ligaments), the radii (biceps), the fibulae (flexor hallucis longus), the patellae (patellar ligament) and the calcanei (Achilles' tendon).

Cranial non-metric traits: R and L highest nuchal lines, R and L parietal foramina, R and L ossicles at pterion, R ossicle at asterion, R and L auditory tori, R and L mastoid foramina extrasutural, R posterior condylar canal open, L open foramen spinosum, L bridging of supraorbital notch, L accessory supraorbital foramen, R and L anterior ethmoid frontal foramina extrasutural, R and L accessory infraorbital foramina.

Post-cranial non-metric traits: R and L plaque, R and L hypotrochanteric fossae, R and L exostoses in trochanteric fossae, R and L accessory sacral facets, R and L acromial articular facets, R and L circumflex sulci, R and L vastus notches, R and L medial talar facets, R and L peroneal tubercles, R posterior atlas bridging, R and L CV1, and R CV6 transverse foramina bipartite.

Dentition:

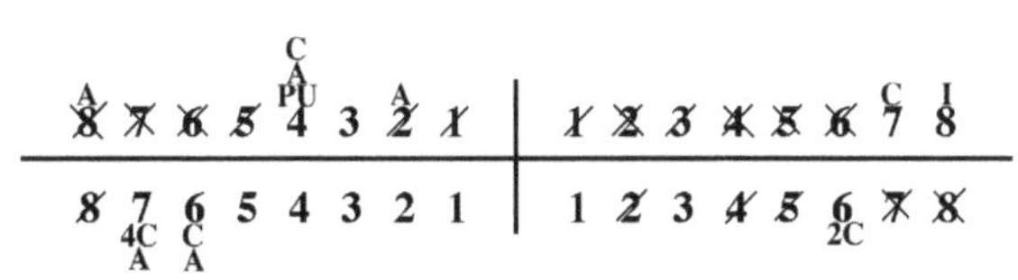

Attrition: Molar wear patterns of Stage 3+ for the 2nd molars and Stage 4 for the 1st molars were apparent. Secondary dentine was exposed in 4⌋.

Calculus: Flecks of calculus were present on the buccal surface of 3⌋, the lingual surfaces of ⌊7 and 6 4⌉ and the buccal and lingual surfaces of 3 1⌉. Slight deposits were visible on the lingual surface of 5⌉ and the buccal and lingual surfaces of 2⌉.

Periodontal disease: Periodontal disease of considerable severity was apparent.

Caries: Small caries were present in the occlusal surface of ⌊7 and the mesial surface of 4⌋. A small caries and a medium sized caries were present in the occlusal surface of ⌈6. Two small occlusal caries, a medium sized distal caries, and a large buccal caries were present in 7⌉. A large caries was evident in the buccal surface of 6⌉.

Abscesses: Small externally draining abscesses were associated with 4⌋ and 6⌉. Medium sized externally draining abscesses were associated with 2⌋ and 7⌉, and a large externally draining abscess was associated with 8⌋.

Ante-mortem tooth loss: 8 7 6⌋ ⌊2 4 5 6 and⌈7 8 had been lost ante-mortem. All sockets were in an advanced state of healing.

Enamel hypoplasia: Hypoplastic grooves were apparent on 4 3⌉ ⌈3 which indicated that the individual had been subject to at least one incident of physiological stress at the age of 3-3.5 years.

Dental anomalies: The L maxillary 3rd molar was impacted.

Context XXXI. 31b

Young adult 17-25 yrs: Male: 167.2 cm, 5'6"

Year of excavation: 1979

Anthropological number: 6863-245 (skeleton)

Archaeological context:

Pit - The remains of a cairn was apparent on the ground surface, beneath which was situated a large trapeziform-shaped pit. The remains of two burials were recovered from deep within the pit. A near-oval shaped pit orientated south-south-east to north-north-west, which contained the remains of a child, had cut the western edge of Burial XXXI. 31a.

Body position - The individual was lying on his back in an extended position, with the head to the north-north-west. The remains of the skull and some disturbed bones were present.

Grave goods - The remains of an iron ring-shaped buckle and an iron disc-shaped object were lying adjacent to the R side of the pelvis.

Visual inventory:

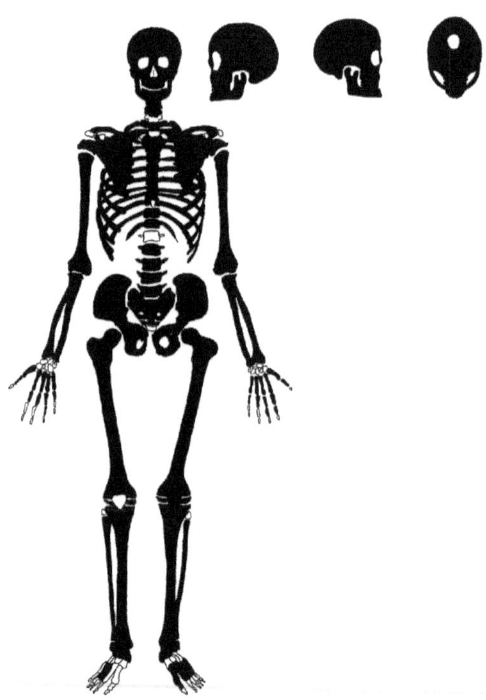

Preservation: The remains were in an excellent state of preservation with 90-100% of bones present.

Post-cranial pathology: The neural arches of S3 and S4 were cleft, and the neural arch of S2 was bifurcated, probably as a result of developmental delay during the early morphogenesis of the spine. A bilateral ossification centre was apparent in the second sternebra, probably due to delay in caudal fusion. The remainder of the sternebrae had a normal morphology.

Erosive lesions were apparent on the distal articular surfaces of all metatarsals of the L foot. The articular surfaces of MT3, MT4 and MT5 had been completely destroyed. The lesions were both marginal and central. A small area of eburnation was present on the distal articular surface of MT1. The proximal ends of the metatarsals were unaffected. The proximal and distal ends of the L proximal 1st phalanx also displayed lytic activity. Plaques of periosteal reactive woven bone were apparent on the lateral surface of the shaft of MT3 and the medial surface of the shaft of MT4. The bones of the R foot and the hands were normal.

Erosive lesions were also apparent on the superior and inferior surfaces of the vertebral bodies of TV7, TV9, TV10, TV11 and TV12. The lesions did not form deep cavities and only affected the exterior surfaces of the superior and inferior bodies. In several vertebrae lytic activity had also eroded the sides of the vertebrae. TV11 had been compressed to form a very distinct wedge-shaped vertebra. Marginal osteophytes were extending from the superior and inferior surfaces of the body at its anterior surface. The wedge-shape of TV11 would have resulted in anterior kyphosis of the spine by approximately forty-five degrees. Inflammatory pitting and slight lytic destruction were apparent on the anterior surfaces of S1. Inflammatory pitting and woven bone formation were visible on the anterior surfaces of S2 and S3.

It is possible that the individual was suffering from Reiter's disease, although other seronegative arthropathies cannot be excluded as feasible aetiologies. The age and sex of the individual are compatible with this diagnosis. In addition, the occurrence of asymmetrical erosive lesions in the L foot, lytic activity in the TV, the normal TV8 (skip vertebra) and periosteal reactive woven bone on the metatarsals all correspond to the characteristic skeletal changes of this disease (after Rogers and Waldron 1995, 73-77). The inflammatory pitting and woven bone on the sacrum could be indicative of a gastrointestinal infection, another feature of Reiter's disease.

Dentition:

~~8~~	7	6	5	4	3	~~2~~	~~1~~	~~1~~	2	~~3~~	4	5	6	7	~~8~~
8	7	6	5	4	3	~~2~~	~~1~~	~~1~~	~~2~~	3	~~4~~	5	6	7	8

Bite: Overbite.

Attrition: Molar wear patterns of Stage 1 for the 3rd molars, Stages 1-2 for the 2nd molars and Stage 3- for the 1st molars were apparent.

Calculus: Flecks of calculus were apparent on the buccal surfaces of 7 6 5 4 3⌋ ⌊6, the lingual surfaces of 7 6⌉ ⌈6, the distal surface of 8⌉ and the buccal and lingual surfaces of ⌊4 5 7 and 5⌉. Slight deposits were present on the buccal surfaces of 4 3⌉, the lingual surface of ⌊2, the buccal and lingual surfaces of ⌈3 and the lingual and distal surfaces of ⌈8.

Periodontal disease: Periodontal disease of slight severity was apparent.

Enamel hypoplasia: Hypoplastic grooves were present on 7 5 4 3⌋ ⌊4 5 7 and 5 4 3⌉ which indicated that the individual had been subject to at least three physiological insults at the ages of 3-3.5, 4-4.5 and 4.5-5 years.

Dental anomalies: The L maxillary lateral incisor was shovel-shaped.

Context XXXI. 40

Adult 25-35 yrs: Female: 156.7 cm, 5'2"

Plate: 24

Year of excavation: 1979

Archaeological context:

Ground pit and wooden coffin - The remains of a poorly preserved stone enclosure were apparent on the ground surface. A rectangular pit orientated south-east to north-west, which contained a wooden board coffin, was situated within the enclosure. Stone slabs were positioned on each end of the coffin lid. The coffin contained a single skeleton.

Body position - The individual was lying in an extended, supine position, with the head to the north-west. Both arms were slightly flexed so that the hands were positioned on the superior aspect of the pelvis. The skull was disturbed.

Grave goods - A clay pot was situated in close proximity to the northwestern corner of the coffin.

Visual inventory:

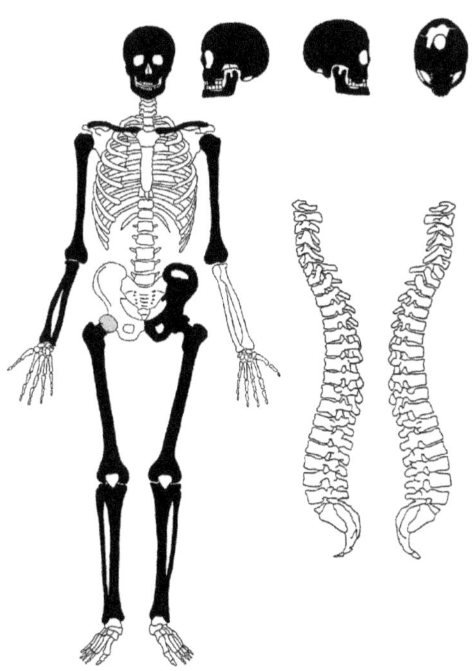

Preservation: The remains were in a fair state of preservation with 40-50% of bones present. Extensive surface erosion was present on the majority of bones.

Cranial pathology: A simple transverse fracture was present at the midpoint of the R nasal bone. The fracture was poorly healed, and the parts of the fracture were ununited. There was no malalignment. A simple transverse fracture was also present at the medial aspect of the L nasal bone. A possible depressed fracture was also apparent at the inferior margin of the R orbit, immediately adjacent to the nasal cavity. It was difficult to assess the situation, but a fracture line appeared to have extended medio-laterally from just inferior to the orbit for approximately 9 mm. The line then extended supero-inferiorly for a further 14.5 mm. It is probable that all three fractures had been obtained in a single incident, and that they had been caused by a direct blow to the medial aspect of the R side of the face. An ivory osteoma, which measured 3.6 mm by 3.6 mm, was present at the midpoint of the L parietal.

Cranial non-metric traits: R and L highest nuchal lines, ossicle at lambda, 3 L and 4 R ossicles in lambdoid suture, R and L parietal foramina, R and L ossicles at pterion, ossicle at asterion, R mastoid foramen extrasutural, L mastoid foramen absent, R posterior condylar canal open, R double anterior condylar canal, R and L open foramina spinosum, R and L accessory supraorbital foramina, R and L anterior ethmoid frontal foramina extrasutural, R and L accessory infraorbital foramina.

Post-cranial non-metric traits: L plaque, R and L exostoses in trochanteric fossae, R and L lateral tibial squatting facets, L accessory sacral facet.

Other observations: The superior aspect of the occipital had a flattened appearance, which may be indicative of the lambdoid form of cranial deformation. Venous-arterial remains were preserved within the cranium.

Dentition:

NP 8	7	C 6	5	4̸	3̸	2̸	1̸	1̸	2̸	3̸	4̸	C 5	Ĉ 6	7̸	B 8
8̸	~~7~~	~~6~~	5̸	4̸	3̸	2̸	1̸	1̸	2̸	3	4̸	5̸	~~6~~	~~7~~	~~8~~

Bite: Overbite.

Attrition: Molar wear patterns of Stage 3- for the 2nd molars and Stage 4 for the 1st molars were apparent.

Calculus: Flecks of calculus were present on the buccal and lingual surfaces of ⌊5, and slight deposits were visible on the buccal surfaces of 7 6⌋ ⌊6.

Periodontal disease: Periodontal disease of considerable severity was evident.

Caries: A small caries was present in the occlusal surface of ⌊5, a medium caries in the occlusal surface of 6⌋ and a large caries in the occlusal surface of ⌊6.

Abscesses: A medium sized externally draining abscess was associated with⌊6.

Ante-mortem tooth loss: 7 6⌉ ⌈6 7 8 had been lost ante-mortem.

Dental anomalies: The R maxillary 3rd molar was genetically absent.

Context XXXI. 42

Adult 35-45 yrs: Male: 172.8 cm, 5'8"

Year of excavation: 1979

Archaeological context:

Ground pit and wooden coffin - A stone cairn and the remains of a circular stone enclosure were apparent on the ground surface. A rectangular pit orientated south-east to north-west, which contained a poorly preserved wooden board coffin, was discovered to the east of the fence. The coffin contained a very poorly preserved skeleton.

Body position - The individual was lying in a supine position, with the head to the north-west. The remains of the skull were situated at the upper part of the northeastern wall.

Grave goods - The area where the pelvis would have originally been located was covered by a piece of birch bark. The remains of bone arrowpoints were situated near the skull, while the remains of an iron knife were recovered from beneath the skull. Fragments of an iron object were situated adjacent to the L femur, and another iron object was recovered from beside the disturbed feet bones. A single sheep bone was situated in the southeastern area of the coffin.

Visual inventory:

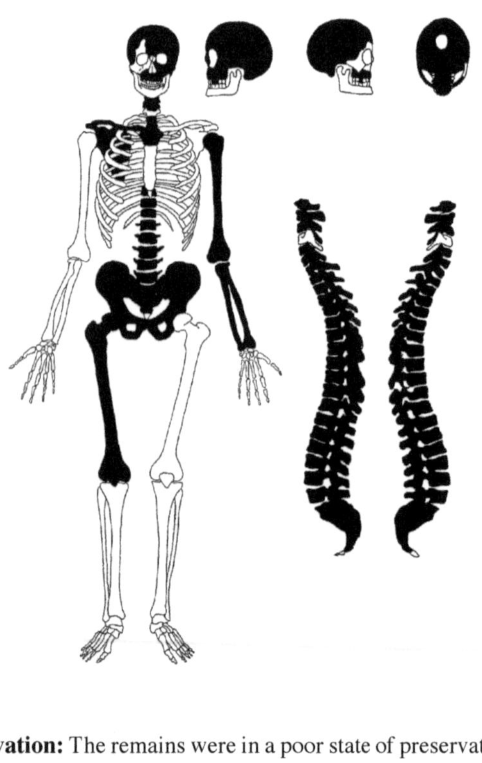

Preservation: The remains were in a poor state of preservation with 40-50% of bones present. Several bones were fragmentary.

Cranial pathology: Porosity was apparent on the superior surface of the frontal and the medial aspects of the parietals, which may be indicative of healed porotic hyperostosis.

Post-cranial pathology: An enthesopathy was present on the R patella (patellar ligament). Vertebral osteophytosis was apparent in TV1, TV3, TV4, TV6-TV12 and LV2-LV5. Schmorl's nodes were visible in TV5-LV4.

Weapon trauma: A transverse cut mark with very sharp margins, which measured approximately 7.5 mm medio-laterally, was present on the posterior surface of the distal third of the shaft of the R femur. The injury had probably been caused by a sword.

Cranial non-metric traits: R and L highest nuchal lines, ossicle at lambda, 5 L and 6 R ossicles in lambdoid suture, R parietal foramen, R ossicle at parietal notch, R and L ossicles at asterion, R and L mastoid foramina extrasutural, L posterior condylar canal open, R and L open foramina spinosum, L accessory supraorbital foramina, L accessory infraorbital foramen.

Post-cranial non-metric traits: R and L acetabular creases, R and L accessory sacral facets, R acromial articular facet, R circumflex sulcus, R and L double atlas facets, R posterior atlas bridging, R CV5, and R and L CV6 transverse foramina bipartite.

Other observations: Orange discoloration (rust) was apparent on several of the vertebrae.

Dentition:

8̸	7 (B)	6	5̸	4̸	3̸	2̸	1̸	1̸	2̸	3̸	4̸	5	6	7̸	8̸
~~8~~	~~7~~	~~6~~	~~5~~	~~4~~	~~3~~	~~2~~	~~1~~	~~1~~	~~2~~	~~3~~	~~4~~	~~5~~	~~6~~	~~7~~	~~8~~

Attrition: Molar wear patterns of Stage 5 were apparent for the 1st molars.

Calculus: Flecks of calculus were visible on the buccal and lingual surfaces of 6⌋ and ⌊5 6.

Periodontal disease: Periodontal disease of slight severity was apparent.

Context XXXI. 44

Juvenile 11 - 12 years

Year of excavation: 1980

Anthropological number: 6863-247

Archaeological context:

Pit and cist - The remains of an oval slab enclosure were apparent on the ground surface. A rectangular pit orientated south-east to north-west, which contained a rectangular stone cist, was situated in the interior of the enclosure. The remains of a single skeleton were recovered from the cist.

Body position - The individual was lying on his or her R side, with the head to the north-west. The arms were extended in front of the body, and the legs were slightly flexed. The L lower leg was disturbed.

Grave goods - The remains of a small wooden dish were situated near the superior aspect of the skull. An iron knife and an iron awl were lying adjacent to the pelvis.

Visual inventory:

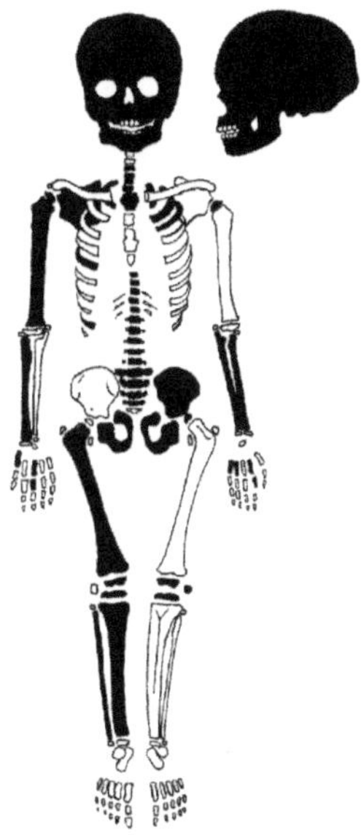

Preservation: The remains were in a fair state of preservation with 40-50% of bones present.

Cranial pathology: Cribra orbitalia of Type 2 was present in Areas 2 and 3 of both orbits.

Weapon trauma: Two small transverse cutmarks were present at the anterior margin of the R inferior articular facet of the CV1. Transverse cutmarks were also present at the midpoint of the anterior surface of CV2. The cutmarks on CV2 displayed bevelled margins indicating that they had been made infero-superiorly. A transverse cutmark was also apparent on the anterior surface of the L transverse foramen of CV3. The cutmarks have a similar morphology and similar location to those observed in other individuals. It is possible that the cutmarks were attained as a consequence of throat cutting.

Cranial non-metric traits: Ossicles in lambdoid suture (full set of traits not recorded).

Other observations: The superior aspect of the occipital had a flattened appearance, which may be indicative of the lambdoid form of cranial deformation.

Dentition:

U	E													E	U
8	7	6	~~5~~	~~4~~	3	~~2~~	~~1~~	~~1~~	~~2~~	~~3~~	4	~~5~~	6	7	8
8	7	6	~~5~~	~~4~~	3	2	~~1~~	~~1~~	~~2~~	3	~~4~~	~~5~~	6	7	8
U															U

Bite: Overbite.

Attrition: Molar wear patterns of Stage 1 for the 2nd molars and Stage 2+ for the 1st molars were apparent.

Calculus: Flecks of calculus were present on the buccal surfaces of 3⌋ ⌊6 and 2⌉ and the buccal and lingual surfaces of 3⌉ ⌈3. Slight deposits were visible on the buccal surfaces of 6⌋ ⌊4 and the lingual surfaces of 6 2⌉.

Enamel hypoplasia: Hypoplastic grooves were present on 6⌋ ⌊4 6 and ⌈3 6 which indicated that the individual had been subject to at least two physiological insults at the ages of 2.5-3 and 3.5-4 years.

Dental anomalies: Carabelli's cusps were present on the maxillary 1st molars.

Context XXXI. 47

Adult 25-35 yrs: Male: 173.4 cm, 5'8"

Year of excavation: 1980

Archaeological context:

Ground pit and cist - No surface signs of the burial were apparent. A rectangular pit orientated south-east to north-west, which contained the remains of a cist, was discovered. The remains of a single skeleton were recovered from the cist.

Body position - The individual was lying on his R side, with the head to the north-west. The R arm was extended, and the L arm was flexed with the hand positioned near the R side of the pelvis. The legs were slightly flexed. The skull was not present.

Grave goods - The poorly preserved remains of an oval wooden dish were located near the northwestern wall. A badly decomposed sheep skull was situated on top of the dish. Fragments of a small wooden mug were situated in close proximity to the dish. The remains of an iron object, two ring-shaped buckles and a plate were situated beneath the lumbar vertebrae. Fragments of a ring-shaped iron buckle, an oval buckle and the remains of an iron knife were positioned below the R side of the pelvis. An iron knife was also located beneath the L side of the pelvis.

Visual inventory:

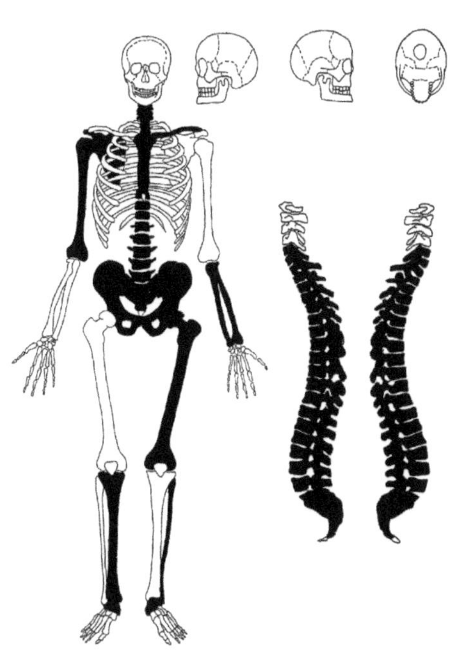

Preservation: The remains were in a fair state of preservation with 40-50% of bones present.

Post-cranial pathology: The manubrio-mesosternal joint was fused, which is a developmental defect due to the failure of development of the fibrous lamina between the primitive manubrium and the first sternebra of the mesosternum. DJD was apparent on the L distal fibula and the trochlear surface of the L talus. The L tibia was not present, so it was not possible to determine if the DJD was secondary to a fracture of the tibia. Osteophytosis was present on the distal R humerus. Enthesopathies were visible on the calcanei (Achilles' tendon). Slight vertebral osteophytosis was apparent in CV5-LV5, and Schmorl's nodes were visible in TV6-LV2.

Weapon trauma: It was recorded in the archive report that the skull had not been present in the cist. In addition, the first four cervical vertebrae had not been recovered. It is feasible, therefore, that the individual had been decapitated prior to burial.

Post-cranial non-metric traits: L plaque, L hypotrochanteric fossa, R lateral tibial squatting facet, R acromial articular facet, R circumflex sulcus, L vastus notch, R medial talar facet, R and L peroneal tubercles, R and L CV5, L CV6, and R and L CV7 transverse foramina bipartite.

Context XXXI. 48

Adult 35-45 yrs: Female: 154.6 cm, 5'2"

Year of excavation: 1980

Archaeological context:

Ground pit and wooden coffin - A circular stone enclosure was apparent on the ground surface. The enclosure had been partially destroyed by a later burial, which was situated at ground level in the northwestern part of the enclosure and covered by a stone cairn. A rectangular pit orientated south-east to north-west, which contained a wooden board coffin covered over by three logs, was located within the enclosure. A single skeleton was recovered from the coffin.

Body position - The individual was lying in an extended, supine position, with the head to the north-west. The R arm was extended and the L arm was flexed so that the hand was positioned on the R side of the pelvis. The legs were flexed to the right.

Grave goods - A clay pot was situated near the frontal bone, and fragments of a clay pot, a round iron plate, and part of a birch bark vessel were recovered near the occipital. A miniature iron knife and fragments of iron tweezers were retrieved from inside the birch bark vessel. A clay spindle whorl was positioned beneath the mandible, and small fragments of a circular iron buckle were situated underneath the lumbar vertebrae. The remains of a circular iron buckle were positioned under the R side of the pelvis, and the remains of an iron buckle were also located just superior to the L side of the pelvis. A lid from a birch bark vessel and some sheep bones were located near the feet of the individual.

Visual inventory:

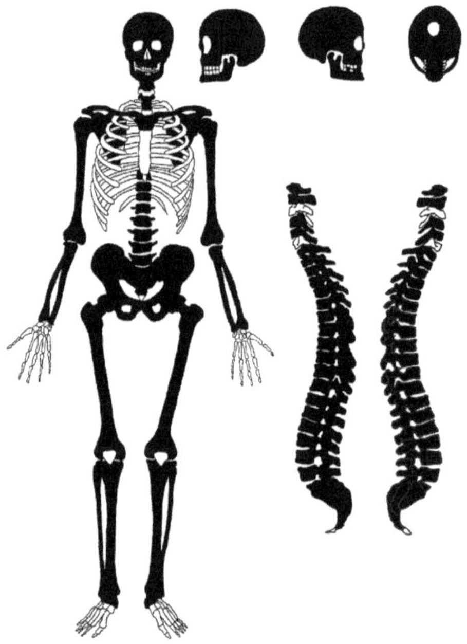

Preservation: The remains were in a very good state of preservation with 70-80% of bones present.

Post-cranial pathology: Partial lumbarization was apparent. The R side of S1 was normal, while the L side of S1 was independent from the remainder of the sacrum and had developed a transverse process. There may have been slight secondary scoliosis of the spine to the left. Secondary osteophytosis was apparent on the superior and inferior surfaces of the body of S1, the superior articular facets of S1 and on the posterior margin of the auricular surface of the R ilium. The neural arches of S1 and S2 were completely cleft. In addition the spinal canal was markedly widened for both vertebrae. It is therefore likely that the spina bifida occulta apparent in this individual was due to a neural tube defect.

A very well healed simple transverse fracture was present at the distal third of the shaft of the L radius (Colles' fracture). Slight posterior displacement of the distal part of the fracture by approximately five degrees was evident. In addition, 100% apposition, no hypertrophy at the fracture site and overlap of the fracture parts by only 6 mm was apparent. Large osteophytes were visible at the proximal end, and osteophytosis was also present at the distal end of the radius. Osteophytosis was present on the distal humeri, the proximal and distal ulnae and the proximal R radius. Vertebral osteophytosis was evident in CV1, CV2, CV4, CV5, TV1-TV3, TV6-TV12 and LV2-LV5. Enthesopathies were apparent on the radii (biceps) and the calcanei (Achilles' tendon).

Cranial non-metric traits: R and L highest nuchal lines, ossicle in L lambdoid suture, L parietal foramen, R ossicle at pterion, R ossicle at parietal notch, R and L mastoid foramina extrasutural, R and L posterior condylar canals open, R open foramen spinosum, R and L accessory supraorbital foramina, R and L anterior ethmoid frontal foramina extrasutural, R and L accessory infraorbital foramina.

Post-cranial non-metric traits: R and L hypotrochanteric fossae, R and L lateral tibial squatting facets, R and L accessory sacral facets, R and L acromial articular facets, R and L circumflex sulci, R medial talar facet, R double inferior articular talar facet, R and L double anterior calcaneal facets, R and L peroneal tubercles, R CV4, R and L CV5 and R CV7 transverse foramina bipartite.

Dentition:

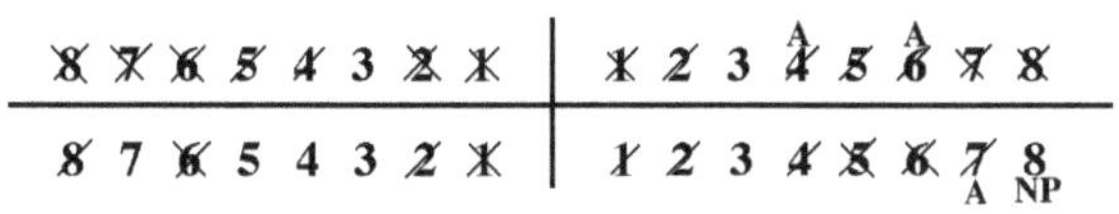

Bite: Overbite.

Attrition: Moderate attrition was apparent on the non-molar teeth, and a molar wear pattern of Stage 3- was present on the R 2nd mandibular molar.

Calculus: Flecks of calculus were apparent on the buccal surfaces of ⌊3 and ⌈3, the lingual surfaces of 5 3⌉ and the buccal and lingual surfaces of 4⌉. Slight deposits were visible on the distal surface of 7⌉.

Periodontal disease: Periodontal disease of considerable severity was evident.

Abscesses: A small externally draining abscess was associated with ⌈7, while medium sized externally draining abscesses were associated with ⌊4 and ⌊6.

Ante-mortem tooth loss: 8 7 6 2 1⌋ ⌊1 7 8 and 6 1⌉ ⌈5 6 had been lost ante-mortem. The sockets for 6⌋ and ⌈5 6 were almost completely remodelled, while those for 8 7⌋ ⌊7 8 were in the later stages of remodelling. The sockets for the remainder of the teeth lost ante-mortem were in the early stages of remodelling.

Dental anomalies: The L mandibular 3rd molar was genetically absent.

Context XXXI. 52

Adult 25-35 yrs: Male: 167 cm, 5'6"

Plate: 25

Year of excavation: 1980

Archaeological context:

Ground pit and cist - A cairn, which may have represented the remains of a circular enclosure, was apparent on the ground surface. A rectangular pit orientated south-east to north-west, which contained a stone cist covered by a layer of thin logs and small slabs, was recovered from beneath the eastern area of the cairn. A single skeleton was retrieved from the cist.

Body position - The individual was lying on his R side, with the head to the north-west. The R arm was extended, and the L arm was flexed so that the hand was positioned near the R elbow. The legs were markedly flexed.

Grave goods - The remains of a wooden cup were recovered from the northwestern corner. An iron awl with a wooden handle, and a bone pin were situated near the superior aspect of the skull. A rectangular fragment of a lacquered wooden cup was situated in close proximity to the awl and pin. A clay vessel and a wooden dish were located near the R humerus, and two circular iron objects were situated near the lumbar vertebrae. Fragments of an iron hook-shaped object associated with some wooden remains, a fragment of an iron knife, and the remains of an iron awl were situated at the R femur. An iron knife, with the remains of a wooden sheath, was lying on the sacrum.

Visual inventory:

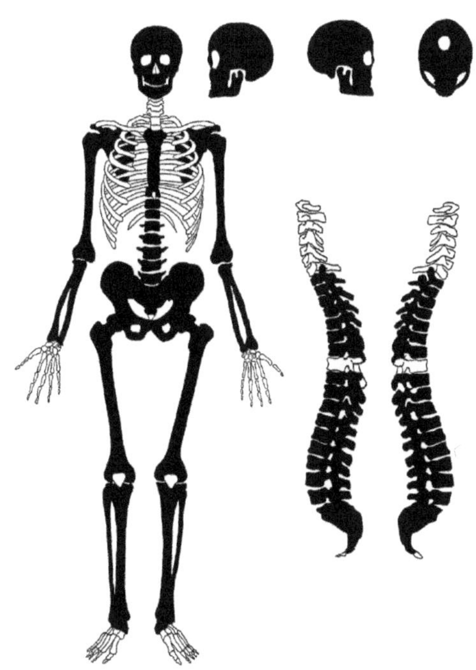

Preservation: The remains were in a very good state of preservation with 70-80% of bones present.

Cranial pathology: Cribra orbitalia of Type 4 was present in Areas 1, 2 and 4 of the R orbit and Areas 1 and 4 of the L orbit. Porosity was apparent on the superior aspect of the frontal, the medial parietals and the superior occipital, which may be indicative of healed porotic hyperostosis. An ivory osteoma, which measured 6 mm by 6 mm, was present at the midpoint of the L parietal. A simple oblique fracture was present on the L nasal bone. The distal part of the fracture was depressed, and a gap was visible between the two parts of the fracture at the medial aspect. The fracture line was clearly evident.

Post-cranial pathology: Os acromiale was present in both scapulae. Osteophytosis was apparent at the glenoid fossa of the R scapula. Vertebral osteophytosis was present in TV4-TV7, TV9-LV2 and LV4. Schmorl's nodes were visible in TV10, TV11, LV1 and LV2.

Enthesopathies were evident on the calcanei (Achilles' tendon).

Cranial non-metric traits: R and L highest nuchal lines, R and L parietal foramina, R and L ossicles in coronal suture, R and L mastoid foramina extrasutural, R posterior condylar canal open, R and L accessory supraorbital foramina, R and L anterior ethmoid frontal foramina extrasutural, R and L accessory infraorbital foramina.

Post-cranial non-metric traits: R medial tibial squatting facet, L lateral tibial squatting facet, R and L acetabular creases, sternal foramen, R and L accessory sacral facets, R and L circumflex sulci, R os trigonum, R and L medial talar facets, R and L double inferior articular talar facets, R and L peroneal tubercles.

Dentition:

8 7 6 5 ~~4~~ 3 2 $\overset{B}{1}$	~~1~~ ~~2~~ 3 4 5 6 7 8
8 7 6 5 4 3 2 1	1 2 3 4 5 6 7 8

Bite: Edge to edge.

Attrition: Molar wear patterns of Stage 3- for the 2nd and 3rd molars and Stages 4-4+ for the 1st molars were apparent.

Calculus: Flecks of calculus were present on the buccal surfaces of ⌊5 6 7 8, the lingual surfaces of ⌊3 and 7 6 5 4⌉ ⌈4 5 6, the buccal and lingual surfaces of 7⌋ and ⌈7, the lingual and distal surfaces of 8⌉ and the buccal, lingual and distal surfaces of ⌈8.

Periodontal disease: Periodontal disease of slight severity was apparent.

Enamel hypoplasia: Hypoplastic grooves were present on 3 2⌋ ⌊3 4 6 and 3⌉ and ⌈3 which indicated that the individual had been subject to at least two incidents of physiological stress at the ages of 2.5-3 and 3.5-4 years.

Dental anomalies: Mandibular tori were present. The R maxillary lateral incisor was shovel-shaped.

Context XXXI. 54

Young adult 17-25 yrs: Male: 153.8 cm, 5'1"

Plates: 42, 43

Year of excavation: 1980

Archaeological context:

Ground pit, birch bark and wicker mat - The poorly preserved remains of a circular enclosure were apparent on the ground surface. A rectangular pit orientated south-east to north-west was situated to the north of the enclosure. A layer of horizontal slabs, under which was a layer of birch bark were situated in the lower levels of the pit. A wicker mat made from reeds was located beneath the layer of birch bark. A skeleton was lying on the layer of birch bark.

Body position - The individual was positioned in an extended, supine position, with the head to the north-west.

Grave goods - A sheep scapula and leg bone were situated near the R lower leg.

Visual inventory:

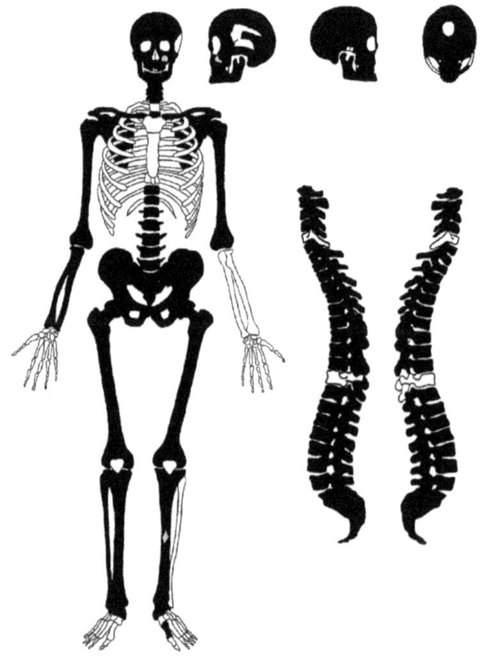

Preservation: The remains were in a good state of preservation with 60-70% of bones present.

Post-cranial pathology: The neural arches for S3 and S4 were cleft, probably as a result of a developmental delay during the early morphogenesis of the vertebral column. Vertebral osteophytosis was evident in TV11 and LV2. Schmorl's nodes were visible in TV7, TV9, TV11, LV1 and LV2. Enthesopathies were present in the L clavicle (costoclavicular ligament), the humeri (deltoid) and the calcanei (Achilles' tendon).

Weapon trauma: A sword chop was apparent on the L parietal and the L side of the frontal. The injury was situated immediately superior to the squamosal suture, and ran parallel to the suture for a length of 102 mm from the most posterior aspect of the temporal to the frontozygomatic suture. The edge of the injury was very sharp and ran at a supero-medial to infero-lateral angle. Small striations, running supero-inferiorly, were apparent in the margins.

Another sword chop was present on the L clavicle. In this case the superior surface of the acromial end had been sliced through nearly horizontally for a medio-lateral length of 17 mm. The injury displayed a slight supero-medial to infero-lateral angle. The acromion of the L scapula also displayed a sword chop. A clean diagonal cutmark was apparent on the superior surface of the acromion, in which the lateral margin of the bone had been removed. The injury measured approximately 42.5 mm medio-laterally and ran at an antero-superior to postero-inferior angle.

It is probable that all three sword chops had been attained in a single incident. The position and angle of each injury suggested that the assailant had been standing in front of the victim, and that he or she had wielded the sword above the victim's head, and then struck downwards at an angle to the right, nearly cleaving the skull in two. The weapon must have been extremely sharp and applied with a great deal of force.

Cranial non-metric traits: R and L parietal foramina, R and L mastoid foramina extrasutural, R bridging of supraorbital notch, L accessory supraorbital foramen, R anterior ethmoid frontal extrasutural, R and L accessory infraorbital foramina.

Post-cranial non-metric traits: R hypotrochanteric fossa, R exostosis in trochanteric fossa, R and L third trochanters, R and L lateral tibial squatting facets, R septal aperture, R and L accessory sacral facets, R acromial articular facet, R and L circumflex sulci, R and L medial talar facets, R and L peroneal tubercles, R and L CV1, R and L CV6, and R and L CV7 transverse foramina bipartite.

Dentition:

		A C PU												C PU		
	8̸	7	6	5	4	3	2	1̸	1̸	2	3	4	5	6	7	8̸
	8	7	6	5	4	3	2	1	1	2	3	4	5	6	7	8
														PU		

Bite: Overbite.

Attrition: Molar wear patterns of Stage 1 for the 3rd molars, Stage 3 for the 2nd molars and Stage 5 for the 1st molars were apparent. Secondary dentine was exposed in 6⌋ ⌊6 and ⌈6.

Calculus: Slight deposits of calculus were present on the buccal surfaces of 6 5 4 2⌋ ⌊2 3 7 and 6⌉ ⌈7, the buccal and lingual surfaces of 7 3⌋ and 7 5 4 3 2 1⌉ ⌈1 2 3 4 5 6, the buccal and distal surfaces of ⌈8 and the buccal, lingual and distal surfaces of 8⌉. Medium deposits were present on the buccal surfaces of ⌊4 6 and the lingual surface of 6⌉.

Periodontal disease: Periodontal disease of slight severity was apparent.

Caries: Medium sized caries were present in the occlusal surface of 6⌋ and the mesial surface of ⌊6.

Abscesses: A medium sized externally draining abscess was associated with 6⌋.

Enamel hypoplasia: Hypoplastic grooves were apparent on 4 3⌉ ⌈3 which indicated that the individual had been subject to at least one incident of physiological stress at the age of 3-3.5 years.

Context XXXI. 59 (i)

Infant 1 - 1.5 years

Plates: 44, 45

Year of excavation: 1980

Anthropological number: 6863-253

Archaeological context:

Pit and wooden coffin - The poorly preserved remains of a circular slab enclosure were apparent on the ground surface. A rectangular pit orientated south-east to north-west, which contained a trapezoid-shaped wooden plank coffin surrounded by several slabs, was situated within the enclosure. The remains of two children were recovered from the coffin.

Body position - Skeleton 1 was lying along the southwestern wall. The individual was probably lying on his or her L side, with the head to the north-west. The bones of the arms and legs were disturbed. Only the skull was available for analysis.

Grave goods - A clay pot was situated at the midpoint of the northwestern wall.

Visual inventory:

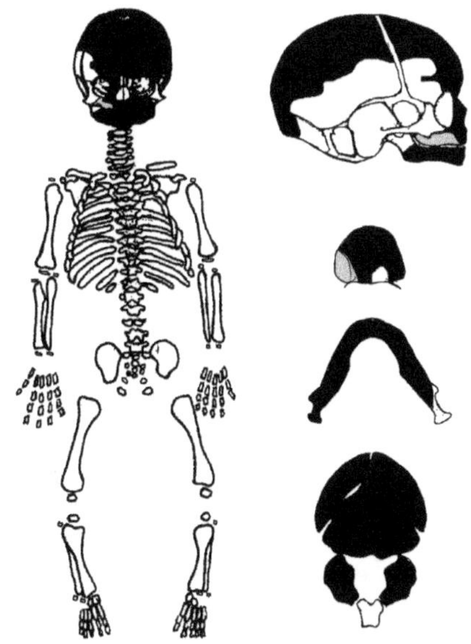

Preservation: Only the skull was present, which was fragmentary.

Weapon trauma: Two definite sword injuries were apparent on the cranium. A transverse sword cut, measuring 90 mm medio-laterally was apparent on the postero-lateral aspect of the L parietal. The cut extended from the superior margin of the temporal to approximately 14 mm lateral of the sagittal suture. The margins of the injury were extremely sharp. The second sword cut was oblique and measured 72 mm. It extended from approximately 10 mm to the R of the posterior aspect of the sagittal suture, through the lambda, and ran for a further 18 mm on the L side of the occipital. The margins of the sword cut were clearly defined and sharp. It is probable that the attacker was standing posterior to the child, and that he or she had wielded the sword in a supero-inferior manner and dealt two successive blows to the posterior aspect of the skull. No evidence of healing was associated with either wound.

Dentition:

U									U
e	d̸	c̸	b̸	a̸	a̸	b̸	c̸	d	e
e̸	d̸	c̸	b̸	a̸	a̸	b̸	c̸	d̸	e
									U

Archaeology & Trauma

Attrition: A molar wear pattern of Stage 1 was present on the 1st deciduous molars.

Context XXXI. 61

Young adult 17-25 yrs: Female: 161.8 cm, 5'4"

Year of excavation: 1980

Archaeological context:

Ground pit and log coffin - The remains of a very poorly preserved circular stone enclosure were apparent on the ground surface. A rectangular pit orientated south-east to north-west, which contained a tree-trunk coffin surrounded by slabs, was situated within the enclosure. A poorly preserved skeleton was recovered from the coffin.

Body position - The individual was lying on her R side, with the head to the north-west. The skull was not present, and the arm and leg bones were disturbed.

Grave goods - A gold earring was recovered from the northwestern corner of the coffin. A base from a birch bark vessel and some fragments of a clay pot were situated near the earring.

Visual inventory:

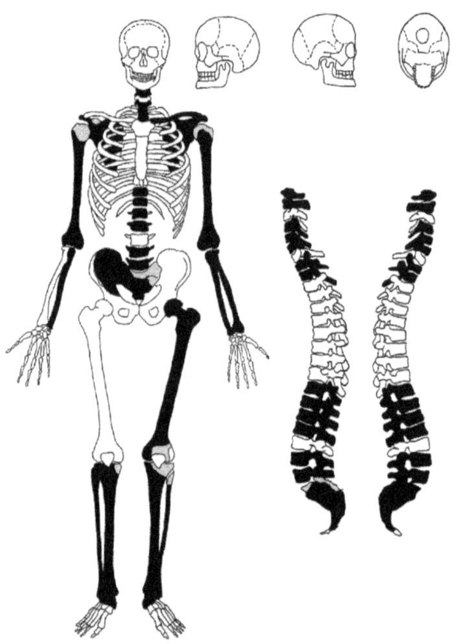

Preservation: The remains were in a poor state of preservation with 40-50% of bones present. The majority of bones were fragmentary and displayed extensive surface erosion.

Post-cranial pathology: The posterior arch of the atlas was cleft, which was probably due to developmental delay during the early morphogenesis of the vertebral column.

Weapon trauma: It was recorded in the archive report that the skull was not present within the tomb. It is possible that the individual had been decapitated prior to burial.

Post-cranial non-metric traits: R and L lateral tibial squatting facets, R and L septal apertures, R and L acromial articular facets, R and L circumflex sulci, R and L medial talar facets, R and L double inferior articular talar facets, L double anterior calcaneal facet, R double atlas facet, R and L CV5 transverse foramina bipartite.

Context XXXI. 64

Adult 25-35 yrs: Female: 148.6 cm, 5'0"

Year of excavation: 1980

Archaeological context:

Ground pit, coffin cover and wooden coffin - The remains of a poorly preserved circular stone enclosure were apparent on the ground surface. A rectangular pit orientated south-east to north-west was present within the enclosure. The pit contained a layer of stones, a layer of slabs, a plank structure which was supported on poles and would have covered the coffin, and a wooden plank coffin. A single skeleton was recovered from the coffin.

Body position - The individual was in a supine position, with the head to the north-west. The R arm was markedly flexed with its hand positioned near the superior aspect of the thorax. The L arm was flexed with its hand situated inferior to the R elbow. The legs were flexed to the right.

Grave goods - Two clay pots were situated to the right of the skull, and a silver earring was recovered from adjacent to the L temporal. A fragment of a stone ring was positioned inferiorly to the mandible, and a birch bark vessel was located at the R forearm. The base of a birch bark vessel was situated near the R side of the pelvis, and the base and side fragments of a birch bark vessel were located near the feet of the individual.

Visual inventory:

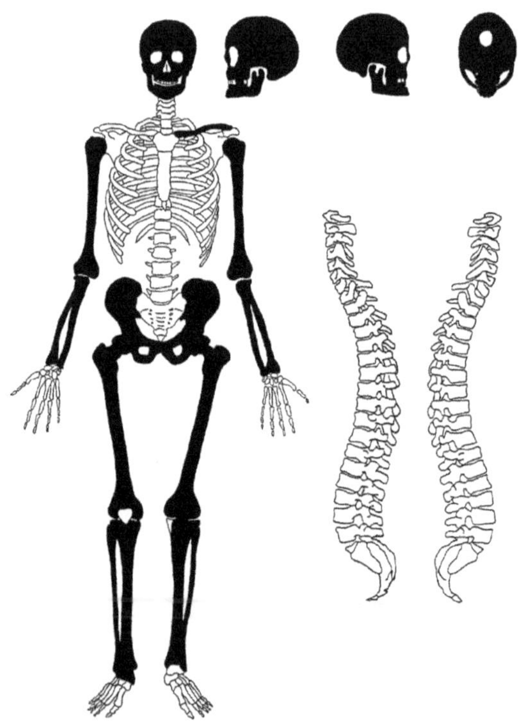

Preservation: The remains were in a good state of preservation with 50-60% of bones present. Slight surface erosion was present on several bones.

Cranial pathology: Cribra orbitalia of Type 1 was present in Areas 1, 2 and 4 of the R orbit, and cribra orbitalia of Type 4 was present in Areas 1, 2 and 4 of the L orbit.

Post-cranial pathology: A bony exostosis was apparent on the lateral surface of the L tibia, immediately inferior to the lateral condyle. It is possible that the exostosis was an enthesophyte of part of the tibialis anterior.

Weapon trauma: A circular depression, which measured 6 mm by 6 mm and had a depth of 2-3 mm, was apparent on the superior surface of the frontal bone immediately to the right of the midpoint. The margins of the lesion were smooth, and it did not perforate the inner table of the vault. It is possible that the lesion is indicative of healed trauma, and it may represent a weapon injury caused by an arrowhead.

Cranial non-metric traits: R and L highest nuchal lines, ossicle at lambda, 2 R and 1 L ossicles in lambdoid suture, R and L ossicles at pterion, R ossicles at parietal notch, R and L mastoid foramina absent, R and L posterior condylar canals open, R double anterior condylar canal, R and L open foramina spinosum, R and L anterior ethmoid frontal foramina extrasutural, R and L accessory infraorbital foramina.

Post-cranial non-metric traits: L hypotrochanteric fossa, R and L exostoses in trochanteric fossae, R medial tibial squatting facet, R and L lateral tibial squatting facets, L septal aperture, R accessory sacral facets, R medial talar facet, R peroneal tubercle.

Dentition:

8	7	6	~~5~~	~~4~~	~~3~~	~~2~~	~~1~~	~~1~~	~~2~~	~~3~~	~~4~~	~~5~~	6	7	~~8~~
8	7	6	5	4	3	2	1	1	2	3	4	5	6	7	8

Bite: Overbite.

Attrition: Molar wear patterns of Stage 2+ for the 3rd molars, Stage 3 for the 2nd molars and Stage 4 for the 1st molars were apparent.

Calculus: Flecks of calculus were present on the buccal surface of 5⏋, the lingual surfaces of 8 4⏋⎡4 5 6, the distal surface of ⎡8 and the buccal and lingual surfaces of 2 1⏋⎡1 2.

Enamel hypoplasia: Single hypoplastic grooves were apparent on 5 4 3 1⏋⎡1 3 5, and double hypoplastic grooves were visible on 2⏋⎡2. The hypoplasia indicates that the individual had been subject to at least two incidents of physiological stress at the ages of 2.5-3 and 3-3.5 years.

Context XXXI. 80

Adult 35-45 yrs: Male: 166.4 cm, 5'5"

Year of excavation: 1980

Archaeological context:

Ground pit, oval cist, log coffin - The poorly preserved remains of a circular stone enclosure were apparent on the ground surface. A rectangular pit orientated east-south-east to west-north-west, which contained an oval cist and a trapezoid-shaped coffin, was situated within the enclosure. A skeleton, which had been partially disturbed by rodents, was recovered from the coffin.

Body position - The individual was lying on his R side, with the head to the north-west. The R arm was extended, and the L arm was markedly flexed with its hand situated near the chin. The pelvis was turned on its side, and the legs were extended. The L lower leg was overlying the R lower leg.

Grave goods - A clay pot and a bone pin were situated at the superior aspect of the skull. Fragments of bowstraps were recovered from posterior to the skull, the spine and the legs. The remains of a circular iron buckle and an iron knife were situated just superior to the R side of the pelvis. An iron knife was also recovered from near the R femur.

Visual inventory:

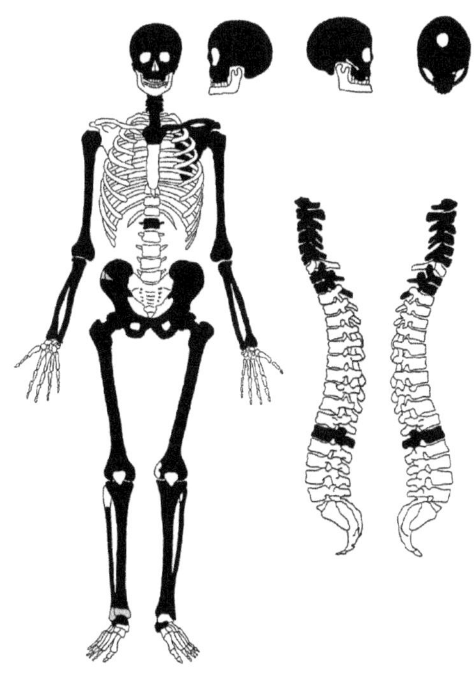

Preservation: The remains were in a good state of preservation with 50-60% of bones present.

Post-cranial pathology: A very well healed simple oblique fracture was present at the midshaft of the L ulna (Night-stick fracture). Very little overlap of the fracture parts and 100% apposition was evident. Hypertrophy was, however, apparent at the fracture site which gave this area a globular appearance. The antero-posterior breadth at the fracture site was 22.7 mm (R=14.6 mm), and the medio-lateral measurement was 17.4 mm (R=12.2 mm). Osteophytosis was apparent on the glenoid fossa and the inferior articular facet of the acromion of the L scapula, the proximal humeri, the distal R femur, the distal tibiae and the proximal R tibia. Rotator cuff disease was evident in the humeri. Vertebral osteophytosis was present in CV3-CV6, TV1 and LV1. Enthesopathies were visible on the L clavicle (costoclavicular ligament), the humeri (teres major, latissimus dorsai, pectoralis major) and the L calcaneus (Achilles' tendon).

Cranial non-metric traits: R and L highest nuchal lines, R and L parietal foramina, R ossicle at pterion, R and L mastoid foramina extrasutural, R and L posterior condylar canals open, R bridging of supraorbital notch, L accessory supraorbital foramen, R and L anterior ethmoid frontal foramina extrasutural, R and L accessory infraorbital foramina.

Post-cranial non-metric traits: L hypotrochanteric fossa, L lateral tibial squatting facet, R septal aperture, L acromial articular facet, L circumflex sulcus, R medial talar facet.

Dentition:

	C PU	A PU										A PU	PU		
8̸	7	6	5	4̸	3̸	2̸	1̸	1̸	2̸	3̸	4	5	6	7	8
~~8~~	~~7~~	~~6~~	~~5~~	~~4~~	~~3~~	~~2~~	~~1~~	~~1~~	~~2~~	~~3~~	~~4~~	~~5~~	~~6~~	~~7~~	~~8~~

Bite: Overbite.

Attrition: Molar wear patterns of Stages 4+-5 for the 2nd molars and Stage 5 for the 1st molars were apparent. The L maxillary 3rd molar displayed a wear pattern of Stage 1 which may indicate that the corresponding mandibular 3rd molar was either genetically absent or had been lost ante-mortem. Secondary dentine was exposed in 7 6⌋ ⌊5 6.

Calculus: Slight deposits of calculus were apparent on the buccal surfaces of 7⌋ ⌊6, the lingual surface of 6⌋, the buccal and lingual surfaces of ⌊4 5 7 and the buccal, lingual and distal surfaces of ⌊8. Medium deposits were visible on the buccal surfaces of 6 5⌋.

Periodontal disease: Periodontal disease of medium severity was evident.

Caries: A large caries was apparent in the occlusal surface of 7⌋.

Abscesses: Medium sized externally draining abscesses were associated with 6⌋ and ⌊5.

Context XXXI. 83

Young adult 17-25 yrs: Male: 161.3 cm, 5'4"

Year of excavation: 1980

Archaeological context:

Ground pit and log coffin - The remains of a poorly preserved circular stone enclosure were apparent on the ground surface. A trapezoid shaped pit orientated south-east to north-west was situated to the north of the enclosure. The fill of the pit consisted of stones, under which were the remains of a tree-trunk coffin which was surrounded by vertical slabs. A skull was lying on the eastern end of the lid of the coffin. The disturbed remains of a single skeleton were recovered from the coffin.

Body position - The individual had probably been lying in a supine position, with the head to the north-west. The R arm was extended, and the L arm was flexed.

Grave goods - An iron object was situated near the eastern wall. The remains of an iron arrowhead were recovered embedded in the posterior aspect of one of the vertebrae, near which was a circular iron object. A fragment of a whetstone was situated between the forearms. The remains of an iron object were recovered near the superior margin of the L side of the pelvis, and an iron arrowhead was situated beneath the pelvis. The remains of a circular iron buckle and an iron knife were positioned on the R side of the pelvis. A circular iron buckle was also situated on the R femur.

Visual inventory:

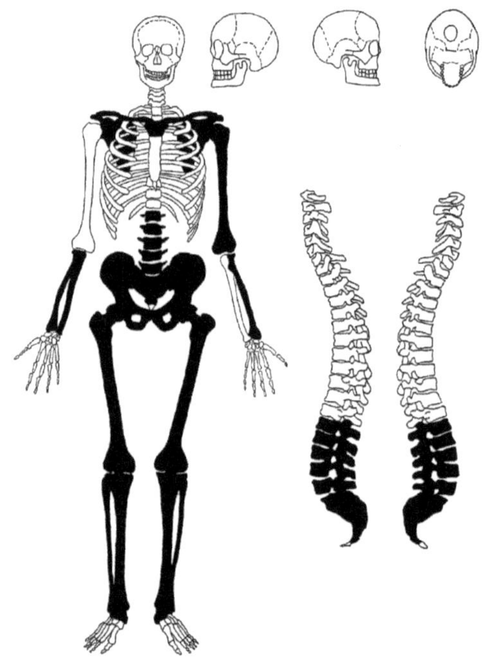

Preservation: The remains were in a good state of preservation with 60-70% of bones present.

Post-cranial pathology: The neural arches of S3 and S4 were cleft, while the neural arch for S2 was partially cleft and partially bifurcated. It is likely that the clefting was due to a developmental delay during the early morphogenesis of the vertebral column. Enthesopathies were apparent on the R clavicle (costoclavicular ligament) and the R fibula (peroneus brevis).

Weapon trauma: A deep transverse cutmark, which measured 11 mm medio-laterally, was present on the posterior surface of the distal third of the shaft of the L radius. Two small cutmarks were apparent on the medial condyle of the R femur. It was reported in the archive report that an arrowhead had been embedded in one of the vertebra. It is feasible, therefore, that the cutmarks, particularly the example on the radius, are indicative of weapon trauma rather than post-mortem defleshing. The morphology of the cutmarks would suggest that they had been made using a small sharp weapon, possibly a knife.

Post-cranial non-metric traits: R and L plaque, R and L lateral tibial squatting facets, R and L acromial articular facets, R and L circumflex sulci, R and L medial talar facets, R and L peroneal tubercles.

Context XXXI. 87. Sk. 1

Young adult 17-25 yrs: Possible male

Plates: 57, 58

Year of excavation: 1980

Archaeological context:

Ground pit and cist - The remains of a poorly preserved circular stone enclosure were apparent on the ground surface. A rectangular pit orientated south-east to north-west, which contained a rectangular cist, was present within the enclosure. Two compact groups of bones, which represented the remains of two individuals, were recovered from the cist.

Body position - Vertebrae, ribs and parts of the pelvis were situated in the lower layer of the group of bones, while a skull, arm and leg bones were recovered from the upper layer. The skull was situated in close proximity to the northwestern wall of the cist.

Grave goods - The remains of a circular iron buckle were situated beneath the R scapula. An iron knife, and two circular iron buckles were located near the centre of the northwestern wall. A circular iron buckle, two iron knives, and fragments of various iron objects were recovered from among the bones.

Visual inventory:

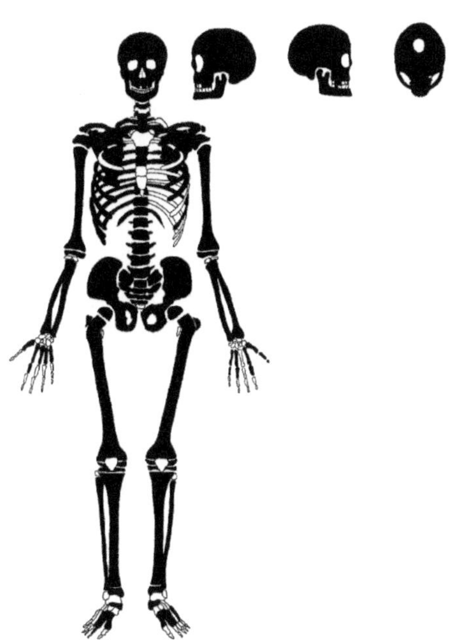

Preservation: The remains were in a very good state of preservation with 80-90% of bones present.

Cranial pathology: Cribra orbitalia of Type 5 was present in Areas 1, 2 and 3 of the R orbit and Areas 1, 2 and 4 of the L orbit. Porosity and trabecular markings were apparent on the superior aspect of the frontal, the parietals and the occipital, which may be indicative of healed porotic hyperostosis.

Scalping: Numerous small sharp cutmarks were apparent on the vault surface. The cutmarks appeared to be following the circumference of the skull, as if the cutter was attempting to remove a circular area of skin from the superior aspect of the cranium. The cutmarks on the occipital and the posterior parietals were running horizontally, parallel to one another. The superior margins of these cutmarks were generally bevelled, suggesting that the person who made the cutmarks was positioned superior to the skull, and that the skin had been peeled off the skull anteriorly. The cutmarks on the frontal and the anterior R parietal also displayed a bevelled superior margin which suggested that these cutmarks were made in the opposite direction. The location and morphology of the cutmarks indicated that the objective was to remove a circular area of the superior and posterior surfaces of the scalp.

Weapon trauma: A sharp cutmark was present on the L side of the neural arch of TV11. The cut mark ran obliquely through the L neural arch and transverse process for a length of approximately 17 mm. The cut mark on TV11 continued onto TV12 where the lateral margin of the L superior articular facet had been chopped off. The cutmarks had probably been made with a sharp sword, which would have been struck supero-inferiorly and to the left. The assailant would have been positioned posterior to the victim.

Dentition:

8̸	7	6	5̸	4̸	3	2̸	1	1	2̸	3	4̸	5̸	6	7	E 8
8̸	7	6̸	5̸	4̸	3̸	2̸	1̸	1̸	2̸	3̸	4̸	5̸	6̸	7̸	8̸

Bite: Overbite.

Attrition: Molar wear patterns of Stage 1 for the 2nd molars and Stage 2+ for the 1st molars were apparent.

Calculus: Flecks of calculus were present on the buccal surface of ⌊6, the lingual surfaces of 3 1⌋ ⌊1 6 and 7⌉ and the buccal and lingual surfaces of 6⌋ ⌊3 7.

Enamel hypoplasia: Single hypoplastic grooves were present on 3 1⌋ ⌊1 3 which indicated that the individual had been subject to at least one incident of physiological stress at the age of 4-4.5 years.

Dental anomalies: A parastyle was present on the mesiobuccal cusp of the R maxillary 2nd molar.

Context XXXI. 87. Sk. 2

Adult 35-45 yrs: Male: 160 cm, 5'3"

Plate: 59

Year of excavation: 1980

Archaeological context:

Ground pit and cist - The remains of a poorly preserved circular stone enclosure were apparent on the ground surface. A rectangular pit orientated south-east to north-west, which contained a rectangular cist, was present within the enclosure. Two compact groups of bones, which represented the remains of two individuals, were recovered from the cist.

Body position - Ribs, vertebrae, pelvic bones, and lower leg bones were recovered from the lower layer of the group of bones, while a skull, arm bones and the femora were present in the upper layer.

Grave goods - A clay pot and an iron spearhead were recovered from the northeastern corner of the tomb. The remains of three circular iron buckles, a knife and fragments of various iron objects were situated amongst the bones. Sheep bones were located near the southeastern wall.

Visual inventory:

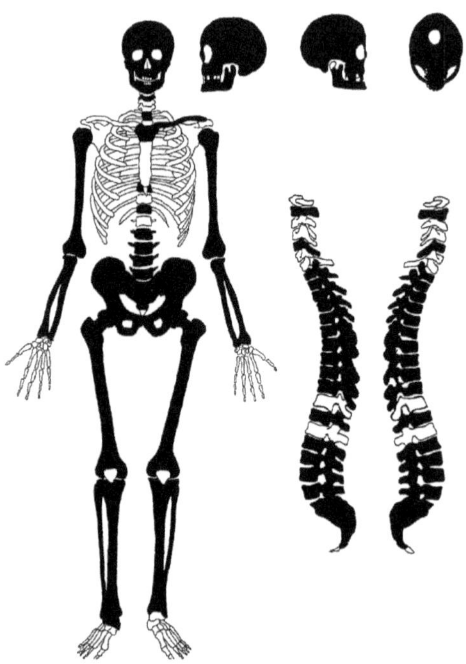

Preservation: The remains were in a good state of preservation with 60-70% of bones present.

Cranial pathology: Porosity was present on the superior surfaces of the frontal and the parietals, which may be indicative of healed porotic hyperostosis. An ivory osteoma, which measured 4 mm by 4 mm, was present on the posterior surface of the R parietal. Porosity was apparent on the articular tubercle of the R temporal, which may be indicative of the early stages of DJD of the R temporomandibular joint.

Post-cranial pathology: Osteophytosis was present on the proximal L ulna. Vertebral osteophytosis was evident in CV2, TV1, TV7, TV10, TV12 and LV2-LV4. Enthesopathies were visible on the L clavicle (costoclavicular ligament), the humeri (pectoralis major), the radii (biceps), the femora (medial head of gastrocnemius) and the R calcaneus (Achilles' tendon).

Scalping: Numerous small sharp cutmarks were apparent on the vault surface. The cutmarks appeared to be following the circumference of the skull, as if the cutter was attempting to remove a circular area of skin from the superior aspect of the cranium. The cutmarks on the occipital and the posterior parietals were running horizontally, parallel to one another. The superior margins of these cutmarks were generally bevelled, suggesting that the person who made the cutmarks was positioned superior to the skull, and that the skin had been peeled off the skull anteriorly. The cutmarks on the frontal and the anterior R parietal also displayed a bevelled superior margin which suggested that these cutmarks were made in the opposite direction. The location and morphology of the cutmarks indicated that the objective was to remove a circular area of the superior and posterior surfaces of the scalp.

Weapon trauma: A sharp diagonal chop mark, which measured approximately 8 mm supero-laterally to infero-medially, was present on the superior aspect of the linea aspera of the R femur. It is likely that the chop mark had been caused by a sword.

Cranial non-metric traits: R and L highest nuchal lines, ossicle at lambda, 4 L and 2 R ossicles in lambdoid suture, R and L parietal foramina, 1 L and 2 R ossicles in coronal suture, 2 ossicles at R parietal notch, R and L ossicles at asterion, R and L mastoid foramina extrasutural, R and L posterior condylar canals open, R bridging of supraorbital notch, L accessory supraorbital foramen, R and L anterior ethmoid frontal foramina extrasutural, R and L accessory infraorbital foramina.

Post-cranial non-metric traits: R and L plaque, R and L lateral tibial squatting facets, R accessory sacral facets, R medial talar facet, R double inferior articular talar facet, R double anterior calcaneal facet.

Dentition:

C 8	7	C 6	~~5~~	C PU 4	PU 3	~~2~~	~~1~~	~~1~~	~~2~~	~~3~~	PU 4	~~5~~	C 6	A ~~7~~	~~8~~
~~8~~	7	6	~~5~~	4	3 B	~~2~~	~~1~~	~~1~~	~~2~~	~~3~~	4	5	6	7	8

Bite: Overbite.

Attrition: Molar wear patterns of Stage 3- for the 3rd molars, Stage 4+ for the 2nd molars and Stages 4+-5 for the 1st molars were apparent. Secondary dentine was exposed in 4 3⌋ ⌊4.

Calculus: Flecks of calculus were present on the buccal surface of 4⌋, the buccal and lingual surfaces of 7⌋ and the buccal, lingual and distal surfaces of 8⌋. Slight deposits were visible on the buccal surfaces of 6⌉ ⌈6, the buccal and lingual surfaces of 4⌉ ⌈4 7 and the buccal, lingual and distal surfaces of ⌊6. Medium deposits were evident on the lingual surfaces of 7 6⌉, the buccal and lingual surfaces of ⌈5 and the buccal, lingual and distal surfaces of ⌈8.

Periodontal disease: Periodontal disease of medium severity was evident.

Caries: Small caries were present in the mesial surfaces of 6⌋ and ⌊6, the distal surface of 4⌋, and the buccal surface of 8⌋.

Abscesses: A small externally draining abscess was associated with ⌊7.

Context XXXI. 97

Adult 25-35 yrs: Female: 161.6 cm, 5'4"

Year of excavation: 1981

Archaeological context:

Ground pit and wooden coffin - A cairn was apparent on the ground surface, associated with which were two parallel slabs fixed firmly into the ground surface. A rectangular pit orientated south-east to north-west, which contained fragments of slabs and a wooden plank coffin with a lid made from three logs, was situated beneath the cairn. The remains of a single skeleton were recovered from the coffin.

Body position - The individual was lying in an extended, supine position, with the head to the north-west. Several of the bones were disturbed.

Grave goods - A sheep scapula lying on an oval wooden dish was situated in the northwestern corner. Fragments of a birch bark vessel were located underneath the wooden dish. A clay pot was situated posterior to the skull, and the remains of a bronze object and a bronze earring were situated to the L of the skull. A poorly preserved disk-shaped iron object was located just inferior to the sternum.

Visual inventory:

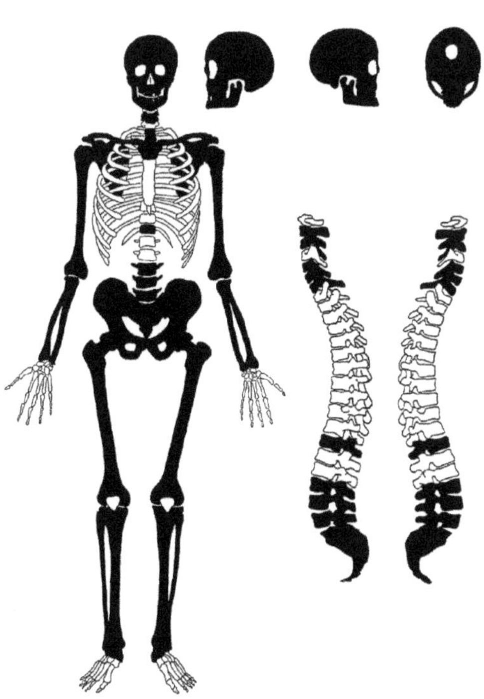

Preservation: The remains were in a good state of preservation with 60-70% of bones present. Several bones displayed surface erosion.

Cranial pathology: Cribra orbitalia of Type 1 was present in Areas 1, 2 and 3 of the R orbit and Areas 3 and 6 of the L orbit.

Post-cranial pathology: The neural arches of S1, S2 and S4 were cleft, while that for S3 was bifurcate. The cleft neural arches were probably due to developmental delay during the early morphogenesis of the vertebral column. The pubic bones, and in particular the pubic symphyses, were hypoplastic. Secondary porosity and eburnation was evident on the superior aspects of the symphyses. An ossified haematoma, which measured 33.4 mm supero-inferiorly by 5 mm antero-posteriorly, was present on the lateral surface of the distal third of the R tibia. Vertebral osteophytosis was evident in CV2, TV12 and LV3-LV5. Enthesopathies were present on the calcanei (Achilles' tendon).

Cranial non-metric traits: R and L highest nuchal lines, 6 ossicles at lambda, 1 R ossicle in lambdoid suture, R parietal foramen, 2 L and 1 R ossicles at pterion, R and L parietal notches, 2 L and 2 R ossicles at asterion, R and L mastoid foramina extrasutural, R and L posterior condylar canals open, L double anterior condylar canal, R and L accessory supraorbital foramina, R and L anterior ethmoid frontal foramina extrasutural, R and L accessory infraorbital foramina.

Post-cranial non-metric traits: R and L hypotrochanteric fossae, R and L lateral tibial squatting facets, R and L acromial articular facets, R and L circumflex sulci, R and L medial talar facets, L CV6 transverse foramen bipartite.

Dentition:

8	7	6	5	4	3̸	2	1	1	2̸	3	4	5	6	7	8
8	7	6	5	4	3	2	1	1̸	2	3	4̸	5	6	7	8 2C A

Bite: Overbite.

Attrition: Molar wear patterns of Stage 1-3- for the 2nd and 3rd molars, and Stage 3+ for the 1st molars were apparent.

Calculus: Flecks of calculus were present on the buccal surfaces of 7⌋ ⌊7 and the lingual surfaces of 6 5 2 1⌋ ⌊3 7. Slight deposits were visible on the buccal surface of 5⌉, the lingual surfaces of 4 1⌋ ⌊1 4 5 and 6 4 2 1⌉ ⌈2 3 5 6, the distal surface of ⌈8, the buccal and lingual surfaces of 7 3⌉, the distal and lingual surfaces of 8⌋ ⌊8 and the buccal, lingual and distal surfaces of 8⌉.

Periodontal disease: Periodontal disease of slight severity was present.

Caries: Both a small and a large caries were present in the occlusal surface of ⌈8.

Abscesses: A medium sized externally draining abscess was associated with ⌈8.

Enamel hypoplasia: Hypoplastic grooves were present on 6 4 2 1⌋⌊1 6 7 and 7 6 2⌉⌈2 which indicated that the individual had been subject to at least three incidents of physiological stress at the ages of 2.5-3, 3-3.5 and 5.5-6 years.

Context XXXI. 99

Adult 25-35 yrs: Female: 148.6 cm, 5'0"

Plate: 29

Year of excavation: 1981

Archaeological context:

Ground pit and cist - A cairn and the remains of a circular stone enclosure were apparent on the ground surface. A rectangular pit, which contained loose stones and a stone cist, was situated beneath the cairn. The remains of an adult and a child were recovered from the cist.

Body position - The individual was lying in an extended, supine position, with the head to the west. Several of the arm bones were displaced. The remains of a poorly preserved child's skeleton were situated beneath the L arm. The child was lying on his/her L side, with the head to the west.

Grave goods - A clay pot was situated near the L forearm, and the remains of fabric were present on the surfaces of the bones of the child.

Visual inventory:

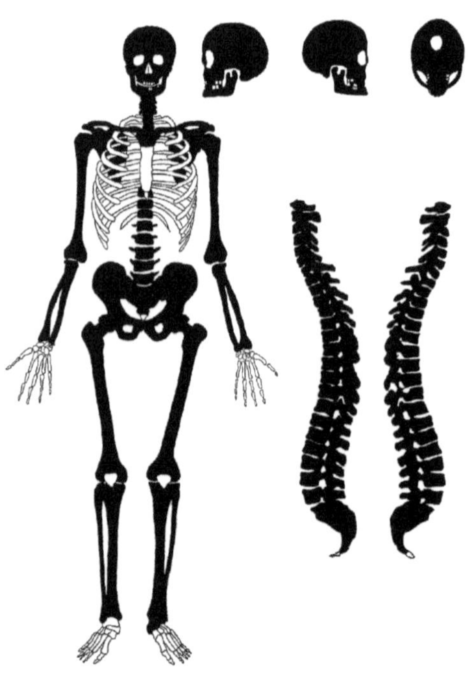

Preservation: The remains were in a very good state of preservation with 70-80% of bones present.

Post-cranial pathology: Possible osteochondritis dissecans was evident on the trochleae of the distal humeri. The lesions were circular, measured approximately 8 mm by 8 mm, consisted of an area of spongy bone with a central disc of thin cortical bone, and were depressed. A plaque of woven bone was present on the posterior surface of the L body of the mandible. No other examples of periosteal reactive bone were evident on the skeleton and it was probably due to trauma. Vertebral osteophytosis was evident in TV5-TV7, LV3 and LV4. An enthesopathy was present on the L calcaneus (Achilles' tendon).

Cranial non-metric traits: R and L highest nuchal lines, ossicle at lambda, R and L parietal foramina, L ossicle at asterion, R and L mastoid foramina extrasutural, L open foramen spinosum, R and L accessory supraorbital foramina, R and L anterior ethmoid frontal foramina extrasutural, R and L accessory infraorbital foramina.

Post-cranial non-metric traits: R and L lateral tibial squatting facets, R and L acromial articular facets, R and L circumflex sulci, L medial talar facet, R CV2, R CV5, R and L CV6 and L CV7 transverse foramina bipartite.

Dentition:

~~8~~	~~7~~	6	5	4	3	~~2~~	~~1~~	~~1~~	~~2~~	~~3~~	4	5	6	~~7~~	~~8~~
~~8~~	7	~~6~~ A	5	4	3	2	1	~~1~~	~~2~~	3	4	5	~~6~~	~~7~~	8 C

Bite: Overbite.

Attrition: Molar wear patterns of Stage 1 for the 2nd and 3rd molars, and Stage 3- for the 1st molars were apparent.

Calculus: Flecks of calculus were present on the buccal surface of ⌈3 and the distal and lingual surfaces of 7⌉. Slight deposits were present on the lingual surface of ⌈3 and the buccal and lingual surfaces of 6 5 4 3⌋⌊4 5 6 and 5 2 1⌉⌈4. Medium deposits were apparent on the buccal and lingual surfaces of 4 3⌉⌈5.

Periodontal disease: Periodontal disease of medium severity was present.

Caries: A small caries was visible in the occlusal surface of ⌈8.

Abscesses: A large externally draining abscess was associated with 6⌉.

Context XXXI. 100. Sk. 1

Adult 35-45 yrs: Female: 149 cm, 5'0"

Year of excavation: 1981

Archaeological context:

Ground pit and wooden coffin - A cairn was present on the ground surface. A rectangular pit orientated south-east to north-west, which contained a wooden plank coffin with stones lying on its lid, was situated beneath the cairn. Two skeletons were recovered from the coffin.

Body position - The southwestern individual was lying on her R side, with the head to the north-west. The R arm was extended, and the L arm was flexed. The hands were situated adjacent to one another, and the legs were flexed.

Grave goods - A sheep skull and an iron knife were lying on the remains of an oval wooden dish which was situated in the northwestern corner. A clay pot, the remains of an iron awl, an iron knife, and a fragment of a bronze object were recovered near the dish. A clay spindle whorl was situated near the L knee.

Visual inventory:

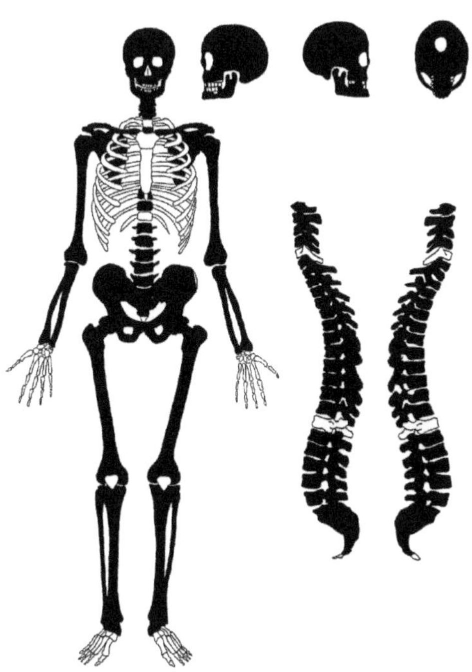

Preservation: The remains were in a very good state of preservation with 70-80% of bones present.

Cranial pathology: Cribra orbitalia of Type 4 was present in Areas 2 and 3 of the R orbit and Areas 1, 2 and 3 of the L orbit. Porosity was apparent on the posterior surfaces of the parietals and the occipital, which may be indicative of healed porotic hyperostosis.

Post-cranial pathology: A Clay Shoveller's fracture was present on TV1, where the tip of the spinous process had separated from the remainder of the process. Spondylolysis was evident in LV4. The distal L humerus was dysplastic with a prominent groove present between the capitulum and the trochlea of the anterior surface. Vertebral osteophytosis was present in CV1, CV2, CV4, CV6, TV2-TV11, LV2 and LV4. A Schmorl's node was visible in TV11. Enthesopathies were apparent on the R clavicle (costoclavicular ligament), the femora (gluteus medius) and the calcanei (Achilles' tendon).

Cranial non-metric traits: R and L highest nuchal lines, ossicle at lambda, R ossicle in lambdoid suture, R and L parietal foramina, R and L mastoid foramina extrasutural, R and L posterior condylar canal open, R and L accessory supraorbital foramina, R and L anterior ethmoid frontal foramina extrasutural, R and L accessory infraorbital foramina.

Post-cranial non-metric traits: R and L hypotrochanteric fossae, R and L third trochanters, R and L lateral tibial squatting facets, R septal aperture, R and L acromial articular facets, R and L circumflex sulci, R and L medial talar facets, R and L double inferior articular talar facets, R and L peroneal tubercles, L CV4, R CV6 and R CV7 transverse foramina bipartite.

Dentition:

NP		PU	PU		PU					PU					
8	~~7~~	6	5	~~4~~	3	~~2~~	~~1~~	~~1~~	~~2~~	3	~~4~~	5	~~6~~	~~7~~	~~8~~
8	7	6	5	~~4~~	3	~~2~~	~~1~~	~~1~~	~~2~~	3	~~4~~	~~5~~	~~6~~	~~7~~	8
C		PU			PU					PU					C

Bite: Overbite.

Attrition: Molar wear patterns of Stage 3 for the 3rd molars, Stage 4 for the 2nd molars and Stage 5 for the 1st molars were apparent. Secondary dentine or pulp cavity exposure was evident in 6 5 3⌋ ⌊3 and 6 3⌉ ⌈3.

Calculus: Flecks of calculus were present on the buccal surfaces of 3⌋ and 3⌉. Slight deposits were visible on the buccal surface of ⌈3, the lingual surfaces of 7⌉ ⌈8, the buccal and lingual surfaces of 5⌋ ⌊3 5 and 6 5⌉ and the buccal, lingual and distal surfaces of 6⌋. Medium deposits were present on the buccal surface of 7⌉ and the buccal, lingual and distal surfaces of 8⌉.

Periodontal disease: Periodontal disease of considerable severity was evident.

Caries: Small caries were visible in the occlusal surface of 8⌉ and the mesial surface of ⌈8.

Ante-mortem tooth loss: ⌊1 7 and 4 1⌉ ⌈1 6 had been lost ante-mortem.

Enamel hypoplasia: Single hypoplastic grooves were present on 3⌉ ⌈3 which indicated that the individual had been subject to at least one physiological insult at the age of 3.5-4 years.

Dental anomalies: The R maxillary 3rd molar was genetically absent.

Context XXXI. 108

Adult 25-35 yrs: Male: 171.5 cm, 5'7"

Year of excavation: 1981

Archaeological context:

Ground pit and ?wooden coffin - A cairn was present on the ground surface and a rectangular pit, orientated south-east to north-west, was partially situated beneath the stones. A layer of thin horizontal logs, followed by a layer of fragments of vertical slabs were recovered from the pit. The remains of a single skeleton were situated in the bottom of the pit. Fragments of a thin log were lying in a transverse position across the skull. It was thought that these fragments may have been the remains of a cover.

Archaeology & Trauma

Body position - The individual was lying in an extended, supine position, with the head to the north-west.

Visual inventory:

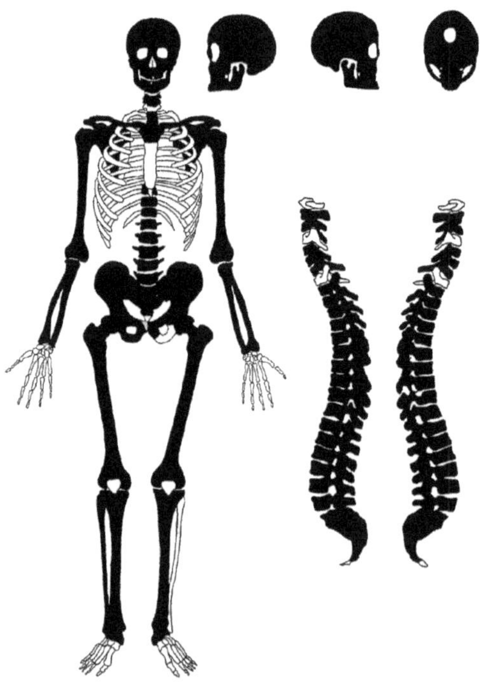

Preservation: The remains were in a good state of preservation with 60-70% of bones present. Slight surface erosion was apparent.

Cranial pathology: Cribra orbitalia of Type 1 was present in Areas 1 of both orbits. A well healed simple oblique fracture was apparent at the distal third of the L nasal bone. The line of the fracture extended onto the maxilla for a few millimetres. A well healed simple oblique fracture was also present on the distal third of the R nasal bone, which extended from the medial margin to the midpoint of the inferior margin. The inferior parts of the fractures were depressed, although all bones were well aligned.

Post-cranial pathology: A well healed simple oblique fracture was present in the distal third of the L ulna (Night-stick fracture). The distal part of the fracture was malaligned laterally by approximately twenty degrees. Hypertrophy was evident at the fracture site, and the parts of the fracture were overlapped by approximately 40 mm. A small sulcus was situated on the posterior surface at the fracture site. Secondary osteophytosis was evident in the L distal humerus, the proximal L radius and the proximal L ulna. An enthesopathy was present in the L ulna (triceps).

Osteophytosis was visible on the acromial facets and glenoid fossae of the scapulae, the proximal R humerus and the proximal and distal R ulna. Rotator cuff disease was apparent in the R humerus. Vertebral osteophytosis was present in CV2, CV3, CV5-CV7 and TV2-LV5. Schmorl's nodes were apparent in TV6-TV12, LV2, LV3 and S1. The body of TV7 was slightly compressed. Enthesopathies were visible in the L humerus (subscapularis), the radii (biceps), the femora (gluteus medius, adductor longus, adductor brevis) and the calcanei (Achilles' tendon).

Weapon trauma: Possible well healed sword cuts were present on the L side of the cranium. A depressed line extended obliquely for 94 mm from the mid parietal area to the greater wing of the sphenoid. Another depressed line was present just inferior to the midpoint of first lesion, which ran parallel to the lesion for a distance of 24.5 mm.

Cranial non-metric traits: R and L highest nuchal lines, L parietal foramen, R and L mastoid foramina extrasutural, R and L posterior condylar canals open, L double anterior condylar canal, R and L accessory supraorbital foramina, R and L anterior ethmoid frontal foramina extrasutural, R and L accessory infraorbital foramina.

Post-cranial non-metric traits: R plaque, R and L exostoses in trochanteric fossae, L third trochanter, R and L acetabular creases, R and L acromial articular facets, R and L circumflex sulci, R and L medial talar facets, L peroneal tubercle, R CV3, R CV5, and R and L CV6 transverse foramina bipartite.

Dentition:

8̸	7	6	5	4	3	2	1	1̸	2	3	4	5	6 (PU)	7	8̸
8 (C)	7	6	5	4	3	2	1̸	1	2	3	4	5	6	7	8

Bite: Overbite.

Attrition: Molar wear patterns of Stage 3 for the 3rd molars, Stage 4 for the 2nd molars and Stage 5 for the 1st molars were apparent on the teeth of the L side of the mouth. Molar wear patterns of Stage 3- for the 3rd molars, Stage 3 for the 2nd molars and Stage 4+ for the 1st molars were apparent on the teeth of the R side of the mouth. The differential wear indicates that the individual predominantly chewed on the L side of the mouth. Secondary dentine was exposed in ⌊6.

Calculus: Flecks of calculus were present on the buccal surfaces of 6 3 2 1⌋ ⌊2 3 4 and 6⌉ ⌈3 4, the lingual surface of 7⌋ and the buccal and lingual surfaces of ⌊6 7 and 8 4 3 2⌉ ⌈1 2 5 6 7 8. Slight deposits were apparent on the distal surface of 7⌋.

Periodontal disease: Periodontal disease of slight severity was evident.

Caries: A small caries was present in the distal surface of 8⌉.

Enamel hypoplasia: Single hypoplastic grooves were present on 7 6 3 2 1⌋ and 3 2⌉ ⌈3 which indicated that the individual had been subject to at least three incidents of physiological stress at the ages of 2.5-3, 3-3.5 and 3.5-4 years.

Context XXXI. 110

Adult 35-45 yrs: Female: 156.1 cm, 5'2"

Year of excavation: 1981

Archaeological context:

Ground pit and cist - The remains of a poorly preserved circular stone enclosure with a central cairn were apparent on the ground surface. A rectangular pit orientated south-east to north-west, which contained loose stones and a stone cist, was situated beneath the cairn. The remains of a single skeleton were recovered from the cist.

Body position - The individual was lying in an extended supine position, with the head to the north-west. The arms were flexed, with the R hand situated near the R scapula, and the L hand positioned above the pelvis. The legs were markedly flexed, with the knees situated near the southwestern wall.

Grave goods - A clay pot was located in the northwestern corner. The remains of an iron knife were recovered from beneath the skull, and a spindle whorl was situated near the L shoulder.

Visual inventory:

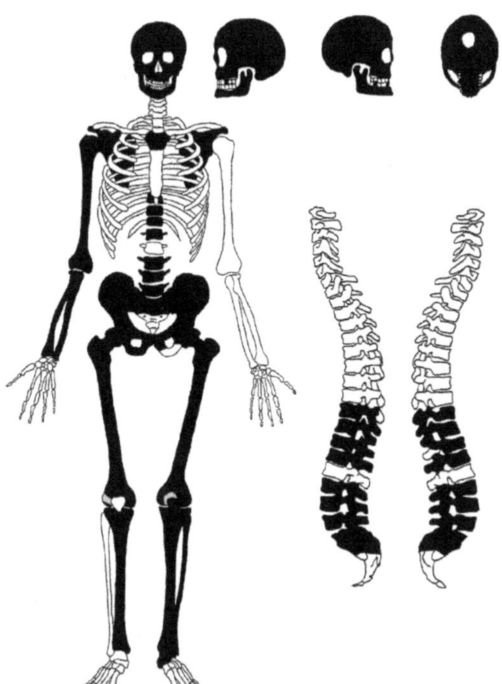

Preservation: The remains were in a good state of preservation with 50-60% of bones present.

Cranial pathology: Porosity was evident on the mandibular condyles, which may be indicative of the early stages of DJD of the temporomandibular joints.

Post-cranial pathology: A very well healed simple oblique fracture was present in the distal third of the L fibula. The distal part of the fracture was deflected anteriorly by approximately ten degrees, and 50% apposition of the fracture parts was evident. Hypertrophy at the site of fracture was visible, and the fracture parts overlapped by 44 mm causing considerable shortening of the bone.

Osteophytosis was present in the glenoid fossae of the scapulae, the distal R humerus, the distal R ulna, the distal R radius, the acetabulae, the distal R femur, the proximal tibiae and the R patella. Rotator cuff disease was present in the R humerus. Vertebral osteophytosis was evident in TV10-LV1 and LV3-LV5. Schmorl's nodes were visible in TV10-LV1, LV3 and LV4. Enthesopathies were apparent in the R radius (biceps), the R patella (rectus femoris) and the R calcaneus (Achilles' tendon).

Cranial non-metric traits: R and L highest nuchal lines, R ossicle in lambdoid suture, R parietal foramen, R and L ossicles at pterion, R and L mastoid foramina extrasutural, R and L posterior condylar canals open, R and L open foramina spinosum, R and L accessory supraorbital foramina, R and L anterior ethmoid frontal foramina extrasutural, R and L accessory infraorbital foramina.

Post-cranial non-metric traits: R and L exostoses in trochanteric fossae, R and L lateral tibial squatting facets, R and L accessory sacral facets, R acromial articular facet, R circumflex sulcus, R vastus fossa, L medial talar facet, L double inferior articular talar facet, L double anterior calcaneal facet.

Other observations: Cutmarks were present on LV4 and the distal L femur, which may be indicative of post-mortem defleshing/ disarticulation.

Dentition:

~~8~~	~~7~~	~~6~~	~~5~~	~~4~~	~~3~~	~~2~~	~~1~~	~~1~~	~~2~~	~~3~~	~~4~~	~~5~~	6 PU	~~7~~	~~8~~
~~8~~	7	~~6~~	~~5~~	4 PU	~~3~~	~~2~~	~~1~~	~~1~~	~~2~~	~~3~~	~~4~~	~~5~~	~~6~~	~~7~~	~~8~~

Bite: Overbite.

Attrition: Molar wear patterns of Stage 5 for the 1st and 2nd molars was apparent. Secondary dentine was exposed in ⌊6 and 4⌉.

Calculus: Slight deposits of calculus were present on the lingual surface of 7⌉ and the buccal and lingual surfaces of ⌊6 and 4⌉.

Periodontal disease: Periodontal disease of considerable severity was evident.

Ante-mortem tooth loss: 8 7 6 4 2 1⌋ ⌊4 7 8 and 6⌉ ⌈5 6 7 8 had been lost ante-mortem. The sockets were generally in the late stages of remodelling.

Context XXXI. 111

Adult 25-35 yrs: Male: 170.9 cm, 5'7"

Year of excavation: 1981

Archaeological context:

Ground pit and wooden coffin - A poorly preserved circular stone enclosure was apparent on the ground surface. A rectangular pit orientated south-west to north-east, which contained a layer of stones and a wooden plank coffin, was situated within the enclosure. The lid of the coffin consisted of three planks, and vertical slabs surrounded the coffin. A single skeleton was recovered from the coffin.

Body position - The individual was lying in an extended position, with the head to the north-east. The R arm was flexed so that the hand was lying on the sacrum.

Grave goods - The remains of a small wooden bowl were located near the southeastern wall. A fragment of an iron object and a whetstone were situated on the L side of the thorax. The remains of a circular iron buckle and fragments of an iron object were positioned superiorly to the L side of the pelvis. A fragment of an iron object was situated above the pelvis, and the remains of an iron knife were located near the L side of the pelvis. A wooden object and an iron object with a wooden handle, were situated between the femora. Fragments of an iron object were located underneath the L femur.

Visual inventory:

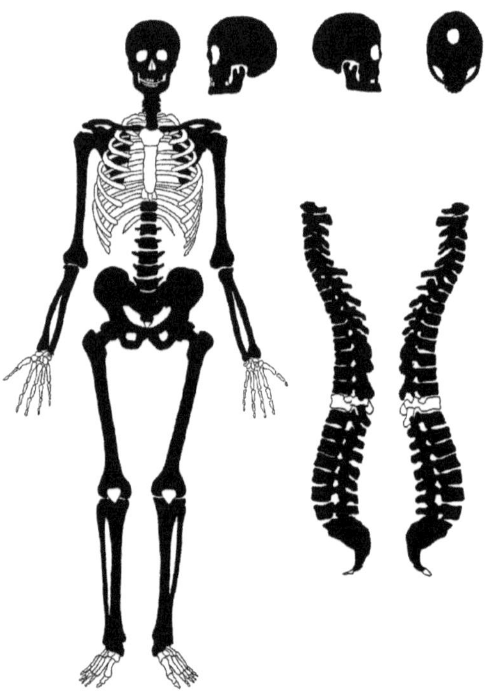

Preservation: The remains were in a very good state of preservation with 70-80% of bones present.

Post-cranial pathology: Os acromiale was present in the L scapula. A circular roughened area of bone apparent on the anterior surface of the capitulum of the R humerus may be indicative of osteochondritis dissecans. Vertebral osteophytosis was evident in TV3-TV6 and TV12-LV2. Schmorl's nodes were visible in TV7-TV9 and TV11-LV4. Enthesopathies were apparent on the clavicles (costoclavicular ligament).

Cranial non-metric traits: R and L highest nuchal lines, L parietal foramen, R and L ossicles at pterion, R and L mastoid foramina extrasutural, R and L posterior condylar canals open, R and L absent zygomaticofacial foramina, R and L accessory supraorbital foramina, R and L anterior ethmoid frontal foramina extrasutural, R and L accessory infraorbital foramina.

Post-cranial non-metric traits: R plaque, R and L hypotrochanteric fossae, R and L lateral tibial squatting facets, R and L acetabular creases, R and L acromial articular facets, R and L circumflex sulci, R and L medial talar facets, R and L double inferior articular talar facets, R and L double anterior calcaneal facets, R CV6 transverse foramen bipartite.

Dentition:

8	7	6	5	4	~~3~~	~~2~~	~~1~~	~~1~~	2	~~3~~	4	5	6	7	8
8	7	6	~~5~~	4	~~3~~	~~2~~	~~1~~	~~1~~	~~2~~	~~3~~	~~4~~	5	6	7	~~8~~

Bite: Overbite.

Attrition: Molar wear patterns of Stage 1 for the 3rd molars, Stage 3- for the 2nd molars and Stage 4+ for the 1st molars were apparent.

Calculus: Flecks of calculus were present on the buccal surfaces of ⌊4 5. Slight deposits were apparent on the buccal surfaces of 5 4⌋ ⌊2, the lingual surfaces of 7⌉ ⌈5, the buccal and lingual surfaces of 7 6⌋ ⌊6 7 and 6 4⌉ ⌈6 and the buccal, lingual and distal surfaces of 8⌋ ⌊8 and 8⌉. Medium deposits were visible on the lingual surface of ⌈7.

Periodontal disease: Periodontal disease of slight severity was evident.

Enamel hypoplasia: Single hypoplastic grooves were present on 7 6 5 4⌋ ⌊4 5 6 7 and 7 4⌉ which indicated that the individual had been subject to at least three incidents of physiological stress at the ages of 2-2.5, 3-3.5 and 5-5.5 years.

Dental anomalies: The L maxillary lateral incisor was shovel-shaped.

Context XXXI. 115

Adult 35-45 yrs: Female: 150.8 cm, 5'0"

Plates: 46, 47, 48

Year of excavation: 1981

Archaeological context:

Ground pit and wooden coffin - A poorly preserved circular stone enclosure was apparent on the ground surface. A rectangular pit orientated south-west to north-east, which contained a rectangular plank coffin with a lid constructed from a single plank, was situated within the enclosure. Several slabs were lying on top of the coffin lid. A single skeleton was recovered from the coffin.

Body position - The individual was lying in an extended, supine position, with the head to the north-east. The R arm was flexed with the hand lying on the L side of the pelvis. The legs were slightly flexed to the left, and the skull was disturbed.

Grave goods - The remains of a birch bark vessel were located beneath and to the posterior of the skull. The remains of a clay pot and an oval wooden dish were situated near the vessel. The remains of an iron knife with a wooden handle were lying on the dish. Fragments of an iron object were recovered from amongst the ribs, and another fragment of an iron object was situated near the inferior aspect of the L side of the thorax. The remains of two wooden objects were lying near the proximal L femur, and a fragment of a wooden object was positioned between the lower legs. The remains of a wooden vessel were situated near the feet, and an iron awl with a wooden handle and a fragment of a wooden object were situated on top of the vessel.

Visual inventory:

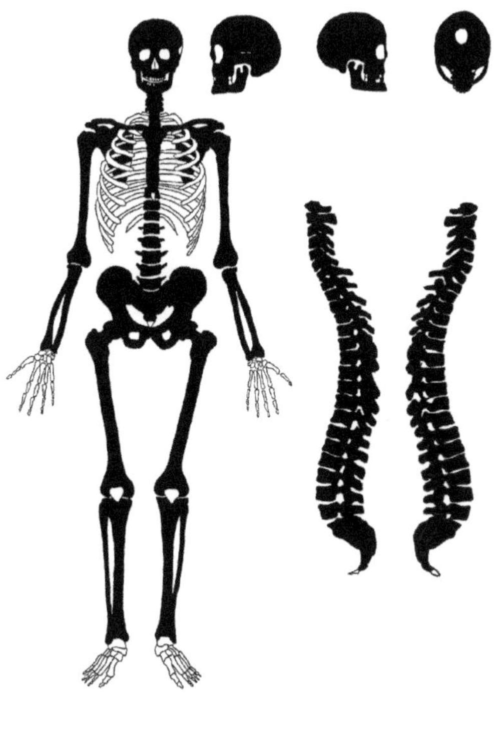

Preservation: The remains were in a very good state of preservation with 70-80% of bones present. A few bones displayed surface erosion and fragmentation.

Post-cranial pathology: The manubrio-mesosternal joint was fused, probably due to a failure of the fibrous lamina to develop between the manubrium and the first sternebra. Osteophytosis was apparent at the auricular surfaces of the R and L ilia and both sides of the sacrum. Enthesopathies were apparent on the ischial tuberosities (adductor magnus, semimembranosus, semitendinosus, long head of biceps).

Weapon trauma: Several examples of sword injuries were apparent in the remains of this woman. Two sword chops were present on the L side of the cranium. The injuries were situated in very close proximity to one another and it was difficult to follow their individual courses. The first chop mark appeared to extend from the supero-posterior border of the temporal for 68 mm to the midpoint of the zygomatic. The chop mark was almost transverse, and as it had cut through the posterior aspect of the zygomatic it had exposed the underlying spongy bone. The injury was probably made by an attacker who was standing just anterior or to the left of the victim, and who had swept the sword anteriorly. In the second injury it appeared that a vertical stroke had chopped through the cortical surface of the anterior zygomatic for approximately 22.5 mm, exposing the underlying spongy bone. In this instance the attacker was probably standing directly anterior to the victim, and had brought the sword down from above his or her head, slicing the victim's cheek in the process.

A very deep sword chop was present on the lateral aspect of the blade of the R scapula. The injury continued straight through the blade, and had cleanly sliced through the midpoint of the spine. The superior aspect of the blade displayed a large oval perforation which may also have been caused by a sword. It is probable that the assailant was standing to the posterior of the victim, and that the sword was raised above the attacker's head and then brought down with a great deal of force onto the victim's R shoulder.

A small, sharp oblique cutmark, which measured 6 mm supero-inferiorly, was present on the antero-lateral margin of the proximal third of the L humerus. Inferior to the cut mark, at the midshaft of the humerus, the bone had been cleaved in two by a sword. A very deep oblique chop mark, which measured approximately 23 mm supero-inferiorly, was apparent on the antero-lateral surface. The cutmark had cleaved through the lateral two-thirds of the bone, causing the bone to fracture at its medial aspect and to separate into two parts. It is probable the attacker had been standing either directly anterior or posterior to the victim, and that the injury was made initially to the lateral surface of the arm. A shallow oblique sword cut, which measured 14 mm supero-inferiorly by 8.5 mm medio-laterally, was present on the posterior surface of the proximal third of the L radius. It appeared that the sword had just nicked the surface of the bone in this injury. The attacker would have been standing either directly posterior or anterior to the victim.

The six different examples of weapon trauma apparent in the remains of the individual suggest that she had been attacked by an individual, or individuals who were not stationary. The variety and quantity of injuries suggests that the woman was actively engaged in the combat, probably also wielding a weapon making it difficult for her attacker to make a clean fatal stroke. The lack of healing around any of the injuries indicates that they were all attained in a single event, and that she had died as a result of the armed conflict.

Cranial non-metric traits: 2 ossicles at lambda, L ossicle in lambdoid suture, R and L parietal foramina, R and L ossicles at pterion, R and L ossicles at parietal notch, R and L ossicles at asterion, R and L mastoid foramina extrasutural, R and L posterior condylar canals open, R double anterior condylar canal, R and L open foramina spinosum, R and L accessory supraorbital foramina, R and L anterior ethmoid frontal foramina extrasutural, R and L accessory infraorbital foramina.

Post-cranial non-metric traits: L Allan's fossa, R and L exostoses in trochanteric fossae, R and L lateral tibial squatting facets, R and L accessory sacral facets, R and L acromial articular facets, R and L circumflex sulci.

Dentition:

8̸	7	6	5	4̸	3̸	2̸	1̸	1̸	2̸	3̸	4	5	6	7	8̸
8 NP	7	6	5̸	4	3̸	2̸	1̸	1̸	2	3	4	5	6	7	8 NP

Bite: Overbite.

Attrition: Molar wear patterns of Stage 1 for the 2nd molars, and Stage 3 for the 1st molars were apparent. The attrition was unusually light for an individual of 35-45 years.

Calculus: Flecks of calculus were present on the buccal surfaces of 7 6⌉and the buccal and lingual surfaces of 4⌉. Slight deposits were visible on the buccal surface of ⌊7, the lingual surfaces of 5⌋ and 1⌉ ⌈2, the buccal and lingual surfaces of 7 6⌋ ⌊5 6 and the buccal, lingual and distal surfaces of ⌈7. Medium deposits were visible on the lingual surfaces of 7 6⌉ ⌈3 4 5 and the buccal and lingual surfaces of ⌈6.

Periodontal disease: Periodontal disease of slight severity was apparent.

Enamel hypoplasia: Single hypoplastic grooves were present on 3⌋and 3⌉ ⌈3 which indicated that the individual had been subject to at least one incident of physiological stress at the age of 3.5-4 years.

Dental anomalies: Both mandibular 3rd molars were genetically absent. The morphology of the socket for the R maxillary 3rd molar indicated that it had been peg-shaped. Enamel pearls were present on the distal aspects of both maxillary 2nd molars.

Context XXXI. 127 (i)

Adult 25-35 yrs: Male: 169.1 cm, 5'6"

Year of excavation: 1981

Archaeological context:

NB: Three individuals were labelled as Context XXXI. 27.

Ground pit and log coffin - A poorly preserved circular enclosure with an oblong-shaped central cairn, was apparent on the ground surface. A rectangular pit orientated south-south-east to north-north-west, which contained a rectangular log coffin surrounded by vertical slabs, was situated beneath the cairn. The lid of the coffin consisted of a single plank. The remains of a single individual were recovered from the coffin.

Body position - The individual was lying on his R side, with the head to the north-north-west. Both arms were flexed, with the R hand located near the western wall opposite the pelvis. The L hand was lying on top of the R hand, and the legs were flexed.

Grave goods - Two clay pots and a bone pin were situated posterior to the skull. A bronze earring was recovered from beneath the skull. The remains of an iron knife with a wooden handle were situated beneath the R hand. Two small circular iron buckles were lying on the lumbar vertebrae, and the remains of an iron knife was located beneath the R femur. A bone bowstrap was situated near the knees.

Visual inventory:

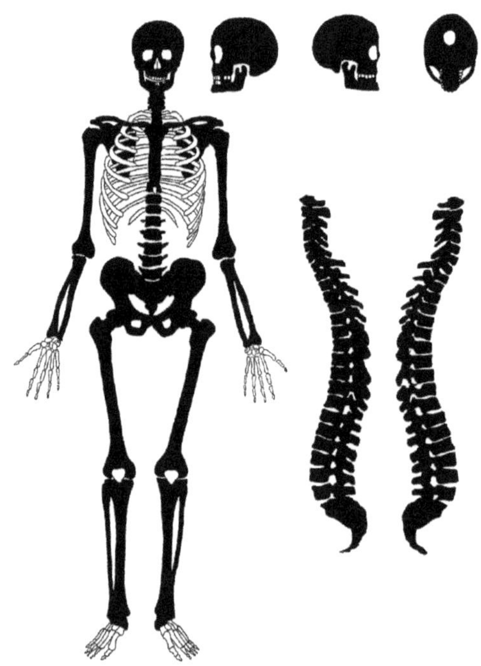

Preservation: The remains were in a very good state of preservation with 70-80% of bones present.

Cranial pathology: Cribra orbitalia of Type 5 was present in Areas 1, 2 and 4 of the R orbit and Areas 1, 2, 4 and 7 of the L orbit. Extensive porosity was apparent on the articular tubercles of the temporals, which may be indicative of DJD of the temporomandibular joints.

Post-cranial pathology: A well healed simple oblique fracture was present at the distal third of the L fibula. No malalignment was evident and the fracture parts displayed 100% apposition. Slight hypertrophy was apparent at the point of fracture. The L talus appeared to have been subject to a crushing fracture at the lateral malleolus of the posterior calcaneal articular surface. The fracture was comminuted and transverse. The greatest breadth of the fractured facet was 44.7 mm, nearly 11 mm greater than the breadth of the corresponding R surface. The trochlear surface displayed extensive secondary osteophytosis. A smaller crushed area was present on the posterior articular facet of the L calcaneus. Osteophytes, porosity and eburnation were apparent on the articular surfaces of the calcaneus. It is possible that the three fractures evident on the L lower leg and the ankle had been attained in a single incident.

A bony exostosis was present on the antero-lateral surface of the midshaft of the R humerus. The exostosis may be indicative of myositis ossificans of pectoralis major. An ossified haematoma, which measured 14.6 mm supero-inferiorly by 3.5 mm medio-laterally, was present on the medial surface of the midshaft of the R tibia. Periosteal reactive woven bone surrounded the nodule.

Osteophytosis was evident on the R clavicle, the acromion of the R scapula, the distal R humerus, the proximal ulnae, the distal R femur, the proximal L tibia, the proximal and distal R tibia and the distal R fibula. Rotator cuff disease was apparent in the humeri. DJD was present in the distal ulnae, the distal radii and the distal L femur. Vertebral osteophytosis was present in CV2-S1. Several vertebrae also displayed DJD. Schmorl's nodes were present in TV10 and TV11. Enthesopathies were visible in the radii (biceps), the femora (gluteus medius, psoas major, iliacus, adductor longus, adductor medius) and the calcanei (Achilles' tendon).

Post-cranial non-metric traits: R and L hypotrochanteric fossae, R and L medial tibial squatting facets, L accessory sacral facets, R and L acromial articular facets, R and L circumflex sulci, R and L medial talar facets, R CV5, R CV6 and R CV7 transverse foramina bipartite.

Other observations: Orange discoloration (rust) was present on the R and L parietals and temporals.

Dentition:

		2A										A		PU	2A	A
	~~8~~	~~7~~	~~6~~	~~5~~	~~4~~	~~3~~	~~2~~	~~1~~	~~1~~	~~2~~	~~3~~	~~4~~	~~5~~	6	~~7~~	~~8~~
	8	7	6	5	4	3	2	~~1~~	~~1~~	2	3	~~4~~	~~5~~	6	7	8
		PU		PU	PU	PU	PU			PU	PU					

Bite: Overbite.

Attrition: Molar wear patterns of Stage 3+ for the 3rd molars, Stage 4+ for the 2nd molars and Stage 5 for the 1st molars were apparent. Secondary dentine or pulp cavity exposure was evident in ⌊6 and 7 5 4 3 2⌉ ⌈2 3.

Calculus: Flecks of calculus were present on the buccal surface of ⌊6 and the buccal and lingual surfaces of ⌈7. Slight deposits were visible on the buccal surfaces of 8 6⌉, the lingual surface of ⌈8 and the buccal and lingual surfaces of 7 4 3 2⌉ ⌈2 3 6. Medium deposits were evident on the lingual surfaces of 6 5⌉, the buccal and distal surfaces of ⌈8 and the lingual and distal surfaces of 8⌉.

Periodontal disease: Periodontal disease of medium severity was apparent.

Abscesses: A medium sized externally draining abscess was associated with ⌊4. A small internally draining abscess, and a medium sized externally draining abscess were both associated with 6⌋. A single large externally draining abscess was associated with ⌊7 8, and a medium sized externally draining abscess was also associated with ⌊7.

Ante-mortem tooth loss: 5⌋ ⌊7 had been lost ante-mortem.

Enamel hypoplasia: Single hypoplastic grooves were present on 3⌉ ⌈3 which indicated that the individual had been subject to at least one incident of physiological stress at the age of 3.5-4 years.

Context XXXI. 127 (iii)

Adult 25-35 yrs: Male: 165.7 cm, 5'5"

Year of excavation: 1981

Archaeological context:

NB: Three individuals were labelled as Context XXXI. 27.

Ground pit and log coffin - A poorly preserved circular enclosure with an oblong-shaped central cairn, was apparent on the ground surface. A rectangular pit orientated south-south-east to north-north-west, which contained a rectangular log coffin surrounded by vertical slabs, was situated beneath the cairn. The lid of the coffin consisted of a single plank. The remains of a single individual were recovered from the coffin.

Body position - The individual was lying on his R side, with the head to the north-north-west. Both arms were flexed, with the R hand located near the western wall opposite the pelvis. The L hand was lying on top of the R hand, and the legs were flexed.

Grave goods - Two clay pots and a bone pin were situated posterior to the skull. A bronze earring was recovered from beneath the skull. The remains of an iron knife with a wooden handle were situated beneath the R hand. Two small circular iron buckles were lying on the lumbar vertebrae, and the remains of an iron knife was located beneath the R femur. A bone bowstrap was situated near the knees.

Visual inventory:

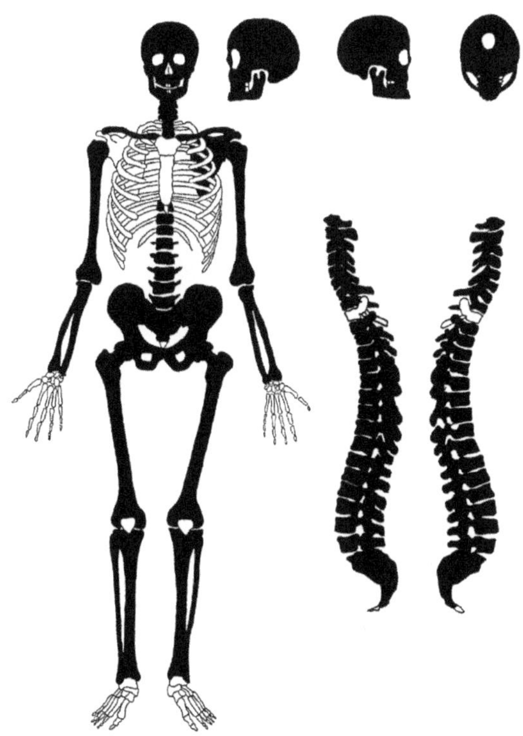

Preservation: The remains were in a good state of preservation with 60-70% of bones present.

Post-cranial pathology: Several CV displayed congenital abnormalities. The transverse process on the R side of CV1 was hypoplastic, and CV2 and CV3 had formed a block vertebra. In addition these vertebrae were dysplastic with the R side of CV3 displaying an abnormal morphology, and the L side being aplastic. CV4 was dysplastic with the L side hyperplastic and the R side normal. CV5 had a normal morphology, and CV6 and CV7 formed a block vertebra. The developmental defects apparent in the cervical vertebrae are paraxial mesoderm field defects which involved errors of segmentation. It is possible that CV3 was a solitary hemivertebral segment which had resulted from hemimetamere shifting, and that the hyperplasia of the L side of CV4 was to compensate for the abnormality. The block vertebrae CV2-CV3 and CV6-CV7 were probably due to the failure of segmentation of the vertebral segments (Type II Klippel-Feil syndrome).

Complete unilateral sacralization was evident with the L transverse process of LV5 fused to the L ala of S1. The R sides of LV5 and the sacrum were normal. A raised area of bone which measured 43 mm supero-inferiorly, and was probably an ossified haematoma, was apparent on the postero-lateral surface of the distal third of the shaft of the R tibia. Vertebral osteophytosis was present in TV10-LV1, and Schmorl's nodes were apparent in TV7-LV1. Enthesopathies were apparent in the clavicles (costoclavicular ligament) and the R humerus (teres major).

Cranial non-metric traits: R and L highest nuchal lines, L ossicle in lambdoid suture, R and L parietal foramina, R ossicle at pterion, R ossicle at parietal notch, R and L ossicles at asterions, R and L mastoid foramina extrasutural, R posterior condylar canals open, R double anterior condylar canal, R and L accessory supraorbital foramina, R and L anterior ethmoid frontal foramina extrasutural, R and L accessory infraorbital foramina.

Post-cranial non-metric traits: R and L plaque, R and L hypotrochanteric fossae, R and L lateral tibial squatting facets, L acromial articular facet, L circumflex sulcus, R and L double atlas facets, L CV1 and L CV6 transverse foramina bipartite.

Dentition:

NP 8	7̸	6	5	4	3	2	1̸	1̸	2	3	4	5	6	7	NP 8
8	7	6	5	4	3	2	1̸	1̸	2	3	4	5	6	7	8 NP

Bite: Overbite.

Attrition: Molar wear patterns of Stage 1 for the 2nd and 3rd molars, and Stage 3 for the 1st molars were apparent.

Calculus: Flecks of calculus were present on the buccal surface of ⌈7, the buccal and lingual surfaces of 6 5 4 3 2⌋ ⌊2 4 5 6 and the buccal, lingual and distal surfaces of ⌊7. Slight deposits were apparent on the buccal and lingual surfaces of ⌊3 and 7 6 5 4 3 2⌉ ⌈2 3 4 5 6 and the buccal, lingual and distal surfaces of 8⌉. Medium deposits were evident on the lingual surface of ⌈7.

Periodontal disease: Periodontal disease of slight severity was evident.

Enamel hypoplasia: Single hypoplastic grooves were present on 5 4 3⌋ ⌊4 5 and 2⌉ which indicated that the individual had been subject to at least one incident of physiological stress at the age of 4.5-5 years.

Dental anomalies: The L mandibular 3rd molar and both maxillary 3rd molars were genetically absent.

Context XXXI. 137

Adult 35-45 yrs: Male: 165.1 cm, 5'5"

Year of excavation: 1982 or 1983

Archaeological context:

The archive reports for the 1982 and 1983 excavations were not available for consultation.

Visual inventory:

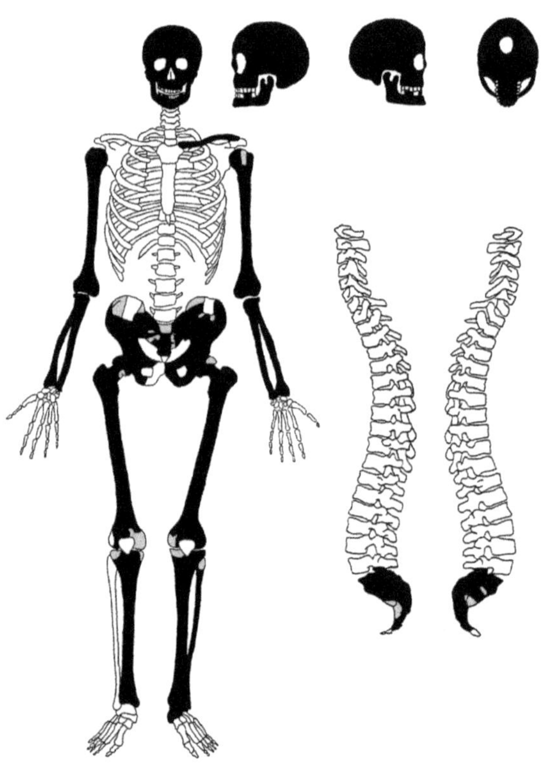

Preservation: The remains were in a fair state of preservation with 50-60% of bones present. Most of the bones displayed surface erosion and were fragmentary.

Cranial pathology: Cribra orbitalia of Type 1 was present in Areas 1, 2 and 4 of the R orbit. The lesion was not present in the L orbit. A poorly healed simple transverse fracture was visible on the distal third of the R nasal bone. An oval area of the medial aspect of the L nasal bone had been fractured in the same event, and the distal aspects of the fractures were depressed. Porosity was apparent on the articular tubercle of the L temporal, which may be indicative of the early stages of DJD of the L temporomandibular joint.

Post-cranial pathology: Osteophytosis was evident in the acromial end of the L clavicle, the proximal humeri, the distal ulnae, the acetabulae and S1. Rotator cuff disease was present in the humeri. Enthesopathies were visible on the L clavicle (costoclavicular ligament), the humeri (deltoid, latissimus dorsai, pectoralis major, teres major, lateral head of triceps), the radii (biceps), the femora (gluteus maximus, adductor longus, adductor medius), the tibiae (soleus) and the L calcaneus (Achilles' tendon).

Preservation: The remains were in a fair state of preservation with 50-60% of bones present. Most of the bones displayed surface erosion and were fragmentary.

Cranial pathology: Cribra orbitalia of Type 1 was present in Areas 1, 2 and 4 of the R orbit. The lesion was not present in the L orbit. A poorly healed simple transverse fracture was visible on the distal third of the R nasal bone. An oval area of the medial aspect of the L nasal bone had been fractured in the same event, and the distal aspects of the fractures were depressed. Porosity was apparent on the articular tubercle of the L temporal, which may be indicative of the early stages of DJD of the L temporomandibular joint.

Post-cranial pathology: Osteophytosis was evident in the acromial end of the L clavicle, the proximal humeri, the distal ulnae, the acetabulae and S1. Rotator cuff disease was present in the humeri. Enthesopathies were visible on the L clavicle (costoclavicular ligament), the humeri (deltoid, latissimus dorsai, pectoralis major, teres major, lateral head of triceps), the radii (biceps), the femora (gluteus maximus, adductor longus, adductor medius), the tibiae (soleus) and the L calcaneus (Achilles' tendon).

Periodontal disease: Periodontal disease of considerable severity was apparent.

Caries: A small caries was present in the distal surface of ⌈7.

Abscesses: Medium sized externally draining abscesses were associated with 4⌋, ⌊3 and ⌊6.

Ante-mortem tooth loss: 7 6 5 3⌋ ⌊4 5 7 and ⌈1 had been lost ante-mortem.

Dental anomalies: The maxillary 3rd molars were genetically absent.

Context XXXI. 138

Adult 25-35 yrs: Male: 172.4 cm, 5'7"

Year of excavation: 1982 or 1983

Archaeological context:

The archive reports for the 1982 and 1983 excavations were not available for consultation.

Visual inventory:

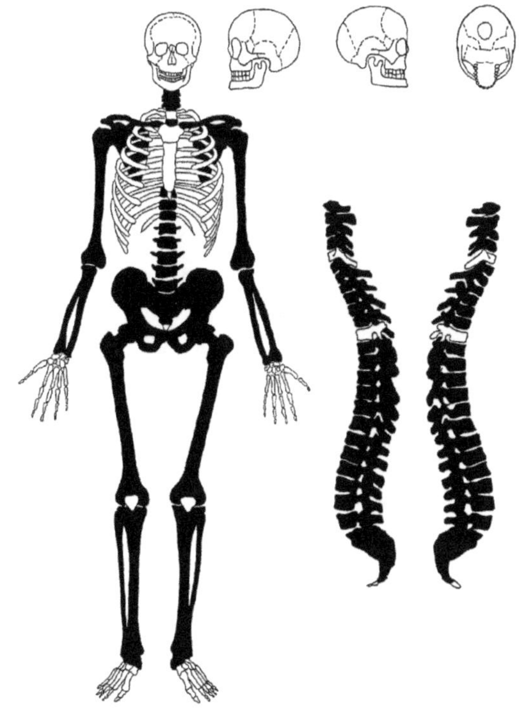

Preservation: The remains were in a good state of preservation with 60-70% of bones present.

Post-cranial pathology: A well healed simple transverse fracture was apparent at the midshaft of the L ulna (Night-stick fracture). Almost fifty percent apposition, and overlap of the fracture parts by approximately 5 mm was evident. The inferior part of the fracture was displaced anteriorly by approximately thirty degrees. Osteophytosis was apparent in the distal L femur, the proximal and distal R femur and the proximal tibiae. Vertebral osteophytosis was visible in CV2-CV4, CV6-TV1, TV3, TV4 and TV8-LV2. Schmorl's nodes were apparent in TV10 and TV11. Enthesopathies were present in the R femur (adductor longus, adductor medius) and the calcanei (Achilles' tendon).

Post-cranial non-metric traits: R and L plaque, R and L hypotrochanteric fossae, R and L lateral tibial squatting facets, R and L acetabular creases, R and L accessory sacral facets, R and L acromial articular facets, R and L circumflex sulci, R and L medial talar facets, R and L double inferior articular talar facets, L double anterior calcaneal facet.

Context XXXI. 139 (i)

Adult 25-35 yrs: Male: 168.6 cm, 5'6"

Year of excavation: 1982 or 1983

Archaeological context:

NB: Two individuals were labelled as Context XXXI. 139.

The archive reports for the 1982 and 1983 excavations were not available for consultation.

Visual inventory:

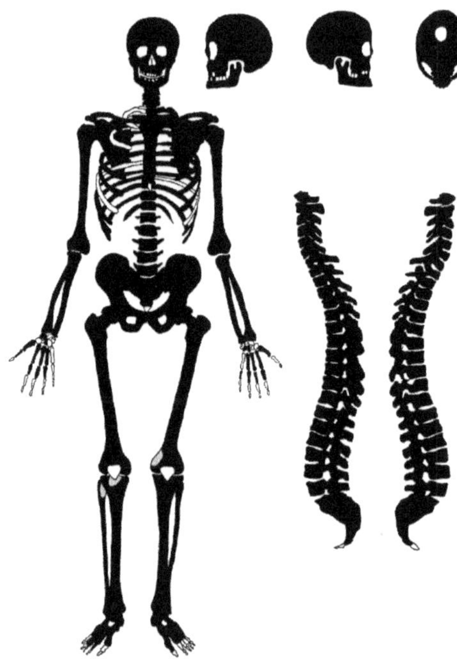

Preservation: The remains were in an excellent state of preservation with 90-100% of bone present. Slight surface erosion was evident on several bones.

Post-cranial pathology: A very well healed simple oblique fracture was present at the midshaft of the L MC4. The bone had a slightly curved appearance, although there was 100% apposition of the fracture parts. Slight hypertrophy was apparent at the site of fracture, and the fracture parts overlapped by approximately 5 mm. The L MC4 had a greatest length of 57.3 mm (R MC4 = 61.1 mm). A possible crushing fracture was present on the proximal articular facet of L MT1. The appearance of the surface indicated that the fracture had been comminuted, and hypertrophy was evident at the site of the fracture. The fracture lines were still prominent. Possible osteochondritis dissecans was apparent on the lateral condyle of the L femur, where a circular depressed area, which measured 24.4 mm by 14.1 mm, was apparent. Evidence for an accessory ossicle was present at the distal end of the R calcaneus.

Vertebral osteophytosis was apparent in CV1, CV2, CV4-TV10, TV12, LV3 and LV5. Schmorl's nodes were present in TV10-LV1 and LV3. Enthesopathies were visible on the clavicles (trapezoid ligament, conoid ligament, costoclavicular ligament), the L humerus (pectoralis major, deltoid), the R humerus (pectoralis major, deltoid, teres major, latissimus dorsai) and the calcanei (Achilles' tendon),

Cranial non-metric traits: R and L highest nuchal lines, 2 L and 1 R ossicles in lambdoid suture, R and L parietal foramina, R and L ossicles at asterions, R and L mastoid foramina extrasutural, R and L posterior condylar canals open, R and L accessory supraorbital foramina, R and L anterior ethmoid frontal foramina extrasutural, R and L accessory infraorbital foramina.

Post-cranial non-metric traits: R and L plaque, R and L lateral tibial squatting facets, R and L accessory sacral facets, R and L acromial articular facets, R and L circumflex sulci, R and L medial talar facets, L double inferior articular talar facet, L double anterior calcaneal facet, R and L peroneal tubercles, L double atlas facet, L CV4, R and L CV5, and R and L CV6 transverse foramina bipartite.

Other observations: Green discoloration was apparent on the L mastoid process and the L side of CV1. Fragments of body tissue were adhering to CV1 and CV2.

Dentition:

8	7	6	~~5~~	~~4~~	3	~~2~~	~~1~~	~~1~~	~~2~~	3	4 (A)	~~5~~	6	7	8
~~8~~	~~7~~	6	5	4	~~3~~	2	~~1~~	1	~~2~~	3	4	~~5~~	6	~~7~~	8 (NP)

Bite: Overbite.

Attrition: Molar wear patterns of Stage 1 for the 3rd molars, Stage 3 for the 2nd molars and Stage 4 for the 1st molars were apparent. The wear on the teeth of the R side of the mouth was less than for those on the L side of the mouth, which may indicate that the individual predominantly chewed using the L side of the mouth.

Calculus: Flecks of calculus were present on the buccal side surface of 6⏋, the lingual surfaces of 4⏋⎾1, the buccal and lingual surfaces of 8 7⏌ ⎿3 and 5⏋ ⎾3 6 and the buccal, lingual and distal surfaces of ⎾4. Slight deposits were visible on the lingual surfaces of 6 3⏌and 2⏋and the buccal and lingual surfaces of ⎿7. Medium deposits were visible on the distal surface of 8⏌and the distal and occlusal surfaces of ⎿8. The presence of calculus on the occlusal surface of ⎿8 may be related to the genetic absence of the corresponding mandibular molar.

Periodontal disease: Periodontal disease of slight severity was apparent.

Abscesses: A medium sized externally draining abscess was associated with ⎿4.

Ante-mortem tooth loss: 7⏋had been lost ante-mortem, and the socket was in the early stages of remodelling.

Enamel hypoplasia: Single hypoplastic grooves were present on 7 6⏌ ⎿3 and ⎾3 which indicated that the individual had been subject to at least one incident of physiological stress at the age of 3.5-4 years.

Dental anomalies: The L mandibular 3rd molar was genetically absent.

Context XXXI. 148

Adult 35-45 yrs: Female: 152.8 cm, 5'1"

Year of excavation: 1982 or 1983

Archaeological context:

The archive reports for the 1982 and 1983 excavations were not available for consultation.

Visual inventory:

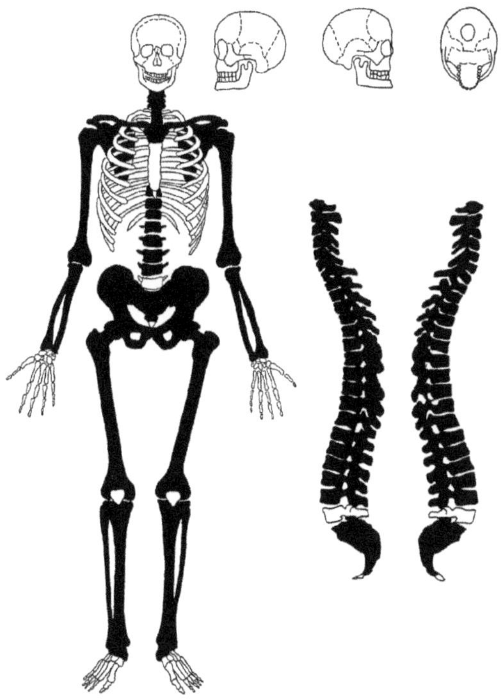

Preservation: The remains were in a good state of preservation with 60-70% of bones present.

Post-cranial pathology: The neural arch of TV12 was bifurcated, which was probably due to a developmental delay during the early morphogenesis of the vertebral column. Spondylolysis was present in LV4. DJD apparent on the inferior surfaces of the inferior articular facets of LV3 was probably secondary to the spondylolysis. Vertebral osteophytosis was present in CV2-CV4, CV6-TV12, LV3 and LV4. Schmorl's nodes were visible in TV7, TV8 and TV11. Enthesopathies were apparent in the R clavicle (trapezoid ligament) and the calcanei (Achilles' tendon).

Post-cranial non-metric traits: L Poirier's fossa, L hypotrochanteric fossa, L third trochanter, R and L lateral tibial squatting facets, R and L septal apertures, R and L acetabular creases, L accessory sacral facets, R and L acromial articular facets, R and L circumflex sulci, R and L medial talar facets, R and L peroneal tubercles, R CV5, R CV6 and R CV7 transverse foramina bipartite.

Context XXXI. 149

Young adult 17-25 yrs: Female: 151 cm, 5'0"

Year of excavation: 1982 or 1983

Archaeological context:

The archive reports for the 1982 and 1983 excavations were not available for consultation.

Visual inventory:

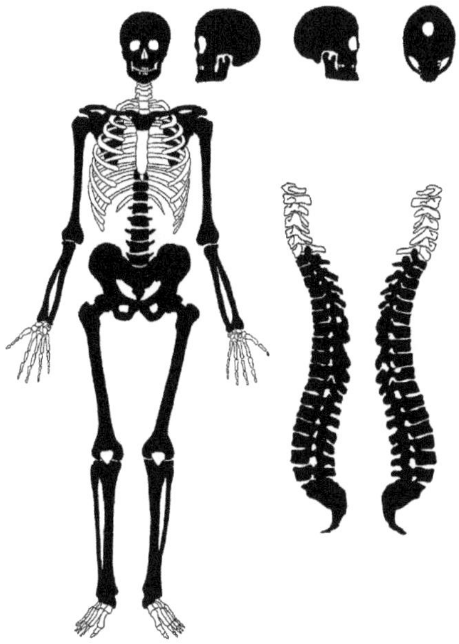

Preservation: The remains were in a very good state of preservation with 70-80% of bones present.

Cranial pathology: A simple transverse fracture in the mid-stages of healing, was apparent on the distal half of the R nasal bone. The distal part of the fracture was slightly depressed. A simple oblique fracture, also in the mid-stages of healing, was present on the distal half of the L nasal bone. No malalignment of the fracture parts was evident. Porosity was apparent on the superior surfaces of the frontal, the parietals and the occipital, which may be indicative of healed porotic hyperostosis.

Post-cranial pathology: Spondylolysis was apparent in LV3.

Cranial non-metric traits: R and L highest nuchal lines, R ossicle in lambdoid suture, 2 R parietal foramen, R and L mastoid foramina extrasutural, R and L posterior condylar canals open, R double anterior condylar canal, R and L accessory supraorbital foramina, R and L anterior ethmoid frontal foramina extrasutural, R and L accessory infraorbital foramina.

Post-cranial non-metric traits: R and L Allan's fossae, R and L lateral tibial squatting facets, R and L acromial articular facets, R and L circumflex sulci, R and L medial talar facets, R and L double inferior articular talar facets, R and L double anterior calcaneal facets, R and L peroneal tubercles.

Dentition:

~~8~~	~~7~~	6	5	~~4~~	3	2	~~1~~	~~1~~	~~2~~	3	~~4~~	~~5~~	6	7	8 (E)
8	7	6	5	4	~~3~~	2	~~1~~	~~1~~	~~2~~	3	4	5	6	7	8

Bite: Overbite.

Attrition: Molar wear patterns of Stage 1 for the 2nd and 3rd molars, and Stage 2+ for the 1st molars were apparent.

Calculus: Flecks of calculus were present on the buccal surfaces of 7 6 5 4⌉ ⌈4 6 and the buccal and lingual surfaces of 3 2⌋ ⌊3. Slight deposits were evident on the buccal surfaces of ⌈3 5 7, the lingual surfaces of 5 4⌉ ⌈4, the distal surface of ⌈8 and the buccal and lingual surfaces of 6 5⌋ ⌊6 7 and 8⌉. Medium deposits were present on the lingual surfaces of 7 6 2⌉ ⌈3 5 6 7.

Periodontal disease: Periodontal disease of slight severity was present.

Enamel hypoplasia: Single hypoplastic grooves were present on 2⌋ ⌊3 and 5 4⌉ ⌈3 4 which indicated that the individual had been subject to at least one incident of physiological stress at the age of 3.5-4 years.

Dental anomalies: The R maxillary lateral incisor was shovel-shaped. Small mandibular tori were apparent.

Context XXXI. 157

Adult 35-45 yrs: Male: 162.4 cm, 5'4"

Plate: 28

Year of excavation: 1982 or 1983

Archaeological context:

The archive reports for the 1982 and 1983 excavations were not available for consultation.

Visual inventory:

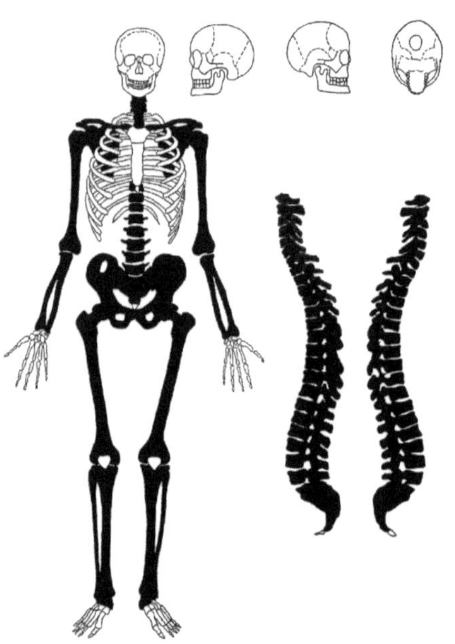

Preservation: The remains were in a good state of preservation with 60-70% of bones present.

Post-cranial pathology: Thirteen thoracic vertebrae were apparent. An ossified haematoma, which measured 23 mm supero-inferiorly by 16.7 mm antero-posteriorly, was present on the medial surface of the midshaft of the R femur. Osteophytosis was apparent on the distal L ulna, the distal R radius, the distal femora and the proximal L tibia. Vertebral osteophytosis was evident in CV2-CV6, TV1, TV2-TV13, LV4 and LV5. Enthesopathies were visible on the humeri (deltoid, teres major, latissimus dorsai, pectoralis major), the L radius (biceps) and the calcanei (Achilles' tendon).

Weapon trauma: A sword nick, which measured 7 mm supero-inferiorly, was present on the L inferior margin of the spinous process of TV4. The spinous process of TV9 had been chopped off, exposing the underlying spongy bone. Evidence of a sharp oblique cut were apparent on the superior margin of the injury. A sharp oblique cutmark, which followed the line of the injury visible on TV9, was present on TV10. The cutmark measured 15.8 mm and extended supero-medially to infero-laterally on the R side of the lamina. The cutmark further extended onto TV10 where an oblique cut, measuring 7.5 mm supero-medially to infero-laterally, was apparent. The assailant had probably been standing directly posterior to the victim, and had brought his or her sword down from a superior position onto the victim's back. No evidence of healing was present.

Post-cranial non-metric traits: R plaque, R and L hypotrochanteric fossae, R and L accessory sacral facets, R and L acromial articular facets, R and L circumflex sulci, R and L medial talar facets, L peroneal tubercle, R and L CV6 and L CV7 transverse foramina bipartite.

Context XXXI. 172

Adult 25-35 yrs: Male: 176.3 cm, 5'9"

Year of excavation: 1982 or 1983

Archaeological context:

The archive reports for the 1982 and 1983 excavations were not available for consultation.

Visual inventory:

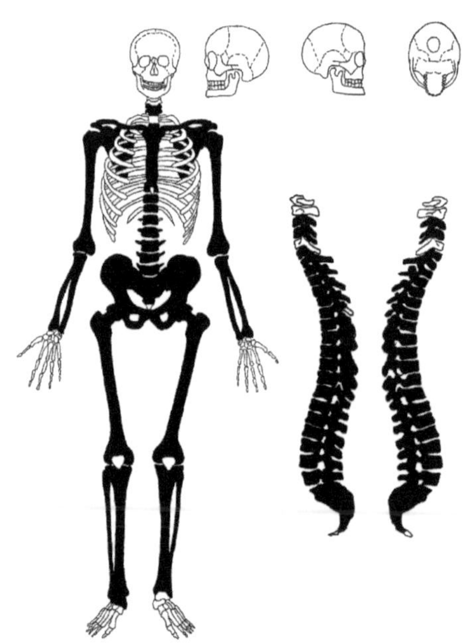

Preservation: The remains were in a good state of preservation with 60-70% of bones present.

Post-cranial pathology: Osteophytosis was present on the distal R humerus, the proximal R ulna, and the proximal R radius. Vertebral osteophytosis was visible in CV4, CV6 and TV6-LV5. Schmorl's nodes were evident in TV6-TV10 and TV12-LV3. Enthesopathies were apparent on the humeri (pectoralis major), the radii (biceps), and the R calcaneus (Achilles' tendon).

Weapon trauma: A very sharp oblique sword cut, which measured 26.5 mm supero-posteriorly to antero-inferiorly, was apparent at the midpoint of the spine of the R scapula. Two fracture lines were present at the inferior margin of the injury, and no sign of healing was evident. It is likely that the attacker had been standing posterior to the victim, and that the sword was brought down from a superior position onto the R shoulder.

Post-cranial non-metric traits: R and L exostoses in trochanteric fossae, R and L lateral tibial squatting facets, R and L acromial articular facets, R and L circumflex sulci, R medial talar facet, R double inferior articular talar facet, R peroneal tubercle, L CV4, R and L CV6 and R and L CV7 transverse foramina bipartite.

Context XXXI. 175

Adolescent 15.5-18.5 yrs: Probable female

Year of excavation: 1982 or 1983

Anthropological number: 6863-273

Archaeological context:

The archive reports for the 1982 and 1983 excavations were not available for consultation.

Visual inventory:

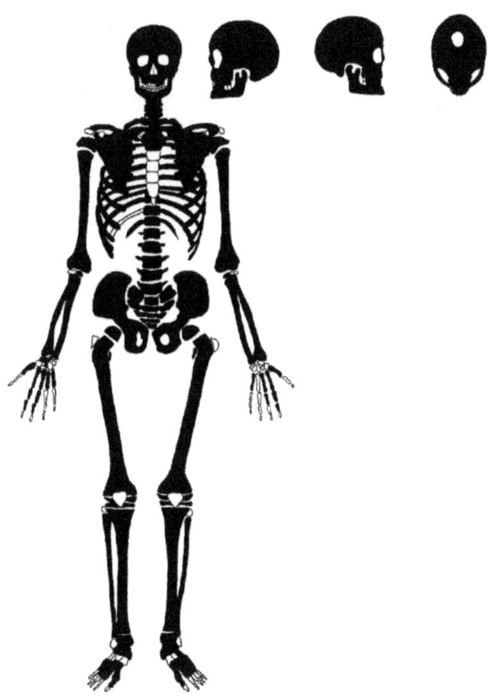

Preservation: The remains were in an excellent state of preservation with 90-100% of bones present.

Post-cranial pathology: Spondylolysis was evident in LV5. Schmorl's nodes were present in TV11 and TV12.

Cranial non-metric traits: Large ossicle at lambda, R and L septal apertures (full set of traits not recorded).

Dentition:

U															U
8	7	6	5	4	3	2	1	1	2	3	4	5	6	7	8
8	7	6	5	4	3	2	1	1	2	3	4	5	X6	7	8
U		PU 2A C											A		U

Bite: Overbite.

Attrition: Molar wear patterns of Stage 1 for the 1st and 2nd molars were apparent. A molar wear pattern of Stage 5 was present on the R mandibular 1st molar. It is possible that the tooth had been broken when the individual was younger, or that the extensive attrition was occupationally related. Secondary dentine was exposed in 6⌉.

Calculus: Flecks of calculus were present on the buccal surfaces of 5⌋ ⌊1 6, the lingual surfaces of 7 2⌋ and 4 3⌉ and the buccal and lingual surfaces of ⌊2 3 4 5 and ⌈5 7. Slight deposits were apparent on the buccal surfaces of 6 4 3 1⌋ and 3⌉, the lingual surfaces of ⌈3 4 and the buccal and lingual surfaces of 5 2 1⌉ ⌈1 2. Medium deposits were apparent on the buccal surface of ⌈4 and the lingual surface of 7⌉.

Periodontal disease: Periodontal disease of medium severity was apparent.

Caries: A large caries was present in the occlusal surface of 6⌉.

Abscesses: Two medium sized externally draining abscesses were associated with 6⌉, and a medium sized externally draining abscess was associated with ⌈6.

Ante-mortem tooth loss: ⌈6 had been lost ante-mortem, and the socket was in the middle stages of remodelling.

Enamel hypoplasia: Hypoplastic grooves were present on 7 1⌋ ⌊1 2 3 and 3 2 1⌉ ⌈1 2 3 which indicated that the individual had been subject to at least four incidents of physiological stress at the ages of 1.5-2, 2-2.5, 3-3.5 and 3.5-4 years.

Context XXXI. 176

Adult 25-35 yrs: Male: 157.7 cm, 5'2"

Plates: 49, 50

Year of excavation: 1982 or 1983

Archaeological context:

The archive reports for the 1982 and 1983 excavations were not available for consultation.

Visual inventory:

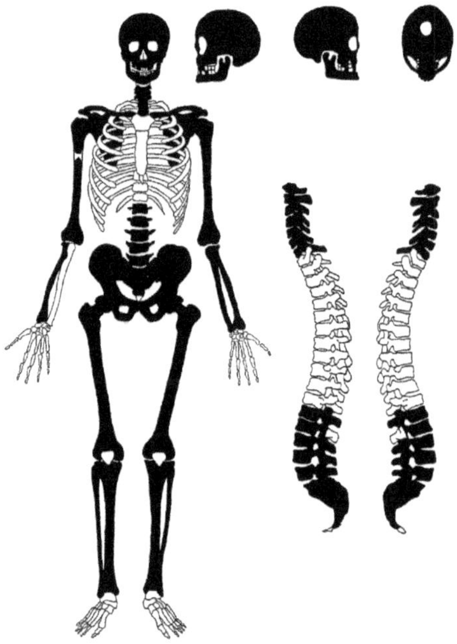

Preservation: The remains were in a good state of preservation with 60-70% of bones present.

Cranial pathology: Cribra orbitalia of Type 2 was present in Areas 1, 2 and 4 of the R orbit and Areas 1, 2, 3 and 4 of the L orbit. Porosity was evident on the superior surfaces of the parietals and the occipital, which may be indicative of healed porotic hyperostosis.

Post-cranial pathology: Osteophytosis was present on the distal L ulna. Vertebral osteophytosis was evident in CV2-CV6 and LV1-LV5. A Schmorl's node was present in LV2. Enthesopathies were visible on the clavicles (costoclavicular ligament).

Weapon trauma: Several examples of sword injuries were apparent in the skeleton. An oblique sword nick, which measured 14 mm antero-posteriorly, was present on the superior aspect of the frontal, just adjacent to the bregma. A similar sword nick, which also measured 14 mm antero-posteriorly, was present on the posterior surface of the R parietal. The nicks had been made using a very sharp weapon which had just scraped the surface of the cranium.

The R humerus had been cleanly chopped into two separate parts. A sword chop with very sharp margins, which extended through the entire diameter of the bone, was apparent at the proximal end of the distal fragment. The fragment which would have linked the proximal and distal parts of the humerus was not present. It was not possible, therefore, to examine the corresponding weapon injuries on the proximal aspect of the bone. The injury would have been made with a very sharp weapon using a great deal of force. The attacker would probably have been standing directly anterior to the victim, and would have initially struck the lateral aspect of the R humerus.

An oblique sword cut, which measured 8 mm antero-posteriorly, was also present on the medial margin of the distal third of the R radius. This injury had probably been made prior to the injury on the R humerus which would have resulted in the amputation, or near amputation, of the lower part of the R arm. The injury had probably been made with a stabbing motion which had pierced the medial margin of the bone.

A transverse blow had been dealt to LV1, which had resulted in the removal of the most inferior aspects of the inferior articular processes. A large transverse chop mark was present on the posterior aspect of LV2. The chop mark had sliced the L inferior articular facet off at the lamina, and had continued transversely, removing the spinous process and the R superior articular facets in the process. The most inferior aspects of the inferior articular processes of LV1 had been removed and they also displayed a chipped appearance. It is probable that these injuries had been caused by the same blow which had affected LV2.

Another large transverse chop mark was present on the posterior surface of LV4. In this case the chop mark had chopped cleanly through and removed the inferior articular processes. It had left a deep cut mark on the posterior surface of the vertebral body. The injuries apparent on the vertebrae had been made with a great deal of force using a slashing motion. The attacker would have been standing posterior to the victim. It is probable that the spinal cord would have been severed as a consequence of this injury.

Several tiny cutmarks were present on the anterior surface of CV2, which had probably been made using a very sharp knife. It is possible that the individual had been stabbed in the throat during the course of the combat. Alternatively, once the victim had been extensively weakened as a result of the sword injuries his attacker may have methodically slit his throat. No signs of healing were associated with any of the sword or knife injuries.

Cranial non-metric traits: R and L highest nuchal lines, R and L parietal foramina, R and L mastoid foramina extrasutural, R open foramen spinosum, L bridging of supraorbital notch, R and L accessory supraorbital foramina, R and L anterior ethmoid frontal foramina extrasutural, R and L accessory infraorbital foramina.

Post-cranial non-metric traits: R and L medial talar squatting facets, R and L acromial articular facets, R and L circumflex sulci, L CV4 and R and L CV6 transverse foramina bipartite.

Dentition:

I	A	A PU	2A	2A							PU	PU	A		A	
8	7	6	~~5~~	~~4~~	3	2	~~1~~	|	~~1~~	~~2~~	3	4	5	~~6~~	~~7~~	~~8~~
~~8~~	~~7~~	6	~~5~~	4	3	~~2~~	~~1~~	|	~~1~~	~~2~~	~~3~~	4	~~5~~	6	~~7~~	~~8~~
		PU	A													

Bite: Overbite.

Attrition: Extensive attrition was apparent on the majority of teeth, with molar wear patterns of Stage 5 for the 1st molars. A molar wear pattern of Stage 1 was apparent on the R maxillary 2nd molar, which indicated that the tooth had not been used in mastication, possibly due to the ante-mortem loss of the corresponding mandibular molar. Secondary dentine or pulp cavity exposure was evident in 6⌋ ⌊3 4 and 6⌉.

Calculus: Slight deposits of calculus were present on the buccal and lingual surfaces of 6 3 2⌋ ⌊3 4 5 and 4 3⌉ ⌈4 6 and the buccal, lingual and distal surfaces of 6⌉. Medium deposits were present on all surfaces of 7⌋, indicating that the tooth had not been used in chewing.

Periodontal disease: Periodontal disease of considerable severity was evident.

Abscesses: Medium sized externally draining abscesses were associated with 5⌉ and ⌊5, and a large externally draining abscess was associated with ⌊7. Single large externally draining and internally draining abscesses were associated with 5 4⌋, and a single large externally draining abscess was associated with 7 6⌋.

Ante-mortem tooth loss: 1⌋ ⌊1 6 7 8 and 8 7⌉ ⌈7 8 had been lost ante-mortem. All associated tooth sockets were in the mid-stages of remodelling.

Dental anomalies: The R maxillary 3rd molar was impacted, with the tooth partially erupted and the occlusal surface wedged against the distal surface of the 2nd molar.

Context XXXI. 182a

Young adult 17-25 yrs: Male: 164.8 cm, 5'5"

Plate: 26

Year of excavation: 1982 or 1983

Archaeological context:

The archive reports for the 1982 and 1983 excavations were not available for consultation.

Visual inventory:

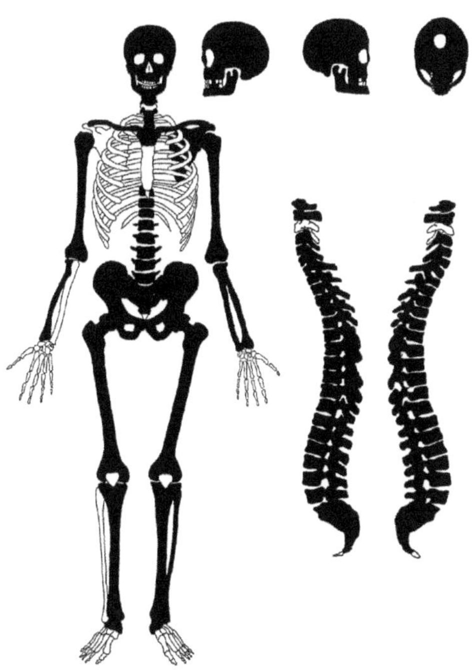

Preservation: The remains were in a good state of preservation with 60-70% of bones present.

Cranial pathology: Porosity was apparent on the superior frontal, the parietals and the occipital, which may be indicative of healed porotic hyperostosis.

Post-cranial pathology: Compression fractures were apparent in the bodies of TV5, TV6 and TV7. The superior aspect of each body surface had been crushed inferiorly, leaving an irregular strip of bone on the supero-anterior surface of each vertebra. The compression fractures were not severe enough to have resulted in the vertebrae becoming wedge-shaped, and would probably have only caused slight anterior kyphosis of the spine. The fracture lines were clearly visible, and it is likely that the fractures were in the mid-stages of healing when the individual had died. The force which would have caused the fractures may have acted from above, and as such may have been caused by a heavy weight falling on the individual when he was in an upright position. Alternatively, the force could have acted from below and may have occurred if the individual had a heavy fall on his feet or buttocks.

Vertebral osteophytosis was evident in TV4-TV8, and Schmorl's nodes were visible in TV7, TV9-TV12 and LV2. Enthesopathies were present on the clavicles (costoclavicular ligament) and the humeri (pectoralis major, latissimus dorsai).

Cranial non-metric traits: R and L highest nuchal lines, R and L mastoid foramina extrasutural, R and L posterior condylar canals open, R and L precondylar tubercles, L double anterior condylar canal, R open foramen spinosum, R and L accessory supraorbital foramina, R and L anterior ethmoid frontal foramina extrasutural, R and L accessory infraorbital foramina.

Post-cranial non-metric traits: R and L hypotrochanteric fossae, L third trochanter, R and L lateral tibial squatting facets, L acetabular crease, R and L accessory sacral facets, L acromial articular facet, L circumflex sulcus, R and L medial talar facets, R and L double inferior articular talar facets, R and L double anterior calcaneal facets, R and L peroneal tubercles, R CV5, R and L CV6, and L CV7 transverse foramina bipartite.

Other observations: Small cutmarks were present on the distal L femur and the L anterior superior iliac spine, which may have been indicative of post-mortem defleshing/disarticulation.

Dentition:

8 (I)	7	6	~~5~~	4	3	~~2~~	~~1~~	~~1~~	~~2~~	~~3~~	~~4~~	~~5~~	6	~~7~~	~~8~~
8 (C)	7	6	~~5~~	~~4~~	~~3~~	~~2~~	~~1~~	~~1~~	2	~~3~~	~~4~~	5	6	7	~~8~~

Bite: Overbite.

Attrition: Molar wear patterns of Stage 1 for the 3rd molars, Stage 3 for the 2nd molars and Stage 4 for the 1st molars were apparent.

Calculus: Flecks of calculus were apparent on the lingual surfaces of ⌊6 and 6⌉ and the buccal and lingual surfaces of 7⌉. Slight deposits were visible on the lingual surfaces of 7⌋ and ⌈2, the distal surfaces of 7⌉ ⌈7 and the distal and lingual surfaces of 8⌉.

Periodontal disease: Periodontal disease of slight severity was evident.

Caries: A small caries was present in the occlusal surface of 8⌉.

Dental anomalies: The R maxillary 3rd molar was impacted. The tooth was partially erupted and the mesial aspect of its occlusal surface was wedged against the distal surface of the 2nd molar.

Context XXXI. 185

Adult 35-45 yrs: Female: 149.4 cm, 5'0"

Year of excavation: 1982 or 1983

Archaeological context:

The archive reports for the 1982 and 1983 excavations were not available for consultation.

Visual inventory:

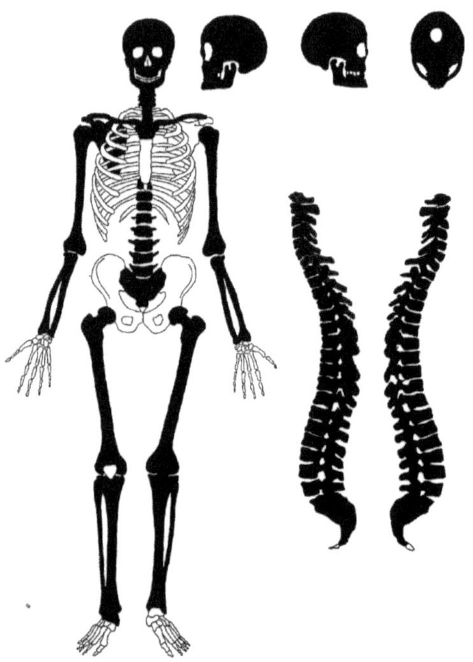

Preservation: The remains were in a good state of preservation with 60-70% of bones present.

Cranial pathology: Cribra orbitalia of Type 1 was present in Areas 1 and 2 of the R orbit, and cribra orbitalia of Type 2 was evident in Areas 1, 2 and 3 of the L orbit.

Post-cranial pathology: A squatting facet, which was completely eburnated, was apparent in the groove for flexor hallucis longus of the distal L tibia. A corresponding eburnated squatting facet was present on the anterior surface of the neck of the L talus. It is possible that the individual was prone to squatting in a position that put particular stress on her L ankle.

Osteophytosis was present on the glenoid fossa of the R scapula, the proximal humeri and the proximal L radius. Rotator cuff disease was present in the humeri. Vertebral osteophytosis was visible in CV2-S1, with DJD present in CV5, CV6, LV5 and S1. A Schmorl's node was evident in LV1. Enthesopathies were apparent on the clavicles (costoclavicular ligament), the humeri (pectoralis major, teres major) and the femora (gluteus medius).

Weapon trauma: A possible sword cut, which ran supero-inferiorly, was present on the medial epicondyle of the R femur. The cut mark measured 23 mm antero-posteriorly by 12 mm supero-inferiorly. The margins did appear to be sharp, but they were somewhat obscured due to the soft spongy nature of the underlying bone.

Decapitation: The woman had been decapitated. A clean, near-transverse sword cut had completely removed the dens of the axis. The posterior margin of the L superior articular facet and the superior surface of the spinous process had also been removed by the blow, exposing the underlying spongy bone. The lateral margin of the R inferior articular facet, and the most supero-lateral point of the R transverse process of CV1 had also been removed in the same blow. The weapon used to decapitate the woman must have been extremely sharp, since there is only evidence of a single blow to the cervical vertebrae. The angle of the sword chop at the base of the dens indicates that the blow was probably struck from the L to the R with the assailant positioned so that he or she had been facing the woman.

Cranial non-metric traits: R and L highest nuchal lines, R ossicle in lambdoid suture, R parietal foramen, R and L mastoid foramina extrasutural, R and L open foramina spinosum, R and L accessory supraorbital foramina, R and L anterior ethmoid frontal foramina extrasutural, R and L accessory infraorbital foramina.

Post-cranial non-metric traits: R and L exostoses in trochanteric fossae, R and L third trochanters, R and L lateral tibial squatting facets, R acromial articular facet, R circumflex sulcus, L medial talar facet, L peroneal tubercle.

Other observations: Orange discoloration (rust) was present on the posterior surfaces of TV3 and TV4.

Dentition:

NP 8	7̸	6̸	5̸	4̸	3	2	1̸	1	2	PU 3	4̸	PU 5	PU 6	PU 7	8̸
8 C	7	6	5	4	3	2	1	1̸ A	2	3	4	5	6	7	8

Bite: Overbite.

Attrition: Extensive attrition was apparent on the majority of teeth, with molar wear patterns of Stages 3+-4+ for the 2nd molars, and Stage 5 for the 1st molars. The wear on the 3rd molars was only slight suggesting that these teeth were not used in mastication. The R maxillary 3rd molar was genetically absent, which would have contributed to the lack of wear apparent on the corresponding mandibular molar. Secondary dentine or pulp cavity exposure was evident in ⌊3 5 6 7.

Calculus: Slight deposits of calculus were present on the buccal surface of ⌊5, the lingual surfaces of ⌊2 and 6⌉, the buccal and lingual surfaces of 3⌋ ⌊1 2 3 6 and 7 3 2 1⌉ ⌈2 3 7, the buccal, lingual and distal surfaces of ⌈8 and the buccal, lingual, and occlusal surfaces of 8⌉. Medium deposits were visible on the buccal surfaces of ⌊7 and 6⌉, the distal surface of 8⌉ and the buccal and lingual surfaces of 5 4 3⌉ ⌈4 5 6.

Periodontal disease: Periodontal disease of considerable severity was apparent.

Caries: A medium sized caries was present in the mesial surface of 8⌉.

Abscesses: A medium sized externally draining abscess was associated with ⌈1.

Ante-mortem tooth loss: 6 4⌋ ⌊4 had been lost ante-mortem, and the sockets had been fully remodelled.

Dental anomalies: The R maxillary 3rd molar was genetically absent.

Context XXXI. 189 (ii)

Young adult 17-25 yrs: Male

Plates: 51, 52, 53, 60, 61

Year of excavation: 1982 or 1983

Anthropological number: 6863-276

Archaeological context:

The archive reports for the 1982 and 1983 excavations were not available for consultation (***NB: Three individuals were labelled as Context XXXI. 189.).***

Visual inventory:

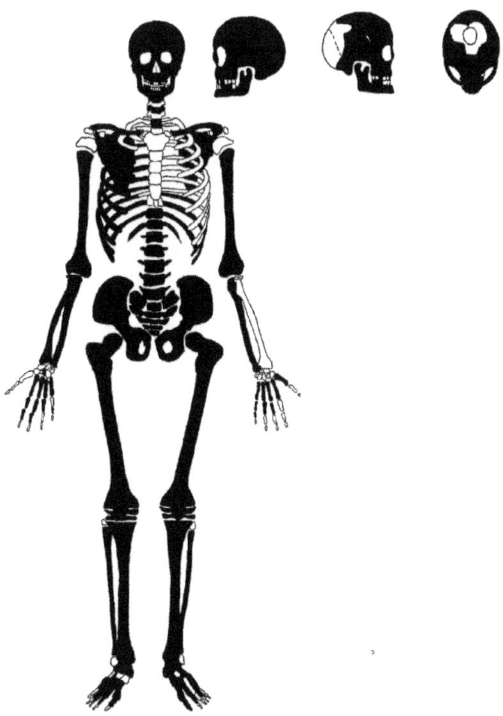

Preservation: The remains were in a very good state of preservation with 80-90% of bones present.

Cranial pathology: A simple transverse fracture was present on the distal half of the L nasal bone. One hundred percent apposition of the fracture parts was evident. The fracture line was barely visible, indicating that the fracture was in the advanced stages of healing.

Post-cranial pathology: Enthesopathies were apparent on the clavicles (costoclavicular ligament).

Weapon trauma:

Cranium - A very large sharp edged chop mark was present on the R side of the cranium. The injury ran from the midpoint of the R side of the frontal bone through the coronal suture to the midpoint of the R parietal. The sword cut measured approximately 91 mm antero-posteriorly. The margins were slightly bevelled and it is probable that the attacker had wielded the weapon supero-medially to infero-laterally.

Mandible - A sharp oblique sword cut was present on the inferior aspect of the L side of the mandible, immediately adjacent to the mental tubercle. Another sharp oblique cut mark was present on the R side of the mandible at the mandibular angle.

R Scapula - A sword cut was present on the posterior surface of the most lateral part of the spine of the R scapula. A small perforation with radiating fracture lines, and a single small sword cut was apparent. It is probable that the point of the sword had penetrated the bone in a stabbing rather than a slashing motion.

L humerus - Weapon trauma had resulted in the diaphysis having separated into two parts. A sharp oblique sword cut was present at the proximal third of the shaft. The cut mark measured approximately 20 mm antero-posteriorly. The full extent of the lesion had been obscured by post-mortem damage. A transverse sword cut was situated approximately 30 mm superior to the first sword cut. The injuries must have been made with a large, sharp edged weapon. It is probable that the combination of the two injuries had weakened the bone and resulted in it fracturing post-mortem.

R humerus - A large transverse sword chop, which measured 21 mm antero-posteriorly, was apparent on the lateral surface of the midpoint of the shaft.

Ribs - A small sharp cut mark was present on the posterior aspect of the R 12th rib. A very sharp sword chop was present at the midpoint of the L 10th rib. The cut mark was oblique, and it had cleanly cut the rib into two separate parts.

Vertebrae - A sharp oblique cut mark, which measured 7.6 mm antero-posteriorly, was present on the infero-medial point of the L transverse process of TV4. A sharp oblique sword cut was visible on the anterior surface of the body of TV12. Small transverse cut marks were present on the posterior surface of LV2 immediately inferior to the superior articular facets.

Sacrum - A very large transverse chop mark was present on the posterior aspect of S4. The sword appeared to have cleanly chopped through the neural arch. The injury would probably have been made by an attacker positioned posterior to the victim who had wielded the sword in a slashing motion.

L femur - A sharp cut mark was present on the lateral surface immediately adjacent to the lesser trochanter. The cut measured 2 mm supero-inferiorly by 21 mm antero-posteriorly, and had been made in a supero-inferior motion.

The presence of numerous examples of weapon trauma on the mid to upper parts of the body indicated that the individual had probably attained the injuries during combat. The lack of healing associated with the injuries implied that the individual had died as a result of the combat.

Peri-mortem trauma: The posterior aspect of the skull appeared to have been broken with a blunt, rounded implement. There was evidence for at least four separate blows. One of the blows had been made to the lambda, and a blow had been dealt to the R and L sides of the injury at the lambda. At least one further blow had been made to the area inferior to the three well defined areas of trauma.

The inferior aspect of the skull in the region of the foramen magnum had been broken. The trauma was oval-shaped and extended from the basilar part of the occipital to the R and L styloid processes, ending at the posterior margin of the foramen magnum.

Seven very sharp cutmarks were apparent at the area of the occipital which extended transversely from the posterior margin of the foramen magnum. The cutmarks each measured approximately 30-35 mm, and ran supero-inferiorly. The margins of the cutmarks were bevelled, and they appeared to have been made by a right handed person in a supero-inferior motion. The regularity of the cutmarks indicates that they were made deliberately and methodically. It is probable that the blunt trauma and the cutmarks at the basioccipital area were made peri-mortem. It is likely that the individual who had broken the skull and made the cutmarks had been attempting to extract the brain from the cranium. The crudeness of the peri-mortem trauma and the appearance of numerous examples of weapon trauma on the skeleton would suggest that the peri-mortem trauma was a post-combat activity probably carried out by an aggressor. The peri-mortem trauma may have been part of a ritual of warfare.

Dentition:

											B				U
~~8~~	~~7~~	6	~~5~~	~~4~~	~~3~~	~~2~~	1	~~1~~	~~2~~	3	4	~~5~~	~~6~~	~~7~~	~~8~~
8	7	6	5	~~4~~	~~3~~	~~2~~	~~1~~	~~1~~	~~2~~	3	~~4~~	~~5~~	6	~~7~~	8
E										B					E

Bite: Overbite.

Attrition: Molar wear patterns of Stage 2+ for the 2nd molars and Stage 3 for the 1st molars were apparent.

Calculus: Flecks of calculus were present on the buccal surface of ⌊3 and the buccal and lingual surfaces of 1⌋, while slight deposits were apparent on the buccal surface of 6⌋.

Context XXXI. 203. Sk. 1

Adult 35-45 yrs: Female: 157.6 cm, 5'3"

Plate: 27

Year of excavation: 1983

Archaeological context:

The archive report for the 1983 excavations was not available for consultation.

Visual inventory:

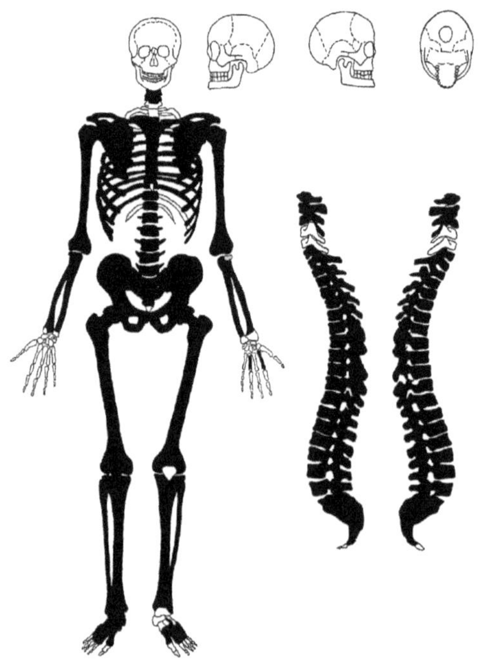

Preservation: The remains were in a good state of preservation with 70-80% of bones present. Extensive surface erosion and fragmentation was evident on the bones. Small canine tooth marks were apparent on the articular ends of the metacarpals.

Post-cranial pathology: Plaques of woven bone were present on the visceral surfaces of the heads of the R 2nd, 3rd, 4th, 5th and 6th ribs. A poorly healed transverse fracture was apparent on the body of the sternum. The fracture ran transversely between the notches for the R and L 6th costal cartilages. The fracture line was clearly visible and callus was present, indicating that the fracture was in the early stages of healing. It is possible the plaques of bone present on the ribs were related to the fractured sternal body and that the woman had suffered a direct blow to her chest, which also affected the underlying organs.

Very pronounced enthesopathies of pectoralis major and the deltoid were apparent on the L humerus. The proximal end of the bone appeared to curve posteriorly. It is possible that the changes apparent in the L humerus were related to her occupation. Alternatively the individual may have attained a greenstick fracture to the anterior surface of the L humerus when she was a child. Vertebral osteophytosis was present in CV7, TV4-TV9, LV4 and LV5. Schmorl's nodes were visible in TV12-LV4.

Post-cranial non-metric traits: R and L hypotrochanteric fossae, R lateral tibial squatting facet, R and L acromial articular facets, R and L circumflex sulci, R medial talar facet, R double inferior articular talar facet, R double anterior calcaneal facet, R and L CV6 and R and L CV7 transverse foramina bipartite.

Context XXXI. 203. Sk. 2

Adult 35-45 yrs: Female: 167.9 cm, 5'6"

Year of excavation: 1983

Archaeological context:

The archive report for the 1983 excavations was not available for consultation.

Visual inventory:

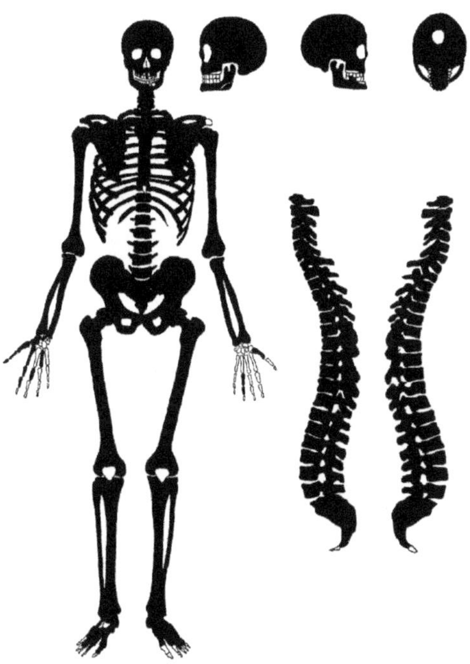

Preservation: The remains were in an excellent state of preservation with 90-100% of bones present.

Post-cranial pathology: Os acromiale was apparent in the L scapula. A very well healed simple transverse fracture was present on the midshaft of the R 3rd rib. Small plaques of woven bone were present on the visceral surfaces of the necks of the R 3rd and the L 8th ribs. Osteophytosis was present on the bones of the L hand (MC1, proximal PH4), the R hand (MC1, MC2, proximal PH1-3, intermediate PH2, trapezium), the L foot (MT1-5, proximal PH4, proximal PH5, cuboid), and the bones of the R foot (MT1-MT3, MT5, proximal PH1, proximal PH4, cuboid, 1st cuneiform). DJD was present on the proximal L MC1, the distal R MC1, the R trapezium, the R distal proximal hand PH3 and the distal R MT1. Osteophytosis was also present on the sternal and acromial ends of the clavicles, the glenoid fossae of the scapulae, the proximal and distal humeri, the proximal ulnae, the proximal and distal radii, the acetabulae, the proximal and distal femora, the proximal and distal tibiae and the distal fibulae. Vertebral osteophytosis was evident in CV1-LV5. Schmorl's nodes were apparent in TV8-TV11.

Enthesopathies were present on the clavicles (costoclavicular ligament), the humeri (subscapularis, deltoid, pectoralis major, latissimus dorsai), the radii (biceps), the femora (adductor longus, adductor magnus, adductor brevis, gluteus medius) and the tibiae (soleus, patellar ligament).

Cranial non-metric traits: R and L highest nuchal lines, ossicle at lambda, R parietal foramen, R and L mastoid foramina extrasutural, R and L posterior condylar canals open, R and L open foramina spinosum, R and L absent zygomaticofacial foramina, R and L accessory supraorbital foramina, R and L anterior ethmoid frontal foramina extrasutural, R and L accessory infraorbital foramina.

Post-cranial non-metric traits: R and L hypotrochanteric fossae, R and L exostoses in trochanteric fossae, R and L lateral tibial squatting facets, R and L acetabular creases, R and L accessory sacral facets, R acromial articular facet, R and L circumflex sulci, R and L medial talar facets, R and L lateral atlas bridging, L CV1 and R CV5 transverse foramina bipartite.

Dentition:

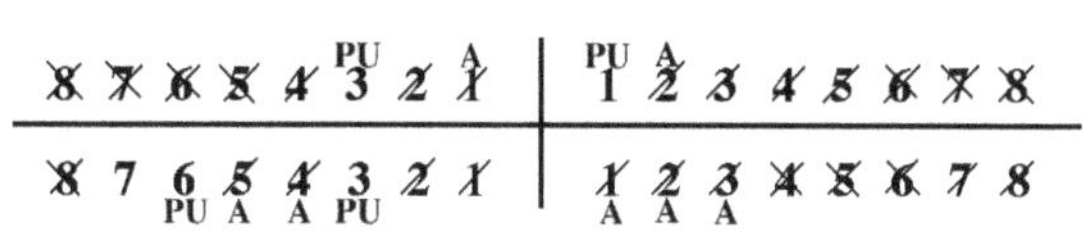

Bite: Overbite.

Attrition: Extensive attrition was evident on the teeth, with molar wear patterns of Stage 4 for the 2nd molars, and Stage 5 for the 1st molars. Secondary dentine was exposed in 3⌋ ⌊1 and 6 3⌉.

Calculus: Flecks of calculus were present on the buccal and lingual surfaces of 6⌉.

Periodontal disease: Periodontal disease of considerable severity was apparent.

Abscesses: A small externally draining abscess was associated with 5⌉, and medium sized externally draining abscesses were associated with 1⌋, ⌊2, 4⌉, and ⌈1. A single large externally draining abscess was associated with ⌈2 3. Plaques of pitted woven bone were apparent on the body of the mandible inferior to 8 7 6 5⌉. The woven bone was probably related to the abscess associated with 5⌉.

Ante-mortem tooth loss: 8 7 6 5⌋ ⌊6 7 8 and 8⌉ ⌈4 5 6 had all been lost ante-mortem. The sockets for the majority of the teeth were in the final stages of remodelling.

Context XXXI. 209

Mature adult 45+ yrs: Male: 156.4 cm, 5'2"

Year of excavation: 1983

Archaeological context:

The archive report for the 1983 excavations was not available for consultation.

Visual inventory:

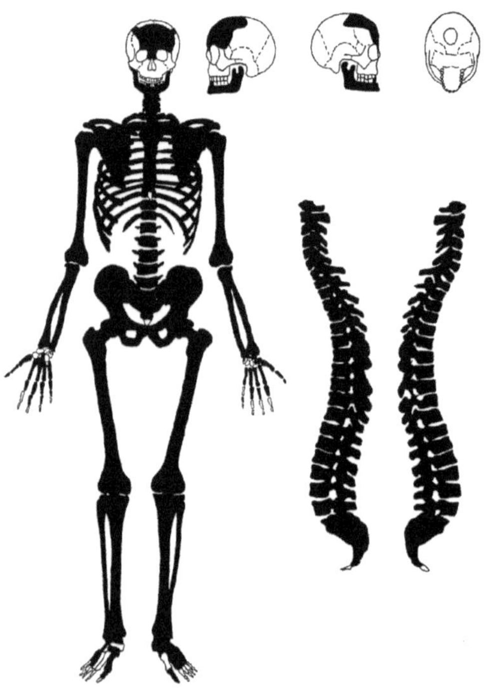

Preservation: The remains were in a very good state of preservation with 80-90% of bones present.

Post-cranial pathology: The neural arches of S1-S4 were dysplastic. The neural arch of S1 was bifurcated, while only a medial strip of bone represented the neural arches of S2-S4. The defects were probably due to delay during the early development of the vertebral column. Well healed transverse fractures were present at the midshafts of the R 12th rib, and the L 5th - 11th ribs. A transverse fracture was apparent at the midpoint of the body of the sternum. There was approximately 14 mm overlap of the fracture parts, and the inferior part of the bone was displaced anteriorly by approximately ten degrees. The bodies of TV4, TV11 and LV1-LV3 were compressed, and had formed wedge-shaped vertebrae. It is possible that all of the bones had been fractured in a single incident.

Osteophytosis was present on the majority of the rib heads, the bones of the R hand (MC1, MC2, MC5, proximal PH1, proximal PH5, intermediate PH3-5, and distal PH5), the bones of the L hand (MC1, MC2, proximal PH1, intermediate PH5, and distal PH1), the bones of the L foot (proximal PH1 and distal PH1), the sternal and the acromial ends of the clavicles, the acromial facets and the glenoid fossae of the scapulae, the distal humeri, the proximal radii, the proximal ulnae, the patellae, the R fibulae and the R talus. DJD was apparent on the distal radii, the distal ulnae, the distal femora, and the proximal tibiae. Rotator cuff disease was present in the humeri. Vertebral osteophytosis was present in CV2-LV5, and DJD was visible in CV3, CV4, CV7, TV1, TV6, TV7, TV11, TV12, LV1 and LV4. Schmorl's nodes were present on TV9, and TV11-LV2. Enthesopathies were evident in the radii (biceps) and the femora (adductor magnus, adductor medius, adductor longus).

Cranial non-metric traits: R and L accessory supraorbital foramina.

Post-cranial non-metric traits: R and L plaque, R and L exostoses in trochanteric fossae, R and L lateral tibial squatting facets, R and L acromial articular facets, R and L circumflex sulci, R and L medial talar facets, R double inferior articular talar facet, L peroneal tubercle, L CV4 and R and L CV5, CV6 and CV7 transverse foramina bipartite.

Dentition:

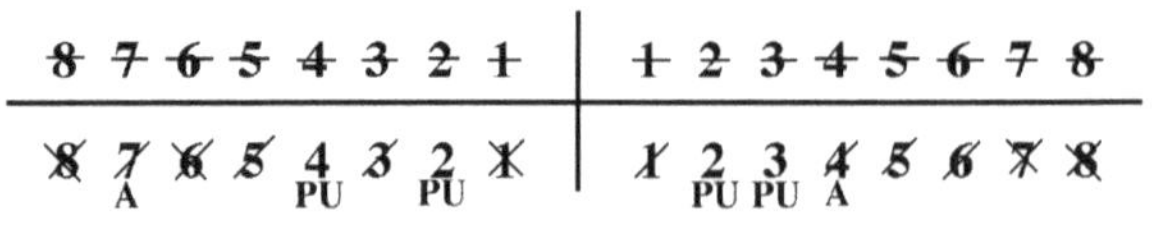

Bite: Overbite.

Attrition: Extensive attrition was apparent on the teeth present. Secondary dentine or pulp cavity exposure was evident in 4 2⏌⎿2 3.

Calculus: Slight deposits of calculus were present on the buccal surfaces of 4 2⏌⎿2 3.

Periodontal disease: Periodontal disease of considerable severity was evident.

Abscesses: Medium sized externally draining abscesses were associated with 7⏌and⎿4.

Ante-mortem tooth loss: 8 6 1⏌⎿7 8 had been lost ante-mortem. The sockets for 1⏌⎿7 8 had been almost completely remodelled, while those for 8 6⏌were in the early stages of remodelling.

Context XXXI. 211A

Adult 35-45 yrs: Male: 171.8 cm, 5'7"

Plate: 23

Year of excavation: 1983

Archaeological context:

The archive report for the 1983 excavations was not available for consultation.

Visual inventory:

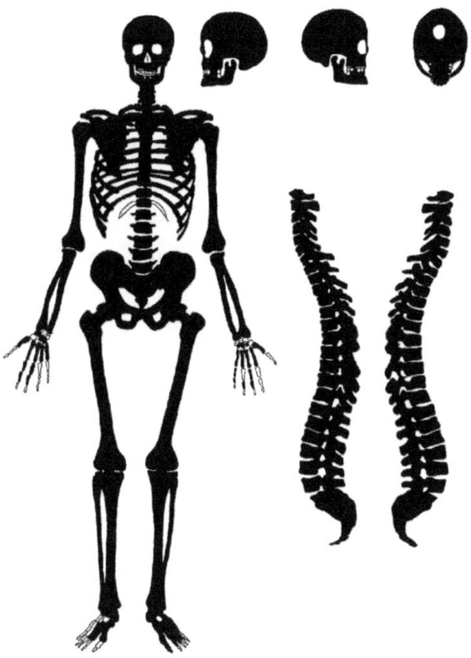

Preservation: The remains were in an excellent state of preservation with 90-100% of bones present.

Cranial pathology: A well healed simple transverse fracture was present on the distal third of the R nasal bone. A well healed simple oblique fracture was present on the distal third of the L nasal bone. The fracture line met the inferior margin of the L nasal bone at its midpoint. The distal parts of both fractures were slightly depressed. Cribra orbitalia of Type 1 was present in Areas 1 and 2 of the R orbit, and cribra orbitalia of Type 2 was apparent in Areas 1, 2 and 4 of the L orbit.

Post-cranial pathology: Sacralization was present with LV5 completely fused to S1. Well healed simple, transverse fractures were present at the midpoints of the L 5th, 6th and 7th ribs. A very well healed simple oblique fracture was present at the acromial end of the L clavicle. There was approximately 27.5 mm overlap of the fracture parts, and almost 100% apposition. The bone was generally hypertrophied and shortened, with a greatest length of 134 mm (R = 155 mm). The acromial end was displaced posteriorly by approximately forty-five degrees.

Osteophytosis was present on all rib heads, the bones of the R hand (intermediate PH2 and PH3), the bones of the L hand (MC1, MC2), the acromial end of the R clavicle, the acromial ends and the glenoid fossae of the scapulae, the proximal humeri, the distal ulnae, the distal radii, the distal femora, the proximal tibiae, the distal L tibia, the patellae, the R fibula, the L talus and the R calcaneus. Rotator cuff disease was present in the humeri. DJD was visible in the R and L MT1s, the R hand intermediate PH2, the L hand intermediate PH3, the R distal tibia and the R talus. Vertebral osteophytosis was evident in CV2-S1, with DJD visible in CV4 and CV5. A Schmorl's nodes was apparent in LV3. Enthesopathies were present on the clavicles (costoclavicular ligament), the humeri (pectoralis major, latissimus dorsai, deltoid), the radii (biceps), the femora (adductor magnus, adductor longus, adductor medius) and the tibiae (soleus, patellar ligament).

Cranial non-metric traits: L plaque, R and L hypotrochanteric fossae, R and L exostoses in trochanteric fossae, R and L acetabular creases, R and L accessory sacral facets, R and L acromial articular facets, R and L circumflex sulci, R and L vastus notches, R and L medial talar facets, R and L double inferior articular talar facets, R and L double anterior calcaneal facets, R and L peroneal tubercles, R and L CV4, CV5, CV6 and CV7 transverse foramina bipartite.

Post-cranial non-metric traits: R and L highest nuchal lines, R and L parietal foramina, R ossicle at asterion, R and L auditory tori, R and L mastoid foramina extrasutural, R posterior condylar canal open, L accessory supraorbital foramen, R and L anterior ethmoid frontal foramina extrasutural, R and L accessory infraorbital foramina.

Dentition:

NP 8	A 7	6	5	4	3	2	1	A 1	2	3	A 4	5	6	7	8
8	7	6	5	4	3	2	1	1	2	3	4	5	6	7	8

Bite: Overbite.

Attrition: Molar wear patterns of Stage 3 for the 2nd molars, and Stage 4+ for the 1st molars were apparent.

Calculus: Flecks of calculus were present on the buccal and lingual surfaces of 4⏋⎡6. Slight deposits were evident on the buccal surface of ⎡5, the lingual surface of 5⏋ and the buccal and lingual surfaces of 7 6 3 1⏋⎡1 3 4. Medium deposits were visible on the lingual surfaces of ⎡5 7.

Periodontal disease: Periodontal disease of medium severity was apparent.

Abscesses: A small externally draining abscess was associated with 7⏌, and medium sized externally draining abscesses were associated with ⎣1 and ⎣4.

Ante-mortem tooth loss: 7 4 3 1⏌ ⎣8 had been lost ante-mortem.

Dental anomalies: The R maxillary 3rd molar was genetically absent.

Context XXXI. 212

Adult 35-45 yrs: Male: 173.5 cm, 5'8"

Year of excavation: 1983

Archaeological context:

The archive report for the 1983 excavations was not available for consultation.

Visual inventory:

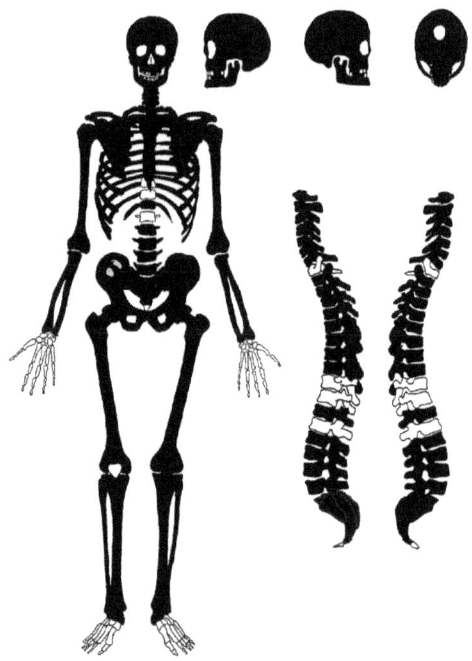

Preservation: The remains were in a very good state of preservation with 80-90% of bones present. Several bones displayed slight surface erosion.

Post-cranial pathology: A very well healed simple oblique fracture was present at the midshaft of the R 9th rib. The anterior aspect of the fracture was displaced anteriorly, and approximately 90% apposition of the fracture parts was evident. A well healed simple oblique fracture was present at the distal half of the shaft of the R radius (Colles' fracture). The fracture parts were very poorly aligned, with only approximately 30% apposition. Overlap of the fracture parts by 31.5 mm was apparent, and the bone had a greatest length of 250 mm (L = 263 mm). Hypertrophy was evident at the fracture site. The distal part of the fracture was displaced postero-laterally by forty-five degrees. Secondary osteophytosis was present on the distal R radius and ulna, which may be indicative of dislocation of the radio-ulnar joint.

Spondylolysis was evident in LV4. DJD was present on the inferior surfaces of the superior articular facets. The superior surfaces of the superior articular facets of LV5 also displayed DJD. Extensive DJD was apparent on the superior and inferior articular facets of the L sides of the cervical vertebrae. As a result the L sides of the cervical vertebrae were compressed, causing this side of the neck to be shorter than the R side. The head may have been tilted slightly posteriorly, and to the left. Vertebral osteophytosis was present in TV2-TV9, TV12 and LV2-S1. The body of LV5 was slightly compressed, and there may have been mild kyphosis of the spine at this point. Schmorl's nodes were visible in TV6-TV9, TV12, LV2 and LV3. Osteophytosis was present on the proximal and distal R MC1, the glenoid fossae of the scapulae, the R distal humerus, the proximal ulnae and the distal femora. Enthesopathies were apparent on the femora (adductor magnus, adductor longus, adductor brevis) and the calcanei (Achilles' tendon).

Cranial non-metric traits: R and L parietal foramina, R and L mastoid foramina extrasutural, R and L posterior condylar canals open, R and L accessory supraorbital foramina, R and L anterior ethmoid frontal foramina extrasutural, R and L accessory infraorbital foramina.

Post-cranial non-metric traits: R and L hypotrochanteric fossae, R and L exostoses in trochanteric fossae, R and L lateral tibial squatting facets, R and L acetabular creases, sternal foramen, R and L accessory sacral facets, L acromial articular facet, R and L circumflex sulci, R and L medial talar facets, R and L double inferior articular talar facets, R and L double anterior calcaneal facets, R and L peroneal tubercles.

Dentition:

8	7	PU 6	~~5~~	~~4~~	3	~~2~~	A ~~1~~	1	2	A ~~3~~	A ~~4~~	A ~~5~~	3A ~~6~~	C 7	8
~~8~~	7 PU	6 PU	5	~~4~~	3	~~2~~	~~1~~	~~1~~	~~2~~	~~3~~	~~4~~	5	6	7	8 PU

Bite: Overbite.

Attrition: Extensive attrition was apparent for the molars, with wear patterns of Stage 5 for the 1st, 2nd and 3rd molars. The non-molar teeth displayed less attrition. Secondary dentine or pulp cavity exposure was evident in 6⌋ and 7 6⌉ ⌈8.

Calculus: Flecks of calculus were present on the buccal surface of 8⌋, the lingual surface of 3⌉ and the buccal and lingual surfaces of ⌊7. Slight deposits were visible on the buccal surface of 6⌋ and the buccal and lingual surfaces of 7⌋ and 7 6⌉ ⌈6 7 8.

Periodontal disease: Periodontal disease of medium severity was apparent.

Caries: A large caries was visible in the distal surface of ⌊7.

Abscesses: A medium sized externally draining abscess was associated with 1⌋, and large externally draining abscesses were associated with ⌊4 and ⌊5. A medium sized internally draining abscess was associated with ⌊3. A small externally draining abscess, and two medium sized externally draining abscesses were associated with ⌊6.

Ante-mortem tooth loss: 4⌋ and 8 1⌉ had been lost ante-mortem. The sockets for 4⌋ and 8⌉ had been completely remodelled, while that for 1⌉ was in the mid-stages of remodelling.

Context XXXI. 214

Adult 35-45 yrs: Female: 156.1 cm, 5'2"

Year of excavation: 1983

Archaeological context:

The archive report for the 1983 excavations was not available for consultation.

Visual inventory:

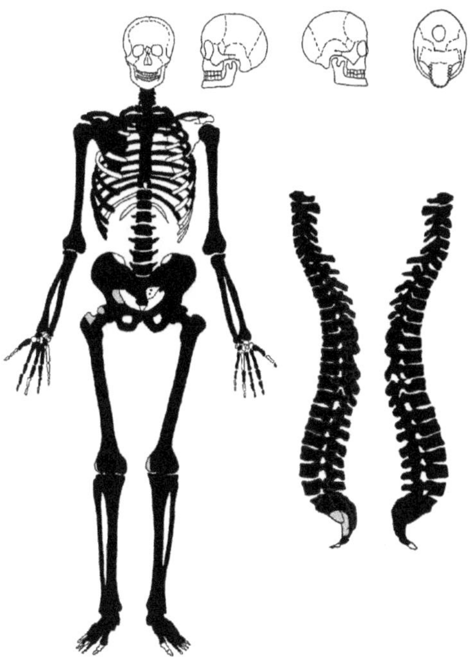

Preservation: The remains were in a very good state of preservation with 70-80% of bones present.

Post-cranial pathology: The metacarpal, proximal phalanx, and intermediate phalanx of the R 4th finger were ankylosed. No signs of degenerative changes at the joints were apparent. An ununited simple transverse fracture was present at the midshaft of the proximal phalanx. The distal part of the fracture was located posteriorly relative to the proximal part of the fracture, and there would have been 0% apposition of the fracture parts. The malalignment and the subsequent ankylosis would have resulted in the finger having a claw shape. It is surprising that the parts of the fracture had not reunited, even out of alignment. It is possible that due to the pain of the fracture that the individual had avoided using the fourth finger, and that it had gradually stiffened due to the ankylosis of the bones.

A very well healed simple oblique fracture, with the fracture line barely visible, was present at the acromial end of the R clavicle. Approximately 12 mm overlap of the fracture parts was visible, and the greatest length of the bone was 117 mm (L = 125 mm). The acromial end of the clavicle was generally hypertrophied. A very well healed simple transverse fracture was present at the distal half of the R radius (Colles' fracture). The distal part of the fracture was malaligned posteriorly by almost ten degrees, but approximately 90% apposition of the fracture parts was evident. A degree of overlap of the fracture parts was apparent, and the greatest length of the bone was 217 mm (R = 223 mm). Secondary osteophytosis was apparent at the ulnar notch of the radius and the distal ulna, indicating that there may have been associated dislocation of the radio-ulnar joint.

Osteophytosis was present on the majority of rib heads, the proximal and distal R MC1, the proximal and distal L MC1, the proximal and distal L ulna and the distal L radius. Vertebral osteophytosis was present in CV1, CV3-CV7 and TV4-S1. Schmorl's nodes were visible in TV5 and TV9-LV4. The bodies of TV12 and LV2-LV5 were wedge-shaped, causing anterior kyphosis of the vertebral column by forty-five degrees.

Post-cranial non-metric traits: R and L hypotrochanteric fossae, R and L lateral tibial squatting facets, R acromial articular facet, R circumflex sulcus, R and L medial talar facets, R and L double inferior articular talar facets, R and L peroneal tubercles.

www.ingramcontent.com/pod-product-compliance
Lightning Source LLC
La Vergne TN
LVHW072121210926
848482LV00028B/732